Neurotransmitters and Anterior Pituitary Function

Neurotransmitters and Anterior Pituitary Function

EUGENIO E. MÜLLER

Department of Pharmacology
University of Milan
Milan, Italy

GIUSEPPE NISTICÒ

Institute of Pharmacology
University of Messina
Messina, Italy

UMBERTO SCAPAGNINI

Department of Pharmacology
University of Catana
Catana, Italy

ACADEMIC PRESS New York San Francisco London 1977

A Subsidiary of Harcourt Brace Jovanovich, Publishers

ACADEMIC PRESS, INC.
111 Fifth Avenue, New York, New York 10003

United Kingdom Edition published by
ACADEMIC PRESS, INC. (LONDON) LTD.
24/28 Oval Road, London NW1

Library of Congress Cataloging in Publication Data

Müller, Eugenio E
Neurotransmitters and anterior pituitary function.

Bibliography: p.
Includes index.
1. Neurotransmitters. 2. Adenohypophysis. 3. Pituitary hormones. 4. Pituitary hormone releasing factors. I. Nisticò, Giuseppe, joint author. II. Scapagnini, Umberto, joint author. III. Title. [DNLM: 1. Neurohumors–Physiology. 2. Pituitary hormones, Anterior–Physiology. WL104 M946n]
QP364.7.M83 599′.01′88 77-74058
ISBN 0–12–510550–9

PRINTED IN THE UNITED STATES OF AMERICA

Contents

Preface

Since the report based on a Neuroscience Research Program Work Session, held May 1968 in Cambridge, Massachusetts appeared (see Wurtman, 1971), no complete survey on the biochemistry and physiology of brain neurotransmitters in relation to anterior pituitary function has been published, and in the last few years there has been a tremendous surge of activity in this field. This book presents the first thorough survey of the most salient constituents of the neural–endocrine communication system in mammals, and represents the first detailed attempt to integrate research findings on neurotransmitter–neurohormone interactions in the control of the anterior pituitary.

Since knowledge of neurotransmitter biochemistry and physiology is essential for an understanding of their role in regulating anterior pituitary secretion, this book presents, initially, a detailed analysis of the biochemistry, physiology, and pharmacology of proved or putative neurotransmitters and describes some of the techniques utilized for determining their synthesis or turnover in the CNS of experimental animals or in man. The principal functions of the most well-known neurotransmitter-containing neurons are also considered, based on up-to-date sophisticated techniques for neurotransmitter measurements. Finally, particular attention is given to both traditional and to a vast series of recently developed drugs that affect both neurotransmitter and neuroendocrine function. The experimental evidence that the brain controls pituitary function via releasing and inhibiting hormones secreted by hypothalamic neuroendocrine cells is then concisely reviewed in relation to the secretion of ACTH, GH, LH-FSH, PRL, TSH, and MSH. Studies on the chemical isolation, identification, and synthesis of hypothalamic neurohormones are reported, as well as the most recent concepts of their mechanism of action at the level of the pituitary cells. Space is also allotted to recent findings in the field of neuroendocrine communication, e.g., radioimmunoassay techniques for neurohormone determination in brain and biological fluids, the topography of the neurosecretory pathways—reviewed in relation to the distribution of the principal neurotransmitter pathways—and the recently described extraendocrine actions of hypothalamic neurohormones.

After elucidation of the neurohumoral and the neurohormonal components of the neural–endocrine communication system, research on the role of specific brain neurotransmitters in controlling pituitary hormone secretions in both experimental animals and in man is considered. The evidence available from the different experimental strategies used is discussed and critically evaluated. A separate section deals with the possible CNS site(s) at which neurotransmitters and neurohormones interact for the control of anterior pituitary secretion. The final part of the book describes the actual or potential import of neuropharmacologic approaches to the diagnosis of and therapy for specific disorders of neuroendocrine function in which neurotransmitter dysfunction may play an etiologic role.

Our major aims have been to interpret and clarify theories derived from different disciplines and to provide not merely a compilation of data but primarily a synthesis of information enabling readers to appreciate the significance of the advances in the field.

This comprehensive work (more than 2000 references and 30 exhaustive tables are included), which has been designed to make the topic understandable to the novice and to appeal to the specialist in the field, will be an invaluable aid to students and research workers not only in the field of neuroendocrinology but also neurobiology, neuropharmacology, and neurophysiology. Clinical neuroendocrinologists will find in it the background essential to the rational and safe use of powerful CNS-acting compounds.

We are grateful to Delia Deriu for providing invaluable help with the literature survey and to Giorgio Marcandalli for original drawings.

Eugenio E. Müller
Giuseppe Nisticò
Umberto Scapagnini

List of Abbreviations

(Other abbreviations used are defined in the text.)

A

AAAD aromatic amino acid decarboxylase
AC adenylate cyclase
Ac acetyl
ACh acetylcholine
AChe acetylcholinesterase
ACTH adrenocorticotropic hormone
AHA anterior hypothalamic area
Ala alanine
AP anterior pituitary
APUD cells amine precursor uptake and decarboxylation cells
ARC n. arcuate nucleus
Arg arginine
Asn asparagine
Asp aspartic acid
ATP adenosine triphosphate
APTase adenosine triphosphatase

B

plasma B plasma corticosterone
BA bioassay
BAL 2,3-dimercaptopropanol
BBB blood-brain barrier
BDB bisdiazotided benzidine
BLA basolateral amygdala

C

cAMP cyclic adenosine monophosphate
3′,5′-cAMP 3′,5′-cyclic adenosine monophosphate
CE cortical extracts
ChAc choline acetyltransferase
Che cholinesterase
CMC carboxymethyl cellulose
CNS central nervous system
CNV contingent negative variation
acetyl-CoA acetyl coenzyme A
COMT catechol-*O*-methyltransferase
CPZ chlorpromazine
CRF corticotropin-releasing factor
CSF cerebrospinal fluid
CTZ chemoreceptor trigger zone
Cy cyproheptadine
Cys cystine

D

DA dopamine
DAO diamine oxidase
db-cAMP dibutyryl cyclic adenosine monophosphate
D-β-H dopamine-β-hydroxylase
DDC dichlorodithiocarbamate

2-DG	2-deoxy-D-glucose
5,6-DHT	5,6-dihydroxytryptamine
5,7-DHT	5,7-dihydroxytryptamine
DMI	desmethylimipramine
dopa	dihydroxyphenylalanine
DOPAC	3,4 dihydroxyphenylacetic acid
DOPS	3,4-dihydroxyphenylserine

E

E	epinephrine
EEG	electroencephalogram
EM	electron microscopy
EP	estrogen-progesterone

F

plasma F	plasma cortisol
FA	fusaric acid or butylpicolinic acid
FSH	follicle-stimulating hormone
FSH-RH	follicle-stimulating hormone-releasing factor

G

GABA	γ-aminobutyric acid
GABA-T	γ-aminobutyric acid transaminase
GAD	glutamate decarboxylase
GAD-I	glutamate decarboxylase I
GAD-II	glutamate decarboxylase II
GH	growth hormone
GH-IF, GIF or GH-RIH	growth hormone-inhibiting factor or somatotropin-release inhibiting factor or somatostatin
GMP	guanosine monophosphate
GRF	growth hormone-releasing factor
Glu	glutamic acid
Gln	glutamine
Gly	glycine

H

H	histamine
hGH	human growth hormone
HHAA	hypothalamohypophyseal-adrenal axis
HHTA	hypothalamohypophyseal-thyroidal axis
5-HIAA	5-hydroxyindoleacetic acid
His	histidine
HIOMT	hydroxyindole-*O*-methyltransferase
5-HT	serotonin
5-HTP	5-hydroxytryptophan
HVA	homovanillic acid

I

IC	intracisternal
IgG	immunoglobulin
INI	isoniazid
IP nucleus	interpeduncular nucleus
ip	intraperitoneally
IR	immunoreactive
Ile	isoleucine
ISO	isoproterenol
iv	intravenous
IVT	intraventricular

L

L-AAAD	aromatic L-amino-acid decarboxylase
L-dopa	L-dihydroxyphenylalanine
Leu	leucine
LH	luteinizing hormone
LHA	lateral hypothalamic area
LH-RH	luteinizing hormone-releasing hormone
LH-RH/FSH-RH	luteinizing hormone-releasing hormone/follicle-stimulating hormone-releasing hormone
L-5-HTP	L-5-hydroxytryptophan
L-TH	L-tyrosine hydroxylase
L-Trp	L-tryptophan
L-Tyr	L-tyrosine
LVP	lysine vasopressin

M

MA's	monoamines
MAO	monoamine oxidase
MAOA	monoamine oxidase A
MAOI	monoamine oxidase inhibitors
MBH	medial basal hypothalamus
α-MD	α-methyldopa
ME	median eminence

Met methionine
MFB medial forebrain bundle
MI melanophore index
MIF melanocyte-inhibiting factor
MIF-I melanocyte-inhibiting factor I
MIF-II melanocyte-inhibiting factor II
α-MmT α-methyl-*m*-tyrosine
α-MNE α-methyl norepinephrine
MPOA medial preoptic area
α-MpT α-methyl *p*-tyrosine
MRF melanocyte-releasing factor
MSH melanocyte-stimulating hormone

N

n. nucleus
NE norepinephrine
NAD nicotinamide-adenine dinucleotide
NAS *N*-acetyl serotonin
NAT *N*-acetyltransferase
NEFA nonesterified fatty acids

O

17-OHCS 17-hydroxycorticosteroids
6-OHDA 6-hydroxydopamine
5-OHDA 5-hydroxydopamine
6-OH-dopa 6-hydroxydopa

P

pCA *p*-chloroamphetamine
pCMA *p*-chloro-*N*-methylamphetamine
pCPA *p*-chlorophenylalanine
PG's prostaglandins
PGDH prostaglandin dehydrogenase
pGlu pyroglutamic acid
PGO pontogeniculooccipital
Phe phenylalanine
Phe-H phenylalanine hydroxylase
PI pars intermedia
PIF prolactin-inhibiting factor
PMS pregnant mare serum
PNMT phenylethanolamine-*N*-methyltransferase
PO preoptic
POA preoptic area
PON preoptic nucleus
PPP polyphloretin phosphate
PRA prolactin-releasing activity
PRL prolactin
Pro proline
PRF prolactin-releasing factor
PS paradoxical sleep
PVN paraventricular nuclei

R

REMs rapid eye movement sleep
RF's releasing factors
RIA radioimmunoassay
RNA ribonucleic acid
mRNA messenger ribonucleic acid
RNase ribonuclease

S

SCN suprachiasmatic nucleus
Ser serine
SLI somatostatin-like immunoreactivity
SME stalk median eminence
SON supraopticus nucleus
SR-IF growth hormone-inhibiting factor
SSA succinic semialdehyde
SWS slow wave sleep

T

$t_{1/2}$ half life
T_3 triiodothyronine
T_4 thyroxine
TH tyrosine hydroxylase
Thr threonine
TIDA tuberoinfundibular dopamine
Trp tryptophan
TPO tryptophan pyrrolase
TRH, TRF thyrotropin-releasing factor
Trp-H tryptophan hydroxylase
TSH thyroid-stimulating hormone
TYA 5,8,11,14-tetraynoic acid
Tyr tyrosine

U

UV ultraviolet

V

Val valine
VMA vanillylmandelic acid
VMN ventromedial nucleus
vs versus
VTP ventral tegmental pathway

I

Neurotransmitters and Neurohormones

A. NEUROCHEMICAL MEDIATION: CURRENT CONCEPTS

The stability of the internal environment, i.e., the plasma and extracellular fluids bathing the cells, depends on the coordinated activity of two major regulatory systems: the endocrine system and the nervous system. The manner in which the two integrative systems communicate with one another in coordinating effectively the body's regulatory activity and the results of this combined activity constitute neuroendocrinology (E. Scharrer, 1966; B. Scharrer, 1967). Communication between neurons and endocrine cells in mammals is usually mediated by two types of chemical signals: neurotransmitters and neurohormones.

The distinction between neurotransmitter and neurohormonal activities is not always clear. Neurotransmitters characteristically are low molecular weight, water-soluble compounds that are charged at physiological pH. These messenger substances elicit strictly localized short-lived responses at an easily identifiable locus, the synapse. Since the effector cells are in close proximity to the respective presynaptic terminals, and a presynaptic neuron makes contact with few cells, minute amounts of neurotransmitters are effectively directed at the appropriate receptor sites and the contacts are well insulated from other cells. A feature peculiar to neurotransmitter substances is the rapidity with which they are inactivated following the completion of the signal, either by return of the active principle to its presynaptic storage site or by enzymatic degradation (see Chapter II, Sections B,9 and 10). Substances of this kind [those most generally responsible for brain function include acetylcholine (ACh) and three monoamines, i.e., dopamine (DA), norepinephrine (NE), and serotonin (5-HT)] lack the essential capabilities of endocrine factors, particularly access to and use of vascular pathways. Further properties of this class of neurochemical mediators will be described later (see Chapter II).

In addition to this conventional type of communication, neural elements can influence effector cells by means of a second class of chemical mediators. The principal feature of these compounds is that they act as blood-borne neurochemical messengers which, in sharp contrast to those characteristic of synaptic function, generate more prolonged signals since they are not as speedily inactivated. Being disseminated by a vascular route, these neurochemical messengers are available simultaneously to multiple effector cells in appropriate concentrations. The specificity of the signal derives from the fact that it is in code, and only few cells are capable of translating the code into understandable "information." This group of compounds can be classified as neurohormones (B. Scharrer, 1969). Typically, neurohormones reach their destination in effective concentration, being delivered from sender to receiver cell either by way of the general circulation or by a more restricted vascular route (see Section B,2). It should be noted that the effector sites of neurohormones are not necessarily always endocrine cells. These hormones may, in fact, act directly on target nonendocrine tissue such as the uterus, mammary gland, or kidney tubules (Fig. 1). A neurohormonal mechanism of this type is that controlled by the posterior lobe hormones, vasopressin and oxytocin.

A basic difference between conventional synaptic transmission and neurohormonal mediation lies in the type of neuronal element from which neurohumoral and neurohormonal substances are derived. The source of neurohormones is not, in fact, a conventional neuronal element but a special "neurosecretory" neuron. This is a neuron which is engaged in secretory activity to a degree which exceeds that of the conventional neuron (see B. Scharrer, 1969). Their capacity for the synthesis of specific secretory products is so highly developed as to dominate all other neuronal functions and to impose a prominent glandular appearance to the neuron. The proteinaceous nature of the secretory products—another characteristic distinguishing them from the classic neurotransmitters—allows for their identification with light and electron microscopy. Characteristic membrane-bound granules of several sizes have been observed in many ultrastructural studies (for further information on the phenomenon of neurosecretion, see B. Scharrer, 1969).

The neurosecretory neurons are usually not so ubiquitous within nervous tissue as are more conventional neurons. They tend to gather in specialized neuronal groups and terminate in close proximity to a general or limited circulatory pathway thus frequently forming neurohemal or-

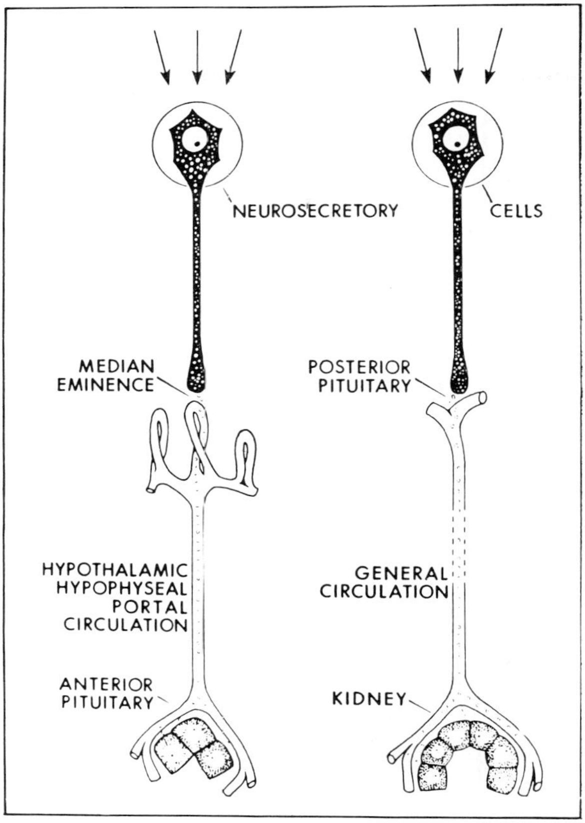

Fig. 1 Release of neurohormones from hypothalamic neurosecretory neurons occurs either in the posterior pituitary, from which they reach terminal target cells by way of the general circulation (right diagram), or in the median eminence of the neurohypophysis, where a special portal system carries neurohormonal signals to endocrine way stations in the adenohypophysis (left diagram). (Adapted from E. Scharrer, 1966.)

gans. Examples of this are the neurohypophysis of vertebrates, the corpus cardiacum of insects, and the sinus gland of crustaceans. These neurohemal organs and the corresponding group of neurosecretory perikarya in which the active material is synthesized constitute neurosecretory systems.

It must be noted, however, that because of the additional knowledge gained in the last few years, the dividing line between conventional and neurosecretory neurons is not so distinct as it appeared to be in the past:

(1) new fibers have been discovered (B fibers), in addition to the "peptidergic" neurosecretory A fibers, in which the secretory product to be discharged into the circulation is of nonpeptide character (see Knowles, 1965; Björklund *et al.*, 1968; B. Scharrer and Weitzman, 1970; these elements will be further discussed below); (2) A and B fibers have been found that do not necessarily end at or near a vascular route since they may establish direct contact with neuronal or endocrine elements or with neither. In the pars intermedia of vertebrates or in the corpus allatum or prothoracic gland of insects (Bargmann *et al.*, 1967; B. Scharrer, 1964a,b) sites exist of these direct contacts (neurosecretomotor junctions) (Bern, 1966). Since the transfer of the chemical message does not involve, in this instance, a vascular route and the site of release is close to the site of action of the messenger substance the latter is not a neurohormone. On the other hand, due to its peptidergic nature, this mediator can be distinguished from conventional neurotransmitters. These peptidergic neurosecretomotor junctions do not end in a synapse, but rather make synaptoid contacts.

The broad spectrum of neuroendocrine mediation and the existence in mammals of peptidergic neurons with neurohormonal and probably neurotransmitter function (see below) create the problem of distinguishing between neurohormones and neurohumors. The two sides of the neurochemical spectrum are neither rigidly uniform nor separated by as clear-cut a demarcation as originally conceived. Data are rapidly emerging for the hypothalamic peptidergic neurons, which control AP function, which show their widespread distribution to extrahypothalamic CNS areas, their localization in nerve terminals and release from nonhypothalamic nerve endings after hypothalamic stimulation and, finally, their important behavioral effects not attributable to action on the pituitary (see Chapter III, Sections C and E). The possibility that the involvement of the hypothalamic peptides in the regulation of the AP function may merely represent a recent evolutionary development of molecules endowed uniquely with neurotransmitter function in the lower phylogenetic species has to be considered (Jackson and Reichlin, 1974; Martin *et al.*, 1975a; Reichlin *et al.*, 1976). The postulated neurotransmitter function of hypothalamic hypophysiotropic peptides seems to have its counterpart in the neurohormone-like behavior of known brain neurotransmitters. Although the evidence shows that catecholamines act predominantly in the CNS (see Chapter IV), several findings suggest that DA, transported by hypophyseal portal blood, may directly affect the AP gland by inhibiting the secretion of prolactin (Chapter IV, Section A,4,a,i).

In summary, the mode of neuroendocrine interaction ranges from conventional information transfer by chemical synaptic transmission to classic neurohormonal signals, and also deals with several intermediate types in which the neurochemical mediator cannot be classified as either hormonal or neurohumoral.

B. NEUROENDOCRINE TRANSDUCTION IN MAMMALS

The widespread diffusion of neurosecretory neurons in living systems underlines their basic function: that of providing a link between the nervous and the endocrine system for the control of stability of the internal environment. It is then apparent that the "vocabulary" of neuroendocrine communication in the animal kingdom is rich and susceptible to further enrichment by new types of "languages." We will now consider the most common type of neuroendocrine communication in mammals.

Mammalian cells differ from bacterial cells in that they are less equipped to cope with hostile environments, and compensate for this lack through the organization of similarly differentiated elements—specialized cells—which react with other groups of cells to regulate the composition of plasma and extracellular fluid constituents. These communication cells have been classified as "neuroendocrine transducers" because they act to translate brain-type signals (i.e., neurotransmitters acting at synapse) to hormonal outputs (Wurtman, 1970, 1973). Communication, therefore, is usually mediated by two types of chemical signals, neurotransmitters and hormones. In this respect, neuroendocrine transducers are said to differ from "endocrine transducers" (thyroid, ovary, adrenal cortex), whose physiological inputs come only from the circulation, since, in addition to specific humoral signals, they also respond to neuronal inputs.

Many groups of neuroendocrine transducer cells have been recognized in mammals, and their characteristics have been described (see Wurtman, 1973, and Table I). At least six such groups of neuroendocrine transducer cells have been demonstrated; additional candidates include the cells of the rodent thyroid and the "amine precursors uptake and decarboxylation cells" (APUD cells) (Pearse, 1968) for which either monoamines or somatostatin or both may represent the input signal(s) (see also Chapter III, Sections A,3,c, and C). We have also proposed that pituitary lactotrophs may act as transducer cells.

The demonstration that a particular cell type mediates neuroendocrine transduction requires that nerve terminals be in close proximity to the cell and that the capacity of the cell to make a secretory response after an appropriate stimulus be damaged when its innervation is interrupted. As for their embryologic derivation, even though most neuroendocrine transducers are of neuroectodermal origin, others (e.g., β-cells of pancreatic

TABLE I

Neuroendocrine Transducer Cells

Cell type	Input signal	Hormonal output signal
Hypothalamic neurohormone cells	Monoamines and other neurotransmitters	Releasing and inhibiting neurohormones secreted into pituitary portal circulation
Cells of the hypothalamic supraoptic nucleus	Acetylcholine or norepinephrine	Vasopressin
Pituitary lactotropes?	Dopamine	Prolactin (inhibition)
Adrenal medulla	Acetylcholine (preganglionic sympathetic neurons)	Epinephrine
Pineal organ	Norepinephrine (postganglionic sympathetic neurons)	Melatonin
Juxtaglomerular cells of the kidney	Norepinephrine (postganglionic sympathetic neurons)	Renin
β-Cells of pancreatic islets	Norepinephrine (postganglionic sympathetic neurons)	Insulin
Thyroid follicular cells?	Monoamines	Thyroid hormones
β- and α_2-cells of pancreas, parafollicular cells of the thyroid, gastrointestinal gland cells?	Somatostatin	Insulin (inhibition), glucagon (inhibition), calcitonin, gastrin, pepsin (inhibition)

islets) are not. Thus, not necessarily all of them have to be classified as true neurosecretory cells. A unique feature of the neuroendocrine transducers thus far identified is the frequent involvement of catecholamines and indoleamines in their communication function. These aminergic links are widespread; the monoamines either transmit messages between cells over long distances as hormones (e.g., epinephrine release from the adrenal medulla) or act over short distances as transmitter substances (neuroendocrine transducers of the hypothalamus). In addition, MA's are normal constituents of many polypeptide hormone-producing cells (see also below), and evidence suggests that activation of the cells to induce hormone release is associated with changes in the amine levels and turnover within the cell (Owman *et al.,* 1973).

1. The Neurohypophysis

In mammals the major neuroendocrine transducers are the supraoptic and paraventricular–hypophyseal neurons which constitute the neurohypophysis, and the neurons of the base of the hypothalamus which secrete hypophysiotropic hormones into the blood supplying the AP gland.

The neurohypophysis consists of neurons whose cell bodies are situated in the supraoptic and paraventricular nuclei of the hypothalamus. The biologically active peptides which they synthesize, vasopressin and oxytocin, are released from the endings of their axons into the posterior pituitary where they are stored. They are then released into the general circulation in response to either suckling or genital stimulation (oxytocin) or changes in extracellular fluid osmolarity and volume (vasopressin).

Two kinds of vesicles are found in the terminals of the supraoptic hypophyseal neurons: the neurosecretory vesicle (about 1500 Å in width), related to hormone synthesis and transport, and a smaller electron-lucent vesicle (about 400 Å) (Palay, 1957), which was termed "synaptic vesicle" and was by analogy with ordinary nerves believed to contain neurotransmitter substances (such as acetylcholine or norepinephrine) that are liberated during stimulation to initiate hormone release. Interestingly enough, these vesicles stain preferentially with the zinc iodide osmic acid (ZIO) procedure as do the classic synaptic vesicles of ordinary nerves (Christ and Back, 1970; Rufener and Dreifuss, 1970). In accordance with the above hypothesis, both morphologic (Shute and Lewis, 1966; Fuxe and Hökfelt, 1967) and electrophysiologic studies (see Cross, 1973) produced evidence for neurotransmitter involvement of the cholinergic and adrenergic systems in the release of neurohypophyseal hormones.

To summarize the results of many studies, some of which refer to the identification of neurosecretory cells *in vivo* by antidromic stimulation in conjunction with iontophoretic injection of drugs into hypothalamic cells (see Cross, 1973), it would appear that the cholinergic afferent connections are mainly responsible for the stimulation of hypothalamic nuclei, while afferent noradrenergic fibers exert inhibitory effects on hormone release.

Koelle (1961) proposed that ACh might be stored in, and released from, hormone- and ACh-containing nerve endings on stimulation of the secretory neurons and then act on the same nerve endings to influence the release of neurohypophyseal hormones. This would produce the unique situation of a neuron's own transmitter substance providing the stimulus for the release of its own endocrine product. Subsequent studies, however, have challenged Koelle's theory of the "double neurohumoral role of acetylcholine" in the neurohypophysis. Under *in vitro* conditions, when the release of hormones can be easily stimulated and measured (Douglas, 1963; Douglas and Poisner, 1964), ACh was shown to stimulate the release of neurohypophyseal hormones (Douglas and Poisner, 1964; Dicker,

1966). Lederis and Livingston (1970), using differential and density-gradient centrifugation followed by bioassay (BA) and electron microscopy (EM), have found ACh in nerve endings other than those containing the neurohypophyseal hormones. Finally, Douglas (see Douglas, 1973) has suggested that the "synaptic" vesicles are merely the result of the retrieval of the neurosecretory granule membrane by vesiculation, following induction of exocytosis and extrusion of granule contents directly to the cell exterior (Fig. 2).

If ACh does not actually exert a neurotransmitter function, it may serve some other function than that proposed by Koelle (1961), e.g., a vasomotor function (Lederis, 1967; Lederis and Livingston, 1968). It has been postulated (Lederis and Livingston, 1970) that since vasodilatation has been shown to occur in the neural lobe after systemic or intraventricular (IVT) administration of ACh (Konstantinova, 1967), an increased blood flow through the neurohypophysis (facilitated by ACh) may coincide with or follow a hormone-releasing stimulus and enhance the early arrival of vasopressin or oxytocin at the target organ.

If this hypothesis is correct, the hypothalamic–neurohypophyseal system would not represent a true neuroendocrine neurotransducer in the sense categorized by Wurtman (1973), and the control of posterior lobe

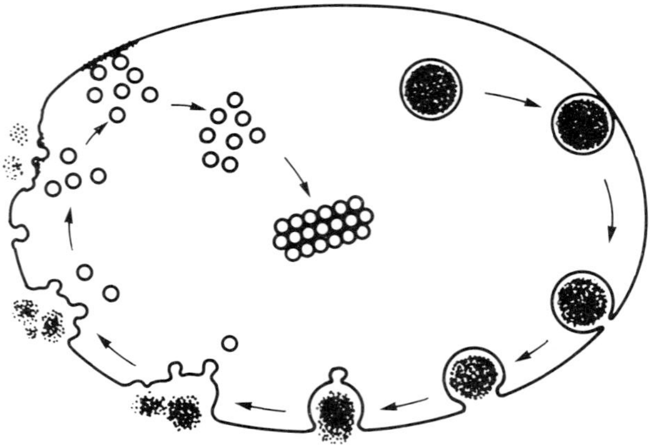

Fig. 2 A scheme depicting the proposed exocytosis–vesiculation sequence and suggested origin of microvesicles ("synaptic" vesicles). In the interest of clarity the "coating mechanism" involved in microvesicles formation is omitted here. The clustered vesicles (center) are probably digested by lysosomal activity. (Reproduced, with permission, from Douglas *et al.*, 1971.)

hormones would be similar to the neural control of exocrine glands (secretomotor innervation) (see Scharrer, 1972).

2. The Hypothalamic–Adenohypophyseal System

The neuroendocrine transducer function of the hypothalamic–adenohypophyseal portal system will be discussed in later chapters. Only some features of this communication system will be considered now.

Extrinsic and intrinsic afferent stimuli are conveyed to extrahypothalamic and hypothalamic neurons, where they are transformed into uniform afferent commands and relayed to the various glandular elements of the pars anterior and pars intermedia by specific neurohormones. For reaching their destination, the hypothalamic neurohormones, instead of using the general circulation, make use of special pathways, particularly the primary plexus of the hypophyseal portal system (see also Chapter III, Section A,1). This classic notion, however must be reevaluated in the light of recently acquired knowledge. In the last few years, the availability of specific antibodies against isolated neurohormones has made possible the immunohistochemical tracing of neuronal systems containing RF's (see Hökfelt *et al.*, 1976c, and Chapter III, Section C). In addition to dense neurohormone innervation of the median eminence (ME), which is in agreement with the postulated role of a humoral factor transported in the portal vessels to the AP, the unexpected finding of dense plexuses of nerve endings scattered in various hypothalamic or extrahypothalamic areas does not seem to indicate a release into blood vessels but instead a transmitter type of release at synapses. This would agree with the reported extraendocrine actions of RF's (see Chapter III,E), and is consistent with the possibility that a single neurosecretory neuron may liberate a product which acts either as a neurohormone when released into the bloodstream, or in close proximity, as a true neurotransmitter (Renaud *et al.*, 1975) (Fig. 3). Evidence for the existence of these bipolar or multipolar neurons has been shown by electrophysiological (Dyer, 1975) (see Chapter II, Section B,14) and immunohistochemical (Alpert *et al.*, 1976) methods.

There is, in addition, evidence for the presence of additional brain mediators, which also participate in the control of anterior pituitary function (B fibers of Knowles, see Section A). Catecholamine- and serotonin-containing neurons and terminals occur in discrete regions of the anterior hypothalamus and ME-arcuate areas. The main CA-neuron systems in the CNS have been mapped by fluorescent histochemistry and by biochemical analysis after inducing selective lesions in the brain (see Chapter II, Sec-

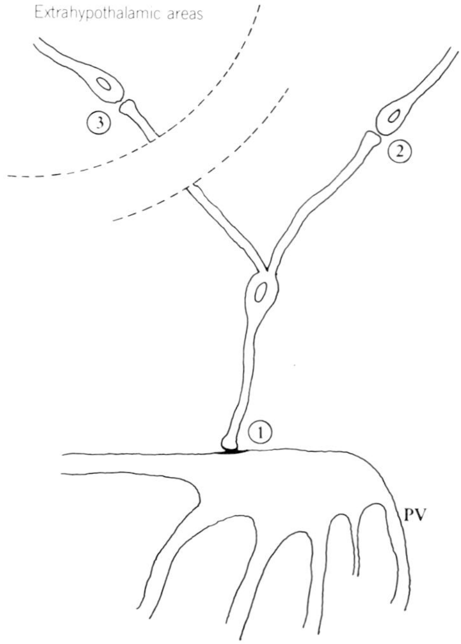

Fig. 3 Schematic drawing of a hypothalamic neurosecretory neuron which, besides innervating the ME (1), sends a collateral axon that terminates locally within the hypothalamus (2), or extends to brain sites remote from the hypothalamus (3). PV, hypophyseal portal vessels.

tion B,1–5). (For the distribution of the serotoninergic system see Chapter II, Sections C,2 and 3.)

Pharmacologic studies and deafferentation experiments have shown that the CA projection to the external layer of the ME is almost entirely derived from the DA tuberoinfundibular projection, whereas the CA projection to the internal layer of the ME is mainly noradrenergic and arises from ascending extrahypothalamic NE pathways.

The massive monoaminergic input to the external layer of the ME (one third of all nerve endings) suggests an important functional role for DA in this region (see Chapter IV). However, the mode and site of action of DA are still unclear, one major problem being whether DA acts at the ME or at the pituitary level (see Chapter IV, Section C,4).

It is worthwhile to consider some features of the tuberoinfundibular dopamine (TIDA) neurons (Shoemaker, 1973). First, the DA cells of the arcuate nucleus are atypical neurons in that they do not form synapses. The EM studies of Ajika and Hökfelt (1973) have shown, in fact, that in the external layer of the ME there are no typical morphological synapses between axon profiles. Rather the efferent cell processes of DA neurons appear to terminate adjacent to the perivascular space (Stoeckart *et al.*, 1972), thus suggesting that DA may influence any of the glial elements in the ME (Hökfelt, 1974). Typical, although few, synaptic profiles have, on the other hand, been described for the DA boutons of the substantia nigra–corpus striatum-containing neurons (Hökfelt, 1968). Second, resistance of the hypothalamic DA neurons to the depleting effect of 6-hydroxydopamine (6-OHDA) has been described in the rat following high doses of 6-OHDA administered into the 3rd ventricle (Cuello *et al.*, 1973). Finally, a DA-stimulated adenyl cyclase which has been equated to the postsynaptic DA receptor (Kebabian *et al.*, 1972) is lacking at the ME level (P. F. Spano, personal communication) and so are DA/neuroleptic receptors in the MBH (G. M. Brown *et al.*, 1976a), and the regulatory mechanisms governing the activity of TIDA neurons appear to differ from those in the other dopaminergic systems (Gudelsky and Moore, 1976).

The unique characteristics of the TIDA neurons raise the question of their actual nature, and of the function of DA as synaptic transmitter at the ME level. The studies of Clementi *et al.* (1970) for isolating different types of nerve terminals by sucrose density-gradient centrifugation of the ME gave evidence for the relationship of the two RF's, the growth hormone-releasing factor (GRF), and the follicle-stimulating hormone-releasing factor (FSH-RF) with the DA-containing nerve terminals. Similarly, Rodriguez (1972) has suggested that MA's and hypothalamic RF's are present in the same neuron, although two reports to the contrary have been presented (Kizer *et al.*, 1975a; Knigge *et al.*, 1977). It might be postulated that DA in the tuberoinfundibular neurons is present in the same neurosecretory neurons in which RF's are present, where it may function other than as a true neurotransmitter (see also Chapter IV, Section C). It is noteworthy that many hormone-producing cells in a variety of peripheral tissues contain large amounts of CA's (Owman *et al.*, 1973). There is no evidence that MA's are synthesized in the cell, but rather that they are taken up and formed on decarboxylation (APUD cells), nor is the specific function of such amines known. As already mentioned, Owman *et al.* (1973) have proposed that amines are normal constituents of polypeptide hormone-producing cells and that the intracellular amines have a function related to the formation, storage, and/or release of the hormone. In the light of present knowledge, the possibility has to be considered that the hypophysiotropic neurons of the CNS, just as the peripheral amine-containing endocrine cells—which incidentally have the same embryologic derivation from the neural crest (Weichert, 1970)—may take up DA with functions in the binding and release of polypeptide hormones.

However, more recent findings do not assign to TIDA neurons and terminals at the ME level the mere function of storage sites for neurotransmitter and neurohormones, but instead an important role in neuroendocrine feedback regulation (Kizer *et al.*, 1976b). It would appear, in fact, that the majority, if not all, of the endocrine responsive catecholaminergic neurons in the ME are dopaminergic (see also Chapter IV, Sections A,5 and B,3).

II

Proved and Putative Neurotransmitters in the Central Nervous System

Despite the growing list of compounds with putative neuroregulatory functions, there are only a small number for which there are both strong suggestive evidence of CNS transmitter activity and histochemical localization of their pathways. These three classes of compounds are catecholamines (NE and DA), 5-HT, and ACh (see Andén *et al.,* 1969; Glowinski and Baldessarini, 1966; Katz, 1969; Marchbanks, 1972; Potter, 1972). Other substances, γ-aminobutyric acid (GABA), aspartate, cyclic AMP (cAMP), glutamate, glycine, histamine, γ-hydroxybutyrate, melatonin, prostaglandins, substance P, taurine, tryptamine, etc., are still considered putative transmitters, not yet having fulfilled criteria reported below for true neurotransmitters.

The identification and estimation of real or putative transmitters, revealed initially by means of classic methods, e.g., biological, colorimetric, and fluorometric, have recently received great impetus after the discovery of more sophisticated and highly sensitive (picamole range) techniques, e.g., enzymatic–isotopic, mass spectroscopy, which enable detection and measurement of these substances in the nuclei of different areas of the brain (Table II).

A. STEPS INVOLVED IN SYNAPTIC TRANSMISSION

The criteria that must be fulfilled for a substance to be considered a neurotransmitter include (1) recovery of the suspected transmitter from the perfusate of an innervated structure after nerve stimulation; (2) iden-

TABLE II

Methods Used for the Identification and Estimation of Proved or Putative Neurotransmitters

Transmitter	Method	References
Catecholamines	Biological assay	See Vane (1966); Gaddum (1959)
	Colorimetric methods	Ehringer and Hornykievicz (1960)
	Fluorometric methods	von Euler (1959); Laverty and Taylor (1968); Montagu (1957); Laverty and Sharman (1965)
	Radioisotopic methods	Saelens *et al.* (1967); Nikodijevic *et al.* (1969)
	Radioenzymatic assay	Ben-Jonathan and Porter (1976)
	Gas chromatographic methods	Clarke *et al.* (1967)
	Mass spectrometry	Knoche *et al.* (1969); Koslow *et al.* (1974)
	Histochemical fluorescence methods and microspectrofluorometry	Jonsson *et al.* (1972); Björklund and Nobin (1973)
	Immunohistofluorescence technique	Goldstein *et al.* (1972); Hökfelt *et al.* (1974a)
	Microdissection and microassay technique	Palkovits (1973); Coyle and Henry (1973)
	Electron microscopy + autoradiography	Aghajanian and Bloom (1967b); see Iversen and Schon (1973); Reivich and Glowinski (1967)
Serotonin	Biological assay	Erspamer (1953); Twarog and Page (1953); see Garattini and Valzelli (1965)
	Spectrophotometric methods	Udenfriend *et al.* (1955)
	Fluorometric methods	Udenfriend *et al.* (1955); Bogdanski *et al.* (1956); Snyder *et al.* (1965); Maickel and Miller (1968); Giacalone and Valzelli (1969); Curzon and Green (1970)
	Radioenzymatic methods	Saavedra *et al.* (1973)
	Autoradiography + EM	See Iversen and Schon (1973); Aghajanian and Bloom (1967a)
	Mass spectrometry	Koslow *et al.* (1974); Abramson *et al.* (1974)
	Radioimmunological methods	Peskar and Spector (1973)
Acethylcholine	Biological assay	Whittaker (1963)
	Colorimetric methods	Friesen *et al.* (1965)
	Fluorometric methods	Fellman (1969)
	Isotopic methods	Collier and MacIntosh (1969); Schubert *et al.* (1969); Feigenson and Saelens (1969); Potter (1970); see Hebb (1972a); Wonnacott and Marchbanks (1975)

TABLE II *(Continued)*

Transmitter	Method	References
	Gas chromatography	Stavinoha and Ryan (1965); Cranmer (1968); Hanin and Jenden (1969); see Jenden (1973)
	Mass spectrometry + gas chromatography	Johnston *et al.* (1968a) Jenden (1973); Koslow *et al.* (1974); Hanin and Schubert (1974)
GABA	Biological assay	See Baxter (1970); Ash and Tucker (1966); Florey (1961); Florey and Elliott (1961)
	Chromatographic methods	Sandman (1962); Baxter and Senoner (1964); Crowshaw *et al.* (1967); Smith (1960)
	Chromatography + radiochemistry or spectrofluorometry	Tarver *et al.* (1975)
	Enzymatic assay	Baxter and Roberts (1961); Graham and Aprison (1966)
	Autoradiography + EM	Iversen and Schon (1973)
	Mass spectrometry	Cattabeni *et al.* (1975)
	Radioreceptor assay	Enna *et al.* (1975)
Prostaglandins	Bioassay	Horton and Main (1967)
	Spectrophotometric method	Bydgeman and Samuelsson (1964, 1966)
	Enzymatic analysis	See Karim and Hillier (1972); Änggård (1971)
	Gas–liquid chromatography	Ramwell *et al.* (1968); Shaw and Ramwell (1969); Jouvenaz *et al.* (1970)
	Gas–liquid chromatography + mass spectrometry	Sweely *et al.* (1966); Samuelsson *et al.* (1970); Thompson *et al.* (1970)
Histamine	Bioassay	Kwaitkowski (1943); Harris *et al.* (1952); Adam (1961)
	Fluorometric method	Shore *et al.* (1959); Kremzner (1966); Medina and Shore (1966)
	Enzymatic–isotopic method	Snyder *et al.* (1966); Taylor and Snyder (1972); Brownstein *et al.* (1974a)

tification of the substance; (3) demonstration that the exogenous application of the suspected transmitter mimics the action of the transmitter released by nerve stimulation; and (4) demonstration that the responses to nerve stimulation and to the injected substance are modified in the same way by various drugs (see Cooper *et al.*, 1974; Goodman and Gilman, 1975).

If a substance acts as a neurotransmitter, it is of particular importance to know the dynamic aspects involved in synaptic transmission (see Fig. 4), since each step may constitute a possible site of action of the various interfering drugs.

1. Synthesis

One or more metabolic steps catalyzed by proper enzymatic activity allow the synthesis of a given transmitter. As will be described in the

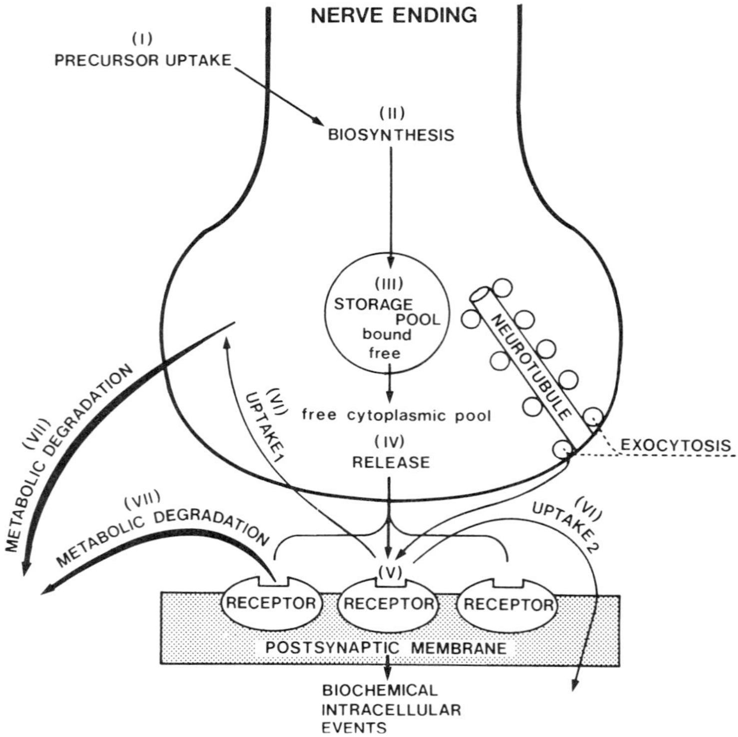

Fig. 4 Schematic illustration of the dynamic steps involved in the transmission of a neural message by an ideal neurotransmitter. (I) Uptake of the precursor into the neuron by a transport system. (II) Biosynthesis of the neurotransmitter. (III) Storage in specific granules where it exists in part as bound and in part as free. (IV) Release by nerve impulse and illustration of the exocytosis mechanism. (V) Effect on specific postsynaptic receptors with activation of membrane and intracellular biochemical events leading to the physiological response. (VI and VII) Termination of the neurotransmitter effects by uptake (VI) into the terminals or into the postsynaptic cells or degradation (VII) at the postsynaptic or presynaptic level or after diffusion into the general circulation (for details see text).

following paragraphs, enzymes or other factors limit the rate of biosynthesis, accelerating or decreasing it (von Euler, 1972). The enzymes involved in transmitter formation are synthesized in the cell body, and by axonal flow are transported to the nerve endings (see Dählström and Häggendal, 1966; Smith and Winkler, 1972; Weiss, 1970). Furthermore, an uptake mechanism which enables transport of the precursor from the plasma to the nerve ending must be operating.

2. Storage

After being synthesized, the transmitter is stored in synaptic vesicles (or granules) where it is protected from metabolic degradation. Small (about 500 Å) and large (about 1000 Å) granular vesicles are present in monoamine neurons (see Grillo and Palay, 1962; Smith and Winkler, 1972). NE, DA, and 5-HT are present not only in the terminals but also in the cell bodies and in the axons. The bulk of the transmitter is present in the small vesicles; in the large ones, which are more abundant in the axons, enzymes such as D-β-H and only relatively small amounts of amines are found (Bloom, 1972; see Hökfelt, 1970). Generally, two pools of transmitter are present within the granules: a free one, more readily available for release, and another bound to ATP or other nucleotides (Douglas and Poisner, 1966). A "small critical" free pool is also present in the cytoplasm (for details see Section B,7).

The introduction of the permanganate fixation method (Hökfelt, 1967a) has provided information on the precise ultrastructural storage sites of brain neurotransmitters. From these studies it was confirmed that, as in the peripheral sympathetic nervous system, small and large dense core vesicles are among the storage sites for amines in the brain.

3. Release

The arrival of a nerve action potential at the axonal terminals produces a synchronous release of several quanta (Katz, 1969) of the transmitter toward the postsynaptic receptors. It is likely that one step involved in the transmitter release is the mobilization of Ca^{2+} which may enter the intraxonal medium, activating the discharge of the transmitter (Douglas, 1968; Rubin, 1970). The various modes of release of CA's from storage sites have been excellently reviewed (Smith, 1973) (see also Section B,8).

4. Interaction with Postsynaptic Receptors

When the transmitter is released from its storage site by the presynaptic action potential or pharmacologically, it acts on some specific postsynaptic receptors, generating either excitatory or inhibitory postsynaptic potentials. In many parts of the peripheral autonomic nervous system and in some sites of the CNS, the cellular mechanism of the response to neurotransmitters and hormones involves the stimulation of adenylate cyclase (AC) with a subsequent increase in 3′5′-cAMP (see Weiss, 1970; Bloom *et al.*, 1973a; Iversen, 1975). The 3′-bond of cAMP is a high-energy one which permits in nervous and nonnervous tissue the phosphorylation of certain key enzymes as a result of an activation of a cAMP-dependent protein kinase (see Fig. 5), thus triggering changes in ion permeability of the membrane, generation of postsynaptic potentials, and other metabolic events leading to a physiologic response (see Iversen, 1975). Several groups of investigators have presented evidence that AC is activated by a variety of hormones and neurotransmitters, principally CA's and 5-HT (see Weiss, 1970). In peripheral tissue the pharmacologic properties of CA-sensitive AC are those typical of a β-type of adrenoreceptor (see Iversen, 1975). Also, in the brain, in which the activity of AC and the concentration of cAMP are higher than in any other organ, NE-sensitive AC, in cerebellar Purkinje cells (see Bloom *et al.*, 1973a), and DA-

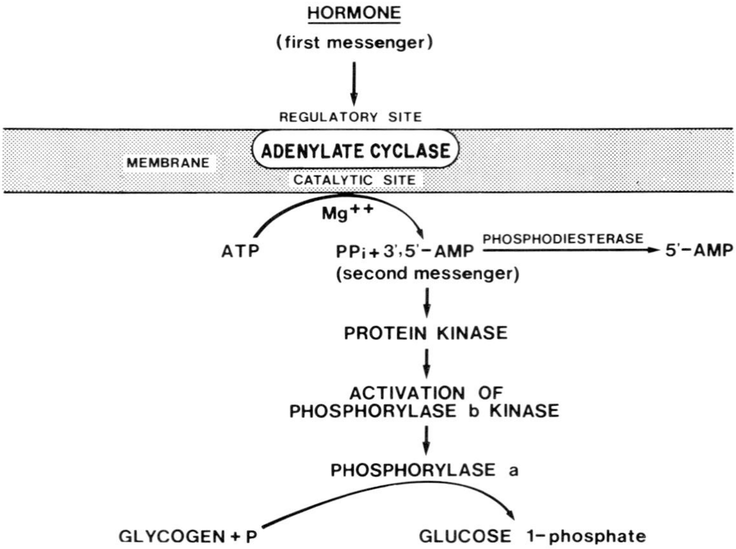

Fig. 5 Sequence of events following neurohormonal activation of adenylate cyclase with synthesis of 3′,5′-AMP (II messenger). This in turn stimulates a protein kinase with consequent activation of a phosphorylase leading to glycogenolysis as ascertained in neural and nonneural tissues (for details see text).

sensitive AC, in retinas (Brown and Makman, 1972) and in the striatum (Kebabian *et al.*, 1972), exist.

5. Termination of the Transmitter Effects

Several efficient mechanisms exist to terminate within a short interval the transmitter action. They include metabolic postsynaptic degradation, e.g., acetylcholinesterase (AChe), diffusion into the circulation and subsequent metabolic degradation in other organs, physical removal (e.g., CA's) by means of recapture into the presynaptic neuron ($Uptake_1$) and into the vesicles, or by recapture into postsynaptic or other cells ($Uptake_2$) followed by metabolic degradation (Iversen, 1967, 1971b).

B. CATECHOLAMINES

1. Occurrence and Regional Distribution of Catecholamines in the CNS

This section, rather than giving an extensive survey of the literature on this topic, for which the reader is referred to the reviews of Fuxe (1965a,b), Ungerstedt (1971a), Holzbauer and Sharman (1972), presents information on the organization of the CA system within the CNS, particularly in the hypothalamus, the median eminence, and the anterior pituitary, areas of unique importance in the control of AP hormone secretion. Emphasis will be given to the fluorescence histochemical data which have allowed the visualization of CA and 5-HT nerve terminals, the study of their topographical distribution, and their relation to other neuronal systems.

Norepinephrine in the brain was first identified by Holtz (1939) and its distribution pattern studied by Vogt (1954), who found particularly high concentrations in the hypothalamus and other areas of central sympathetic representation when compared to other brain areas. More NE was generally found in the gray than in the white matter. The nonuniform distribution of NE, which does not coincide with the density of blood vessels present in a given brain area, and its characteristic regional localization within mammalian brain suggested that NE might have specialized functions, perhaps as a central neurotransmitter. This view was reinforced by the finding that the relative distribution of NE was quite similar in most mammalian species (Table III). Epinephrine concentration in mammalian CNS was relatively low, being approximately 5–17% (by bioassay) of NE content (Cooper *et al.*, 1974).

TABLE III

Concentration ($\mu g/gm$) of DA and NE in the Hypothalamus of Several Animals Species[a]

Amine	Chicken	Pigeon	Rat	Rabbit	Cat	Sheep	Pig	Dog	Monkey	Man
DA	0.08	0.15	0.14–1.3	0.20	0.27–0.75	0.19	0.88	0.26, 0.20	1.69	0.8 –1.12
NE	1.42	1.50	1.29–1.79	1.07	1.40–2.01	1.15	0.83	1.0	7.10	1.11–1.25

[a] Data from Hornykiewicz (1966), Iversen (1967), Juorio and Vogt (1967), Holzbauer and Sharman (1972). Further details and references on the distribution of E, NE, and DA in several areas of the brain of mammalian and nonmammalian species can be found in Holzbauer and Sharman (1972).

Since NE was shown to be present in mammalian brain, the occurrence in the brain of small amounts of DA, its immediate precursor, was to be expected. Montagu (1957) and Weil-Malherbe and Bone (1957) first presented evidence that DA is a normal constituent of the mammalian brain. Since then, DA has been found in the brain of all mammalian species examined so far (see Hornykiewicz, 1966, 1973). In 1959, Bertler and Rosengren, as well as Carlsson, reported that the regional distributions of DA and NE in the mammalian brain were markedly different. Little DA was found in the NE-rich hypothalamus, the highest amounts of this amine (approximately 80% of total brain DA) being present in the caudate nucleus and putamen regions. This observation has since been confirmed and extended to the substantia nigra and the globus pallidus. In the caudate nucleus, which is nearly devoid of NE (0.1 μg/gm or less), the concentration of DA is on the average 10 μg/gm of fresh tissue. In contrast, DA concentration in the hypothalamus (see Table III) rarely exceeds 10% of the concentration of NE, the latter ranging between 1 and 2 μg/gm in this region. DA is present in relatively high concentration in the ME and in the infundibular stem as recorded by chemical assay (Laverty and Sharman, 1965; Rinne and Sonninen, 1968; Björklund *et al.*, 1970; Fuxe and Hökfelt, 1969). There are findings to suggest that DA also may be the predominant CA in the retina and tuberculum olfactorium (Häggendal and Malmfors, 1965; Nichols *et al.*, 1967). Reports on the occurrence of DA in the spinal cord are controversial. McGeer and McGeer (1962) found it to be the main CA in the spinal cord of the rat, rabbit, ox, cat, dog, and man. Other authors, however, were unable to find substantial amounts of this amine in the spinal cord (see Hornykiewicz, 1966).

2. Fluorescent Pathways of Norepinephrine and Epinephrine

As mentioned before, the development of a histochemical fluorescence technique of high sensitivity and sufficient specificity has permitted the study of the intraneuronal CA distribution (Falck *et al.*, 1962). The method relies upon the fact that certain monoamines, and some of their precursors, can be converted to fluorescent products by treatment with paraformaldehyde gas provided they are enclosed in a dry protein layer. The catecholamine fluorescence appears green, whereas 5-HT fluorescence is yellow to yellow-green. The detailed mapping of monoamine pathways in the CNS has been presented in excellent reviews (for details, see Dählström and Fuxe, 1965; Fuxe, 1965a,b; Hillarp *et al.*, 1966; Ungerstedt, 1971a; Hökfelt *et al.*, 1974a) and, recently, a topographic atlas of CA neurons has also been published (Palkovits and Jacobowitz, 1974).

The classic work of Fuxe (1965a,b) indicates that 7 out of 10 NE cell

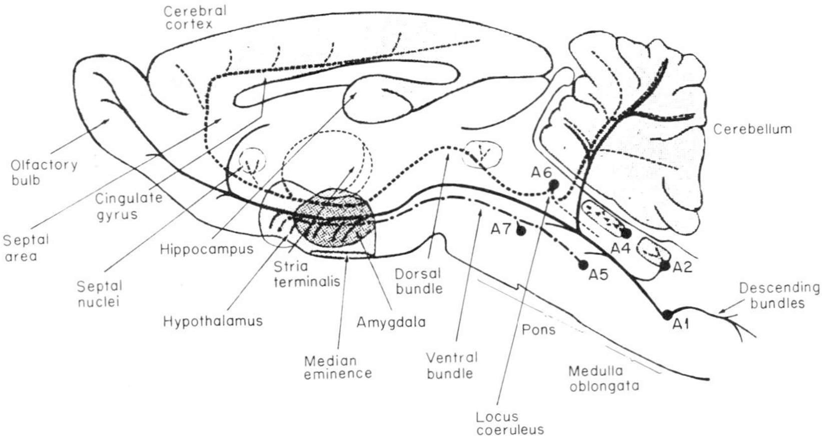

Fig. 6 Schematic diagram of a sagittal section of the rat brain showing central NE cell groups and ascending and descending NE pathways. Shaded areas indicate regions of NE terminals (for details see text). (Reproduced, with permission, from Livett, 1973.)

body groups are present in the pons and medulla oblongata. The A1, A2, A5, and A7 cell groups give rise to one ("ventral pathway") of the two major ascending NE pathways (Fig. 6). The axons ascend to the mid-reticular formation, turn ventromedially along the lemniscus medialis, and continue rostrally, mainly within the medial forebrain bundle (MFB). This pathway gives rise to NE nerve terminals in the lower brainstem, the mesencephalon, and the diencephalon. In the medulla oblongata and pons, the ventral NE pathway and the dorsal pathway, originating in the locus coeruleus, overlap and both contribute to the terminal areas; however, in the mesencephalon and the diencephalon the majority of NE nerve terminals seem to derive from the ventral pathway. In particular, the ventral NE pathway innervates the whole hypothalamus (e.g., the n. dorsomedialis hypothalami, the n. periventricularis, the area ventral to the fornix, the ARC n. and the internal layer of the ME, the retrochiasmatic area, the n. paraventricularis, the SON, and the POA).

The NE cell bodies in the locus coeruleus (A6) (see an example depicted in Fig. 22a and b), besides a dorsal NE pathway, projecting to the cerebral cortex, the hippocampus, and the anterior hypothalamus, give rise to a descending pathway that innervates the lower brainstem nuclei and to a lateral pathway that enters and innervates the cerebellum medial to the pedunculus cerebellaris medius (Olson and Fuxe, 1972). It is generally agreed that part of the hippocampal innervation probably also derives from direct hypothalamic connections while other ascending axons follow the cingulum caudally and then enter the hippocampus.

Recently, by measuring NE concentration by radioisotopic assay in discrete brain regions of the rat following unilateral electrolytic lesions of the locus coeruleus (Kobayashi *et al.,* 1974), a significant reduction was found in NE content ipsilateral to the lesion in the following areas: all portions of the cerebral cortex examined (entorhinal, hippocampus, cingulate, parietal, and occipital areas), anterior half of the cerebellar cortex, hypothalamic periventricular and paraventricular nuclei, anterior ventral thalamic nucleus, ventral thalamic nucleus, and habenula. This demonstrates that these regions receive unilateral innervation by axons originating from the locus coeruleus. No reduction in NE concentration occurred in the medial preoptic nucleus, n. interstitialis striae terminalis (ventralis), dorsomedial hypothalamic n. and MFB.

The use of an immunohistochemical method has made possible the demonstration in mammalian CNS of neurons containing epinephrine (Hökfelt *et al.,* 1974a). In particular, by means of antibodies against PNMT, the enzyme converting NE into E, it has been clearly demonstrated that certain areas of the brain, such as the olfactory bulb, olfactory tubercle, and the hypothalamus, contain substantial amounts of the enzyme and therefore are capable of forming E *in vivo.* The E-cell bodies have been found in the reticular formation of the medulla oblongata; they send descending axons to the sympathetic lateral column and ascending axons into the hypothalamus, the periventricular gray, and the locus coeruleus. Furthermore, there is a dense E innervation of the n. dorsalis motorius (i.e., n. dorsalis motorius n. vagi) n. vagi, and the n. tractus solitarium (Fig. 7). In a detailed report on the localization of PNMT in the rat brain nuclei, this enzyme was found to be unevenly distributed in the

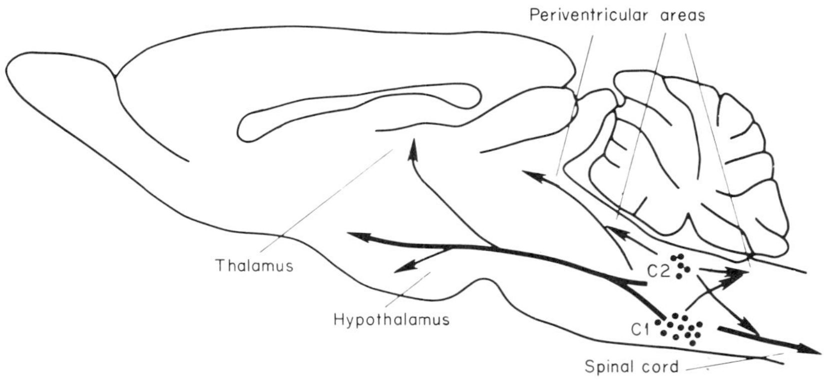

Fig. 7 Schematic diagram of a sagittal section of the rat brain showing central E neurons. C1 and C2: PNMT-positive cell groups from which hypothetical ascending and descending axons arise (for details see text). (Reproduced, with permission, from Hökfelt *et al.,* 1974a.)

hypothalamus, the largest concentration being in the MBH, in the ARC and paraventricular nuclei, and in the ME (Saavedra *et al.,* 1974a,b). Recently, PNMT has been found to be induced in the rat medulla and hypothalamus with continued dexamethasone treatment (Moore and Phillipson, 1975a). Since E nerve terminals are particularly rich in regions involved in vasomotor control, the hypothesis has been formulated (Bölme *et al.,* 1974) that central E neurons constitute a vasodepressor system and that the antihypertensive agent clonidine acts as an E agonist in the brain. At present, the precise significance of the occurrence of epinephrine in other areas of the brain is not completely known, but evidence is accumulating that suggests a role for hypothalamic E in the neuroendocrine control of gonadotropin secretion (see Chapter IV, Section A,3).

3. Fluorescent Pathways of Dopamine

Dopaminergic neurons in CNS are organized in several systems, e.g., nigrostriatal, mesolimbic, tuberoinfundibular, CTZ, and retinal. Some DA-pathways are represented in Fig. 8. The A9 DA cell group in the zona compacta of the substantia nigra and its rostromedial extension in the area ventralis tegmenti give rise to a large pathway ("the nigrostriatal pathway") that ascends in the lateral hypothalamus, enters the crus cerebri in the mid-hypothalamus, intermingles with the myelinated bundles in the

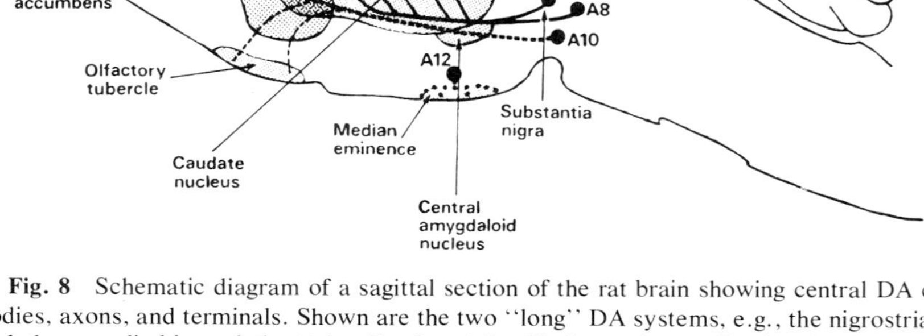

Fig. 8 Schematic diagram of a sagittal section of the rat brain showing central DA cell bodies, axons, and terminals. Shown are the two "long" DA systems, e.g., the nigrostriatal and the mesolimbic and the "short" tuberoinfundibular pathway. Depicted are also the mesocortical dopaminergic projections. Shaded areas indicate regions of DA terminals (for details see text). (Modified from Livett, 1973.)

capsula interna, and provides terminals to the n. caudatus-putamen. The DA nerve terminals in the n. amigdaloideus seem to be an extension of the DA terminals of the n. caudatus-putamen and originate from axons running lateroventrally from the capsula interna (Ungerstedt, 1971a). The axons from the A8 DA cell bodies caudal to the substantia nigra, dorsal to the lemniscus medialis, probably join the nigrostriatal DA system as they show retrograde degeneration after lesions of the corpus striatum.

The axons from the DA cell bodies, surrounding the interpeduncular nucleus (A10), form the mesolimbic system which ascends together with the axons of the nigrostriatal DA system. They follow a more medial route and never enter the crus cerebri but ascend just dorsal to the MFB. At the level of the commissura anterior one branch enters the n. accumbens and the n. interstitialis striae terminalis dorsal part, and another turns lateroventrally to enter the tuberculum olfactorium.

New types of DA nerve endings have been recently described in the amygdaloid complex, the septal area, and in the limbic cortex. These terminals probably arise from axons originating mainly from the A10 DA cell group in the ventral mesencephalon (Fuxe *et al.*, 1975a). In addition, other groups of DA neurons have been identified in the hypothalamus by the same authors (see below).

4. Catecholamine Innervation of the Hypothalamus

In view of the rich distribution of CA's in the hypothalamus and the essential role this area exerts in the control of neuroendocrine mechanisms, a great deal of effort has also been made to determine CA distribution in individual nuclei of the hypothalamus, using histochemical, ultrastructural, and enzymatic–isotopic microassay techniques. As already pointed out, the NE nerve terminals in the hypothalamus arise to a large extent in the MFB, mostly originating from cell bodies in the pons and medulla oblongata (see Fuxe and Hökfelt, 1969). The distribution of NE nerve terminals in the hypothalamus is uneven. High to very high fluorescence intensity was found in the following areas: n. supraopticus, n. paraventricularis, n. dorsomedialis, and retrochiasmatic area. Low to medium intensity was observed in mammillary bodies, in the AHA and in the LHA (Fuxe, 1965a,b). Since some of the NE cell bodies lie in the reticular formation of the lower brainstem, it is possible that this system may influence the hypothalamus; mention has already been made of the NE connections among the hippocampus, the amygdaloid cortex, and the hypothalamus.

In the hypothalamus, the CA innervation of the ME appears to be of particular interest (see also Chapter I,B,2). As a result of an extensive series of deafferentation and neuropharmacological studies (Fuxe, 1963, 1964; Fuxe and Hökfelt, 1966; see Björklund *et al.*, 1973a), it was evident that CA fluorescence is localized in nerve terminals present in the external layer of the ME originating from DA cell bodies present in the ARC n. (A12), the anterior periventricular n. (ventral part) and in a group (A13) dorsolateral to the n. dorsomedialis hypothalami pars medialis (tuberoinfundibular DA system). In addition to the fairly dense collection of CA nerve terminals in the external layer, a smaller number of coarse varicose NE nerve terminals was also recognized in the internal layer (Rethelyi and Halàsz, 1970; Ungerstedt, 1971a; Jonsson *et al.*, 1972; Björklund *et al.*, 1973a) (Fig. 9). Although it was initially postulated that NE terminals

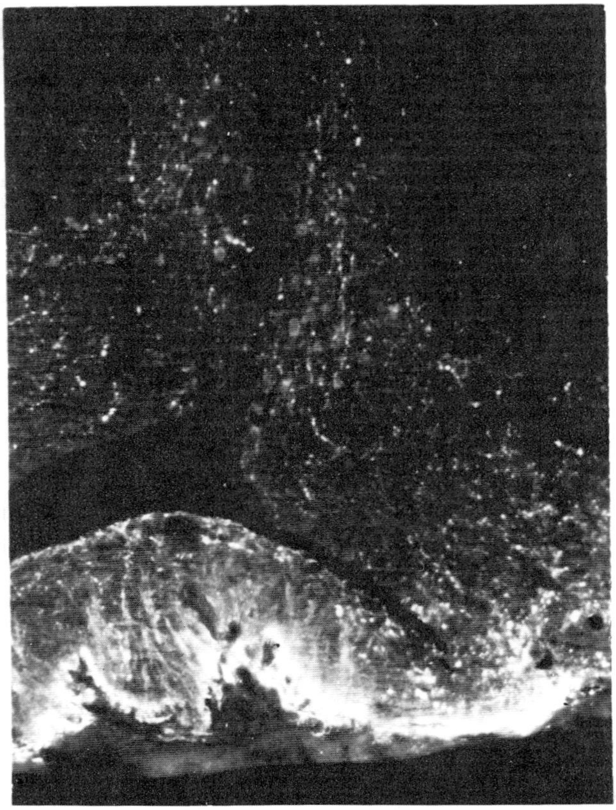

Fig. 9 Transverse section through the hypothalamus–ME region. A dense accumulation of fluorescent nerves is observed in close proximity to the capillary plexus in the external layer. Fewer more discrete varicose fibers are observed in the internal layer of the ME. Low fluorescence ARC nucleus cell bodies are present (arrows). (×135). (Reproduced, with permission from Jacobowitz, 1973.)

project also to the external layer of the ME at an extent similar to that of DA (Björklund *et al.*, 1970), the consensus has now been reached that NE terminals are mainly localized within the internal and DA terminals within the external layer of the ME (Jonsson *et al.*, 1972; Björklund and Nobin, 1973). Only a small projection of NE terminals involves the external layer of the ME as implied by localization of D-β-H (Goldstein *et al.*, 1973) or by studies with 6-OHDA (Cuello *et al.*, 1974). A pattern of CA fluorescence in the hypothalamus and ME similar to that present in the rat has been described in rhesus (Hoffman *et al.*, 1976) and squirrel monkeys (Felten, 1976). In primates, however, CA cell bodies do not fluoresce spontaneously, and only with prior administration of α-MNE, which can enter a CA cell and is not degraded, can the fluorescent neurons of the hypothalamus be visualized.

Measurements of CA turnover in nerve terminals in the central part of the rat ME by means of fluorescence histochemistry in combination with microfluorimetry have shown that DA turnover in the medial external layer is significantly higher than that in the lateral external layer, a difference which may reflect a subdivision of the TIDA neurons. NE terminals in the subependymal layer appear to have the lowest turnover of the CA terminal systems of the ME (Fuxe *et al.*, 1974).

Besides the ARC n. cell bodies (A12) (Fuxe, 1963, 1964), approximately 80% of the hypothalamic DA was localized outside the ARC n.–pituitary region (see Björklund *et al.*, 1975a). In the rat, DA-containing cell bodies have been demonstrated in the posterior hypothalamus and the medial zona incerta, designated as incertohypothalamic system (A11 and A13 according to Fuxe *et al.*, 1969; Björklund and Nobin, 1973; Lidbrink *et al.*, 1974), and in the rostral periventricular hypothalamus (group A14 of Björklund and Nobin, 1973). In a recent study, evidence has been obtained that the incertohypothalamic fibers are the projection of short intradiencephalic DA neurons, the cell bodies of which are located in the A11, A13, and A14 cell groups (Björklund *et al.*, 1975a). More important, according to these authors, the incertohypothalamic system could be subdivided into two parts: a caudal part, originating from A11 and A13 and projecting to anterior and dorsal hypothalamic areas and to the zona incerta, and a rostral part (originating from A14), which seems to constitute a periventricular–preoptic system giving rise to terminals also in septal regions. While TIDA neurons allow for DA-mediated control at the ME level (see Chapter IV, Section C), the incertohypothalamic system provides a morphological substrate for a dopaminergic influence on the hypophysiotropic neurons at higher levels of the hypothalamohypophyseal axis.

Confirmation of these findings on the distribution of brain CA's has been achieved by immunohistofluorescence localization of CA-synthesizing enzymes, e.g., D-β-H and L-AAAD, used to detect the presence of NE or DA neurons, respectively (see Goldstein *et al.,* 1972; Hökfelt *et al.,* 1974a).

Recently, using a very sensitive enzymatic–isotopic technique, the concentration of NE and DA in several rat hypothalamic nuclei or nuclear subdivisions (Table IV) has been determined (Palkovits *et al.,* 1974a). This information can now be added to that already acquired on CA distribution in CNS areas involved in the control of AP function in different animal species (Table V). The hypothalamic nuclei contain 3–20 times more NE and DA in comparison to the cortex, and within the hypothalamus the DA content of the ME is among the highest measured in the brain, whereas NE in this region is about half that of DA.

The nerve terminals of the TIDA system in the ME have also been identified at the ultrastructural level by the permanganate fixation method

TABLE IV

Distribution of Norepinephrine and Dopamine in Hypothalamic Nuclei[a]

Brain area	Norepinephrine (ng/mg protein)	Dopamine (ng/mg protein)
Nucleus periventricularis	33.5	7.1
Nucleus supraopticus	23.6	3.7
Nucleus paraventricularis	51.0	10.0
Nucleus hypothalamicus anterior	16.2	5.0
Nucleus preopticus suprachiasmatis	39.5	11.7
Nucleus suprachiasmatis		
Internal	20.5	8.5
External	29.2	9.5
Area retrochiasmatica	48.0	15.1
Nucleus arcuatus	100.5	73.0
Nucleus ventromedialis	111.8	33.8
Nucleus dorsomedialis	76.4	20.4
Nucleus perifornicalis	17.9	6.0
Nucleus premammillaris	14.2	3.9
Nucleus premammillaris ventralis	16.3	3.4
Nucleus hypothalamicus posterior	13.9	4.3
Medial forebrain bundle		
Anterior	16.9	6.1
Posterior	20.2	11.0
Median eminence	29.5	65.0
	4.4 μg/gm[b]	28.9 μg/gm[b]
Cortex	4.2	1.3

[a] From Palkovits *et al.* (1974a).
[b] From Chiocchio *et al.* (1976).

TABLE V

Distribution of NE and DA in Some Brain Areas Involved in the Control of AP Hormone Secretion

Species	Brain area	Concentration (μg/gm) NE	DA	Reference
Pig	Infundibular stem and ME	0.78	0.85	Rinne and Sonninen (1968)
	Mediobasal hypothalamus	1.97	0.14	Rinne and Sonninen (1968)
	Remainder of the hypothalamus	0.61	0.13	Rinne and Sonninen (1968)
	Anterior pituitary	0.01	0.01	Björklund *et al.* (1967)
Sheep	Pituitary stalk and ME	0.32	5.05	Laverty and Sharman (1965)
Goat	Pituitary stalk and ME	0.16	2.0	Laverty and Sharman (1965)
Cat	Tuber cinereum	2.42–3.73	—	Reis *et al.* (1968)
	Anterior hypothalamus	3.6	—	Kuntzman *et al.* (1961)
	Posterior hypothalamus	2.0	—	Kuntzman *et al.* (1961)
Man	Anterior hypothalamus	0.96	0.18	See Holzbauer and Sharman (1972)
	Intermediate hypothalamus	1.19	0.14	See Holzbauer and Sharman (1972)
	Posterior hypothalamus	0.31	0.22	See Holzbauer and Sharman (1972)

(Hökfelt, 1967a,b, 1968; Ajika and Hökfelt, 1973). The presence of boutons of small (about 500 Å) and large (about 1000 Å) granular vesicles has been shown in the internal layer of the ME, in the ARC n. and in the periventricular region, areas assumed to contain mainly NE axon terminals. The smaller granular vesicles can make axodendritic junctions (Hökfelt, 1968). In addition, boutons with small granular vesicles have been identified in the external layer of the ME which contain predominantly DA terminals (Ajika and Hökfelt, 1973). The density of monoamine boutons appeared to be highest in the superficial, lateral parts of the ME (about 33% of all boutons), whereas in the ARC n., the corresponding values were 2.6% (Fig. 10). These results inferentially demonstrate that in the neurohemal contact zone about two-thirds of all boutons belong to non-monoamine peptidergic neurons and are possibly sites of store and release of hypothalamic neurohormones controlling hormone secretion from the AP (see Chapter IV, Section C). Although DA neurons form axoaxonic contacts (Hökfelt, 1967a), no typical "synapses" with peptidergic neurons have been identified here (Ajika and Hökfelt, 1973) (see also Chapter I, Section B,2).

5. Catecholamine Innervation of the Pituitary Gland

The presence of MA's in the anterior, intermediate, and posterior lobe of the pituitary gland is well documented, but, although an involvement in

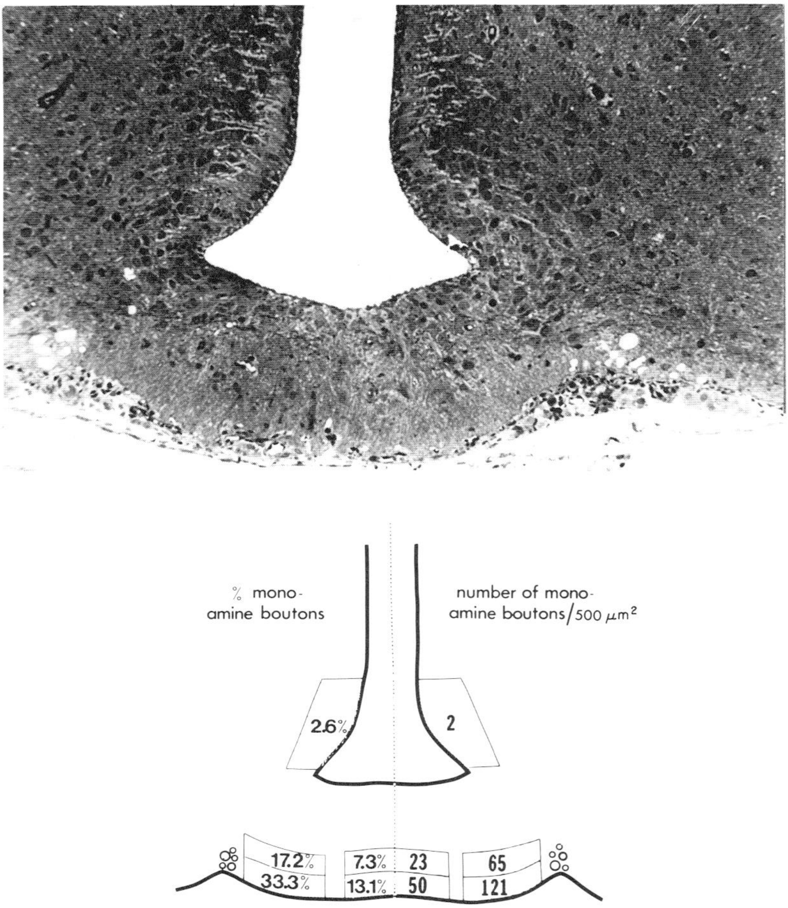

Fig. 10 At the top: light micrograph of the basal hypothalamus including the ME. The lower figure represents a schematic drawing of the same area showing on the left the percentage of monoamine boutons and on the right the number of monoamine boutons per square unit. The highest density is present in the superficial lateral part of the external layer. (Reproduced, with permission, from Ajika and Hökfelt, 1973.)

the control of vasomotor activity has been supposed, their precise role is still obscure (Hyyppä and Wurtman, 1973). The distribution pattern of the MA's in the different lobes of the pituitary is not uniform and varies in the different species studied. Catecholamine terminals have been shown by histofluorescence in the neural and intermediate lobe (see Jacobowitz, 1973) but not in the AP (Dählström and Fuxe, 1966; Fuxe and Hökfelt,

1969). In particular, DA is the predominant catecholamine in the neural lobe and in the pars intermedia of pig (Björklund *et al.*, 1967), rat (Björklund *et al.*, 1970), cat (Björklund and Falck, 1969), horse (Iwata and Ishii, 1969), and is contained in the anterior lobe only in very low amounts (Saavedra *et al.*, 1975). Throughout the gland, NE concentrations are low, about one-tenth that of DA. Epinephrine was not detected in significant concentrations in the pituitary gland (Björklund and Falck, 1969; Björklund *et al.*, 1970; Iwata and Ishii, 1969), a finding consistent with the reported lack of PNMT activity (Saavedra *et al.*, 1975). Table VI reports the distribution of some biogenic amines in the rat pituitary gland. In the anterior lobe of the pituitary strong green to yellow-green CA fluorescence is present in several PAS-positive cells of cats, dogs, pigs, but not rats (Dählström and Fuxe, 1966; Fuxe and Hökfelt, 1969). In contrast to the CA fluorescence of adrenergic axons and terminals, the AP fluorescence is not depleted by reserpine, possibly because CA's are stored in the glycoprotein granules of the PAS-positive cells by a mechanism insensitive to this agent (Fuxe and Hökfelt, 1969).

The recent development of specific and sensitive microassay methods for the quantitative determination of enzymes involved in NE metabolism has also permitted study of their distribution in the AP (see Saavedra *et al.*, 1975). Tyrosine hydroxylase was detected in the posterior and intermediate lobes, indicating local formation of DA in these areas, but was not detected in the anterior lobe. The apparent absence of D-β-H was consistent with the very low NE levels detected (see above). Although it had been postulated that under certain circumstances *O*-methylation was an important pathway in CA inactivation by the pituitary (Ciaranello and Barchas, 1973), it would appear that only the AP lobe possesses some degree of this COMT activity. The high MAO levels detected in all pituitary lobes indicate that oxidative deamination rather than *O*-methylation is the physiological metabolic route for CA in this organ

TABLE VI

Biogenic Amines in the Rat Pituitary Gland[a]

Gland area	Amine content			
	Norepinephrine	Dopamine	Serotonin	Histamine
Anterior lobe	0 (5)	0.20 ± 0.02 (5)	1.85 ± 0.3 (16)	3.03 ± 0.67 (8)
Intermediate lobe	0.15 ± 0.10 (5)	1.78 ± 0.79 (5)	3.69 ± 0.44 (10)	17.15 ± 4.5 (8)
Posterior lobe	0.80 ± 0.28 (5)	7.16 ± 1.91 (5)	4.05 ± 1.0 (14)	13.75 ± 2.06 (8)

[a] From Saavedra *et al.* (1975). Results as ng/mg protein ± SEM. Numbers in parentheses indicate *n* values.

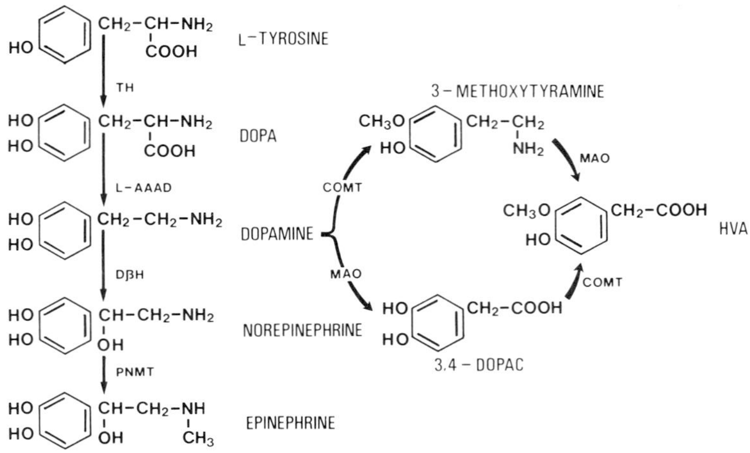

Fig. 11 Enzymatic steps involved in CA biosynthesis and DA metabolism. TH: tyrosine hydroxylase; L-AAAD: aromatic L-amino-acid decarboxylase; D-β-H: dopamine-β-hydroxylase; PNMT: phenylethanolamine-*N*-methyltransferase; HVA: homovanillic acid; 3,4-DOPAC: 3,4 Dihydroxyphenylacetic acid; MAO: monoamine oxidase; COMT: catechol-*O*-methyltransferase.

(Saavedra *et al.*, 1975). Collectively, the presence of biogenic amines and related enzymes in posterior, intermediate and anterior lobes of the pituitary gland (see also Section C,3) suggests their direct participation in the control of pituitary gland secretions. Quantitative determination of amines and enzymes in individual pituitary lobes with different drug treatment and/or after endocrine manipulation could help clarify the extent of this neurohumoral regulation.

6. Biosynthesis and Its Regulation

The metabolism of CA's and its regulation have been extensively studied and summarized in a series of reviews (von Euler, 1972; Axelrod, 1974; Iversen, 1973). This section will focus on the major steps in CA biosynthesis which are essential for understanding CA–neurohormonal interactions in the control of AP function.

Synthesis of CA's occurs *in vivo* from phenylalanine and tyrosine; the main mammalian pathway is illustrated in Fig. 11, while Fig. 12 illustrates the mechanisms involved in NE neurotransmission and the distribution within the neuron of enzymes for synthesis.

The first step is represented by the transformation of L-tyrosine to dopa, a reaction catalyzed by the enzyme tyrosine hydroxylase (TH) which is present in the axoplasm (Nagatsu *et al.*, 1964). A stereospecific "carrier" mechanism common to L-Trp and other amino acids is required for the entrance of tyrosine into the axon (Wurtman and Fernström, 1972). Dopamine formed from L-dopa by L-AAAD is subsequently oxidized to NE, a reaction catalyzed by D-β-H, an enzyme which is present in the storage particles. In CNS several factors can influence the CA rate of biosynthesis, such as uptake of L-tyrosine from plasma into the adrenergic neurons, availability of the precursor, and activity or amount of TH (Wurtman and Fernström, 1972).

From a functional viewpoint TH is the key enzyme in CA biosynthesis

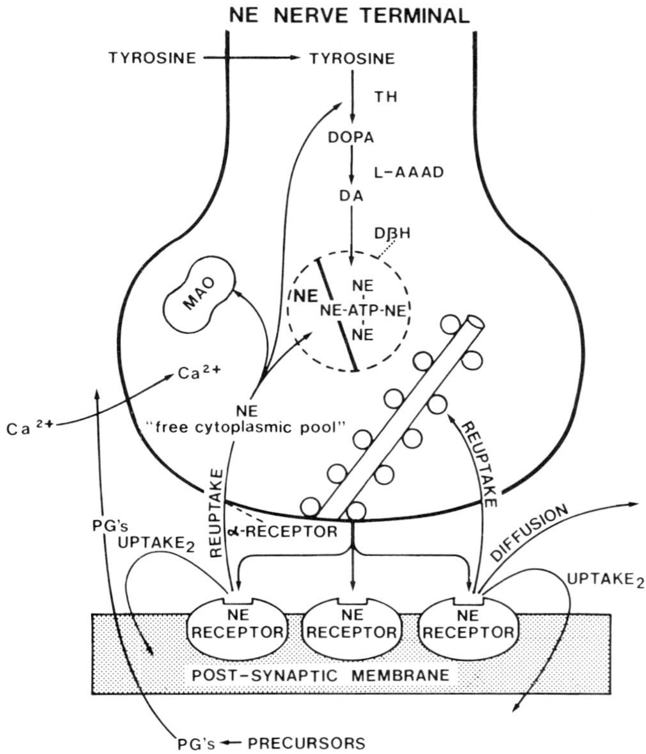

Fig. 12 Schematic illustration of the mechanisms (tyrosine uptake into the neuron, NE biosynthesis and storage within the granules, free cytoplasmic pool, release, effect on α-postsynaptic and presynaptic adrenergic receptors, and termination of the effects by Uptake$_1$, Uptake$_2$, and metabolism) involved in norepinephrinergic neurotransmission and distribution within the neuron of the enzymes involved in CA synthesis. The role of prostaglandins (PG's) in regulating NE release from the terminal by inhibiting Ca^{2+} influx is also shown (for details, see text).

since it catalyzes the rate-limiting step (Spector *et al.*, 1963; Levitt *et al.*, 1965). In comparison to the other enzymes involved in CA synthesis (L-AAAD and D-β-H), TH concentrations are 100–1000 times lower. The intraneuronal localization of TH has been demonstrated by experiments in which 6-OHDA-induced degeneration of sympathetic neurons (see Section D) was accompanied by disappearance of this enzyme from sympathetic innervated organs (Mueller *et al.*, 1969a). TH is stereospecific, requires O_2 and Fe^{2+} and has tetrahydropteridine as cofactor. The levels of pteridine are probably low, and this may influence CA synthesis (Wurtman and Fernström, 1972).

Tyrosine hydroxylase can be inhibited by NE, and such inhibition (end-product regulation) is an important controlling mechanism for the synthesis in order to assure the constant concentration of tissue CA. Inhibition of the enzyme by CA depends upon competition for the cofactor in its reduced form and not with its substrate, tyrosine (Ikeda *et al.*, 1966). Also, drugs which elevate CA levels such as MAOI (see Section C,10,a) are able to block TH (Axelrod, 1974). A list of drugs capable of inhibiting TH activity is given in Table VII. As will emerge from the following chapters, these drugs represent a very useful mode of depleting brain CA's.

The second step in the biosynthesis of the adrenergic neurotransmitter, the formation of DA from L-dopa, is catalyzed by L-AAAD, which is localized in the same subcellular organelle as TH (Udenfriend *et al.*, 1966). This enzyme is not specific for L-dopa in that it decarboxylates a variety of L-aromatic amino acids (tryptophan, 5-hydroxytryptophan, histidine, tyrosine, glutamic acid, etc.).

3,4-Dihydroxyphenylserine (DOPS) is directly decarboxylated by this enzyme to NE, and offers a pharmacological means of increasing NE without affecting DA concentration. Among the four stereoisomers of DOPS, (+)-*erythro*-DOPS is the most effective in inducing increased brain NE levels (Puig *et al.*, 1974). DOPS administered in association with α-MpT allows selective depletion of brain DA levels in the presence of unchanged or increased brain NE; this pharmacological manipulation has been found useful in assessing the importance of NE rather than DA, or vice versa, in the control of hypothalamic–pituitary hormone secretion (see Chapter IV). However, interpretation of data obtained by the use of DOPS must be tempered with the reservation that this drug enters different brain structures nonspecifically, and NE can also be formed in DA and 5-HT nerve endings (Constantinidis *et al.*, 1975).

L-Aromatic amino-acid decarboxylase requires pyridoxal phosphate as a cofactor, and is tightly bound to the apoenzyme as a Schiff base. It has been suggested that under circumstances in which a limited concentration of the cofactor occurs, such as decreased availability of dietary pyridoxine or of enzymes converting pyridoxine to pyridoxal phosphate, L-AAAD may be rate-limiting (Wurtman and Fernström, 1972). Several compounds listed in Table VII are able to inhibit this enzymatic activity.

D-β-H hydroxylates DA on the β-carbon to form NE. It is a mixed function oxidase containing 2 moles of Cu^{2+}, which lacks specificity and can β-hydroxylate a variety of phenylethylamines. D-β-H is localized in the membrane of NE storage nerve vesicles (Potter and Axelrod, 1963) and in chromaffin granules of the adrenal medulla (Kirshner, 1957).

The epinephrine-forming enzyme, PNMT, is highly localized in the cytoplasm of mammalian adrenal medulla (Axelrod, 1962), and is present in the sympathetic nerves of amphibians (Wurtman *et al.*, 1968a) and in mammalian CNS (Hökfelt *et al.*, 1974a). It methylates NE as well as β-hydroxylated phenylethanolamine derivatives, *S*-adenosylmethionine being the methyl donor. The regulation of PNMT has been reviewed by Ciaranello (1973).

It is generally assumed that the enzymatic activity for CA synthesis is formed in the cell body and is transported by axoplasmic flow to the nerve endings (see Weiss, 1970; Kopin, 1973). According to Axelrod (1974), the levels of NE within the sympathetic neuron can be regulated at several sites, i.e., cell body via preganglionic nerves, the axon which transports the biosynthesis enzymes, the cytoplasm and storage vesicles of the nerve terminal, the neuronal membrane, and the postsynaptic membrane.

Several experimental manipulations can accelerate CA synthesis. Thus, after electrical stimulation of the sympathetic nerves, there is a rapid or short-term (within minutes) increase of CA synthesis which may be dependent on the release of the small critical pool of CA (see below) and therefore on the elimination of the end product negative feedback mechanism (Alouisi and Weiner, 1966; Sedvall *et al.*, 1968). Another mechanism by which CA synthesis in the periphery and in the CNS is increased is by a transsynaptic induction of TH and D-β-H; in other words, in conditions in which sympathetic activity is overstimulated (stress, blockade of α-receptors, reserpine, 6-OHDA) (see following sections) there is an increase in the synthesis of TH and D-β-H, and, in turn, an increase in CA synthesis. This mechanism operates more slowly (within days) (see Axelrod, 1974).

Several studies have shown that α-adrenergic blocking agents cause an increase in the release of the neurotransmitter on nerve stimulation (Brown and Gillespie, 1957; Hertting *et al.*, 1961), while, conversely, large amounts of endogenous NE inhibit the discharge of NE from nerves (Stärke, 1971). These observations suggest that another regulatory site in CA biosynthesis is the inhibitory α-adrenergic receptor on the presynaptic membrane or the soma (Stärke, 1971, 1972; Farnebo and Hamberger,

TABLE VII

Inhibitors of CA Synthesis

Class	Drug	Observations[a]	References to main action
TH inhibitors	α-Methyl-*p*-tyrosine (α-MpT)	Conversion to α-methyl-DA and α-methyl-NE	Udenfriend *et al*. (1965); Spector *et al*. (1965)
	α-MpT methyl ester hydrochloride (H44/68)	Soluble compound	Hanson (1965); Corrodi and Hanson (1966)
	3-Iodo-L-tyrosine and halogenated tyrosines	Short half-life in the circulation (Stanbury *et al*., 1956); postulated DA receptors blockade (Smythe *et al*., 1975a)	Udenfriend *et al*. (1965); Goldstein and Weiss (1965); Spector *et al*. (1965)
	Tryptophan and derivatives (5-halotryptophans)	Competition with transport of aromatic amino acids	McGeer *et al*. (1967)
	α-Methyl-5-HTP	Decarboxylated *in vivo* to α-methyl-5-HT	Kopin (1971); Muscholl (1972)
	Catechols, pyrogallols	Competition with tetrahydropteridine cofactor (feedback inhibition of TH)	Udenfriend *et al*. (1965); Levitt *et al*. (1967)
	Chelating agents, e.g., α′,α′-dipyridyl	D-β-H inhibition	Nagatsu *et al*. (1964)
	Troponolones		Goldstein *et al*. (1967)
	Aquamycin		Ayukawa *et al*. (1968)
	Pyrroloisoxazoles		Taylor *et al*. (1968)
L-AAAD inhibitors	α-Methyl-α-hydrazino-3,4-dihydroxyphenylpropionic acid	Inhibition mostly at peripheral sites	Porter *et al*. (1962)
	N′-(3-Hydroxybenzyl)-*N*′-methylhydrazine (NS-1034)		Brodie *et al*. (1962); Reid and Shepherd (1963)
	L-α-Hydrazine methyldopa (MK-486) (Carbidopa)	Inhibition mostly at peripheral sites	Pletscher (1973)
	N-(DL-Seryl)-*N*′-(2,3,4)-trihydroxybenzylhydrazine (Ro 4-4602) (Benseraside)	Inhibition mostly at peripheral sites (Bartholini and Pletscher, 1969)	Burkard *et al*. (1964)

	2,3,4-Trihydroxybenzylhydrazine (Ro 4-5127)		Porter (1971)
	3-Hydroxybenzyloxyamine (NS-1024)	Inhibition at peripheral and central sites	Reid and Shepherd (1963)
	α-Methyldopa (α-MD)	Hypotensive effect; stimulation of medullary α-receptors through formation of α-MNE (Laverty, 1973)	Porter *et al.* (1962); Reid and Shepherd (1963); Pletscher *et al.* (1965)
	Decaborane		Merritt *et al.* (1965)
DA-β-H inhibitors	Disulfiram (bisdiethyldithiocarbamate disulfide)	Copper chelating agent	Goldstein *et al.* (1964)
	Diethyldithiocarbamate (DDC)	Formation *in vivo* from reduction of disulfiram; reduction of brain NE and slight increase in DA (Musacchio *et al.*, 1966; Carlsson *et al.*, 1967)	Collins (1965); Musacchio *et al.* (1964)
	Picolinic acid		
	Fusaric acid (FA) (5-butylpicolinic acid)	Hypotensive effect (Hidaka *et al.*, 1973); antistress ulcer effect (Hidaka, 1973)	
	[bis(4-methyl-1-homopiperazinylthio-carbonyl] disulfide (FLA-63)	Increase in DA in the striatum (Nyback, 1971; Svensson, 1973a,b)	Carlsson *et al.* (1967)
	Troponolone		Goldstein *et al.* (1964)
	α',α'-Dipyridyl		Goldstein *et al.* (1964)
	(1-phenyl-3,2-thiazolyl)-2-thiourea (U14.624 Upjohn)	Increased 5-HT turnover (Johnson *et al.*, 1972).	Johnson *et al.* (1970)

[a] Unless otherwise specified, references of this column are the same as those indicated under the column referring to the main action. The same applies to Tables VIII–XI, XIV, XV, XVII, XIX, and XX.

1971; Kirpekar and Puig, 1971; Langer, 1973; Langer *et al.*, 1971; Enero *et al.*, 1972), and that these presynaptic or "autoreceptors" are part of a negative feedback control mechanism through which the transmitter may inhibit its own release (Fig. 12). According to this hypothesis it might be expected that this mechanism will operate most effectively when the quantity of transmitter released by each impulse is high. More recently, the presence of presynaptic receptors has also been postulated for dopaminergic neurons (see Section B,11).

Prostaglandins (PG's) of the E type (PGE's) also regulate transmitter release from adrenergic nerves. The PGE's are present in most mammalian tissue and in a variety of sympathetically innervated organs; increased nerve activity is associated with increased formation and release of PG's (see Bergström *et al.*, 1968; Hedqvist, 1970, 1973a,b; and Section G). It has been shown that in several sympathetic innervated organs the outflow of NE in response to nerve stimulation is depressed by small doses of PGE's (Hedqvist, 1973a,b). The mechanism by which PGE's act prejunctionally on sympathetic neuroeffector transmission seems to be an inhibition of Ca^{2+} influx into neurons (see Bergström *et al.*, 1968; Hedqvist, 1973a,b) and subsequent inhibition of CA release (Fig. 12). Therefore, locally formed PGE's represent a significant mechanism for feedback control of transmitter release from adrenergic nerves.

a. Biosynthesis Inhibitors

The rate of synthesis of CA's can be pharmacologically controlled, a subject which has been reviewed by Creveling and Daly (1971) and von Euler (1972). Several inhibitors of the enzymes involved in the synthesis have been discovered; the most commonly used are represented in Table VII. Systemic or intracerebral administration of these drugs represents an approach often used in the investigation of the role played by CA's in the neuroendocrine regulation of AP function (see Chapter IV).

Of the three enzymatic steps, the formation of dopa by TH is the slowest one and therefore rate-limiting. Consequently, inhibition of this step is the most useful method for reducing NE synthesis. α-Methyl-*p*-tyrosine (α-MpT), particularly as soluble methyl ester (Spector *et al.*, 1965; Corrodi and Hanson, 1966), inhibits TH. The inhibition of TH by α-MpT may cause severe NE deficiency (Spector *et al.*, 1965). The rate of NE and DA depletion after inhibition of TH is dependent on the rate of CA utilization, and so is more rapid in tissues with a fast turnover (Udenfriend *et al.*, 1965) (see Section B,12). In the brain, maximal CA depletion is reached after about 8 hours from a single injection of α-MpT (Spector *et al.*, 1965).

Since there is only a slow exchange between storage and functional pools (Crout *et al.,* 1962) and repletion of the functional pool by reuptake is probably less than that resulting from synthesis (Kopin *et al.,* 1965) (see Section B,6 and 9), administration of α-MpT is capable of decreasing the function of adrenergic neurons without consistent CA depletion. Effects similar to those of α-MpT have been shown for 3-iodo-L-tyrosine (Goldstein and Weiss, 1965), which has been also postulated as blocker of DA receptors (see Chapter IV, Section A,2).

Strong inhibitory action *in vivo* on the L-AAAD is exerted by hydrazine and hydroxylamine derivatives e.g., *N*-methyl-*N*-(3-hydroxybenzyl)hydrazine (Brodie *et al.,* 1962) (Table VII). These compounds, although up to 100 times more potent than α-methyldopa (α-MD), which has also been shown capable of inhibiting L-AAAD *in vitro* and *in vivo* (Pletscher *et al.,* 1966), are devoid of the blood-pressure lowering and psychotropic effects of α-MD, even in high doses. The introduction of selective peripheral inhibitors, such as DL-seryltrihydroxybenzylhydrazine (Hoffman, La Roche, Ro4-4602) and L-α-methyldopahydrazine (Merck Sharp and Dohme, MK-486), has enabled substantial reduction in the dosage of amine precursors (e.g., L-dopa or L-5-HTP) required for pharmacological effect while concurrently resulting in an enhanced clinical response (see Chapters IV and V). The action of the hydrazine inhibitors is due to a peripheral inhibition of the decarboxylation of the amine precursor in extracerebral tissues, thus allowing larger amounts of it to penetrate into the brain parenchyma, where decarboxylation occurs since cerebral AAAD remains active (Bartholini and Pletscher, 1968). However, at appropriately high doses (e.g., for Ro4-4602 in the rat, $\geq$50 mg/kg) hydrazine inhibitors also inhibit CNS decarboxylase and effectively decrease brain neurotransmitter concentrations (see Porter, 1971), so that caution must be used in interpreting the pharmacological consequences of their administration (see also Chapter IV, Sections A,2,a and A,4,a).

Another inhibitor of L-AAAD activity is decaborane ($B_{10}H_{14}$) (Merritt *et al.,* 1965). The effect of decaborane appears to depend on the inhibition of the cofactor pyridoxal 5′-phosphate, and can be counteracted by the injection of pyridoxine (Wykes and Landez, 1967). Since L-AAAD is not rate-limiting (Hess *et al.,* 1961; Udenfriend *et al.,* 1966), potent inhibitors of this activity do not reduce CA concentrations (Brodie *et al.,* 1962) unless very high doses are administered (see above). A substrate inhibition of the L-AAAD enzymatic step with subsequent decrease in CA synthesis can result from administration of 5-HTP, since this amino acid is transformed to 5-HT in all the sites where the enzyme is present (Wurtman and Fernström, 1972).

Dopamine-β-hydroxylase is a copper-containing enzyme, and is inhib-

TABLE VIII

Drugs Blocking NE Neurotransmission by Mechanisms Other Than Synthesis Inhibition

Class	Drug	Observations	References to main action
Depleting agents	Reserpine	Leakage of CA from granules and inhibition of granular uptake (ATPase Mg^{2+} dependent); same effect on 5-HT terminals (Brodie *et al.*, 1957; Shore, 1962); chronic administration: increased TH and D-β-H (Levitt *et al.*, 1965; Segal *et al.*, 1971; Dairman and Udenfriend, 1970)	Kopin and Gordon (1963); Glowinski and Axelrod (1965); Glowinski *et al.* (1966b)
	Guanethidine	Biphasic action (initial massive CA release followed by depletion by reserpine-like action); no penetration into BBB	Chang *et al.* (1965); Bhagat and Shideman (1963); Boura and Green (1965)
	Tetrabenazine	Short duration of action and more selectivity for the brain	Pletscher *et al.* (1962)
Release inhibitors	Bretylium	Ca^{2+} chelating action with impermeabilization of granular membrane; local anesthetic activity, antihistaminic activity	Boura and Green (1959); see Meyers *et al.* (1974)
	Lithium ions	Antagonism of NE release from brain slices; stimulation of NE reuptake (Schildkraut *et al.*, 1966; Colburn *et al.*, 1967); interference with Na^+ pump and decrease in intracellular Na^+ (Schou, 1968)	Katz *et al.* (1969)
	Bromide	Antagonism of NE release from brain slices (neurosedative effect)	Goodwin *et al.* (1969)
	Colchicine and Vincristine	Microtubules poisoning compounds	Thoa *et al.* (1972)
	Cytochalasin B	Interference with microfilaments	Thoa *et al.* (1972)

False precursors or transmitters	α-Methyldopa (α-MD)	Transformation into α-MNE; hypotensive action due to activation of central α-adrenergic receptors (Laverty, 1973); decrease in brain 5-HT and 5-HIAA (Sharman and Smith, 1962; Roos and Werdinius, 1963; Pletscher *et al.*, 1964); inhibition of L-AAAD (Sourkes, 1954)	Carlsson and Lindqvist (1962); Day and Rand (1963)
	α-Methyl-*m*-tyrosine (α-MmT)	Transformation to metaraminol; TH inhibition; inhibition of L-AAAD (Andén and Magnusson, 1964)	Carlsson and Lindqvist (1962); Andén (1964); Shore *et al.* (1964)
	Metaraminol	Replacement of NE; direct, although low, stimulating effect on adrenergic receptors (Andén and Magnusson, 1964)	Shore *et al.* (1964)
	Octopamine	Formed by β-hydroxylation of tyramine (Poch and Kopin, 1966); less firmly bound to storage sites and preferentially released in comparison to NE (Poch and Kopin, 1966); major amine which accumulates in tissues after chronic MAO inhibition (Kakimoto and Armstrong, 1962)	
	Hydroxyamphetamine (paredrine, α-methyltyramine)	Partial conversion to *p*-hydroxynorephedrine (α-methyloctopamine)	Fischer *et al.* (1965); Thoenen *et al.* (1965)
Selective neurotoxic agents of CA neurons	6-OH-dopa	Crossing of BBB	Ong *et al.* (1969)
	6-OHDA (see Table XI)	Not crossing of BBB (IVT, IC, intracerebral infusion)	Ungerstedt (1971b); Uretsky and Iversen (1969)

ited by various compounds including benzylhydrazines, benzyloxyamines, and derivatives of picolinic acid (Table VII). Some of them, e.g., disulfiram and tropolone, probably act by copper chelation (Pletscher, 1973). Compounds able to inhibit D-β-H have been used to obtain an increase in brain DA and a parallel decrease in brain NE concentration (Goldstein *et al.*, 1964; Hidaka, 1973) (see Chapter IV). However, a caveat related to some inhibitors of D-β-H seems warranted; coincident with the inhibition of D-β-H, there may also be a nonspecific inhibition of TH (Taylor *et al.*, 1968) and/or an increase in brain 5-hydroxyindoleacetic acid (5-HIAA) levels, probably reflecting accelerated 5-HT biosynthesis (Johnson *et al.*, 1972).

b. False Neurotransmitters

The processes for synthesis, storage, release, uptake, and metabolism of NE are not entirely specific so that structurally related compounds may displace and replace NE and be released from terminals; as a consequence there is a decreased release of NE by nerve impulses. As these compounds are usually less effective than NE, they have been called false neurotransmitters (Kopin, 1971; Muscholl, 1972). Some compounds (e.g., α-MD, α-MmT) may act as false precursors inasmuch as they follow the same metabolic transformation of the natural precursors of NE and give rise to false transmitters. A list of false neurotransmitters is given in Table VIII. Besides the displacement and replacement of NE, false transmitters may interfere with NE transmission by means of other mechanisms such as inhibition of NE synthesis, inhibition of DA uptake by the granules, and NE receptor antagonism (Kopin, 1971).

False transmitters which are CA's, e.g., α-methyl norepinephrine (α-MNE), may have direct inhibitory action on TH, while those which displace NE from its storage sites in the cytoplasm (e.g., metaraminol, octopamine) inhibit TH only for a short interval. When the release of NE is complete, and levels of cytoplasmic CA are diminished, the rate of TH formation becomes normal. However, depending on the time after replacement of NE by a false transmitter and the susceptibility of the latter to degradation by MAO, a decrease, no change, or an increase in the rate of tyrosine hydroxylation may occur (see Kopin, 1971).

Uptake and retention of 5-HT or 5-HTP by CA neurons in CNS was demonstrated by Lichtensteiger *et al.* (1967). It is likely that in the central and peripheral nervous system, indoleamines accumulated in CA neurons may also be released by nerve stimulation and act as false adrenergic transmitters (Kopin, 1972). Similarly, after administration of large doses of L-dopa, DA formed by L-AAAD in serotoninergic neurons will replace 5-HT and act as a false transmitter (Bartholini *et al.*, 1968).

7. Storage

In nerve endings NE is bound to a small subcellular particle fraction obtained in the sediment of homogenates of sympathetic nerves of NE-innervated organs by ultracentrifugation (von Euler and Hillarp, 1956). By EM studies on this fraction, two types of vesicles have been found, small granular (400–600 Å) and large granular (800–1200 Å) vesicles. Both vesicles store CA and protect them from enzymatic degradation (Fillenz, 1971; Tranzer and Thoenen, 1972). Vesicles are synthesized in CA-cell body and transported to the terminals by axonal flow (Dählström *et al.*, 1973). They possess a limiting membrane and D-β-H, the enzymatic activity responsible for the transformation of DA to NE. It has been also shown that within the granules NE is in part free and in part bound to ATP in a ratio 4 : 1 (Iversen, 1967, 1971b).

From a functional viewpoint the existence of two pools of NE has been suggested within the nerve terminal, one of which, the more readily available one, is present in vesicles closest to the presynaptic membrane, and the other larger one is located at a distance much further from the neuronal membrane. The two pools seem to be linked in a dynamic way in the sense that the larger pool might serve as a reservoir for the more readily releasable transmitter (Axelrod, 1974).

Besides the two intragranular pools, the existence has been suggested of a functional free cytoplasmic NE pool; this pool seems to be more readily released by drugs such as D-amphetamine or by nerve impulse. Evidence exists that under certain circumstances newly synthesized or newly stored NE is more readily released by nerve stimulation in comparison to older stored NE, both in the periphery and in the CNS (Kopin *et al.*, 1968; Glowinski, 1972; Schildkraut *et al.*, 1971). The marked ability of α-MpT to inhibit amphetamine-induced locomotor activity and stereotyped behavior without complete depletion of brain CA stores supports the hypothesis that newly synthesized NE is uniquely important in neurotransmission (see Shore, 1962, and Chapter IV).

Several drugs, as reported in Table VIII, interfere with CA storage within the terminals. A classic example is represented by reserpine and *Rauwolfia* alkaloids, the action of which on monoaminergic transmission is based schematically on two fundamental mechanisms: the induction of slow but constant release (leaking) of the transmitter from the storage granules and the inhibition of Mg^{2+}-ATP-dependent granular uptake (Fig. 13). As a result of such mechanisms there is a long-lasting depletion of amine stores in the periphery and CNS (Holzbauer and Vogt, 1956; Carlsson *et al.*, 1957), and the tonus mediated by the specific monoamine is decreased since the degrading enzymes, MAO (see Section B, 10), pro-

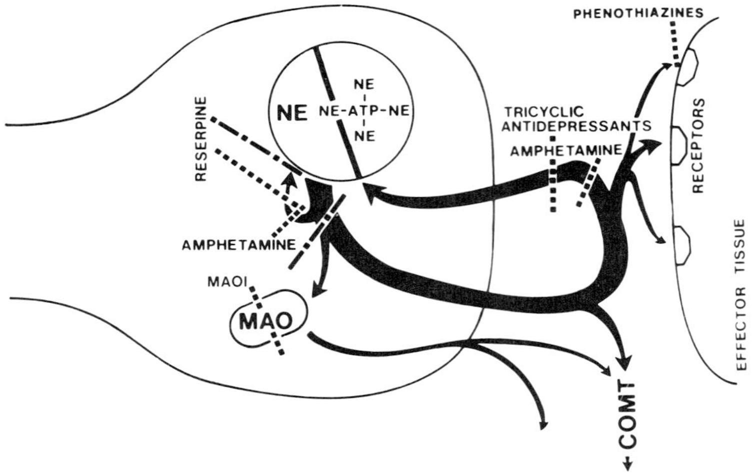

Fig. 13 Schematic representation of a norepinephrinergic nerve ending, synaptic cleft, postsynaptic membrane, and main mechanisms of action of drugs interfering at various levels with NE neurotransmission. Reserpine: depletion of NE as consequence of leakage of NE from the granules and inhibition of granular uptake Mg^{2+}-dependent ATPase mechanism. Amphetamine: release of NE, inhibition of neuronal uptake ($Uptake_1$); MAOI: inhibitors of MAO. Tricyclic antidepressants: inhibition of $Uptake_1$; phenothiazines: NE receptor blockade (for details see text).

gressively inactivate the transmitter leaking from the granules, as indicated by a striking increase in deaminated metabolites (Kopin and Gordon, 1962; Glowinski and Axelrod, 1966; Glowinski *et al.*, 1966a). In the reserpinized brain, exogenous NE is no longer accumulated and stored in CA neurons unless MAO activity is inhibited (Fuxe and Hillarp, 1964; Glowinski *et al.*, 1966a).

It has been found that during continuous reserpine treatment, small daily variations in the NE content of severely depleted amine stores in the brain correlated well with variations in animal behavior. This has suggested that, similar to peripheral adrenergic nerves, the presence of only a very small pool of NE is required for normal brain functioning (Häggendal and Lindqvist, 1964). The recovery from the pronounced sedative effects of a single dose of the drug occurs between 24 and 48 hours after drug administration at a time when brain CA levels remain severely depressed (Segal *et al.*, 1971). A rapid recovery occurs at this time in the ability of the central CA-containing neurons to take up and retain exogenous NE (Glowinski *et al.*, 1966a).

After treatment with reserpine, there is a small reserpine-resistant NE store (Sedvall, 1964; Häggendal and Lindqvist, 1964; Glowinski *et al.*,

1966a) which may be due to NE retention in particles which are still able to take up certain amounts of newly synthesized NE and release it slowly. In this case, an additional dose of reserpine does not always affect this amount which might represent the balance between small uptake and slow release (Iversen, 1965). Reserpine also produces a severe and long-lasting depletion of brain 5-HT stores (Brodie *et al.*, 1957), and it is still unclear whether behavioral effects due to reserpine are dependent on depletion of CA and/or serotonin, or whether either system is involved. As mentioned before (Section B,6), reserpine given continuously is able to induce TH (Levitt *et al.*, 1965; Mueller *et al.*, 1969b; Mandell and Morgan, 1970; Segal *et al.*, 1971; Scapagnini *et al.*, 1974a), an effect which appears to be a result of increased sympathetic nerve activity (Dairman and Udenfriend, 1970).

After the discovery of the CA-lowering properties of reserpine, a systematic research for compounds with a similar type of action led to the introduction of various derivatives of 1,2,3,4,6,7-hexahydro-11b*H*-benzo[*a*]quinolizine, e.g., tetrabenazine and Ro-4-1284. These compounds act similarly to reserpine, but are less potent, of shorter duration of action, and probably more selective for the brain (Pletscher *et al.*, 1962).

Another drug affecting CA stores, guanethidine, possesses a typical biphasic action. Initially, it releases massive amounts of CA from the free intragranular pool and simultaneously inhibits NE reuptake at the neuronal membrane, thus provoking sympathomimetic effects. Thereafter, a second phase follows quickly during which NE depletion occurs by the same mechanisms of reserpine (Boura and Green, 1965; Feilden and Green, 1967).

8. Release

There are two principal ways by which NE stored in vesicles can be released to the extracellular space: (1) NE could first pass from the vesicles into the cytosol and from there cross the presynaptic membrane; (2) the vesicle might release its content directly into the extracellular space (Fig. 14) (for further details, see Smith, 1973). According to Smith (1973), the release of NE from a pool into the cytosol is possible, but this occurs only spontaneously or after the administration of drugs that act on the vesicle; most, if not all, of the NE released by nerve impulse does not pass through the cytosol.

With regard to the direct release of NE from the vesicle into the extracellular space, the hypothetical mechanisms are the following: (a) vesicle and neuronal membrane come into close contact and NE passes across the two membranes; (b) expulsion of the whole vesicle; (c) exocytosis

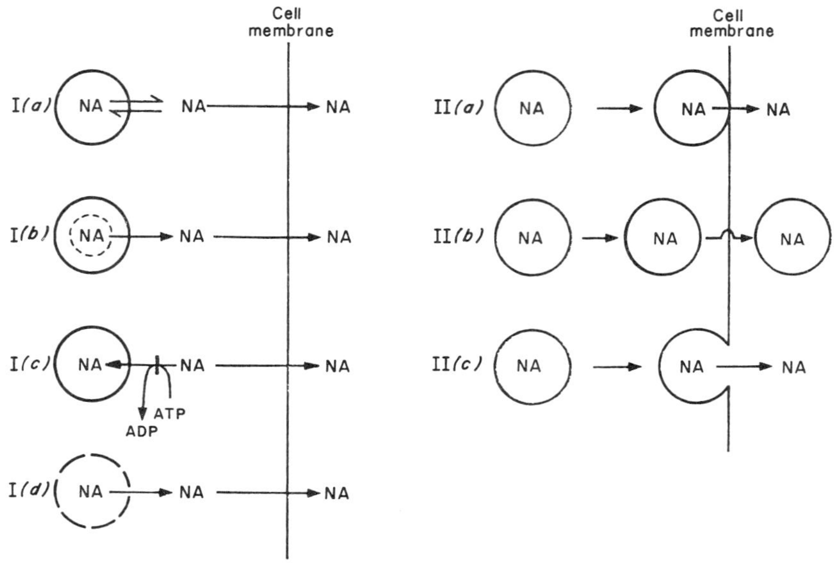

Fig. 14 Mechanisms of type I involve the diffusion of noradrenaline (NA) out of the vesicle, into the cytosol, and across the cell membrane. In mechanisms of type II, the vesicle itself moves and releases NA directly into the extracellular space. ADP, adenosine 5′-pyrophosphate; ATP, adenosine 5′-triphosphate. (Reproduced, with permission, from Smith, 1973.)

(Fig. 14). The latter mechanism consists of the fusion of vesicles and neuronal membrane and rearrangement to form an opening through which the content of the vesicle can pass into the extracellular space. The finding that electrical stimulation of the splenic nerve evokes the release of two vesicle proteins, chromogranin and D-β-H, favors exocytosis as the mechanism of CA release (Douglas and Poisner, 1966; De Potter *et al.*, 1969; Geffen *et al.*, 1969). Arrival of vesicles from the axoplasm to the plasma membrane involves transport by microtubules, since mitotic poisons, like colchicine and vincristine, which destroy microtubules, inhibit NE and D-β-H release (Thoa *et al.*, 1972). Once the two membranes fuse, some contractile mechanism would operate in order to allow the passage of the content of the vesicles into the extracellular space. In fact, cytochalasin B, a compound which interferes with the function of microfilaments (Table IV), has been shown to inhibit NE release (Thoa *et al.*, 1972). NE release from nerve terminals is a Ca^{2+}-dependent process (Hukovic and Muscholl, 1962; Rubin, 1970; Smith and Winkler, 1972); despite much speculations, nothing is known about the role of Ca^{2+} in the events that occur within the terminal before release (Rubin, 1970).

A schematic representation of drugs interfering with CA release is given in Fig. 13, and these drugs with their mechanisms(s) of action are listed in Table VIII.

After being released, NE acts on specific postsynaptic receptors and gives rise to the biochemical events responsible for the physiologic response. The division of adrenoreceptors into two distinct classes, α and β, is by now considered classic (Ahlqvist, 1967), and β-adrenoreceptors have been subdivided into at least two types, β_1 and β_2 (Lands *et al.*, 1967; Furchgott, 1967). A list of drugs interfering at the α- or β-level as selective agonists or antagonists is given in Table IX.

The way in which NE is released by indirect action of sympathomimetic amines such as tyramine differs from the mode of release by nerve impulses (see Smith, 1973). Tyramine, in fact, causes the release of NE from stores which are unavailable for release by nerve stimulation; one of these stores is the "free" cytoplasmic pool as found in animals pretreated with reserpine (which inhibits granular reuptake) and a MAO inhibitor (Furchgott and Sanchez-Garcia, 1968). In particular, exogenous administered tyramine to release NE is taken up in the sympathetic endings by the Uptake$_1$ mechanism (see following section). Once in the terminal, tyramine either displaces NE from the vesicles into the cytosol (e.g., by dissociating or making less stable the storage complex of NE and ATP, according to Berneis *et al.*, 1970) or increases the spontaneous rate of loss of NE from the cytosol to the extracellular space or both. However, it is evident that tyramine does not evoke NE release by exocytosis. Another drug which releases NE from terminals is *d*-amphetamine; however, the mechanism of action of *d*-amphetamine is more complex since it is also able to block Uptake$_1$ mechanism and MAO (see Table X and Fig. 13) (Glowinski and Axelrod, 1966; Glowinski *et al.*, 1966b). In animals pretreated with a MAO inhibitor, a massive release of CA's is observed after administration of reserpine, prenylamine, a phenylethylamine with reserpine-like actions, and some foodstuffs containing tyramine (Marley and Blackwell, 1970; Marley and Nisticò, 1975a). Since NE release is a Ca^{2+}-dependent process, drugs able to chelate Ca^{2+}, such as bretylium, induce inhibition of NE release. Several other mechanisms regulating synthesis and release of NE, such as end-product regulation, presynaptic α-receptors, prostaglandins, have been already considered in Section B,6.

9. Uptake Mechanisms

Several mechanisms are involved in the termination of CA effects; they are schematically illustrated in Fig. 12. They include uptake into the neuron (Uptake$_1$) or into extraneuronal sites (Uptake$_2$), removal by the

TABLE IX

Drugs Interfering with E and NE Receptors

Class	Drug		Observations	References to main action
Receptor agonists	Noradrenaline	Do not cross the BBB	Stimulation of α-adrenoceptors	See Goodman and Gilman (1975); Meyers *et al*. (1974); Jenkinson (1973); Furchgott (1972)
	Phenylephrine		Stimulation of α-adrenoceptors	
	Adrenaline		Stimulation of α- and β-adrenoceptors	
	Isoprenaline		Stimulation of β_1- and β_2-adrenergic receptors	
	Methoxamine		Stimulation of α-adrenergic receptors, slight β-blocking activity	
	Clonidine		Stimulation of central and peripheral α receptors; stimulation of cerebral histamine H_2 receptors (Audigier *et al*., 1976)	See Connolly (1970); Callingham (1971); Laverty (1973)
	Oxymetazoline		Stimulation of central and peripheral α-receptors	
	Salbutamol		β_2-stimulation	Brittain *et al*. (1970)
Receptor antagonists	Ergotamine		α-Adrenergic receptor blockade; direct stimulant action of smooth muscles	
	Phentolamine		Short-lasting α-adrenergic receptor blockade; α-presynaptic receptor blockade	
	Tolazoline		Short-lasting α-adrenergic receptor blockade; blockade of NE receptors (Keller *et al*., 1973); increase NE turnover (Lloyd and Bartholini, 1974)	
	Methiothepin		Blockade of 5-HT (Monachon *et al*., 1972) and DA receptors (Lloyd and Bartholini, 1974)	

Phenoxybenzamine	Long-lasting α-adrenergic receptor blockade; uptake inhibition; blockade response to almost all transmitters	
Piperoxane	α-Adrenergic receptor blockade	
Yohimbine	α-Adrenergic receptor blockade; blockade of 5-HT receptors	
Chlorpromazine	NE and DA receptor blockade; sensitization to β-adrenergic stimulation	
Propranolol	β_1,β_2-receptor antagonist (Dunlop and Shanks, 1968)	Black *et al*. (1965); Black and Prichard (1973)
Practolol	β_1-receptor antagonist; does not cross the BBB (Scales and Cosgrove, 1970)	Black and Prichard (1973)

circulation, metabolic degradation by MAO or/and COMT. One of the most important mechanisms by which the actions of released neurotransmitters on postsynaptic receptors are rapidly terminated is the physical removal of the free transmitter from the synaptic cleft, and this can be catalyzed by a variety of different uptake systems. Only the most usual features of these systems will be described here; more details on their existence and function can be found in several reviews by Iversen (1967, 1971a, 1971b, 1973).

Most of the uptake mechanisms known now are located in the membrane of the presynaptic nerve terminal ($Uptake_1$), so that removal of transmitter is also equivalent to "recapture" in the terminal with subsequent storage by a specific Mg^{2+}-ATP "granular" uptake mechanism and reutilization of some molecules. However, the cellular localization of such removal mechanisms might be at any site of the synaptic region. There are also some examples of transmitter uptake occurring in sites other than the prejunctional terminals ($Uptake_2$).

$Uptake_1$ has been studied most, and is a stereochemical-specific transport system for which 1-NE is the preferred substrate. The uptake process is saturable with a high affinity constant, which means that it can work very efficiently in the presence of concentrations of NE likely to be encountered in the extracellular space under physiological conditions. This process appears to require the maintenance of a normal Na^+ gradient concentration between the intra- and extracellular compartments and can be inhibited by drugs that prevent normal functioning of the Na^+ pump. The uptake system for NE exists in adrenergic nerve endings in both peripheral and central nervous systems (Iversen, 1973).

NE uptake in the CNS can be easily studied in homogenates where amine uptake into synaptosomes readily occurs when such preparations are incubated *in vitro* (Snyder and Coyle, 1968; Coyle and Snyder, 1969).

The importance of the NE uptake system for adrenergic neurotransmission is illustrated by the reported potentiation and prolongation of the responses of innervated tissues to adrenergic nerve stimulation by drugs which inhibit the uptake process. Cocaine, phenoxybenzamine, and tricyclic antidepressants of the imipramine and amitriptyline groups, namely, desmethylimipramine (DMI), are among the most potent inhibitory drugs (Iversen, 1973) (see Table X). In addition, NE uptake is also inhibited by some antihistaminic drugs such as chlorpheniramine and tripelennamine, adrenergic neuron-blocking drugs (guanethidine and bretylium), and several MAOI. The exact proportion of released NE recaptured in this way varies from tissue to tissue and also depends upon the frequency of firing of impulses in the adrenergic terminals, but it is probably in the range 50–90% (Iversen, 1973; Langer, 1970; Häggendal, 1970).

TABLE X

Drugs Potentiating Catecholaminergic Mechanisms

Class	Drug	Observations	References to main action
Precursors	L-Dopa	Increase in synthesis of DA and NE; initial displacement of 5-HT and inhibition of 5-HT synthesis after long-term treatment (Bartholini *et al.*, 1968; Everett and Borcherding, 1970; Karobath *et al.*, 1972). Decrease in brain tryptophan (Karobath *et al.*, 1971)	See Hornykiewicz (1966); Bartholini and Pletscher (1968)
	L-Dopa + peripheral L-AAAD inhibition	Increase in brain CA synthesis	Bartholini and Pletscher (1968); Tissot *et al.* (1969)
	DOPS	Direct decarboxylation to NE; the most active is + (*erythro*-DOPS)	Creveling and Daly (1971); Puig *et al.* (1974)
Releasing agents	*d*-Amphetamine	Inhibition of uptake (Iversen, 1971b); MAO inhibition (Blaschko *et al.*, 1937); increase in 5-HT turnover (Reid, 1970); release of ACh from cerebral cortex (Pepeu and Bartholini, 1968); stimulation of 5-HT receptors (Vane, 1960)	Glowinski and Axelrod (1966); Glowinski *et al.* (1966b); Creveling and Daly (1971)
	Tyramine	Good substrate for MAO (Smith, 1966); partial conversion to octopamine (Musacchio *et al.*, 1966)	See Trendelenburg (1972); Burn and Rand (1958)
	Ephedrine		
MAO inhibitors			
Hydrazides and hydrazines	Iproniazid	Irreversible noncompetitive inhibition	Zeller *et al.* (1952); see Goodman and Gilman (1975); Meyers *et al.* (1974); Creveling and Daly (1971); Marley and Blackwell (1970)
	Nialamide	Reduction in brain CA turnover (Neff *et al.*, 1969)	Zeller *et al.* (1952); see Goodman and Gilman (1975); Meyers *et al.* (1974); Creveling and Daly (1971); Marley and Blackwell (1970)

TABLE X *(Continued)*

Class	Drug	Observations	References to main action
	Isocarboxazide		Zeller *et al.* (1952); see Goodman and Gilman (1975); Meyers *et al.* (1974); Creveling and Daly (1971); Marley and Blackwell (1970)
	Iproclozide		Zeller *et al.* (1952); see Goodman and Gilman (1975); Meyers *et al.* (1974); Creveling and Daly (1971); Marley and Blackwell (1970)
	Pheniprazine (α-methylphenylhydrazine)		Zeller *et al.* (1952); see Goodman and Gilman (1975); Meyers *et al.* (1974); Creveling and Daly (1971); Marley and Blackwell (1970)
	Phenelzine (phenethylhydrazine)		Zeller *et al.* (1952); see Goodman and Gilman (1975); Meyers *et al.* (1974); Creveling and Daly (1971); Marley and Blackwell (1970)
	Mebanazine (α-methylbenzylhydrazine)		Zeller *et al.* (1952); see Goodman and Gilman (1975); Meyers *et al.* (1974); Creveling and Daly (1971); Marley and Blackwell (1970)
	Phenoxypropazine (1-methyl-2-phenoxyethylhydrazine)		Zeller *et al.* (1952); see Goodman and Gilman (1975); Meyers *et al.* (1974); Creveling and Daly (1971); Marley and Blackwell (1970)

Amines	*d*-Amphetamine	Reversible inhibitors of rather low potency	Grana and Lilla (1959); Greig *et al.* (1959)
	α-Ethyltryptamine (Etryptamine)		Grana and Lilla (1959); Greig *et al.* (1959)
	Pargyline (*N*-methyl-*n*-2-propinylbenzylamine)	Containing an acetylenic group	Creveling and Daly (1971)
	Tranylcypromine	Amphetamine-like effects (Carlsson *et al.*, 1960)	Mills *et al.* (1968)
	Harmine, Harmaline		Mills *et al.* (1968)
	Yohimbine	α-Adrenergic receptor blockade	McIsaac and Estevez (1966)
COMT inhibitors	Pyrogallol	Inhibition also of D-β-H	Axelrod and Laroche (1959); Belleau and Burba (1961)
	Troponolone		Axelrod and Laroche (1959); Belleau and Burba (1961)
	Dopacetamide		
	Estrogens		Ball *et al.* (1972)
Uptake$_1$ inhibitors	DMI	Preferential NE inhibition	See Iversen (1967, 1971a, 1971b, 1974)
	Imipramine	Inhibition of 5-HT uptake	
	Amitriptyline	Inhibition of 5-HT uptake	
	Nortriptyline	Inhibition of 5-HT uptake	
	l-Metaraminol		
	Cocaine	Increased turnover of striatal DA (Costa *et al.*, 1972)	
	Phenoxybenzamine	Blockade of α-adrenoceptors; inhibition of uptake$_2$	
	d-Amphetamine	Inhibition of DA uptake	
	Maprotiline	Inhibition of 5-HT uptake	Maitre *et al.* (1974)
Uptake$_2$ inhibitors	*dl*-Metanephrine		Iversen (1971a, 1971b)
	Glucocorticoids		
	Estrogens		

Besides the NE uptake system, there is a similar system in nerve terminals of the DA-containing neurons in the basal ganglia of the CNS (see Section B,11).

In addition to the neuronal "recapture" mechanisms, other uptake systems are present in extraneuronal sites that could similarly accelerate disappearance of transmitter from the extracellular fluid. In various peripheral tissues innervated by the sympathetic nervous system, notably in cardiac muscle and in smooth muscle such as in spleen and blood vessels, NE is taken up by a special system known as "$Uptake_2$" (Iversen, 1971a). This transport system has properties quite distinct from that of the neuronal $Uptake_1$ system. It has relatively lower affinity of NE than for E as substrate, and it is potently inhibited by phenoxybenzamine and certain structurally related compounds, by a variety of steroids (corticosterone, β-estradiol, etc.) and by the *O*-methylated CA metabolites metanephrine and normetanephrine (see Table X and Fig. 13). The inhibitory effects of the steroids might explain the reported potentiating effect that certain of these compounds have on the adrenergic mechanisms (Iversen and Salt, 1970, 1972). Drugs such as imipramine and cocaine, which are potent inhibitors of the neuronal recapture system, are without effect on $Uptake_2$. The NE accumulated by the $Uptake_2$ mechanism in the various tissues examined is not retained but instead rapidly exposed to MAO or/and COMT degradation (Lightman and Iversen, 1969) (see below).

10. Metabolic Degradation

Oxidative deamination by MAO (see Costa and Sandler, 1972; Kopin, 1972) and *O*-methylation by COMT are the major enzymatic pathways of CA's catabolism in the brain (see Marley and Stephenson, 1972). Indirect evidence suggests that MAO, present in brain mitochondria (Bogdanski *et al.,* 1957; Weil-Malherbe *et al.,* 1961a; Nukada *et al.,* 1963), mainly at synaptic endings (Rodriguez de Lores Arnaiz and De Robertis, 1962), functions intraneuronally whereas COMT, present in a soluble supernatant fraction (Axelrod and Tomchick, 1958) associated with synaptosomes (Alberici *et al.,* 1965), functions extraneuronally. Thus, following iv injection of dopa, dihydroxyphenylacetic acid appears in brain before homovanillic acid, suggesting that deamination occurs first, close to the site of dopamine formation (Carlsson and Hillarp, 1962). Evidence exists that MAO is not a single enzyme but rather a family of enzymes with different substrate specificity. In mammalian tissues at least two forms of MAO (A and B) have been identified; types A and B enzyme have different characteristics and drugs are known which can, in the proper concentrations, selectively block one or the other enzyme system (see Costa and Sandler, 1972).

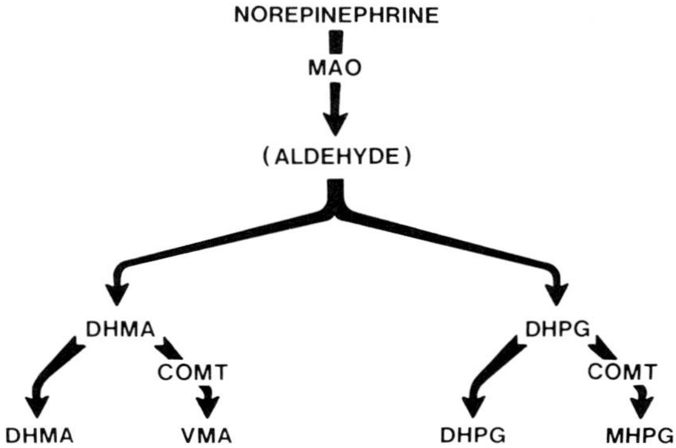

Fig. 15 Metabolism of NE in CNS. Urinary metabolites are shown in the bottom line. Excretion products may be conjugated with glucuronide or sulfate. DHMA, dihydroxymandelic acid; DHPG, dihydroxyphenylglycol; VMA, 3-methoxy-4-hydroxymandelic acid; MHPG, methoxyhydroxyphenylglycol. (Modified from Meyers *et al.*, 1974.)

When endogenous CA's reach the circulation they are degraded mainly by *O*-methylation in the liver and kidney. When they are taken up into the neuron (Uptake$_1$), the storage in the vesicles mostly prevents destruction by MAO, unless the storage is blocked (e.g., by reserpine). Uptake into extraneuronal sites (Uptake$_2$) is followed by metabolism, mainly by COMT but also by MAO. The major catabolites of NE formed in the brain or peripherally are shown in Figs. 15 and 16.

a. Inhibitors of Metabolic Degradation

Since MAO and COMT are the major enzymatic activities involved in CA metabolism in the brain, efforts were made to synthesize compounds able to enhance central CA mechanisms by inhibiting their metabolic inactivation. A list of the most common MAO and COMT inhibitors is given in Table X, which also presents a series of drugs capable of enhancing catecholaminergic neurotransmission by acting at different steps of the synaptic events.

In several animal species, pretreatment with MAOI increases brain concentration of endogenous monoamines, enhances the accumulation of exogenous ^{3}H-NE following its intracerebroventricular administration and the increase in CA's following L-dopa (see Marley and Stephenson, 1972). Also behavioral, electrocortical, and hypothermic effects of NE and DA after their intrahypothalamic infusion can be potentiated by MAOI (Marley and Nisticò, 1972). However, interpretation of results obtained after MAOI is not always easy as these drugs possess other actions which directly or indirectly interfere with monoaminergic neurotransmission,

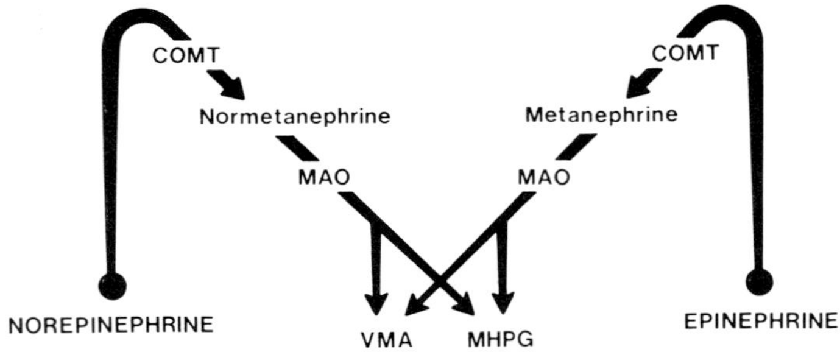

Fig. 16 Metabolism of free catecholamines. Urinary metabolites are shown in the bottom line. VMA, 3-methoxy-4-hydroxymandelic acid; MPHG, methoxyhydroxyphenylglycol. Excretion products may be conjugated with glucuronide or sulfate. (Modified from Meyers *et al.*, 1974.)

e.g., decrease in CA turnover (Neff and Costa, 1966), inhibition of TH as a consequence of increased CA concentration (Udenfriend *et al.*, 1965), and some of them, like phenylisopropylhydrazine and tranylcypromine, exert amphetamine-like effects (Carlsson *et al.*, 1960; Eltherington and Horita, 1960; Pirch and Norton, 1965). Tranylcypromine has also been found capable of blocking NE Uptake$_1$ in rat heart (Iversen, 1965) and rat cerebral cortex (Hendley and Snyder, 1968). In addition, regional differences in MAO inhibition have been shown; thus iproniazid inhibits MAO in the hypothalamus but not in the pituitary, whereas pheniprazine is without effect in both cases (Weil-Malherbe *et al.*, 1961b; Hertting *et al.*, 1961).

Following MAO inhibition, usually, brain 5-HT concentration is increased more than that of CA's, and in some species, such as cats and dogs, NE is not increased at all (Vogt, 1954, 1959; Spector *et al.*, 1960; Mannarino *et al.*, 1963). The increase in brain DA and NE concentration is associated with an increase in their respective *O*-methylated catabolites, 3-methoxytyramine and normetanephrine (Axelrod, 1958; Carlsson and Waldeck, 1964; Glowinski and Iversen, 1966).

Although there are many selective inhibitors of MAO there are none for COMT; consequently, the study of the role played by COMT degradation has been slow. Although a variety of compounds are able to inhibit COMT *in vitro* there are only few effective *in vivo* (see Table X), e.g., pyrogallol, troponolones, 2′,3′,-trihydroxyacetophenone, dopacetamide derivatives (Carlsson *et al.*, 1963; Carlsson, 1964; Ross and Haljasmaa, 1964). With some compounds, such as troponolones, inhibition of peripheral COMT is greater than that of cerebral COMT (Ross and Haljasmaa, 1964), possibly because of the poor penetration into the brain. However, after troponolone (Goldstein, 1964) or α-ethoxydopacetamide (Carlsson, 1964) an increased accumulation of brain CA's from dopa has been obtained. The lack of specificity of these compounds is that some of them, such as dopacetamide derivatives, are also able to inhibit CA and 5-HT synthesis (Carlsson *et al.*, 1963; Roos and Werdinius, 1963) and troponolone inhibits D-β-H *in vitro* (Goldstein *et al.*, 1964).

11. Metabolism of Dopamine

Synthesis and degradation of DA in the brain occur in those areas in which DA is present as neurotransmitter, such as the nigrostriatal system, mesolimbic system, tuberoinfundibular system, CTZ, and retina. The DA-catabolizing enzymes (MAO and COMT) and the DA-synthesizing enzymes (L-TH and L-AAAD) are present in highest concentration in the same areas (Hornykievicz, 1973).

Like NE, dopamine is stored in presynaptic vesicles, is released by nerve impulses, by electrical stimulation and pharmacologically (e.g., *d*-amphetamine, reserpine) (Glowinski, 1972). It exerts its effects through the stimulation of specific receptors, and its effects are terminated by uptake mechanisms and by metabolic degradation (Iversen, 1974). As for NE, the synthesis of DA in dopaminergic terminals seems to be modulated by a mechanism of negative feedback so that increased DA in terminals appears to inhibit TH activity, the rate-limiting step (Javoy and Glowinski, 1971a).

It would appear that in DA nerve terminals two storage compartments exist, a small, functional, readily releasable one and a main storage compartment. Such an assumption is based first on the fact that DA newly synthesized from ^{3}H-tyrosine is preferentially released from caudate terminals (Besson *et al.*, 1971), and, second, because DA levels in the striatum do not decline monophasically after DA synthesis inhibition, but in two distinct phases, an initial rapid phase lasting 15 min and a slower one starting 30 min after the drug injection and lasting for a few hours (Javoy and Glowinski, 1971b). According to the latter authors, the spontaneous release of DA seems to be confined mainly to the newly synthesized transmitter. However, the compartmentalization of DA in the striatum as proposed by Javoy and Glowinski (1971b) has been criticized and, on the basis of kinetic analysis of turnover rate of striatal DA, experimental data have been presented that do not support the presence of two DA compartments (Doteuchi *et al.*, 1974). Quite recently, however, the evaluation of the specific activity of DA and its metabolites at various times after IVT injection of labeled tyrosine has added strength to the proposal of the existence of two (or multiple) DA compartments in the striatum of the rat (Groppetti *et al.*, 1977).

In vitro and *in vivo* release of DA from the striatum have been obtained with several drugs such as *d*-amphetamine, some MAOI (pheniprazine and pargyline), and with an anticholinergic drug, benztropine (Besson *et al.*, 1971) (Table XI). DA release from the striatum has also been obtained by electric stimulation of substantia nigra which contains DA cell bodies (Poirier and Sourkes, 1965; Portig and Vogt, 1969; Riddell and Szerb, 1971; Glowinski, 1972; Andén *et al.*, 1964b).

After being released DA acts on specific receptors of the postsynaptic membrane. The presence of postsynaptic DA receptors in CNS has been demonstrated by experiments in which central DA effects are blocked by specific antagonists (see Marley and Nisticò, 1972), such as neuroleptic drugs (see Table XI). As a result of DA receptor-blocking effects, neuroleptic drugs have been found to increase reflexively DA synthesis and turnover in the striatum (Andén *et al.*, 1964; Roos, 1965; Sharman, 1966; Nyback and Sedvall, 1969). Recently, it has been reported that butyrophenones and other antipsychotic drugs potently inhibit stimulated release of DA from terminals in striatal slices (Seeman and Lee, 1975).

It is now evident that feedback mechanisms involved in the regulation of dopaminergic neurons are far more complex than originally supposed (Carlsson *et al.*, 1972). In addition to the existence of a postsynaptic neuronal feedback circuit responsible for modulating the rate of impulse flow within the dopaminergic pathway, recent biochemical and neurophysiological studies suggest the existence of a "local" mechanism modulating dopaminergic function (Aghajanian and Bunney, 1974; Bunney and Aghajanian, 1975). This would be mediated by presynaptic DA receptors present on the entire dopaminergic neuron, including both soma and terminals. In fact, stimulation of the presumed presynaptic DA receptors (e.g., through microiontophoresis of DA or apomorphine) induced an inhibitory response in the dopaminergic cell soma (Fig. 17), suggesting that if such presynaptic receptors exist on the dopaminergic terminals they may be concerned with a feedback inhibition of DA synthesis and release. In contrast to the direct effects of DA and apomorphine upon dopaminergic neurons, microiontophoretically applied *d*-amphetamine was found to produce only a minimal slowing of DA neurons or to have no effect at all (Bunney *et al.*, 1973). However, *d*-amphetamine had a marked inhibitory effect on the rate of firing of dopaminergic neurons when administered systemically in low doses (Fig. 18, upper trace). In addition, low systemic doses of apomorphine induced inhibition of the rate of firing of dopaminergic neurons in animals lacking anatomic communication between the pre- and postsynaptic neurons for a diencephalic transection separating the substantia nigra from the neostriatum, while systemically administered *d*-amphetamine was ineffective (Fig. 18, lower trace).

TABLE XI

Drugs Interfering with Dopaminergic Neurotransmission

Class	Drug	Observations	References to main action
Receptor agonists	Apomorphine	Stimulation of DA-sensitive AC system of the rat striatum (Kebabian *et al.*, 1972); decrease in DA turnover in nigrostriatal system; presynaptic DA receptor stimulation with inhibition of TH in synaptosomes of rat striatum (Christiansen and Squires, 1974; Iversen *et al.*, 1975); central emetic action; diesters of apomorphine (diacetyl, dipropionyl, diisobutyryl, dipivaloyl, dibenzoyl) possess prolonged apomorphine-like effects (Baldessarini *et al.*, 1976); increase in brain 5-HT and 5-HIAA (Grabowska *et al.*, 1973)	Andén *et al.* (1967); Ernst (1967)
	Epinine		Iversen (1975)
	ADTN (2-amino-6,7-dihydroxytetrahydronaphthalene)	Stimulation of DA-sensitive AC system (Miller and Iversen, 1974)	Woodruff (1971); Iversen (1975)
	Piribedil (ET 495) 1-(2-pyrimidil-4-piperonylpiperazine)	Peripheral vasodilatation (Regnier *et al.*, 1968); no stimulation of DA sensitive AC system (Miller and Iversen, 1974); S 584: catechol metabolite of Piribedil capable to stimulate striatal AC (Miller and Iversen, 1974); no stimulation by Piribedil or S 584 of retinal AC (Schorderet, 1975); increase in striatal ACh (Consolo *et al.*, 1974)	Corrodi *et al.* (1971)
	S 3608 (1-coumaran-5-yl-methyl)-4,2-thiazolyl)piperazine	No cardiovascular properties	Poignant *et al.* (1975)
	Amantadine	No stimulation of DA-sensitive AC system (Karobath, 1974)	Papeschi (1974)
	D 145 (1,3-dimethyl-5-aminoadamantane)	No stimulation of DA-sensitive AC system (Karobath, 1974); stimulation of DA release (Svensson, 1973a,b)	Svensson (1974a,b)
	Ergocornine	Stimulation of 5-HT receptors (Corrodi *et al.*, 1975)	Corrodi *et al.* (1973)
	Ergometrine		Corrodi *et al.* (1973)

TABLE XI *(Continued)*

Class	Drug	Observations	References to main action
	2-Br-α-ergocryptine (CB 154) (Bromocriptine)	Stimulation of DA-sensitive AC system only *in vivo* (Trabucchi *et al.*, 1976a); action depending in part by brain CA stores (Johnson *et al.*, 1976)	Corrodi *et al.* (1973)
	Lergotrile (2-Chloro-6-methyleagoline-8β-acetonitrile)		J. A. Clemens *et al.* (1974)
	Lisuride hydrogen maleate [*N*-(D-6-)methyl-8-isoergolenyld]-*N′*,*N′*-diethylcarbamide	Peripheral antiserotoninergic action (Aùskovà *et al.*, 1974); stimulation of brain 5-HT receptors (Kehr, 1977)	Horowski *et al.* (1975)
Precursors	L-Tyrosine		See von Euler (1972)
	L-Dopa	Often associated with peripheral inhibitors of L-AAAD	Carlsson (1959)
Releasing agents	*d*-Amphetamine (see Table X)	Increase in striatal DA turnover (Costa *et al.*, 1972)	Carlsson and Waldeck (1966); Ernst (1967)
	Amantadine		Carr and Moore (1970); Scatton *et al.* (1970)
Blockers of D-β-H	Disulfiram		
	FLA-63 (see Table VII)		
Uptake inhibitors	*d*-Amphetamine		
	Amantadine	Stimulation of DA release (Carr and Moore, 1970)	Coyle and Snyder (1969); Scatton *et al.* (1970)
	Benztropine	Anticholinergic activity. Release of DA from the striatum (Goodale and Moore, 1975)	Coyle and Snyder (1969); Farnebo *et al.* (1970)
	Diethazine	Anticholinergic activity	
	Orphenadrine	Anticholinergic activity	
	Nomifensine (8-amino-2-methyl-4-phenyl-1,2,3,4-tetrahydroisoquinoline)	Weak anticholinergic activity; inhibition of NE uptake	Hunt *et al.* (1974)
Receptor blockers	Phenothiazines	Increased turnover of striatal DA (Andén *et al.*, 1969)	Horn and Snyder (1971); Carlsson and Lindqvist (1963); Andén *et al.* (1970); Janssen (1965); Jassen *et al.* (1975)

	Methiothepin	NE and 5-HT receptor blockade (Monachon *et al.*, 1972) and increase in striatal DA turnover (Lloyd and Bartholini, 1974)	Lloyd and Bartholini (1974)
	Haloperidol		
	Pimozide	Increased NE turnover at high doses	Andén et al. (1970)
	Clopimozide	Long duration of action	Janssen *et al.* (1975)
	Domperidone	Does not readily cross the BBB (Janssen Research Product Information Service)	
	Penfluridol		
	Spiroperidol		
	α-Flupentixol		
	Metoclopramide	No antagonism of *in vitro* DA stimulated striatal AC (Peringer *et al.*, 1976)	Costall and Naylor (1973); Dolphin *et al.* (1975)
	Sulpiride	No antagonism of *in vivo* and *in vitro* DA stimulated striatal AC (Trabucchi *et al.*, 1976b)	Justine-Besançon *et al.* (1967)
	Clozapine	Inhibition of DA stimulated striatal AC; inhibition of DA release (Miller *et al.*, 1974); strong anticholinergic properties (Stille and Hippius, 1971)	Matz *et al.* (1974)
	DMPEA (3,4-dimethyloxyphenylethylamine or 3,4-di-*O*-methyldopamine)	Postulated DA receptor blockade (Smythe and Lazarus, 1973b)	
Synthesis inhibitors	See NE synthesis inhibitors (Table VII)		
Storage inhibitors	Reserpine (see Table VIII)		
Neurotoxic agents	6-OH-dopa	Passage through the BBB	Jacobowitz and Kostrzewa (1971); Sachs and Johnsson (1972)
	6-OHDA	In association with DMI or protriptyline: selective effect on DA neurons (Evetts and Iversen, 1970; Breese and Traylor, 1971)	

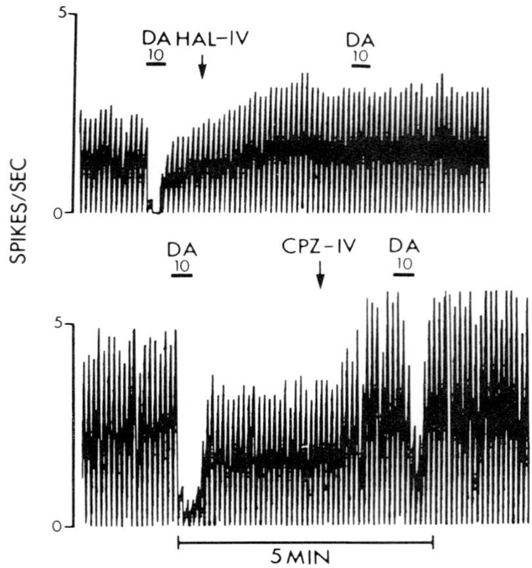

Fig. 17 The depression of firing of dopaminergic neurons by microiontophoretically applied dopamine (DA) and the blockade of this effect by intravenous haloperidol (HAL, upper trace) or chlorpromazine (CPZ, lower trace). Both cells are inhibited by DA ejected at a low iontophoretic current (10 nA). HAL, in a dose of 0.25 mg/kg, causes an increase in the baseline rate of firing (upper trace). Following the HAL, microiontophoretic DA is no longer effective in producing inhibition. CPZ, at a dose of 5.0 mg/kg, attenuates but does not completely block the DA inhibition (lower trace). The bars above tracings indicate duration of microiontophoresis and numbers above bars refer to ejecting currents in nA. The concentration of DA in the 5-barreled microiontophoretic electrodes was 0.2 *M* (pH 4.0). (Reproduced, with permission, from Aghajanian and Bunney, 1974.)

Collectively, these data would indicate that systemically administered apomorphine, like the iontophoretically applied drug, can directly inhibit dopaminergic neurons by acting upon presynaptic DA receptors, while *d*-amphetamine-induced depression of firing of dopaminergic neurons following systemic injection is mediated by the postsynaptic feedback loop. If one accepts that there are both local and postsynaptic feedback mechanisms modulating dopaminergic function, the crucial question deals with the relative importance of these two mechanisms under different physiological conditions and drug treatments. It would appear that in a low dose range the presynaptic action of a drug may predominate, resulting in diminished dopaminergic function. Such a mechanism may explain the biphasic behavioral effect of apomorphine in rats. There is evidence that at low doses it reduces motor activity while at high doses it increases motor activity (Carlsson, 1977).

Activation of postsynaptic DA receptors in the brain is followed by an increase in 3′,5′-cAMP. This has led to the hypothesis that a DA-stimulated adenylate cyclase could well represent DA receptors (Kebabian *et al.*, 1972). On homogenates of striatal tissue, after denervation, adenylate cyclase sensitive to DA is considerably increased (Mishra *et al.*, 1974), thus adding further support to the hypothesis that DA-sensitive adenylate cyclase represents the mechanism by which DA act postsynaptically in the CNS. In addition, powerful stimulants of adenylate cyclase, such as cholera toxin injected into the n. accumbens and the striatum, mimic DA, and dibutyryl-cAMP (db-cAMP) induced increase in motor ac-

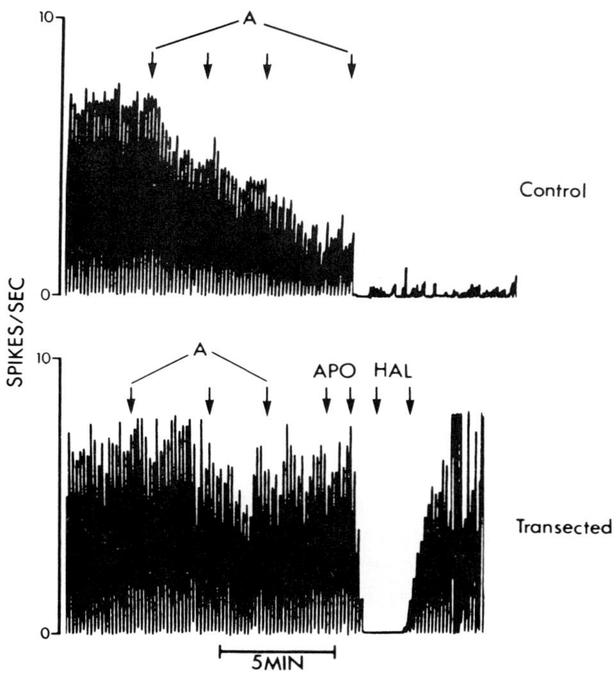

Fig. 18 Effects of intravenous *d*-amphetamine (A) on the rate of firing of dopaminergic neurons in the zona compacta (substantia nigra) in a control rat (upper trace) and in a rat with a diencephalic transection (lower trace) which separates the substantia nigra from the neostriatum. In upper trace, the typical depressant effect of A on dopaminergic cell firing can be seen. An initial dose of 0.25 mg/kg (iv) produces a partial slowing; a total inhibition occurs after a cumulative dose of 2.0 mg/kg has been given. In the transected preparation (lower trace) amphetamine in a cumulative dose of 3.2 mg/kg (iv) fails to produce significant slowing. However, intravenous apomorphine in a cumulative dose of 0.4 mg/kg totally inhibits firing of the cell. Haloperidol (0.2 mg/kg, iv) rapidly reverses the apomorphine-induced depression. (Reproduced, with permission, from Aghajanian and Bunney, 1974.)

tivity (Miller and Kelly, 1975; Nisticò *et al.,* 1976). DA-sensitive adenylate cyclase is inhibited in a dose-dependent fashion by several (e.g., haloperidol, pimozide, α-flupentixol, fluphenazine) but not all neuroleptics (e.g., sulpiride) (Iversen, 1975, and Table XI).

Similar to NE, it is likely that termination of central effects of DA is due to a physical recapture of the transmitter within the DA terminals (Iversen, 1974). The features of the uptake mechanism for DA are similar to those of NE, e.g., Na^+-dependency, stereospecificity, highest affinity for DA. Several drugs able to potentiate dopaminergic transmission in CNS, among them (+)-amphetamine, but not antidepressants, are known to be inhibitors of DA uptake (Coyle and Snyder, 1969).

As a corollary, a pharmacological tool which allows selective degeneration of DA neurons in CNS is represented by the IVT administration of 6-OHDA (see Section D) in animals pretreated with DMI; the latter drug prevents uptake of 6-OHDA in the NE neurons, whereas uptake into DA neurons is unaffected.

The end product of DA catabolism is HVA, and its concentrations in the caudate nucleus and putamen parallel that of DA;, in contrast, in the substantia nigra and in the globus pallidus, the levels of HVA are 5- to 10-fold higher than those of DA. The metabolic rate of DA in the fibers of the nigrostriatal pathways seems to be particularly high (Hornykiewicz, 1973). A schematic representation of DA metabolic pathways is given in Fig. 11.

12. Quantitative Estimation of Turnover

In order to investigate the functional role played by CA's, turnover studies have been carried out using kinetics. Such an approach provides an index of the physiological functions of CA neurons since the steady-state concentrations do not show per se whether a rise in brain CA's reflects an increase in their synthesis or decrease in the amount of CA's released and degraded. Current views tend to describe CA release by nerve impulse as involving a small metabolic pool (Javoy and Glowinski, 1971a,b; Kopin *et al.,* 1968). However, this view is not readily explained on the basis of new concepts regarding the release of CA's by means of an exocytosis process (De Potter *et al.,* 1969; Geffen *et al.,* 1969; Weinshilboum *et al.,* 1971). Several simple models of adrenergic function include two pools of CA's, a storage and a metabolic one, which function as two distinct entities and mix with one another at a very slow rate (see Costa *et al.,* 1972). The regulation of CA synthesis, as mentioned above in Section B,6, also depends on the rate of axonal depolarization (Neff *et al.,* 1969).

Steady-state and nonsteady-state methods to measure turnover rate have been described (Anton-Tay and Wurtman, 1971; Costa, 1972; Costa *et al.,* 1972). In the steady-state methods, trace doses of labeled tyrosine or CA which do not disturb the steady state are administered. The accumulation of brain ^{3}H-CA following systemic or IVT injection of ^{3}H-tyrosine or dopa is taken as an index of CA biosynthesis. The disappearance of ^{3}H-CA molecules from the brain during the first few hours after IVT or IC injection of ^{3}H-CA is taken as an index of CA turnover.

In the nonsteady-state methods, the decline in brain CA levels after treatment with α-MpT, the increase in brain CA after treatment with a MAO-inhibitor, or the accumulation of acid DA metabolites after inhibition with probenecid are measured.

With both methods certain limitations are evident, although it appears that more reliable indications on the functional activity of CA neurons can be obtained by steady-state isotopic methods. The principal characteristics and main drawbacks of the methods used for the evaluation of brain CA turnover are summarized in Table XII. The recent availability of microdissection techniques (Palkovits *et al.,* 1974a) and of highly sensitive radiometric methods for assay of biogenic amines (Coyle and Henry, 1973) has made it possible to estimate CA turnover rate in individual nuclei of the rat brain by the use of α-MpT (Kizer *et al.,* 1975b). In addition to these methods, the functional activity of dopaminergic neurons may be evaluated by the determination of 3-methoxytyramine (3-MT) concentration in brain areas rich in DA-nerve terminals (Groppetti *et al.,* 1977). Since COMT, which is the enzyme that catalyzes the conversion of DA to 3-MT, is most likely localized only extraneuronally, measurement of 3-MT provides a reliable index of DA released from presynaptic nerve junctions.

The turnover rate of a brain monoamine can be affected by several experimental manipulations (Costa *et al.,* 1972; Costa, 1973). Thus, it may be increased by several drugs which block the postsynaptic receptors, block the reuptake of presynaptic terminals, decrease the monoamine concentration if this can control its own biosynthesis by product inhibition, increase the activity of the rate-limiting enzyme by transsynaptic activation (see above Section B,6), decrease the afferent synaptic input generating inhibitory postsynaptic potentials, and increase the afferent synaptic input generating excitatory postsynaptic potentials. The turnover rate of a brain monoamine can be decreased by drugs which inhibit the rate-limiting enzyme for the biosynthesis of the monoamine, act as agonist at the postsynaptic receptors, decrease the afferent excitatory input or increase the afferent inhibitory input in the aminergic system studied, reduce the release of transmitter by activating regulatory receptors in the nerve ending, and increase the monoamine concentration if inhibition by end product is operating for the considered transmitter.

TABLE XII

Methods for the Evaluation of Brain CA Turnover

Method	Observations and main drawbacks	References
A. *Isotopic steady-state methods*		
1. Labeling brain (or discrete areas) CA stores with systemic, IVT or IC injection of labeled precursors (tyrosine, dopa)	Assumption that labeled precursors enter only endogenous amine stores; erroneous estimation of NE biosynthesis because newly synthesized NE molecules mix with the active pool and are released from the neuron; the specific activity of labeled tyrosine within the noradrenergic neurons is not known; endocrine manipulations could increase uptake or decrease CA release with consequent increased accumulation in labeled CA with no change in CA synthesis	Neff *et al.* (1969, 1971); Fuxe and Hökfelt (1969); Antòn-Tay and Wurtman (1971); Javoy and Glowinski (1971b)
2. Decline of labeled brain (or discrete areas) CA stores after IVT or IC injection of labeled DA or NE	Only the periventricular areas are reached by the amines and other CA neurons cannot be studied; labeled amines are taken up also by noncatecholaminergic neurons (e.g., ependymal cells)	Iversen and Glowinski (1966); Fuxe and Hökfelt (1969); Antòn-Tay and Wurtman (1971)
B. *Nonisotopic nonsteady-state methods*		
1. Decline in brain CA stores after synthesis inhibition (α-MpT)	Compensatory mechanisms evoked by CA depletion and consequent nonphysiologic change in the neuronal activity; difficulty in measuring very low concentrations of NE after α-MpT; assumption that α-MpT produces complete inhibition of TH which is independent from the endocrine manipulations	Fuxe and Hökfelt (1969); Neff *et al.* (1969); Javoy and Glowinski (1972); Costa (1973); Doteuchi *et al.* (1974)
2. Increase in brain CA stores after MAO inhibition	Compensatory mechanisms evoked by CA accumulation and consequent nonphysiologic change in the neuronal activity; feedback inhibition of CA biosynthesis by end-product accumulation	
3. Accumulation of acidic DA metabolites (3,4-DOPAC and HVA) in the CSF after inhibition with probenecid of the transport system	Acidic metabolites probably reflect only activity of neurons lying close to the ventricles; no acidic metabolites of NE are formed in the brain and the methods for evaluating brain NE metabolites are not very sensitive	

13. Brain Metabolism and Aging

Several age-related changes in brain CA metabolism have been reported in senescent mice, e.g., reduced levels of striatal DA; reduced conversion of labeled tyrosine or L-dopa to catecholamines in the cerebellum, brainstem, striatum, and hypothalamus; slowed degradation of total NE in the hypothalamus and of total DA in the striatum (Finch, 1973). These changes, because of the key role of catecholaminergic mechanisms, can, with aging, influence neuroendocrine function (see Chapter IV, Section A,3,a and Chapter V, Section A,1). In addition to changes in neuroendocrine function, age-related changes in catecholaminergic-dependent systems include rapid eye movement (REM) sleep and state IV sleep, which are reduced (Feinberg, 1969); number of spontaneous awakenings in humans which is increased (Feinberg, 1969); spontaneous locomotor activity (Goodrick, 1971), which is reduced; and thermoregulation, which is impaired (Finch *et al.,* 1969; Rapaport, 1969).

14. Behavioral and Electrocortical Changes

Catecholamines are involved in the control of several important physiological and behavioral functions which include sleep and arousal, body temperature, food and water intake, and blood pressure (see Iversen and Iversen, 1975; Jouvet, 1972; Marley and Stephenson, 1972). Only two aspects of CA interaction with the CNS will be considered here: electrocortical and behavioral changes evoked by CA's or drugs affecting CA neurotransmission after intracerebroventricular infusion or systemic injection and the effects of microiontophoretic-applied CA's. Both are of importance in relation to the role played by brain neurotransmitters in the control of hypothalamic-pituitary hormones.

The role of neurochemical changes in the brain in affecting sleep and arousal mechanisms has been assessed by means of several experimental strategies, including electrical stimulation of specific brain areas, pharmacological manipulations intracerebral or IVT infusion of transmitter, evaluation of transmitter release in conscious, behaving animals, and measurement of other biochemical parameters (transmitter synthesis, uptake, storage and catabolism) in animals killed at various times during various behavioral states.

The relation between electrolytically or pharmacologically induced degeneration of CA neurons, as well as between increased or decreased CA neuronal activity and sleep-arousal mechanisms, has been periodically reviewed (Jouvet, 1969, 1972, 1973; Putkonen, 1974; Holman *et al.,* 1975). Although in many of the early studies, in addition to catecholaminergic systems, other neuronal elements were sometimes affected by the experimental approach, the conclusion seems to emerge that lesions of brainstem CA neurons by electrocoagulation (Jouvet, 1972; Petitjean and Jouvet, 1970) or induced chemically with 6-OHDA (Laguzzi *et al.,* 1971, 1972) produced in the cat a decrease in cortical arousal associated with an increase in slow wave sleep (SWS). These results were in line with the classic experiments of Moruzzi and Magoun (1949) which showed a marked behavioral arousal after stimulation of the dorsal NE system. According to Jouvet (1972), the results of many neuropharmacological manipulations of the CA system are consistent with the NE theory of cortical arousal, inasmuch as in cats increased arousal or increased sedation and SWS, respectively, were obtained in conditions of increased (L-dopa treatment) or decreased (α-MpT treatment) catecholaminergic activity.

Besides the role played in evoking cortical arousal, NE has been also implicated in the neurophysiological control of paradoxical sleep (PS or REM sleep). Both phasic (PGO-waves and REM's) and tonic (cortical desynchronization and muscle atonia) aspects of PS were suppressed by lesions of the locus coeruleus and subcoeruleus (Jouvet and Delorme, 1965), structures rich in catecholaminergic innervation (see Section B,2). However, other authors have been unable to duplicate the effects on sleep of total locus coeruleus lesions (see Holman *et al.,* 1975).

In cats, IVT injection of 6-OHDA has been shown to produce an initial excitation followed by sedation and appearance of continuous PGO discharges (Laguzzi *et al.,* 1972; Panksepp *et al.,* 1973). The excitation was inhibited if pretreatment with α-MpT depleted cerebral NE (Laguzzi *et al.,* 1972). In animals pretreated with chlorimipramine, in which 6-OHDA produced selective damage of CA neurons (see Section D), there was an initial period of excitation followed by a decrease in both PS and in the frequency of PGO waves during PS (Laguzzi *et al.,* 1972).

Other evidence favoring a role of NE in PS was derived from experiments in which an increased NE turnover in the rat brain was present during a rebound of PS after its selective deprivation (Pujol *et al.,* 1968). On the other hand, single-unit recording from locus coeruleus neurons showed enhanced activity during waking and PS (Chu and Bloom, 1973).

Other indirect evidence for a correlation between brain CA content and increase in arousal was given by treatment with L-dopa alone or in conjunction with a peripheral L-AAAD inhibitor (Bartholini *et al.,* 1967; Reis *et al.,* 1970); however, such treatment is not specific and these results were not always confirmed (Kadzielawa and Widy-Tyszkiewicz, 1970). In humans, the majority of studies showed an increase in REM sleep, but some investigators also reported no changes or decrease in REM sleep with no alterations in non-REM sleep, although high doses of L-dopa could produce insomnia (see Holman *et al.,* 1975).

Contrasting results were also obtained with α-MpT. This drug reportedly reduced waking in cats (Stein *et al.,* 1974) and both waking and REM sleep in rabbits (Fujimori and Himwich, 1971) and monkeys (Weitzman *et al.,* 1969). In the rat, no changes (Marantz and Rechtschaffen, 1967) or decrease (Torda, 1968) in REM sleep have been reported after systemic injection of α-MpT; other authors, however, found an increase in REM sleep when α-MpT was given orally (Hartmann *et al.,* 1971a,b). Associated with NE depletion, an increase in REM sleep after α-MpT was also observed in cats (King and Jewett, 1971; Henriksen *et al.,* 1973; Stern and Morgane, 1973). In man, α-MpT produced an increase in REM sleep with no effect on non-REM sleep (Wyatt *et al.,* 1971a,b; Vaughan *et al.,* 1972).

Intriguing also, were the results of experiments in which the involvement of central CA systems was assessed by direct CA administration. Depending on the dose used and the personal experiences and beliefs of various investigators, a role both in arousal and sleep has been attributed to CA's in the same animal species.

CA's, for a long time, have been regarded exclusively as central excitants, a belief initially based on experiments in which CA's administered systemically in adult animals quite unlikely crossed the BBB (Weil-Malherbe *et al.,* 1961a). Injections of CA's in immature animals (chicks) in which the BBB is not fully developed, therefore allowing penetration in the CNS (Waelsch, 1955; Bakay, 1956; Lajtha, 1957), produced behavioral and electrocortical sleep (Key and Marley, 1961; Dewhurst and Marley, 1965; Spooner and Winters, 1965). This was also true for central (IVT, IC, or intrahypothalamic) injection of CA's in adult fowl, mice, rats, cats, sheep, dogs, monkeys, and man (Marley and Nisticò, 1972; Marley and Stephenson, 1972).

Additional support for the theory that CA's have central depressant effects derived from experiments in which similar electrocortical responses were elicited in mammalian and avian species by systemic or central injection of an α-adrenergic agonist, clonidine (Zaimis, 1970; Holman *et al.,* 1971; Marley and Nisticò, 1975b). Clonidine readily crosses the BBB, and sleep elicited in adult cats and rats after its systemic injection is of particular importance because of the controversy as to the significance of sleep produced by NE administered into the brain. That clonidine-induced SWS actually resulted from activation of central α-adrenoceptors is implied by the antagonistic effect of centrally injected phentolamine (Marley and Nisticò, 1975b). A typical example of electrocortical synchronization after IVT injection of clonidine is given in Fig. 19.

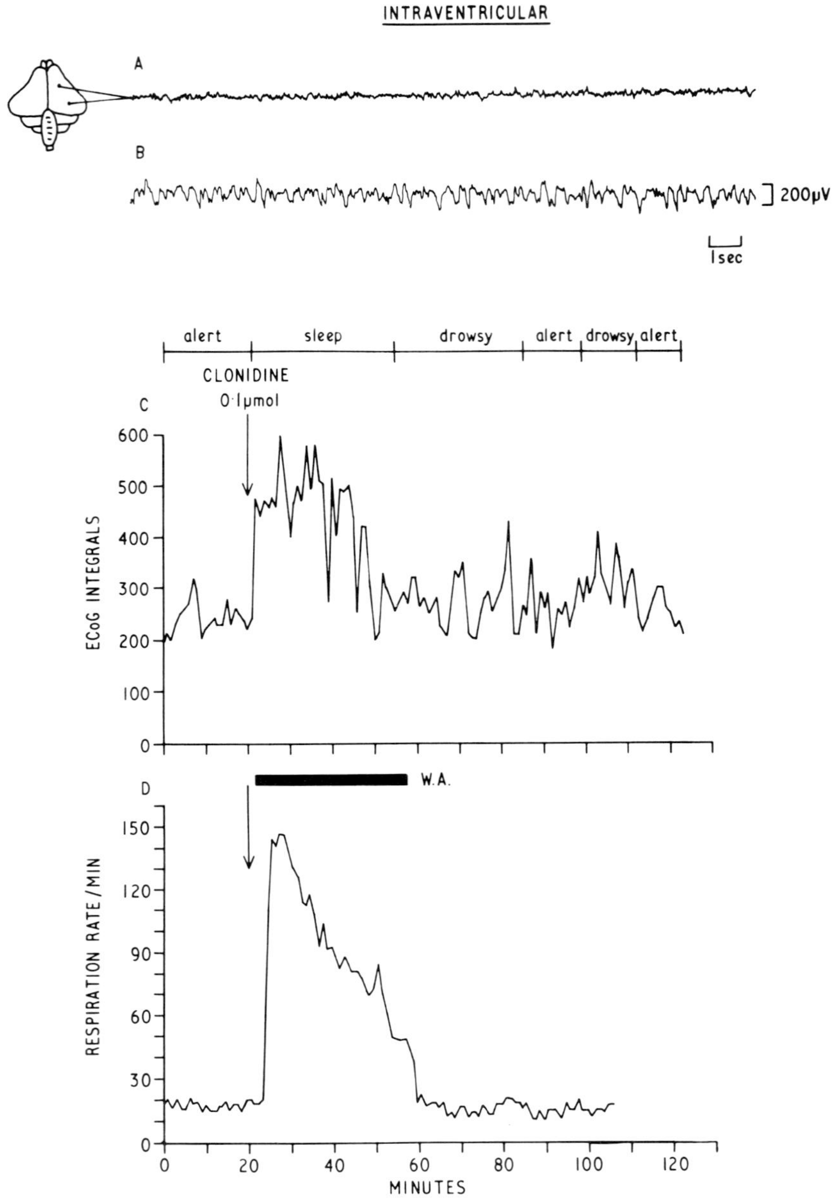

Fig. 19 Records of electrocortical activity (A,B), histogram of integrated electrocortical activity (C) in an unanesthetized adult fowl. (A) Control electrocortical activity. (B) Slow frequency (4–7 Hz), large amplitude electrocortical potentials 15 min after clonidine, 0.1 μmole, infused into the third cerebral ventricle. (C) Electrocortical integrals showing a marked increase (up to a peak of 600/min) following IVT infusion of clonidine and associated with behavioral sleep. (Reproduced, with permission, from Marley and Nisticò, 1975b.)

However, these findings do not necessarily permit the conclusion that NE released from nerve endings within the brain is an inhibitory transmitter since at other brain loci CA's could be excitatory. Other experiments suggest, in fact, that CA's may also possess central excitant effects. However, most of these studies were carried out in *encéphale isolé* preparations and not in intact unanesthetized animals (Bradley, 1960; Key, 1975; also see Marley and Stephenson, 1972). The few studies in which CA's infused into the brain of unanesthetized animals were found to possess central excitant effects (Cordeau *et al.*, 1963) are questionable since a very excessive infusion volume was used.

15. Iontophoretic Application

Microiontophoresis, with properly controlled pH and current, and no mechanical damage, represents a good method for the evaluation of qualitative effects of putative or real neurotransmitters on single neurons (see Bradley, 1968; Bloom *et al.*, 1973a). However, one of the main limitations of this method is that the functional role of a discrete population of neuronal elements can hardly be anticipated, since, invariably, within the same population there are elements which respond positively and others negatively to the same neurotransmitter.

In general, inhibition is the most common response of single neurons to iontophoretically applied NE, although some neurons do exhibit excitation. Thus, both excitatory and inhibitory effects were shown on cells within pontine and medullary reticular formation (Bradley and Wolstencroft, 1962; Couch, 1970; Boakes *et al.*, 1971). The presence of one type of response in one brain region does not exclude that in other brain areas NE could elicit the opposite effect (Bloom *et al.*, 1973a). The interpretation of the data becomes even more difficult since responses qualitatively opposite on similar neurons, under similar conditions, have been reported by different groups of workers. Thus, excitatory responses of cortical neurons to NE were reported by Johnson *et al.* (1969a,b), whereas earlier works of Krnjević and Phillis (1963a;b) indicated generally depressant effects for NE.

A summary of the effects of iontophoretically applied CA's on different brain regions is given here. Mitral cells of the olfactory bulb were found to be consistently depressed by NE (Bloom *et al.*, 1964) as were SO neurons identified by antidromic stimulation of the posterior pituitary (Barker *et al.*, 1971). Neurons of the cat and rat raphe nuclei, in which 5-HT cell bodies have been found (see Section C,2), exhibited less qualitative responses; feline median raphe cells were frequently excited (Crayton and Bloom, 1969), whereas only about half of the neurons in the rat dorsal raphe (Aghajanian and Haigler, quoted by Bloom *et al.,* 1973a) and pontine raphe (Couch, 1970) exhibited this response. The pontine raphe cells, excited by 5-HT, usually were depressed by NE and vice versa (Couch, 1970). More than 90% of the polysensory cortical neurons responded to NE and 5-HT by depression of spontaneous or induced discharge (Nelson *et al.,* quoted by Bloom *et al.,* 1973a). These experiments suggest that in the cortex, as in the other CNS regions, the main cellular action of NE is inhibition. The inhibitory effects of NE on rat cerebellar Purkinje cells were demonstrated to be mediated by cAMP (Siggins *et al.,* 1969; Hoffer *et al.,* 1971), supporting a direct postsynaptic activation of AC as one step in the molecular mechanism by which NE causes inhibition (Siggins *et al.,* 1971).

Dopamine was found to be strongly inhibitory to certain hypothalamic neurons (20%) and to neurons of the n. caudatus and gracilis (Steiner and Pieri, 1969). Neuronal activity after application of DA was inhibited almost instantaneously, and the recovery time was also short (1–5 sec). Dopaminergic neurons of the zona compacta of the substantia nigra typically showed extracellular action potentials with a regular rhythm and a slow rate of firing (2–6 spikes/sec) in unanesthetized rats, but an increase in firing rate and a bursting pattern was registered in anesthetized animals. In contrast, neurons of the zona reticulata of the substantia nigra exhibited a relatively rapid rate of firing and were either unaffected or depressed by anesthesia (Aghajanian and Bunney, 1973). By using the microiontophoretic technique, the same authors found further differences among dopaminergic neurons of the zona compacta of the substantia nigra and the adjacent ventral tegmental area and the nondopaminergic neurons of the zona reticulata. Dopaminergic cells showed little or no response to ACh, whereas neurons of the zona reticulata were stimulated by microiontophoretically applied ACh.

Recently, microiontophoresis and antidromic identification have been utilized to localize the tuberoinfundibular neurons and to determine their sensitivity to DA, NE, and glutamate (Moss *et al.,* 1975a). This study showed that in antidromically identified units of ARC n., two distinct populations of neurons exist so far as the chemical sensitivity to neurotransmitter is concerned: one which displayed excitation to NE and inhibition or no response to DA, and the second which was excited by DA and not responsive to or inhibited by NE. In addition, in a population of neurons of the ARC n., not identified antidromically, heterogeneous responses to NE and DA were recorded, so that some neurons were excited by both NE and DA, some were inhibited by both, and still others were excited by one and inhibited by the other. Glutamate had a powerful excitation action on nearly all the ARC n. neurons tested. In preliminary experiments, utilizing receptor-blocking agents, it was possible to block the DA excitation but not the NE excitation with iontophoretically applied pimozide, while phentolamine blocked both the DA and the NE excitation of the antidromically activated ARC neurons. Responses of a single antidromically identified arcuate neuron to iontophoretically applied NE, DA, and glutamate are illustrated in Fig. 20 (Moss *et al.*, 1975a). Collectively, these findings suggest that the NE and DA neurons present in the MBH may modulate the neuronal activity of ARC neurons whose axons project to the ME, thus giving strong support to the physiological role of brain CA's in the neuroendocrine control of AP hormones.

In addition to antidromic activation, another means of characterizing, both morphologically and functionally, neurons projecting axons to the ME was provided by the retrograde axonal transport of a marker fluorescent dye (Procion yellow) applied by electrophoresis to the cut end of ME terminals and axons (see Dyer, 1975). With this technique, most of the cells projecting to the ME were found in a basal periventricular zone extending up to the main body of the paraventricular n. and as far forward as the suprachiasmatic nuclei. Furthermore, 82% of the tested cells were no more than 0.5 mm lateral to the third ventricle (Blackmore *et al.*, 1974).

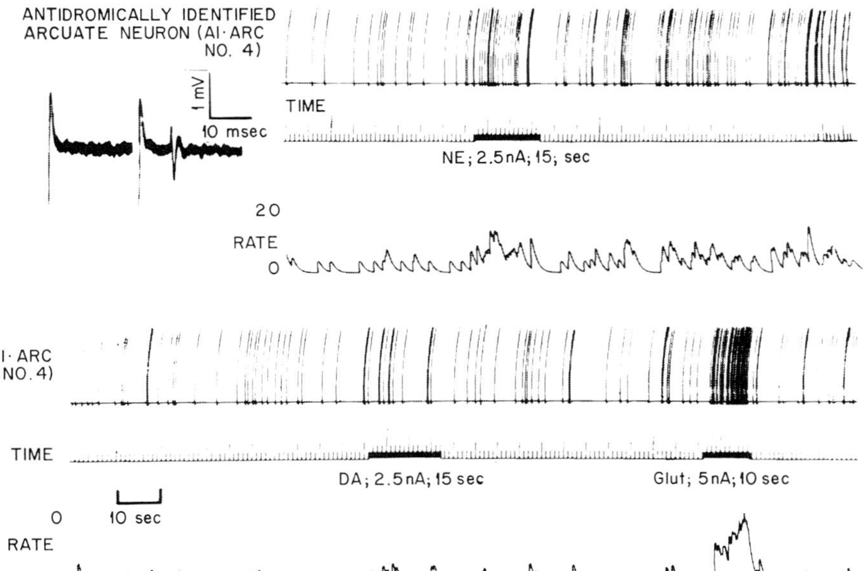

Fig. 20 Responses of a single antidromically identified (AI) arcuate (ARC) neuron to iontophoretically applied norepinephrine (NE), dopamine (DA) and glutamate (GLUT). (Reproduced, with permission from Moss *et al.*, 1975a.)

A particular feature of many of the stained cells is that they are bi- or multipolar, more frequently bipolar with one axon projecting toward the site of the lesion and the other upward toward the more dorsal zones of the rostral hypothalamus (Dyer, 1975).

At the ultrastructural level, these cells were found to contain a population of dense-cored granules with a diameter (800–1600 Å) within the size range found in ME terminals. These periventricular cells are distinguishable from classic neurosecretory cells and adjacent hypothalamic neurons in that they contain significantly more dense-cored granules than the other two cell types (see Dyer, 1975). Recently, some periventricular cells antidromically activated from the ME were also found to be activated by stimulation of the preoptic area (M. C. Harris *et al.* and Wuttke *et al.*, quoted by Dyer, 1975). On the other hand, neurons projecting to the ME also send axons to limbic structures (Wuttke, quoted by Dyer, 1975).

At the present time, the supposition that the bipolar cells are endocrine neurons containing hypothalamic neurohormones requires confirmation, but it is appealing and would fit in well with an extraendocrine role for these neurons in addition to the classic hypophysiotrophic function at the ME level (see Chapter I, Section B,2).

Recently, after iontophoretic application of DA and NE, a reduction was found in the spontaneous discharge of about one-half of the antidromically identified rostral hypothalamic cells. This effect was resistant to α- or β-adrenergic blockade (Dyball *et al.*, 1974). Also, the response of identified VMN neurons to microiontophoretically applied DA was basically depression, although DA was never as potent in action as either GABA or glycine (Renaud, 1975).

C. SEROTONIN

1. Occurrence in the CNS

In mammals, 5-HT has been found in the CNS, the pineal gland, the enterochromaffin cells of the intestinal tract, the mast cells, and the platelets. The first evidence of the presence of 5-HT in the CNS was provided by means of biochemical methods (Amin *et al.*, 1954; Twarog and Page, 1953). In these studies, the distribution pattern of 5-HT in discrete areas of the brain was described; 5-HT was shown to be present in particularly high concentration in the hypothalamus, midbrain, and limbic system (see Garattini and Valzelli, 1965; Erspamer, 1966; Welsh, 1968). More recently, determinations made in individual nuclei of the rat hypothalamus revealed high concentrations of 5-HT in the SCN, the ME, the ARC n., and MFB (Saavedra *et al.*, 1974b) (Table XIII).

TABLE XIII

5-HT Content in Individual Nuclei of Rat Hypothalamus[a]

Area	5-HT content (ng/mg protein)
Nucleus preopticus medialis	11.6
Nucleus preopticus lateralis	14.4
Nucleus preopticus periventricularis	7.4
Nucleus preopticus suprachiasmatis	24.6
Nucleus suprachiasmatis	
Total	25.4
Internal	37.2
External	17.0
Area retrochiasmatica	15.9
Nucleus periventricularis	10.9
Nucleus supraopticus	9.5
Nucleus paraventricularis	13.5
Nucleus hypothalamicus anterior	10.2
Nucleus arcuatus	36.4
Nucleus ventromedialis	8.5
Nucleus dorsomedialis	13.6
Nucleus perifornicalis	30.0
Nucleus hypothalamicus posterior	24.5
Nucleus premammillaris dorsalis	22.9
Nucleus premammillaris ventralis	18.3
Medial forebrain bundle	
Anterior	21.6
Posterior	30.7
Median eminence	15.3

[a] From Saavedra *et al.* (1974b).

Histofluorescence techniques have enabled investigators to isolate, visualize, and map in the CNS 5-HT cell bodies, their axons, and their terminal endings. In this section, the anatomic distribution of CNS 5-HT pathways, as shown by histofluorescence, will be discussed, with particular attention given to the innervation of the hypothalamus.

2. Fluorescent Pathways

Like the NE pathways, the 5-HT pathways in the CNS are organized in two main systems, one ascending from the lower brainstem and innervating cortical and subcortical areas, and the other descending to the spinal cord. We shall confine ourselves mainly to the major elements; further details can be found in Dählström and Fuxe (1964a,b, 1965), Ungerstedt

(1971a), Baumgarten *et al.* (1972a,b), Dählström *et al.* (1973), and Fuxe and Jonsson (1974). In comparison to the detection of CA's by fluorescent histochemistry, many methodological difficulties are involved in the study of the precise distribution and connections of the indoleamine neurons (lower fluorescent yield on a molar base, rapid fading of the fluorophores upon UV irradiation, and extreme fineness of the terminals), although, recently, improved filter systems combined with administration of selective neurotoxic drugs made it possible to distinguish the 5-HT from the CA fluorescence (Ajika and Okinaga, 1976).

Several methods have been used to improve the visualization of 5-HT neurons, viz., pretreatment of animals with a MAO inhibitor alone or associated with systemic or intracerebroventricular injection of L-Trp, infusion into the cerebral ventricles or into different areas of the brain of neurotoxic 5-HT derivatives such as 5,6-DHT or 5,7-DHT, which produce an almost selective degeneration of 5-HT neurons (see Section D) or mechanical lesions of 5-HT axons. After lesions are formed, the amine accumulates on the proximal side of the lesion rendering visible axons otherwise not detectable; distal to the lesion, the nerve terminals and axons degenerate, causing a decrease in 5-HT fluorescence in the innervated areas (Dählström and Fuxe, 1964a,b; Dählström and Häggendal, 1966; Ungerstedt, 1971a).

A schematic representation of the central 5-HT neurons is given in Fig. 21. Two main ascending bundles of 5-HT axons have been localized:

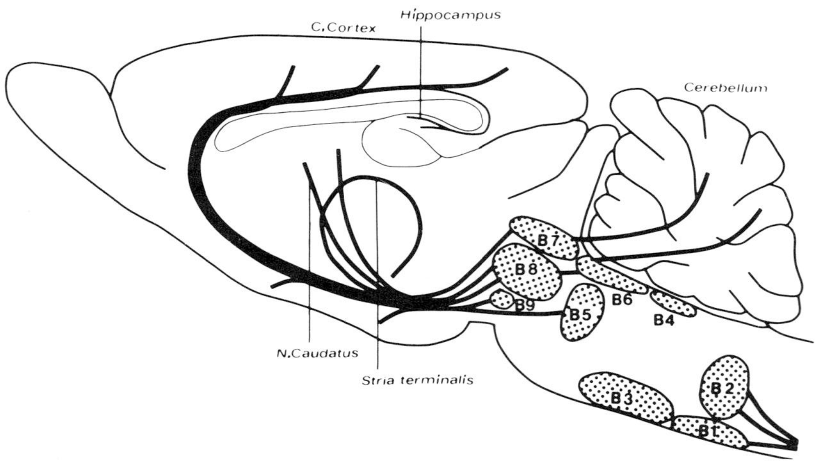

Fig. 21 Schematic diagram of a sagittal section of the rat brain showing central 5-HT cell groups, axons, and terminals. (for details see text). (Reproduced, with permission, from Fuxe and Jonsson, 1974.)

medial and lateral ascending 5-HT pathways. They originate from cell bodies present in the mesencephalic raphe (B7, B8, B9) and pontine raphe (B5, B6) nuclei, run near the midline in the MFB, and give rise to terminals innervating the hypothalamus, ME, preoptic area, limbic system, septal area, striatum, and cerebral cortex (Fuxe, 1965a; Fuxe and Jonsson, 1974; Ungerstedt, 1971a; Baumgarten and Schlossberger, 1973).

As noted above, a great improvement in the understanding of the distribution of 5-HT axons and terminals has been achieved by IVT administration of 5,6-DHT (Björklund *et al.*, 1974a). One to three days after an injection of this compound into the MFB in the middle and posterior hypothalamus, yellow fluorescent material could be traced caudally up to about 4 mm from the site of the injection to the cell bodies in the mesencephalic and pontine reticular formation. Of particular import was the failure of 5,6-DHT injected in the tegmentum to remove the indoleamine fibers in the pretectal region. Since several adjacent structures (habenula, subcommissural organ, pretectal septum, globus pallidus) were denervated instead, the existence of a short epithalamic 5-HT neuronal system, which could constitute a link between the habenula and possibly the pineal gland and the pretectal region, has been postulated (Björklund *et al.*, 1974a). Other specific 5-HT pathways visualized by the intracerebral injection of 5,6-DHT, such as a mesencephalic cortical and a pontine–hypothalamic 5-HT pathway, have been described in detail recently (Fuxe and Jonsson, 1974). As a result of histochemical and electron microscopic studies, the presence has also been shown of supraependymal nerve endings containing either 5-HT or a closely related indoleamine (Richards and Tranzer, 1974), originating in the dorsal and median raphe nuclei (Aghajanian and Gallager, 1975).

Combined evaluation by means of histofluorescence and EM has made it possible to visualize 5-HT terminals and axons in the ME of rats (Baumgarten and Lachenmayer, 1974) and fowl (Gargiulo and Nisticò, 1977) after IVT pretreatment with 5,6-DHT. A typical histofluorescence picture showing 5-HT nerve terminals in the ME of the fowl before and after 5,6-DHT is represented in Fig. 22c and d. Collectively, these observations demonstrate that 5-HT terminals, like DA terminals, are localized mainly in the external layer of the ME in contrast to the predominant distribution of NE terminals in the internal and subependymal layer (see Section B,4).

Fluorescence micrographs suggest that generally the 5-HT terminals may make synaptic contact with nonfluorescent (peptidergic or not) nerve cell bodies and cell processes. With electron microscopy, the 5-HT terminals have an appearance similar to that of NE and DA terminals, with both large and small dense-core vesicles (Hökfelt, 1968; Hökfelt and Ungerstedt, 1969). At the ultrastructural level, the synaptic relations between monoamine terminals seem to be common at least in hypothalamic areas;

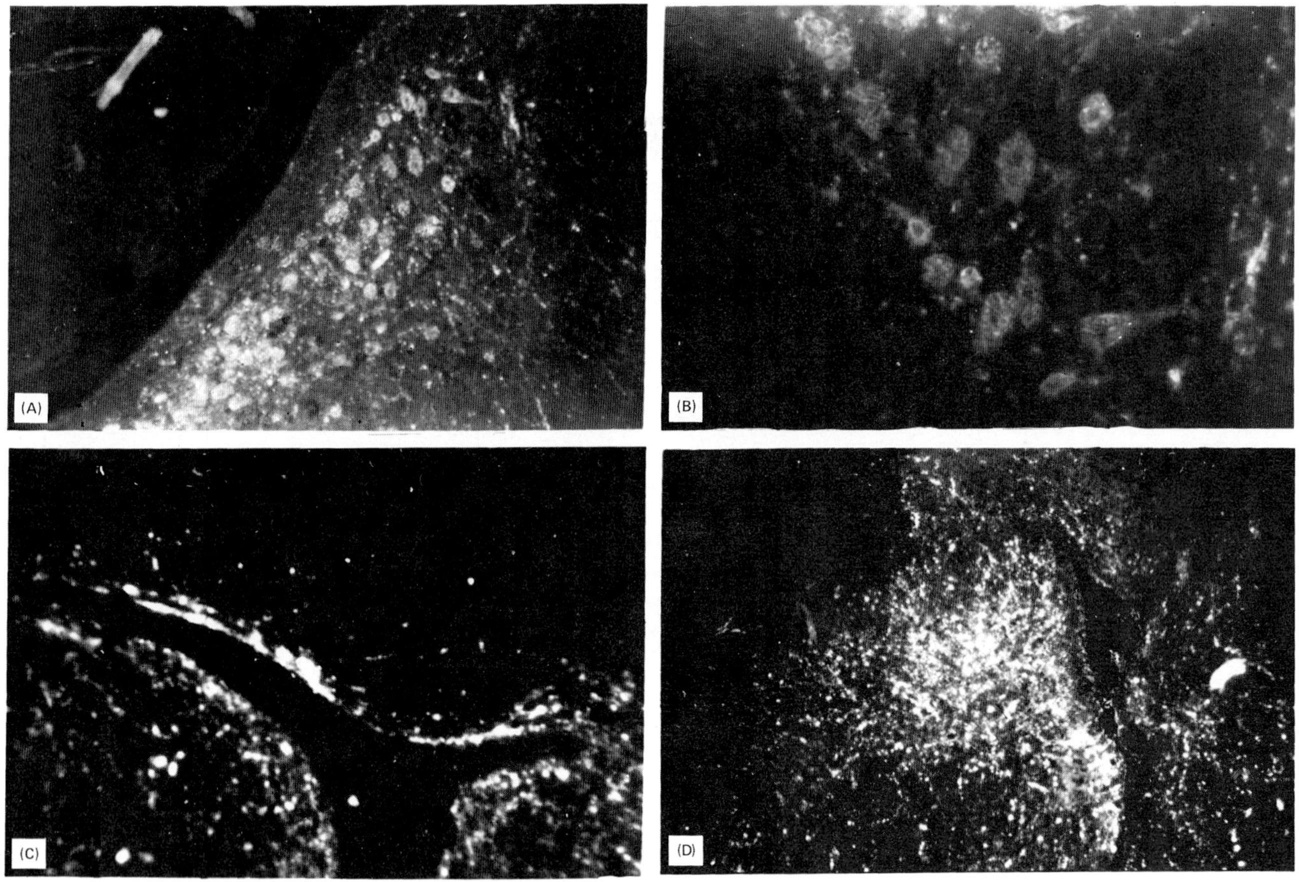

Fig. 22 (A)–(D)

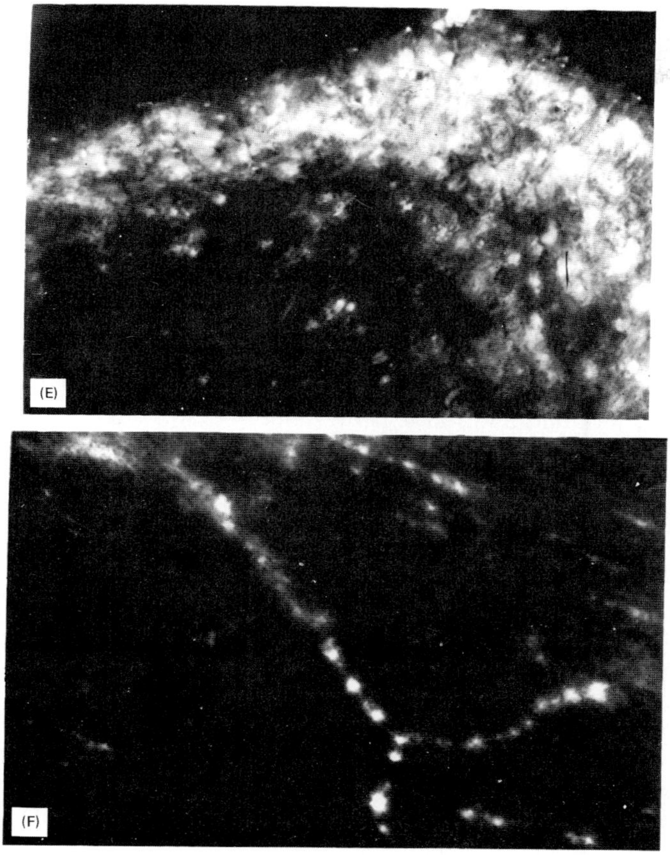

Fig. 22 Typical histofluorescence images (examples) of CA and tryptaminergic neurons in various areas of the brain in fowl. (A) Transverse section of the mesencephalic tegmentum of an untreated fowl, showing the presence of cell bodies and terminals displaying a strong green fluorescence in the nucleus tegmentalis lateralis pars dorsalis (locus coeruleus) (×55). (B) Particular of the green fluorescent perikarya of the locus coeruleus at higher magnification (×140). (C) Medium to strong green and yellow fluorescence of the ME after infusion into the third ventricle of L-Trp (0.5 μmole) in a fowl pretreated with a MAOI, mebanatine (100 μmole/kg 18 hr and 1 hr before) (×55). (D) Marked increase in yellow fluorescence of the ME, 4 days after an infusion into the third cerebral ventricle of 5,6-DHT (100 μg in 10 μl) (×55). (E) Marked increase in yellow fluorescence and swelling of tryptaminergic terminals of the POA 4 days after IVT injection of 5,6 DHT (100 μg) (×55). (F) Typical distorted yellow fluorescent fiber in the hypothalamic periventricular region 4 days after an IVT injection of 5,6-DHT (100 μg). (Reproduced, with permission, from Gargiulo and Nisticò, 1975, 1977,)

in fact, the smaller granular vesicles can be seen aggregated at the presynaptic surface of synaptic axodendritic contacts (Hökfelt and Ungerstedt, 1969; Aghajanian *et al.*, 1969; Hökfelt, 1970).

3. Indoleamine Innervation of the Pituitary Gland

The existence of serotoninergic mechanisms in the mammalian pituitary gland has been postulated by Hyyppä and Wurtman (1973). Evidence for the presence of 5-HT in the pituitary gland, with highest levels in the intermediate and posterior lobe, had been obtained earlier (see Björklund *et al.*, 1974a; Saavedra *et al.*, 1975; and Table VI), and a wide range of 5-HT concentrations was detected among the different mammalian species. Presence of tryptamine-like indoleamines, although reported in the AP of mammals (Björklund and Falck, 1969; Martin *et al.*, 1971), has not been confirmed by determinations made with a specific and sensitive microassay (Saavedra *et al.*, 1975). Similarly, it had been postulated that a portion of the pituitary fluorescence might derive from peptides having tryptophan in the NH_2 terminal position (Håkanson and Sundler, 1971a,b), but no definite proof was given for their presence in the pituitary.

The cellular localization of 5-HT in the pituitary is not presently known. At least part of the 5-HT in the intermediate and posterior lobes could be located within 5-HT-rich mast cells (Björklund *et al.*, 1974b); however, its presence in nerve terminals within the gland cannot be excluded.

With regard to 5-HT biosynthesis, some PAS-positive cells in the AP take up 5-HTP and decarboxylate it to 5-HT. The ability to take up 5-HTP (and L-dopa) is not an exclusive property of the pituitary cell; all cells take them up by the same membrane transport system that mediates the uptake of the aromatic amino acids which form protein (e.g., tyrosine, tryptophan). Moreover, L-AAAD appears to be present in most cells. However, the finding of Trp-H activity in all three pituitary lobes (Saavedra *et al.*, 1975) at levels which are of the same order of magnitude as those reported for ME and other regions of the hypothalamus of the rat (Kizer *et al.*, 1975c) suggests that, in addition, 5-HT can be formed locally. Finally, the very high capacity for deamination of 5-HT in all lobes of the pituitary (Saavedra *et al.*, 1975) indicates the presence of large amounts of MAOA (Goridis and Neff, 1971) (see also Section B,5).

4. Biosynthesis

Subcellular localization studies indicate that 5-HT is present predominantly in nerve terminals (Michaelson and Whittaker, 1963; De Robertis, 1967; Kuhar *et al.*, 1971), where the enzymatic activity regulating the synthesis occurs. The synthesis of 5-HT involves several mechanisms: uptake of L-Trp by the terminals, hydroxylation of Trp to 5-HTP, and decarboxylation to 5-HT (Figs. 23 and 24).

Since the concentration of the L-AAAD enzyme is about 75 times greater than that of Trp-H (Peters *et al.*, 1968), it is unlikely that the regulation of 5-HT synthesis depends on the decarboxylation of 5-HT. On the contrary, the rate-limiting enzyme in the synthesis is represented by Trp-H, the activity of which is relatively low (see Garattini and Valzelli, 1965; Lovenberg *et al.*, 1968; Wurtman and Fernström, 1972; Gàl, 1974; Hamon and Glowinski, 1974). Tryptophan hydroxylase appears to exist in two forms: a soluble and a particulate fraction, each exhibiting different characteristics (Ichiyama *et al.*, 1970; Knapp and Mandell, 1972a,b). As does TH, tryptophan hydroxylase prepared from rat brainstem requires oxygen, Fe^{2+} and a reduced pteridine cofactor for optimal activity (Lovenberg *et al.*, 1968). In the brain Trp-H appears to reside primarily in nerve endings (Ichiyama *et al.*, 1970), and its distribution parallels that of 5-HT (Kizer *et al.*, 1975c). This enzymatic activity is not saturated by the

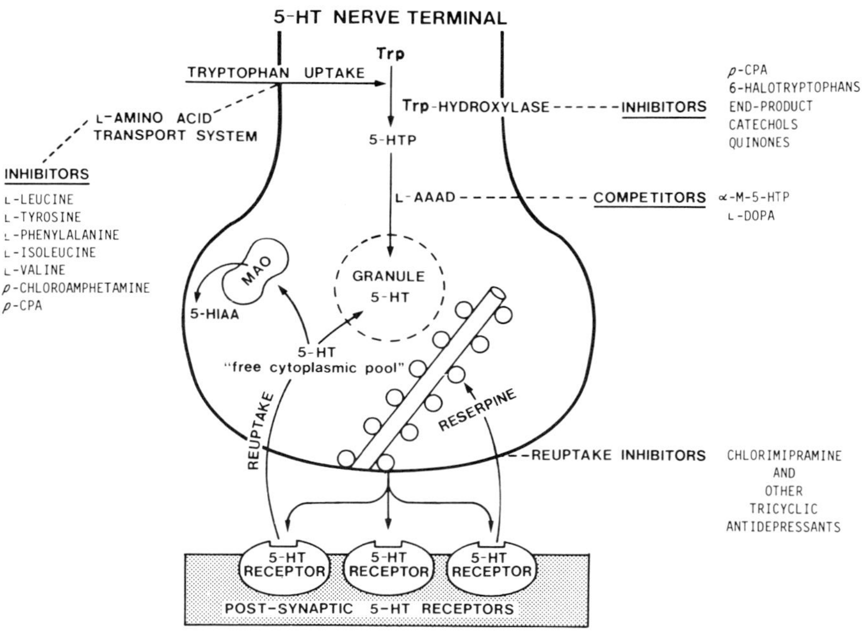

Fig. 23 Schematic illustration of the mechanisms (tryptophan uptake, 5-HT synthesis, storage, free cytoplasmic pool, release, effect on specific postsynaptic receptors, and termination of the effects by uptake and metabolism) involved in the serotoninergic neurotransmission and site of action of the main interfering drugs (for details see text).

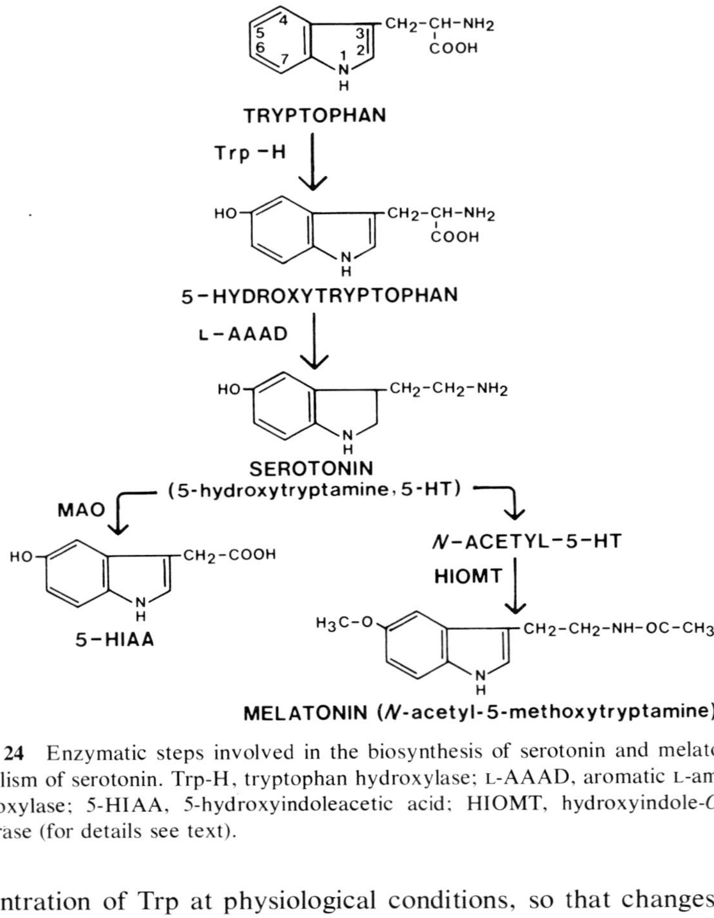

Fig. 24 Enzymatic steps involved in the biosynthesis of serotonin and melatonin and metabolism of serotonin. Trp-H, tryptophan hydroxylase; L-AAAD, aromatic L-amino-acid decarboxylase; 5-HIAA, 5-hydroxyindoleacetic acid; HIOMT, hydroxyindole-*O*-methyl transferase (for details see text).

concentration of Trp at physiological conditions, so that changes in the availability of Trp affect the rate of synthesis of 5-HT in the brain. Accordingly, peripheral injections of Trp (Moir and Eccleston, 1968; Fernström and Wurtman, 1971) or its direct IVT administration (Consolo *et al.*, 1965) increased brain 5-HT and 5-HIAA levels, whereas brain 5-HT levels were reduced in rats fed a Trp-deficient diet (Gàl and Drewes, 1962; Gil-Ad *et al.*, 1976a).

Tryptophan is the only essential amino acid partially bound to albumin in plasma (McMenamy and Oncley, 1958), in contrast to tyrosine, which is entirely free. The amount of Trp which penetrates into the brain seems closely dependent on the small plasma pool of "free" Trp (10–20%) but not on the total one (Knott and Curzon, 1972; Tagliamonte *et al.*, 1973). From the plasma pool, L-Trp is taken up by the brain apparently by the same stereospecific transport system which operates for L-tyrosine, phenylalanine, and other aromatic amino acids (see Fig. 25). A high affinity transport system for Trp has also been discovered in synaptosomal preparations (Belin and Pujol, 1972; Knapp and Mandell, 1972a).

A variety of lipid-soluble compounds which bind to albumin in the blood (e.g., hormones, drugs, NEFA) may displace Trp from binding sites to plasma proteins, increase the free concentration of plasma Trp and, finally, brain Trp and 5-HT synthesis in the brain (see Tagliamonte *et al.*, 1973; Curzon, 1974). However, in some studies no correlation was found between plasma free Trp levels and brain Trp and 5-HT concentrations, whereas the factor limiting the availability of plasma Trp to the brain appeared to be the ratio between total Trp in plasma and other competing neutral amino acids such as L-leucine, L-phenylalanine, L-isoleucine, L-tyrosine, L-valine, and L-methionine transported to the brain by the same uptake mechanism (see Figs. 24 and 25) (Fernström and Wurtman, 1972a,b; Madras *et al.*, 1973, 1974; Fernström, 1976). It has been suggested by these authors that the secretion of insulin increases the effec-

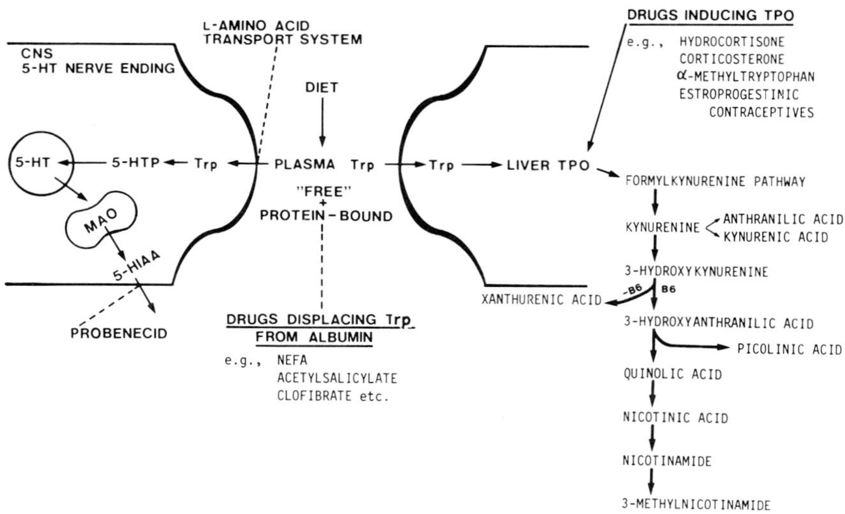

Fig. 25 Schematic illustration of the relations between plasma Trp (free and bound) levels, brain and liver Trp metabolic pathways. On the left: 5-HT nerve terminal in the CNS. An L-amino acid transport system permits the uptake of Trp into the 5-HT neuron in which 5-HT synthesis and metabolic degradation to 5-HIAA by MAO occurs. In the middle: plasma Trp levels (free and albumin-bound) and drugs able to displace Trp from albumins. On the right: metabolic degradation of Trp by liver tryptophan pyrrolase (TPO) to the formylkynurenine pathway and drugs inducing TPO, able therefore of producing a shift of Trp from the brain to the liver pathway (for details see text).

tiveness of Trp uptake by brain because it decreases plasma concentration of neutral amino acids competing with Trp for uptake. According to the same investigators the binding of Trp in plasma may lead to its storage there so that it remains available to the brain. Under some experimental conditions, brain 5-HT turnover did not correlate with brain Trp levels (Curzon, 1974; Hamon *et al.*, 1974), which were not always proportional to either free or total serum Trp concentrations (Gessa and Tagliamonte, 1974). However, more recent evidence has been given explaining apparent discrepancies and in favor of a consistent relationship between plasma-free Trp and brain Trp concentrations (Fernando *et al.*, 1976). According to these authors, it seems that after feeding rats with high fat diets, in contrast to an increase in serum NEFA and free Trp, no changes in brain Trp levels (Madras *et al.*, 1974) were observed since *in vitro*, during either the preparation of serum or in the dialysis at 37°C for 3½ hr in order to separate free Trp, lipolysis may have prevented such changes. Furthermore, it has been pointed out that both plasma free Trp and insulin are determinant factors in regulating the accessibility of plasma Trp to the rat brain (Fernando *et al.*, 1976). However, since it is surprising that the metabolism of a substance of such importance for brain function as 5-HT should be influenced by the manifold factors that can alter brain Trp concentrations, it has been suggested (Grahame-Smith, 1974) that those factors may normally have little influence on the amount of brain 5-HT at the receptor site. Evidence that brain 5-HT exists in two or more pools (Shields and Eccleston, 1973) implies the possibility that total brain 5-HT concentration and functional brain 5-HT are not related (Green and Grahame-Smith, 1976) (see also Section C,7).

Another regulator of 5-HT synthesis is liver tryptophan pyrrolase (TPO), an enzyme system which can be induced by corticosteroids (Curzon and Green, 1969, 1971; Nisticò, and Preziosi, 1969; Scapagnini *et al.*, 1969) or, indirectly, by contraceptives (Nisticò and Preziosi, 1970), with ensuing shift of Trp to the liver kynurenine pathway and decrease in brain 5-HT synthesis (Fig. 25). While pyrrolase activity may influence brain 5-HT metabolism in various drug experiments, it is not yet clear how significant it is in physiological or pathological circumstances. 5-HT synthesis in the brain is not only dependent on the plasma Trp levels but also on the activity of the Trp carrier system (Hamon and Glowinski, 1974). Since the affinity of the carrier molecule for Trp seems to be greater in glial cells than in synaptosomes, glia may regulate in some way the delivery of Trp into 5-HT neurons (Hamon *et al.*, 1974).

Another regulator factor in 5-HT synthesis is the rate of tryptophan hydroxylation under certain conditions such as diethyl ether anesthesia (Bourgoin *et al.*, 1973) or high temperature (Squires, 1974). The rapid changes in the rate of tryptophan hydroxylation cannot be attributed to changes in Trp-H concentration, but could be due to changes in the levels of pteridine cofactor (Gàl, 1974). An interaction with the reduced pteridine cofactor in a manner similar to that observed with TH could also explain the inhibitory effects *in vitro* by NE.

In contrast to the rapid regulatory mechanisms, the sustained changes in the rate of 5-HT synthesis have been related to increase in the concentration of Trp-H, as shown after corticosterone treatment in adrenalectomized rats (Azmitia and McEwen, 1969) or after reserpine treatment (Zivkovic *et al.*, 1973).

Unlike CA's, which inhibit TH, 5-HT has no effect on the activity of soluble Trp-H (Jequier *et al.*, 1969). However, end-product regulation of 5-HT synthesis may occur following MAO inhibition when 5-HT levels are very high (Macon *et al.*, 1971), although other studies do not support the importance of such feedback inhibition (Jequier *et al.*, 1969; Millard *et al.*, 1972; Gàl, 1974).

a. Biosynthesis Inhibitors

Serotonin synthesis can also be controlled pharmacologically. A list of drugs inhibiting 5-HT synthesis is given in Table XIV. Inhibitors of Trp-H in *in vitro* systems include catechols such as α-*n*-propyl dopacetamide (competition with the pteridine cofactor), chelating agents, such as α,α-dipyridyl, quinones, such as aquayamycin, which act aspecifically by easy oxidation-reduction, and amino acid analogs (see McGeer, 1973). Inhibitors of the first two groups have only minimal effects on brain 5-HT *in vitro* and are not specific, being active also against tyrosine and phenylalanine hydroxylase.

One of the more widely employed drugs for inhibiting *in vivo* and *in vitro* brain 5-HT synthesis is *p*-chlorophenylalanine (pCPA) (Koe and Weissman, 1966). However, pCPA possesses several other actions which might contribute to its effects (see Table XIV) (Wurtman and Fernström, 1972; Rechtschaffen *et al.*, 1973; Marley and Whelan, 1976).

Some Trp analogs have been used to decrease brain 5-HT. 6-Fluorotryptophan has been found capable of decreasing *in vivo* brain 5-HT concentrations to the same extent as pCPA; however, the recovery of 5-HT levels is quicker (McGeer, 1973). α-Methyl-Trp also causes profound changes in the metabolism of L-Trp. This synthetic amino acid can decrease brain 5-HT levels probably through a shift of Trp from the brain to the liver metabolic pathway since it induces liver TPO (Curzon and Green, 1969). Moreover, α-methyl-Trp is converted in the brain to α-methyl-5-HT, and this compound may act as a false neurotransmitter (Wurtman, 1966), thus concurring to 5-HT depletion. Inhibition of Trp-H

TABLE XIV

Drugs Inhibiting 5-HT Neurotransmission by Various Mechanisms

Class	Drug	Observations	References to main action
Synthesis inhibitors	*p*-Chlorphenylalanine (pCPA)	Trp-H inhibition; blockade of Phe-H and build-up of Phe and metabolites (Weissman, 1973); CA depletion (Miller *et al.*, 1970; Weissman, 1973). Reduction of NE turnover (Stolk *et al.*, 1969); reduction of DA synthesis (Tagliamonte *et al.*, 1973); reduction in intestinal adsorption of Trp and Phe (Weissman, 1973); inhibition of Trp uptake in synaptosomes and glial cells (Bauman *et al.*, 1974); as pCPA methylester stimulation of Trp uptake in glial cells (Bauman *et al.*, 1974); 5-HT-like action on rat stomach strip, rat uterus, and cat adrenal medulla (Marley and Whelan, 1976)	See McGeer (1973)
	α-*n*-Propyldopacetamide	Trp-H inhibition	
	p-Chloroamphetamine	Trp uptake inhibition (Sanders-Bush and Sulser, 1970); release of 5-HT and perhaps inhibition of Trp-H (Pletscher *et al.*, 1968); amphetamine-like action (Strada *et al.*, 1970); increase in 5-HT turnover (Costa *et al.*, 1971)	
	α,α'-Dipyridyl	Trp-H inhibition	
	6-Halotryptophans	Trp-H inhibition	
	L-Aromatic amino acids	Competition at Trp brain uptake	
	α-Methyltryptophan	Conversion to α-methyl-5-HT (false transmitter); conversion to α-methyl-5-HTP and blockade of L-AAAD; liver TPO induction (Curzon and Green, 1971)	
	α-Methyl-5-HTP	Inhibition of L-AAAD (Moran and Sourkes, 1963)	

Receptor antagonists	Methysergide	Inhibition of 5-HT uptake into synaptosomes (Fiszer and De Robertis, 1969); no change in brain 5-HT turnover (D'Amico *et al.*, 1976)	Vane (1960); Roberts and Straughan (1967)
	2-Bromo-LSD	Blockade of 5-HT release (Chase *et al.*, 1967)	Boakes *et al.* (1970)
	LSD-25	Decrease in 5-HT turnover (Lin *et al.*, 1969); 5-HT-like action (Andén *et al.*, 1968; Aghajanian and Freedman, 1968); agonist at striatal DA receptors (Pieri *et al.*, 1974); stimulation of AC in rat striatum and blockade of DA-induced activation of AC (von Hungen *et al.*, 1974)	Roberts and Straughan (1967); Bloom (1968); Aghajanian *et al.* (1968)
	Yohimbine	α-Receptor blockade	
	Psylocibin	Hallucinogen action	
	Cyproheptadine	Antihistaminic and anticholinergic activity (Stone *et al.*, 1961)	Stone *et al.* (1961)
	Pizotifen (BS-105)		
	Cinanserin		Barasi and Roberts (1975)
	MJ 1999		York (1970)
	Methergoline (1-methyl-8β-carbobenzyloxy-aminomethyl-10α-ergoline)[a]	DA uptake inhibition (?) (Ferrini and Glässer, 1965) Blockade of DA-sensitive AC system in rat striatum (P. F. Spans, personal communication)	Beretta *et al.* (1965)
	Melatonin	Postulated blockade of 5-HT receptors (Smythe and Lazarus, 1974)	
Neurotoxic agents	5,6-Dihydroxytryptamine (5,6-DHT)		Baumgarten *et al.* (1971, 1972a,b)
	5,7-Dihydroxytryptamine (5,7-DHT)	Small depletion of brain NE and DA levels; in association with DMI selective effect on 5-HT neurons (Björklund *et al.*, 1975b)	

[a] The international nonproprietary name recommended by WHO for methergoline (Liserdol, Farmitalia) is metergoline.

or L-AAAD does not seem to be important in explaining depletion in brain 5-HT observed following α-methyl-Trp administration (see Wurtman and Fernström, 1972), however, α-methyl-5-HTP, which is formed *in vivo* from α-methyl-Trp, is a potent inhibitor of L-AAAD (Moran and Sourkes, 1963).

p-Chloroamphetamine and *p*-chloro-*N*-methylamphetamine have been proposed as *in vivo* inhibitors of Trp-H since they decrease brain 5-HT and 5-HIAA levels (Pfeifer and Galambos, 1967; Miller *et al.*, 1970). However, these compounds also exhibit amphetamine-like activity on brain CA's (Strada *et al.*, 1970). pCA and pCMA do not inhibit Trp-H *in vitro* (Pletscher *et al.*, 1968) but can decrease the amount of labeled 5-HT formed from IVT ^{14}C-tryptophan; hence, it seems possible that they inhibit Trp uptake (Sanders-Bush and Sulser, 1970). Therefore, the effects of pCA on serotoninergic tone could be the consequence of a combination of several mechanisms such as release and inhibition of 5-HT uptake and perhaps inhibition of Trp-H (see Sanders-Bush *et al.*, 1974). An enhancement in catecholaminergic activity associated with a decrease in 5-HT synthesis has been achieved by administering L-dopa, which acts as competitor for L-AAAD, thus decreasing the rate of conversion of 5-HTP to 5-HT (see Creveling and Daly, 1971).

Another way to decrease brain 5-HT levels is by the administration of synthetic diets lacking L-Trp and containing neutral amino acids (see above) which compete with Trp for the transport system from plasma into the brain (Wurtman and Fernström, 1972).

5. Storage, Release, and Uptake

Density gradient techniques applied to brain tissue have shown that 5-HT is present in synaptosomes (Michaelson and Whittaker, 1963; De Robertis, 1967; Whittaker, 1969a,b,c). The early ultrastructural studies showed good correlation between large granular vesicles and NE and 5-HT content, but subsequent experiments showed that the main storage sites for NE and 5-HT seem to be the small granular vesicles (Hökfelt, 1968) (see Section A,2). Labeled 5-HT injected intraventricularly accumulates in nerve endings containing a small number of large granular vesicles (800–1200 Å) (Aghajanian and Bloom, 1967a; Bloom and Aghajanian, 1968). Evidence exists that exogenous 5-HT is taken up by the same compartments that contain endogenous 5-HT (Kuhar *et al.*, 1971, 1974).

Blood platelets of several animal species and man store 5-HT together with nucleotides (especially ATP) and bivalent cations (Ca^{2+} and Mg^{2+}) in specific subcellular organelles (vesicle-like structures) possessing a very dense osmiophilic core (Tranzer *et al.*, 1966). These are spherical, 1000–

2000 Å in diameter, bound by a typical unit membrane containing a Mg^{2+}-dependent ATPase (see Da Prada and Pletscher, 1974). In synaptosomes from the hypothalamus and spinal cord, two different serotonin-binding proteins have been recently isolated by affinity chromatography (Chen Shih *et al.,* 1974). Storage in vesicular granules protects 5-HT from degradation by MAO.

Release of 5-HT from nerve terminals can be initiated by nerve impulse, electrical stimulation, and drugs (see Table XV). A relation between nerve impulse flow and turnover of 5-HT has been shown (Carlsson *et al.,* 1972); e.g., turnover is decreased by disconnecting axons from cell bodies (Andén *et al.,* 1966) and is increased after electrical stimulation of raphe nuclei cell bodies (Sheard and Aghajanian, 1968).

As for NE, 5-HT is also taken up by a specific high affinity Na^+-dependent mechanism that appears to be associated with 5-HT neurons in mammalian brain (see Snyder *et al.,* 1970; Iversen, 1974). Such a mechanism is inhibited by various tricyclic antidepressants of the imipramine and amitriptyline group (see Table XV). In contrast to the inhibition of NE uptake, imipramine and amitriptyline are more potent inhibitors of 5-HT uptake than their desmethyl derivatives, DMI and nortriptyline. The most potent inhibitor so far described is 3-chloroimipramine (Iversen, 1974). Recently, it has also been shown that the neurotoxic agents 5,6-DHT and 5,7-DHT, are taken up by the 5-HT neurons by the same mechanism (see Baumgarten and Schlossberger, 1973; Björklund *et al.,* 1974a) (Section D).

As for NE and DA, central 5-HT effects can be mimicked or blocked by a variety of drugs acting by different mechanisms. Serotoninergic activity can be decreased by drugs able to block, by various mechanisms, 5-HT biosynthesis (see above); to inhibit 5-HT storage within the nerve endings, e.g., by reserpine (Grahame-Smith, 1971); to act as 5-HT receptor antagonists (methysergide, LSD25, cyproheptadine, etc.); to induce a rather selective degeneration of 5-HT neurons, e.g., by 5,6-DHT and 5,7-DHT (see Table XIV). Enhancement of 5-HT neuronal function can be obtained with the administration of the precursors L-Trp or 5-HTP alone or following pretreatment with a MAO or L-AAAD inhibitor, respectively, or with drugs which either release 5-HT, e.g., D-amphetamine, inhibit its uptake, or are agonists at 5-HT receptor sites (see Table XV).

It has been shown that loading doses of 5-HTP not only lead to 5-HT accumulation in cells that do not ordinarily contain this indoleamine but also interfere with CA's and their precursors for transport, storage, and metabolism within the CNS (Johnson *et al.,* 1968; Fuxe *et al.,* 1971; Ng *et al.,* 1972). Combined administration of 5-HTP and an AAAD inhibitor seems to enable a more consistent enhancement of 5-HT-mediated neuronal transmission (Henning and Rubenson, 1971; Porter 1971).

TABLE XV

List of Drugs Directly or Indirectly Potentiating Serotoninergic Mechanisms

Class	Drug	Observations	References to main action
Receptor agonists	Tryptamine		Bradley and Marley (1965); Dewhurst and Marley (1965)
	α-Methyltryptamine		
	N-Dimethyl-5-HT (Bufotenin)	Hallucinogen (Jacobsen, 1968)	
	N-Dimethyltryptamine	Hallucinogen (Szara, 1956)	
	Quipazine [2-(1-piperazinyl) quinoline maleate]	DA agonist action? (Grabowska *et al.*, 1974)	Hong *et al.* (1969)
Precursors	L-Tryptophan (Trp)	Associated with MAOI (Coppen *et al.*, 1963; Pare, 1963); possible induction of liver TPO (Mahler and Cordes, 1966)	Carroll (1971)
	5-Hydroxytryptophan (5-HTP)	Associated with L-AAAD inhibitors (Henning and Rubenson, 1971); decrease by displacement of brain NE levels with no change in brain DA (Everett, 1974); displacement by 5-HT of NE and DA (Fuxe *et al.*, 1971)	Moir and Eccleston (1968)
5-HT uptake inhibitors	Chlorimipramine	The most potent uptake inhibitor; slight effect on neuronal NE uptake	Iversen (1967, 1971a, 1971b, 1974)
	Imipramine	More active on 5-HT uptake than its desmethyl derivative	
	Amitriptyline		
	Maprotiline	Inhibition of NE uptake	Maitre *et al.*, (1974)
	(+) *trans*-3-[4-methoxy) methyl-1-methyl]-4-phenyl-piperidine hydrochloride (FG 4963)	Almost equipotent to chlorimipramine; central inhibition of neuronal NE uptake at high concentrations; only weak inhibition of peripheral NE uptake	Buus Lassen *et al.*, (1975)
	3-(*p*-Trifluoromethylphenoxy)-*N*-methyl-3-phenylpropylamine hydrochloride (Lilly 110140) or fluoxetine	Almost equipotent to chlorimipramine; no blockade of NE uptake *in vivo* (Wong *et al.*, 1975)	Fuller and Perry (1974)

MAO inhibitors			
Hydrazides and hydrazines	Iproniazid	Interaction also with brain CA	See Goodman and Gilman (1975)
	Nialamide		
	Isocarboxazide		
	Iproclozide		
	Pheniprazine (α-methylphenylhydrazine)		
	Phenelzine (phenethylhydrazine)		
	Mebanazine (α-methylbenzylhydrazine)		
	Phenoxypropazine (1-methyl-2-phenoxyethylhydrazine)		
Amines	*d*-Amphetamine		
	α-Ethyltryptamine (etryptamine)		
	Pargyline (*N*-methyl-*n*-2-propinyl-benzylamine)		
	Tranylcypromine		
	Harmine, Harmaline		
	Yohimbine		
Releasing agents	Amphetamine	Release also of NE and DA; blockade of their uptake, and of MAO; stimulation of 5-HT receptors (Vane, 1960)	
	Fenfluramine	Costa *et al.* (1971)	

6. Metabolic Degradation

Besides the presence of an active stereospecific uptake mechanism in the nerve terminals for 5-HT, another mechanism by which 5-HT effects are terminated is by metabolic degradation via MAO (Fig. 24). The end product of 5-HT metabolic degradation is 5-HIAA, whose distribution in the brain is similar to that of 5-HT (Roos, 1962). The cerebrospinal fluid concentration of 5-HIAA is similar to that of the brain rather than of the blood (Eccleston, 1968). It has been known for a long time (Ashcroft and Sharman, 1960) that 5-HIAA concentration is reduced in the CSF of depressed patients. In cases of severe depression, low values of 5-HIAA in CSF are always observed, but in less severe cases values may fall in the normal range. Generally, probenecid, an acid compound known to compete with the transport of acidic metabolites from the brain to plasma, e.g., 5-HIAA, HVA (Neff *et al.,* 1964; Werdinius, 1966; Guldberg *et al.,* 1966), is administered before the assay of these metabolites in the CSF. The rate at which 5-HIAA accumulates in the CSF in response to probenecid is regarded as an index of the rate of utilization of brain 5-HT, whereas the formation of labeled 5-HIAA after administration of ^{3}H-Trp is considered an index of the functional activity of 5-HT neurons (Van Praag and Korf, 1973) (see the following section and Table XVI).

7. Quantitative Estimation of Turnover

As for CA's, several methods have been devised to study the brain 5-HT turnover rate, which is a better indicator of the activity of 5-HT neurons than absolute 5-HT concentrations.

Turnover rate of brain 5-HT can be measured by following the rise of 5-HT or the decline in 5-HIAA after MAO inhibition, or by measuring the rise in 5-HIAA in CSF or in the brain after blocking the acid transport system with probenecid (Neff and Tozer, 1968). In other methods the accumulation of labeled 5-HT and 5-HIAA after systemic or IVT injection of ^{14}C-Trp is taken as an index of 5-HT biosynthesis and degradation. In many studies, the turnover of brain 5-HT has been determined on the basis of the overall steady-state system where synthesis of 5-HT is equal to its metabolism to 5-HIAA, and it has been assumed that the 5-HIAA which is produced reflects the amount of 5-HT used in functional activity in the brain (see Table XVI).

The relation between total turnover of brain 5-HT and functional activity of 5-HT has been discussed (Grahame-Smith, 1973). It would appear that the increase in brain 5-HIAA which follows L-Trp loading does not truly reflect the activity of central 5-HT neurons since behavioral changes

TABLE XVI

Methods for the Evaluation of Brain 5-HT Turnover

Method	Observations and main drawbacks	References
A. *Isotopic steady-state methods*		
1. Labeling brain (or discrete areas) 5-HT stores with labeled precursors (tryptophan, 5-HTP)	Precursors given systemically are taken up into platelets; 5-HTP is decarboxylated in sites other than 5-HT neurons	Neff and Tozer (1968); Neff *et al.*, (1971)
2. Labeling brain (or discrete areas) 5-HT stores after IVT or IC injection of labeled 5-HT	Nonspecific labeling of brain tissue and great variation in the disappearance rate of 5-HT	
B. *Non-isotopic, nonsteady-state methods*		
1. Accumulation of brain 5-HT stores after MAO inhibition 2. Decline in brain 5-HIAA after MAO inhibition	Assumption that MAO inhibition is immediate and total; assumption that 5-HT is not lost from the brain by diffusion; increase in brain 5-HT stores after MAO inhibition with consequent changes in 5-HT neuronal activity	Neff *et al.*, (1967); Neff and Tozer (1968); Lin *et al.*, (1969)
3. Accumulation of 5-HIAA after probenecid	See also Table XII	

do not always occur under these conditions. On the contrary, after pretreatment with a MAOI, small doses of Trp produce hyperactivity and hyperpyrexia related to the accumulation of 5-HT in the brain. Therefore, it seems that the level of the function pool of 5-HT within the neurons is regulated by the controlled activity of MAO, the ability of 5-HT to get to this enzyme, and the control of the storage compartment. On the basis of the evidence presented, it seems hard to conceive that the functional pool of neuronal 5-HT depends on the fine regulation of Trp-H. The latter in fact is, in turn, dependent on brain Trp concentration which is not finely controlled. Therefore, measurements of total turnover rate may in no way reflect the turnover of a functionally active compartment since it also estimates the functionally inactive compartment (in fact both active and inactive pools metabolize 5-HT to form 5-HIAA).

8. Electrocortical Changes

Several studies recently reviewed (Holman *et al.,* 1975) indicate that 5-HT brings about or facilitates SWS in many mammalian species. This hypothesis is supported by the observation of insomnia after brain 5-HT depletion in cats (Delorme *et al.,* 1966; Koella and Czicman, 1966) and monkeys (Weitzman *et al.,* 1968) and, conversely, SWS facilitation after treatment with 5-HTP (Jouvet, 1967) or Trp (Hartmann *et al.,* 1971b). In addition, in favor of a positive 5-HT involvement in the control of sleep, experiments have shown that sleep can result from manipulations in which serotoninergic activity is increased, such as electrical stimulation of the 5-HT perikarya in the raphe nuclei (Gumulka *et al.,* 1971; Kostowski *et al.,* 1969; Jouvet, 1972). Conversely, both SWS and REM sleep followed lesions of the raphe nuclei (Jouvet, 1972).

Consistent with the 5-HT theory of sleep induction are also the experiments with pCPA; this drug, however, induces depletion of 5-HT levels in several animal species, but is also capable of producing several other effects which make the interpretation of results difficult (see Table XIV). Insomnia (suppression of both SWS and REM sleep), paralleling 5-HT depletion, followed pCPA treatment (Koella *et al.,* 1968; Jouvet, 1967, 1972). The administration of 5-HTP in pCPA-treated animals restored qualitatively and quantitatively normal sleep (SWS and REM) (Koella *et al.,* 1968; Jouvet, 1967, 1972).

In cats, IVT microinjection of 5-HT produced sedation (Feldberg and Sherwood, 1953, 1954), the disappearance of spontaneous activity (Schain, 1962), lethargy (Gaddum and Vogt, 1956), and prolonged tremors (Domer and Feldberg, 1960). In the same animal species, intracarotid injection of 5-HT was followed, after an initial phase of arousal, by a prolonged period of EEG synchronization and mydriasis (Koella and Czicman, 1963, 1966). Moreover, since, after severing the posterior brainstem from the forebrain, 5-HT induced only arousal, it was suggested by the same authors that the synchronizing effect of 5-HT is due to an action in the posterior brainstem (area postrema) from which the hypnogenic influence is transmitted to the forebrain. Topical application of 5-HT to the posterior region of the fourth ventricle produced an increase in the recruiting response, an increase of spindle bursts, and an increase in slow-wave output lasting at least 15 min. These experiments were later confirmed by Roth *et al.* (1970) and by Bronzino *et al.* (1972).

Recently, by microinfusing 5-HT into the third ventricle and into the anterior hypothalamus of fowl, it was shown that this amine produces behavioral and electrocortical SWS with consistent increase in electrocortical integrals (Fig. 26), lasting at least 30–60 min (Marley and Nisticò, 1975a). These experiments are in agreement with the suggestion that a release of 5-HT in the preoptic anterior hypothalamus, an area rich in

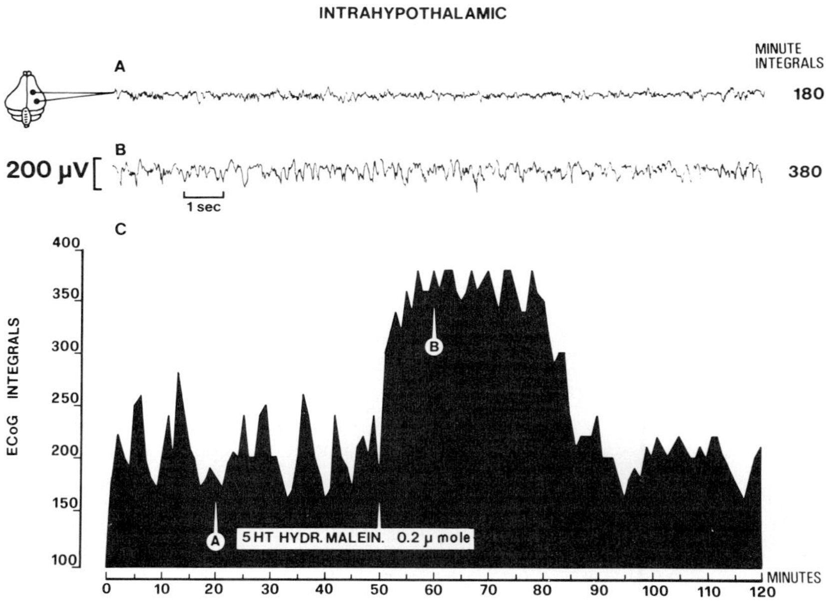

Fig. 26 Records of electrocortical activity (A, B) and histogram of integrated electrocortical activity (C) in an unanesthetized adult fowl. (A) Central electrocortical activity. (B) Slow frequency, large amplitude potentials after 5-HT-hydrogen maleinate (0.2 μmole) infused into the hypothalamus. (C) Electrocortical integrals showing a sustained increase lasting about 30 min after intrahypothalamic infusion of 5-HT (Marley and Nisticò, 1975a.)

5-HT terminals (Gargiulo and Nisticò, 1975), would facilitate or modulate the postsynaptic synchronizing mechanisms which are apparently triggered in this region (Sterman and Clemente, 1962; Yamaguchi *et al.*, 1963).

Results of experiments with IVT or intracerebral administration of 5,6-DHT are also in favor of serotoninergic involvement in both SWS and REM sleep. Immediately after IVT administration of the drug, a state of prostration with electrocortical slow waves occurred in avian and mammalian species (Jouvet and Pujol, 1974; Longo *et al.*, 1974; Nisticò *et al.*, 1975), and in the following days a significant decrease in SWS and REM sleep was observed with a parallel decrease in brain 5-HT and 5-HIAA levels (Jouvet and Pujol, 1974).

However, several reports provide evidence inconsistent with this model. In pCPA-treated rats, an 85–90% depletion of cerebral 5-HT resulted in little or no change in sleep time (Rechtschaffen *et al.*, 1973), and particularly intriguing was the finding that cats kept chronically depleted of 5-HT with repeated injections of pCPA showed a return to essentially normal SWS and REM sleep, despite continued low brain 5-HT concentration (Dement *et al.*, 1972). Possible explanations for this discrepancy include persistence after pCPA inhibition of a functional micropool of 5-HT or, alternatively, increase in receptor sensitivity, for which indirect evidence has been presented (W. C. Dement and J. D. Barchas, unpublished observations).

In addition, besides a regulatory role played by 5-HT in REM sleep, other components of this state, as PGO spikes, seem to be inhibited; in fact, a striking parallel between depletion of 5-HT by pCPA and appearance of PGO activity in the waking state (Henriksen *et al.*, 1974) indicates that 5-HT may be involved in suppression of the PGO activity from SWS and waking (Holman *et al.*, 1975).

Also, extrapolation of animal experiments to humans has yielded inconsistent findings. An increase in REM sleep has in fact been shown after iv injection of 5-HTP (Mendell *et al.*, 1965) or oral doses of Trp (Hartmann, 1967) in normal adults. Conversely, when pCPA was used to suppress the synthesis of 5-HT, REM sleep was reduced (Wyatt *et al.*, 1969). A study on the relationship between plasma free Trp levels and electroencephalographic sleep in normal young women revealed a positive correlation of free Trp with REM sleep and a negative correlation with non-REM sleep (C. N. Chen *et al.*, 1974). As plasma free Trp levels may determine brain Trp and 5-HT concentrations (Section C,4), these and the above results lead to the supposition that in man REM sleep is closely related to indoleamines. In this context, it is also of great interest that in the rat, unit neuronal activity of 5-HT-containing cells in the n. raphe magnus was highest during REM sleep and lowest during SWS (Bloom *et al.*, 1973a).

9. Iontophoretic Application

When applied iontophoretically, 5-HT can influence the discharge of neurons throughout the CNS. Generally, the majority of the cells (telencephalic, raphe nuclei, cerebellar Purkinje cells) respond by decreasing their discharge rate (Cooper *et al.*, 1974; Bloom *et al.*, 1973a,b; Boakes *et al.*, 1969, 1970; Couch, 1970; Galindo *et al.*, 1967; Hoffer *et al.*, 1969; Aghajanian, 1972). However, in certain areas of the brain, particularly the limbic cortex and spinal cord, 5-HT produces a marked activation of discharge rate. Since the most common response to 5-HT is inhibition, the observed activation could well derive from inhibitory neurons.

In contrast to earlier reports indicating no effect of 5-HT in the cortex (Krnjevic and Phillis, 1963a,b; Phillis, 1965; Bloom, 1968) and spinal cord (Curtis *et al.*, 1961), reproducible depressant effects of 5-HT have now been found in these areas (Johnson *et al.*, 1969b; Weight and Salmoiraghi, 1966, 1968; Gershenfeld, 1971). In the brainstem, in an initial study, no sensitive neurons were reported (Curtis and Koizumi, 1961), but in a subsequent study consistent effectiveness of 5-HT was found: 40% of the sensitive neurons were excited and 49% depressed (Bradley and Wolstencroft, 1965). In unrestrained, unanesthetized cats, raphe nuclei 5-HT neurons exhibited at least two different spontaneous patterns of discharge; the majority of units exhibited moderate discharge rates (3–20/sec) during quiet behavior when the animal was awake, and an increase or decrease when the animal entered SWS or subsequent REM stages (Sheu *et al.*, 1971). In addition, according to these authors, several raphe magnus neurons exhibited an extremely rapid pattern during REM sleep stage (see above). Depression of spontaneous activity was the prevalent response of hypothalamic neurons to 5-HT (Bloom *et al.*, 1963). Twenty-four out of 171 units studied in the cat hypothalamus responded to 5-HT; the rate of discharge of 22 neurons was depressed by 5-HT and only 2 neurons were activated. In addition, in more recent studies, a reduction was reported in the spontaneous discharge of about one-half of the rostral hypothalamic neurons after 5-HT application (Dyball and Dyer, 1974), thus confirming previous data that in other hypothalamic nuclei (e.g., cat suprachiasmatic nuclei, SON) 5-HT possesses a depressant effect (Bloom *et al.*, 1972).

D. SELECTIVE NEUROTOXIC DRUGS FOR MONOAMINE NEURONS IN THE CNS

1. 6-Hydroxydopamine and 6-Hydroxydopa

Selective degeneration of CA or 5-HT neurons in the CNS achieved by means of neurotoxic agents has a definite advantage over mechanically or electrolytically induced lesions which inevitably involve different neuronal and nonneuronal elements.

As shown by histofluorescence and electron microscopy, several studies indicate that IVT or intracerebral injections of 6-OHDA produce a rather selective, dose-dependent degeneration of catecholaminergic neurons without affecting other transmitters or other neuronal and nonneuronal elements (Bloom *et al.*, 1969; Bartholini *et al.*, 1970, 1971; Breese and Traylor, 1970; Ungerstedt, 1968, 1971b). In the CNS, there is neither morphological (Bloom, 1971; Ungerstedt, 1971b) nor biochemical (Uretsky and Iversen, 1970) evidence for regeneration of destroyed CA nerve terminals even after more than two years (Ungerstedt, 1971b).

6-OHDA is taken up by CA terminals where initially it displaces the transmitters from the storage vesicles, and may be responsible for activation of the CA system (see Kostrzewa and Jacobowitz, 1974). When critical intraneuronal concentrations are achieved, degenerative processes are initiated (Thoenen and Tranzer, 1968). As target sites of 6-OHDA action, cell membrane, endoplasmic reticulum, mitochondria, and storage granules have been suggested (see Kostrzewa and Jacobowitz, 1974). Certainly, cellular enzymes and energy-producing cytochromes, or related elements of the respiratory transport chain, are destroyed by 6-OHDA, which, being a strong reducing agent, acts as an oxidative phosphorylation uncoupler (Wagner, 1971). Consequently, depletion of brain CA's and impairment of biochemical and electrophysiological processes occur in CA neurons. In particular, IVT and intracerebral injections of 6-OHDA produce a dose-dependent decrease in NE and DA in different areas of the brain (Uretsky and Iversen, 1969). Up to a certain dose (50 μg) only NE is decreased whereas higher doses decrease NE and DA simultaneously (Evetts *et al.*, 1970; Iversen and Uretsky, 1970).

Various methods have been devised for producing greater selective degeneration of one of the two CA's without affecting the other. Pretreatment with tricyclic antidepressants such as imipramine and DMI, has been shown capable of inhibiting the uptake of 6-OHDA by the NE neurons and of producing selective degeneration of central DA neurons (Evetts and Iversen, 1970; Breese and Traylor, 1971). Repeated intracerebroventricular injection of low doses of 6-OHDA selectively depleted NE, leaving DA stores unchanged (Breese and Traylor, 1971; Simmonds and Uretsky, 1970) (see Table VIII). Selective degeneration of CA neurons leaving 5-HT neurons unaffected was obtained by pretreating animals with chlorimipramine (Laguzzi *et al.*, 1972) (for different sensitivity of discrete brain areas to 6-OHDA see Chapter I, Section B,2).

A loss of enzymes which regulate CA synthesis, such as TH and D-β-H, associated with a decreased CA turnover (Bloom *et al.*, 1969; Bartholini *et al.*, 1971; Iversen and Uretsky, 1970) and an impairment in the uptake capacity for CA's (Uretsky and Iversen, 1969; Iversen and Uretsky, 1971), have also been demonstrated following 6-OHDA administration. However, a higher rate of NE turnover seems to be present in the surviving CA neurons which may provide reestablishment of a baseline function (N. J. Uretsky *et al.*, 1971). Development of presynaptic (due to decreased uptake mechanisms for NE) and postsynaptic (due to a proliferation of adrenergic receptors after denervation) supersensitivity has been found to occur after 6-OHDA chemical sympathectomy (see Kostrzewa and Jacobowitz, 1974). Destruction by 6-OHDA of DA-containing nigrostriatal pathways resulted in supersensitivity to DA agonists (Schoenfeld and Uretsky, 1972). When this projection was destroyed unilaterally, the animals developed a characteristic contralateral turning response to DA agonists, e.g., apomorphine, L-dopa, and ET 495 (Ungerstedt, 1971b; Corrodi *et al.*, 1971; von Voigtlander and Moore, 1973). Supersensitivity to dopaminergic and/or noradrenergic agonists has also been produced by inhibition of CA synthesis (Dominic and Moore, 1969; Gianutsos *et al.*, 1974; Tarsy and Baldessarini, 1974; Gudelsky *et al.*, 1975), depletion of CA stores (Stolk and Rech, 1968) or blockade of postsynaptic receptors by neuroleptics (Gianutsos *et al.*, 1974; Tarsy and Baldessarini, 1974; Moore and Thornburg, 1975; Dunstan and Jackson, 1976). A detailed description of behavioral, electrocortical, body temperature, and endocrine effects after intracerebroventricular infusion of 6-OHDA can be found in a review by Kostrzewa and Jacobowitz (1974).

After the discovery that 6-OHDA does not easily cross the BBB (Bloom *et al.*, 1969; Uretsky and Iversen, 1969, 1970; Ungerstedt, 1971b), the effects on brain DA and NE neurons were studied after systemic injection of the carboxyl analog of 6-hydroxydopamine, e.g., 6-OH-dopa, which enters the brain. However, disappointing results were obtained in trying to selectively destroy NE and DA neurons in CNS by 6-OH-dopa (see Thoenen, 1972). The amino acid is more toxic (severe renal and hepatic lesions) than the respective amine, and brain NE depletion is less marked than that which follows IVT-6-OHDA. The use of smaller doses of the drug in animals pretreated with a MAO inhibitor (Jacobowitz and Kostrzewa, 1971; Sachs and Jonsson, 1972) avoided severe toxic effects. Under these conditions, 6-OH-dopa appears to be more selective for NE neurons since neither DA nor 5-HT neurons were affected. A relatively selective (neocortex, hippocampus, and spinal cord) long-term alteration of central NE stores has been reported in the rat after neonatal treatment with either 6-OHDA (Liew and Laverty, 1975) or 6-OH-dopa (Kostrzewa, 1975).

2. 5,6 and 5,7-Dihydroxytryptamines

The biochemical counterpart to 6-OHDA for the indoleaminergic system is 5,6-DHT. 5,6-DHT given in some dose regimens is taken up rather selectively by 5-HT neurons, leaving largely unaffected adrenergic neurons. When critical concentrations are reached, degenerative lesions become evident (Baumgarten *et al.*, 1971, 1972a; Björklund *et al.*, 1974a), and yellow fluorescent material accumulates in 5-HT severed axons and terminals making visible 5-HT terminals and axons otherwise not detectable (Fig. 23d–f) (see also Section C,2).

In several animal species, such as fowl (Nisticò *et al.*, 1975), mice, rats (Longo *et al.*, 1974), and cats (Jouvet and Pujol, 1974), the immediate central effects of 5,6-DHT after IVT administration mimic those evoked by 5-HT. This fact is likely accounted for by the displacement and release of 5-HT from the vesicle stores (Björklund *et al.*, 1974a), without limiting the possibility of a direct action on 5-HT receptor sites (Barzaghi *et al.*, 1973; Carruba, 1974), or an inhibition of 5-HT reuptake (Björklund *et al.*, 1974a). Six to 12 hours after a single IVT injection of 5,6-DHT, a drop occurs in 5-HT content of several brain areas such as septum, striatum, and hypothalamus. Long-term evaluation of 5-HT concentrations in various areas of the rat brain revealed a significant recovery of 5-HT with axonal regeneration in all regions except the spinal cord between 1 and 6 months following IVT injection of 5,6-DHT (see Baumgarten and Schlossberger, 1973).

Another hydroxylated derivative of tryptamine, 5,7-DHT administered centrally, appears to be less specific for the 5-HT system than 5,6-DHT since it produces significant degenerative lesions also on CA neurons (Björklund *et al.*, 1974a). However, because of its lower general toxicity, 5,7-DHT can be administered IVT in much higher doses than 5,6-DHT, and after, intracerebral infusion is more efficient, although less selective, in inducing destruction of 5-HT axons. Recently, combined treatment with 5,7-DHT and DMI, which is more efficient in blocking the membrane pump of NE than of 5-HT neurons (Table X), improved the selectivity of 5,7-DHT action for serotonin neurons (Björklund *et al.*, 1975b).

On the basis of present evidence, 6-OHDA, 6-OH-dopa, 5,6-DHT, and 5,7-DHT administered IVT, IC, or into selected areas of the brain appear to be useful for investigating the functional role played by MA systems in neuroendocrine regulation.

E. MELATONIN

The biosynthesis of melatonin, secreted by the pineal parenchymal cells, is initiated by the uptake of L-Trp by the pineal cells from the circulating blood and, as in 5-HT synthesis, it is likely that the concentration of this amino acid in the plasma may act as an important regulatory mechanism in melatonin synthesis (Fig. 24) (Wurtman and Fernström,

1972). The uptake of Trp by the pineal cells appears to be stimulated by NE (Wurtman *et al.*, 1969).

In the first two steps 5-HT is synthesized, and is then converted by *N*-acetyltransferase (NAT) to *N*-acetyl serotonin (NAS). *N*-Acetyl serotonin is then converted to melatonin in a step catalyzed by hydroxylindole-*O*-methyltransferase (HIOMT), which is present in the body only in this gland (Axelrod *et al.*, 1961). The mammalian pineal gland is on the peripheral side of the BBB, and is stimulated by the superior cervical sympathetic ganglia (Cooper *et al.*, 1974). It functions as a neuroendocrine transducer, which converts an adrenergic neural input (e.g., NE released from nerve endings) into a hormonal output (melatonin and possibly other methoxyindoles) (Wurtman *et al.*, 1969) (see Chapter I,B).

Cyclic daily rhythms of both 5-HT and melatonin in the pineal appear to be directly related to environmental light patterns (Axelrod and Wurtman, 1968). Pineal biosynthesis and probably secretion of melatonin are augmented during the night in most mammals (see Brownstein, 1975). Reports of a nocturnal elevation of plasma and urinary melatonin in the human with lower or undetectable levels during the day have been recently published (Arendt *et al.*, 1975; Vaughan *et al.*, 1976). A marked daily rhythm in the activity of NAT exists (Klein and Weller, 1970). Conditions which lead to an increase in NAT result in a fall of 5-HT (Brownstein *et al.*, 1973a), its substrate, and an increase in NAS, its product (Brownstein *et al.*, 1973b). Conversely, conditions which cause a decrease in the activity of NAT are associated with an increase in 5-HT and a decrease in NAS.

The time course of the change in NAT is reflected in the time course of changes in its substrate and product. Apparently cycles in the activity of NAT are responsible for the cycles observed in 5-HT, NAS, and melatonin (see Brownstein, 1975). The NAT activity appears to be regulated by the action of NE on β-adrenergic receptors (Deguchi and Axelrod, 1972a,b). The effect of NE, in turn, appears to be mediated by cAMP (Klein *et al.*, 1970; Deguchi, 1973). In fact, cAMP of pineal glands exposed to β-adrenergic agonists *in vivo* or *in vitro* increases rapidly. The elevation in cAMP and in NAT levels are blocked by β-antagonist drugs (Strada *et al.*, 1972; Deguchi, 1973). In addition, theophylline, which inhibits phosphodiesterase and prolongs the life of cAMP, potentiates the induction of NAT by CA's (Strada *et al.*, 1972). It has also been shown that stimulation of the superior cervical ganglion, which presumably causes an increase in the release of NE in the pineal gland, causes an increase in the activity of pineal NAT (Volkman and Heller, 1971) (Fig. 27).

Interactions between melatonin and serotoninergic mechanisms at the hypothalamic level have been observed. An increase in the hypothalamic content of both 5-HT and GABA was found after systemic or IVT injection of melatonin (Anton-Tay, 1971) (see also Chapter IV,A,3,b). In addi-

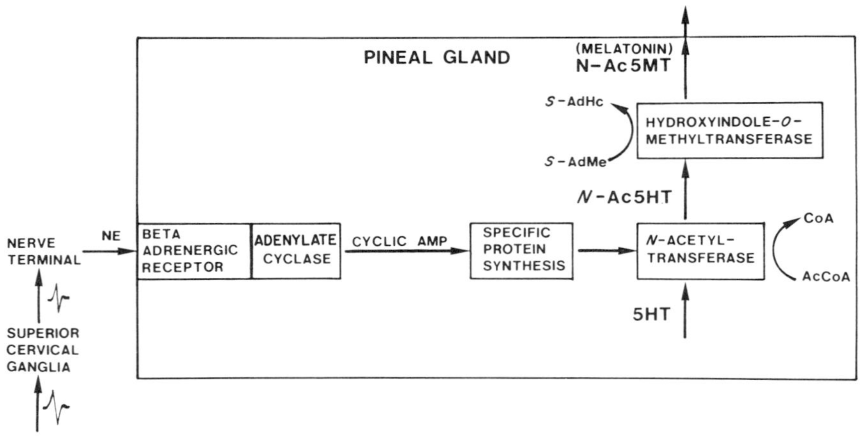

Fig. 27 Schematic illustration of the adrenergic control of melatonin synthesis in the pineal gland. AcCoA, acetyl-CoA; CoA, coenzyme A; *N*-Ac 5-HT, *N*-acetyl-5-hydroxytryptamine; *S*-AdMe, *S*-adenosylmethionine; *S*-AdHc, *S*-adenosylhomocysteine; *N*-acetyl-5-MT, *N*-acetyl-5-methoxytryptamine (melatonin) (for details see text). (Modified from Klein, 1973.)

tion, systemic administration of melatonin markedly reduced hypothalamic 5-HT uptake, but did not modify the uptake of DA, NE, or glutamate; however, a similar effect on the uptake of these neurotransmitters was present 15 days after pinealectomy or bilateral ablation of the superior cervical ganglion (Cardinali, 1975).

Relevant to a discussion of pineal–brain interactions is the observation that melatonin induces sleep. Intravenous injections of melatonin into chicks whose BBB is relatively undeveloped (Hishikawa *et al.*, 1969) or its microinjection into the hypothalamus of cats (Marczynski *et al.*, 1964) produced behavioral and electrocortical SWS mimicking the effects (in chicks) of 5-HT (Jalfre *et al.*, 1974). In humans, although some studies indicated that administration of melatonin is followed during the next 20 minutes by EEG synchronization and sleep (Anton-Tay *et al.*, 1971), others found that it decreased the mean latency of night sleep onset but did not alter total sleep and the percentage of sleep stages I to IV or REM sleep (Cramer *et al.*, 1974).

F. ACETYLCHOLINE

The requirement for a convenient assay for ACh of suitable specificity and sensitivity to replace the tedious bioassay on several test muscles (see

Potter, 1972) is indicated by the large number of methods published recently (see Hanin, 1974) (see Table II). These include colorimetric methods and gas chromatographic and radiometric analyses. In most radiometric assays the acetyl group of the substrate (acetyl-CoA) is labeled, and during the course of the reaction some of it is transferred to choline and can be recovered as labeled ACh (see Hebb, 1972a). A radioenzymatic assay for ACh and the problems found during its development and application have been recently published (Wonnacott and Marchbanks, 1975).

1. Occurrence and Regional Distribution in the CNS

Biochemical and electrophysiological studies have localized ACh in several areas of the CNS, e.g., brainstem, hypothalamus, basal ganglia, limbic system, cerebellum, and cerebral cortex (Quastel, 1962; Votava, 1967; Krnjevic, 1965; Phillis, 1965; Bradley, 1968; Curtis and Crawford, 1969). ACh has been found in higher concentration in gray matter than in white matter (Feldberg and Vogt, 1948), and its distribution in the CNS parallels the distribution of the enzymatic activity responsible for its biosynthesis (choline acetyltransferase or ChAc) (Quastel, 1962). Recently, ChAc levels have been measured in specific nuclei of the rat diencephalon. Thalamic nuclei contain higher concentrations of the enzymes than hypothalamic and preoptic nuclei (Brownstein *et al.*, 1975a). Mappings of central cholinergic pathways have been published by Shute and Lewis (1966) and Palkovits and Jacobowitz (1974).

2. Cholinergic Pathways

Using a version of the thiocholine method (Koelle and Friedenwald, 1949) for the detection of cholinesterase (Che), two main Che-containing pathways have been identified in the rat forebrain (Shute and Lewis, 1966). The dorsal tegmental pathway (DTP) consists of fibers containing Che as well as acetylcholinesterase (AChe), which arise from the n. cuneiformis, situated in the dorsolateral part of the mesencephalic reticular formation, and project to the tectum, pretectal area, geniculate bodies, the thalamus, and the limbic system. The fibers of the ventral tegmental pathway (VTP), originating from the substantia nigra and ventral tegmental area of the midbrain, contain AChe only, and traverse the hypothalamus and subthalamus to reach the basal forebrain areas from which more AChe-containing neurons project into all regions of the cerebral cortex and the olfactory bulb. The existence of AChe-containing corticopetal pathways has been also demonstrated in the cat (Krnjević and

Silver, 1965, 1966). This ascending cholinergic reticular system probably forms the basis of electrocortical arousal, and so is identical with the "ascending reticular activating system" described by neurophysiologists (Shute and Lewis, 1966). This hypothesis is confirmed by the finding that during arousal there is a marked increase in the release of ACh from the cerebral cortex (Kanai and Szerb, 1965) (see also below).

In the hypothalamus and adjacent areas a dense group of cholinergic cells seems to constitute the supramammillary nuclei, whereas a scattering of AChe-containing cells is found in all parts of the posterior and lateral hypothalamic areas, except at the rostral end of the latter.

Particularly rich in AChe-containing cells is the lateral preoptic area. A group of AChe-containing cells in the dorsal part of the lateral preoptic area sends fibers via the stria terminalis to the amygdala. In the dorsal hypothalamic area, the AChe-containing cells form a compact, densely stained island. Their axons can be traced rostrally to the globus pallidus and to the ventral part of the reticular thalamic nucleus. Ventral and periventricular hypothalamic areas, e.g., the SCN, the periventricular hypothalamic n., and the ARC n., do not stain at all. In the mammillary n., the SON, and the paraventricular n., some neurons contain small or moderate amounts of AChe in their cell bodies, but none in their axons. Serious doubts exist, however, as to whether the Che-staining technique can be considered a definite method for the study of the molecular nature of the transmitter in a given neuron (Silver, 1967). Hence, ChAc has been also used as a marker to localize cholinergic neurons (Fahn and Coté, 1968; Yamamura *et al.*, 1974). Since ChAc cannot be measured histochemically and it is not rate-limiting for the synthesis of ACh, its activity measured in small tissue samples may reflect neither the density of cholinergic axons nor the concentration of ACh.

The instability of ACh and the lack of specificity and/or sensitivity of bioassay and gas chromatographic methods prevented, for a while, the study of the ACh concentrations in discrete brain nuclei. Recently, mass fragmentography in conjunction with gas chromatography (Hanin and Schubert, 1974) has enabled measurement of subpicamole quantities of ACh with a minimum of unit mass resolution (Koslow *et al.*, 1974). By means of this method, the concentration of ACh and ChAc in discrete nuclei of the rat brain has been reported (Cheney *et al.*, 1975a), confirming the presence of ACh in some areas of the limbic system and the hypothalamus which may be involved in neuroendocrine regulation. In this context, it is of note that only small changes in ChAc have been found in the ARC n. and VMN after surgical isolation of the MBH; no change at all was found in the ME (Brownstein *et al.*, 1976). This strongly suggests that there is a cholinergic tuberoinfundibular pathway similar to the dopaminergic one, which may be responsible for the neuroendocrine effects of cholinergic agents which have been observed (see Chapter IV).

3. Acetylcholine in the Synaptic Region

In general, the biochemical mechanisms on which cholinergic neurotransmission depends are very similar to those of CA's and 5-HT. Mechanisms responsible for the synthesis, storage, and release of the transmitter to specific receptors of the postsynaptic membrane are present in the nerve terminals, and those for degradation and termination of its effects are present within the synaptic region (Fig. 28).

Because of the lack of histochemical methods available for the observation of ACh in nerve terminals, the study of the compartments in which ACh is located has been accomplished by subcellular fractionation methods (Marchbanks, 1972). In several areas of the brain, ACh is not the

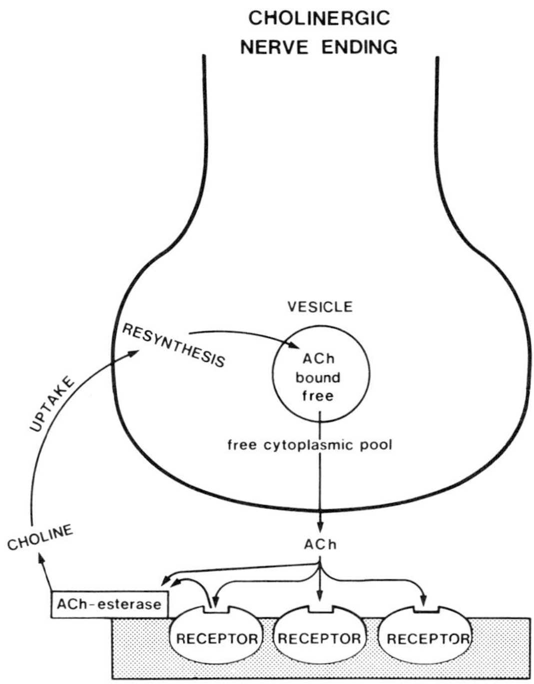

Fig. 28 Schematic illustration of the mechanisms (choline uptake, ACh synthesis, storage, and release) involved in cholinergic neurotransmission. After being released ACh acts on specific ACh receptors and is rapidly degraded by AChe, giving rise to choline which is taken up into the neuron and is used for ACh resynthesis (for details see text).

only neurotransmitter present, so that the synaptosome and synaptic vesicle populations are heterogeneous with respect to chemical transmitters. A pure cholinergic tissue is represented by the electric organ of *Torpedo marmorata* for which fractionation methods have been developed (see Marchbanks, 1972). About two-thirds of the ACh of the cerebral cortex can be isolated in nerve terminals; the proportion is higher in the caudate nucleus and lower in the spinal cord (Whittaker, 1965). When synaptosomes are osmotically lysed about half of their ACh content can be recovered in association with synaptic vesicles (De Robertis *et al.,* 1963; Whittaker *et al.,* 1964). According to Whittaker (1969a,b,c) three pools of ACh exist in peripheral cholinergic nerve endings. One is the labile, extracellular, or released ACh that is normally hydrolyzed by AChe, but can be recovered in the supernatant fraction if an anti-Che agent is added. The second is the labile-bound or cytoplasmic or extravesicular pool that is released from synaptosomes when they are disrupted by osmotic or mechanical means. The third pool is formed by the stable-bound ACh, present in the synaptic vesicles, a large proportion of which can be isolated in single particulate fraction when preparations of disrupted nerve endings are centrifuged on a discontinuous density gradient. The existence of similar pools of ACh in the CNS awaits demonstration.

4. Biosynthesis and Metabolic Degradation

Several highly specialized processes occur in cholinergic nerve terminals that lead to ACh synthesis. They include a facilitated transport mechanism for the uptake of the precursor choline from the interstitial fluid, one enzymatic step (nearly unique to cholinergic nerves) for the synthesis of ACh, a mechanism of storage in synaptic vesicles, a mechanism for a very rapid release of quanta of ACh onto the postsynaptic membrane and, finally, a rapid metabolic hydrolysis by AChe (see Fig. 28).

Both central and peripheral ACh neurons require supplies of extraneuronal choline for maintaining normal tissue levels of ACh and high release rates of the transmitter. The utilization of either endogenous or exogenous choline is inhibited by a choline analog, hemicholinium, which inhibits choline uptake but not ChAc (Hebb, 1972b).

The mechanism of choline uptake by isolated nerve terminals (Marchbanks, 1968; Potter, 1968; Diamond and Kennedy, 1969), brain slices (Schubert *et al.,* 1969), and squid axons (Hodgkin and Martin, 1965) appears to be the same. Choline is taken up by a saturable carrier and by passive diffusion. Uptake does not require energy or specific ions and is unaffected by ouabain. At low, but physiological concentrations, choline is concentrated about fourfold by synaptosomes (Marchbanks, 1968). The concentration of free choline in the neuroplasm of cholinergic nerve terminals appears to be crucial; a low level of choline could limit ACh synthesis since the enzyme ChAc is not saturated by the concentration of choline in the neuroplasm under normal conditions (Glover and Potter, 1971).

Little is known about the origin and the stability of free choline in the interstitial fluid of the CNS. Choline crosses the BBB with difficulty (Potter *et al.,* 1968; Chakrin and Whittaker, 1969; Ansell and Spanner, 1970), but there is evidence suggesting that a phosphocholine compound, most likely phosphatidylcholine (lecithin), may cross the BBB and be partially degraded to provide choline (Ansell and Spanner, 1970). Much of the choline produced by the hydrolysis of ACh is apparently recovered and reutilized by nerve terminals as precursor for ACh synthesis (Fig. 29). This mechanism of reuse of choline seems to be of particular importance in the CNS where choline input and output are more limited (Potter, 1972).

The next step for the synthesis of ACh is the acetylation of choline which is catalyzed by the enzyme ChAc, found in particularly high concentration in nerve terminals (Hebb *et al.,* 1964). The enzyme is synthesized in the cholinergic cell bodies and transported down to the axon terminals (Hebb, 1972b). Subcellular fractionation studies have shown that ChAc is a soluble extravesicular enzyme, a fact in favor of the existence of an extravesicular compartment of ACh (Marchbanks, 1972). According to this theory, extravesicular ACh represents recently synthesized ACh, awaiting to be stored in the vesicles. It is surprising that ACh causes little product inhibition of the enzyme and does not fully inhibit its synthesis at any level. However, the fact that ACh does not compete significantly with either choline or acetyl-CoA can be an advantage in that it permits rapid ACh synthesis to millimolar levels in the presence of only micromolar concentrations of these substrates (Potter, 1972). The rate of ACh synthesis depends on the availability of active enzyme, the concentration of choline and acetyl-CoA and the ratio of CoA to acetyl-CoA. It seems that ACh synthesis is dependent on the equilibrium of ChAc with its substrates and products. The advantage of this mechanism is that it responds to the neuroplasmic levels of ACh, which is in equilibrium with the content of the vesicles.

Some halogenated derivatives of ACh, e.g., chloro- and, to a lesser extent, bromo-, and iodoacetylcholine, have been found to be powerful inhibitors of ChAc (Chase and Tubbs, 1966; Morris and Grewaal, 1969) (see Table XVII); however, they are not completely specific since they also inhibit AChe (Chiou and Sastry, 1968). Another group of ChAc inhibitors are the styrylpyridine analogs (Smith *et al.,* 1967) (see Table XVII) which being quaternary compounds are unable to cross the BBB and hence are not useful for the study of cholinergic transmission in the CNS *in vivo*. A new series of substituted styrylpyridines without quaternary groups has been synthesized (Baker and Gibson, 1971).

5. Storage

The isolation of synaptic vesicles and the demonstration that they contain ACh provided confirmation of their postulated storage function (De Robertis *et al.*, 1962; Whittaker, 1963). With suitable correction for recovery, the estimated vesicular content is more than 80% of the total in the nerve cells (Israël *et al.*, 1968; Israël and Gautron, 1969). In mammals, the number of ACh molecules in each synaptic vesicle seems to be around 1400–9400. Recent studies indicate that in vesicles, ACh is compartmentalized in two pools, a large bound pool and a small pool of recent synthesis (Marchbanks and Israël, 1973). In cholinergic terminals the existence of an extravesicular compartment has been shown ("surplus ACh") (Birks and MacIntosh, 1961; Collier, 1970; Potter, 1970). The mechanism by which the vesicles take up and concentrate ACh from the neuroplasm is not known, but it seems that the vesicles do not directly receive the major proportion of the newly synthesized ACh from extracellular choline (Marchbanks, 1972) (Section F,4).

Two possible mechanisms for uptake may be envisaged (Potter, 1972). The first is that by which neuroplasmic ACh is exchanged for intravesicular Na^+ by a non-energy-requiring carrier in the vesicle membrane. The only enzyme known to be present in vesicle preparations is a Mg^{2+}-activated ATPase (Whittaker, 1969a,b,c). A second mechanism of uptake and concentration of ACh within the vesicles could operate if ACh were bound by constituents within them in the same manner as CA's are bound to ATP. Nucleotides and soluble proteins are present in cholinergic vesicles (Potter, 1971), but the formation of complexes between ACh and nucleotides or proteins remains to be demonstrated (Potter, 1972).

Little is known about the source and turnover of synaptic vesicles (see Section F,6). The presence of some vesicles of the right size in preterminal axons (Zelená, 1969) suggests that some if not all may be manufactured in the cell body and transported by axonal flow to the nerve endings in the same manner as some adrenergic and serotoninergic vesicles are transported (Dählstrom and Häggendal, 1970).

6. Quantitative Estimation of Turnover

Turnover of cytoplasmic ACh in resting nerve terminals is relatively rapid, whereas that in vesicles is slow (Chakrin and Whittaker, 1969; Marchbanks, 1969). Also, when synaptosomes are exposed to labeled

ACh and some enters their cytoplasm, the vesicular content of ACh remains predominantly unlabeled (Marchbanks, 1968). This indicates that in the majority of vesicles ACh is inert metabolically. That a type of equilibrium exists between vesicular and extravesicular compartments of ACh has been suggested (Potter, 1972). The turnover of ACh in vesicles during nerve stimulation is rapid. In fact, most of the total store becomes labeled during exposure to radioactive choline, and the specific activity of ACh in vesicles must approach that of the choline used (Collier and MacIntosh, 1969; Potter, 1970). The rate of turnover of ACh in vesicles under such conditions is presumably comparable to the rate of release (Birks and MacIntosh, 1961). Apparently a small part of the total store of ACh in nerve terminals, perhaps that contained in vesicles nearest the presynaptic membrane, turns over more rapidly than the rest (Potter, 1972). When ACh synthesis is inhibited by hemicolinium, the release of ACh declines in proportion to the decreasing store of ACh (Birks and MacIntosh, 1961; Potter, 1970).

In addition to determinations made on isolated synaptosome preparations, the turnover rate of ACh has been estimated in the brain by applying steady-state kinetics to the change, with time, in brain ACh and choline-specific radioactivity after labeling brain choline *in vivo* (Cheney *et al.*, 1975b). Brain choline was initially labeled by iv injection of radioactive choline (Schubert *et al.*, 1970). However, this method provided questionable results since the amount of choline injected was high in comparison to plasma choline concentrations and the system was not at steady-state. This procedure was disturbed by the low specific radioactivity of the commercially available redioactive choline, which made it impossible to use true tracer doses of radioactive choline. However, this obstacle was overcome by using pulse injection of radioactive phosphorylcholine to label the pools of choline since this compound is rapidly converted into choline (Hanin *et al.*, 1973). For estimating brain ACh turnover rate, it is assumed that at steady-state the concentration of brain ACh is maintained by a balance between its rate of synthesis and metabolism. By a simple kinetic approach, the turnover rate of ACh in mouse brain was calculated to be 0.36 μmole ACh synthesized per gram per hour (Cheney *et al.*, 1975b). Similar results have been obtained using another radioisotopic method (Schubert *et al.*, 1970) and a gas chromatographic–mass spectrometric approach (Jenden, 1973). Drugs, such as oxotremorine and physostigmine, which increase cholinergic activity, have been shown capable of decreasing, in mice, the brain ACh turnover rate (Schubert *et al.*, 1969; Consolo *et al.*, 1972; Trabucchi *et al.*, 1974). The reduced ACh turnover rate associated with increased ACh neurotransmission is not likely due to feedback control by end-product inhibition (see Section F,4);

it could instead be associated with the presence of interneuronal loops regulating the cholinergic function, whose activity may be triggered by cholinergic receptor activation (Trabucchi *et al.*, 1974).

7. Release

The quantal nature of transmission at chemical synapses has been known for a long time (Del Castillo and Katz, 1954). This subject has been periodically reviewed (Katz, 1966, 1969; Hebb, 1972b; Ceccarelli and Hurlbut, 1975). Several experiments at the neuromuscular junction have shown that the release of quanta varies with the degree of depolarization of the presynaptic membrane and as a function of extracellular Ca^{2+} concentration (see Hubbard, 1970). Sodium ions compete with Ca^{2+} on the outside and probably on the inside of the presynaptic membrane.

The "vesicular hypothesis" (Del Castillo and Katz, 1955; De Robertis and Bennett, 1955; Palay, 1956) generally assumes that the ACh quantum which produces one "miniature end-plate potential" comes from only one synaptic vesicle, although it is conceivable that the coordinated and simultaneous release of the content of several vesicles is necessary. This hypothesis raised the problem of the modes by which ACh molecules reach the postsynaptic receptors (see Hubbard, 1970; Potter, 1972; Ceccarelli and Hurlbut, 1975). The most accepted hypothesis, that of the exocytosis (Del Castillo and Katz, 1957), is similar to the one already described for CA's (see Section B,8). A precise line-up of vesicles along neurotubules in preterminal and terminal cholinergic axons of some lower vertebrates has been documented (Smith *et al.*, 1970). However, it is difficult to assess whether the axonal transport of vesicles can supply a sufficient number of new vesicles to match the number of quanta released. If release took place through exocytosis, the vesicle membrane would be recovered. This could occur if each vesicle fused only momentarily with the presynaptic membrane and then reformed within the terminal or if the vesicles were reformed after their total fusion with the membrane. The fact that vesicles have a higher lipid and a lower ganglioside content than other synaptosomal membranes and no Na^+-K^+-activated ATPase (Whittaker, 1969a,b,c) suggests that the vesicular and the presynaptic membranes are different. The possibility must also be considered that the presynaptic membrane could be a mosaic of pieces of vesicle-type membrane which can "flow" to sites of new vesicle formation. Some electron micrographs of nerve terminals suggest that total vesicle fusion may occur (Nickel and Potter, 1970); others of coated vesicles have been interpreted as evidence for vesicle reformation by pinocytosis (Kadota and Kanaseki,

1969), and there is morphological evidence that the number of vesicles near the presynaptic membrane changes during stimulation (Hubbard and Kwanbunbumpen, 1968; Jones and Kwanbunbumpen, 1970a,b). There are two classes of ACh receptors, one designated muscarinic and the other nicotinic. These two classes of receptors have distinct anatomic distributions and physiological functions. They can be distinguished from one another by drugs that either mimic or antagonize the various actions of ACh (see Goodman and Gilman, 1975). There is evidence that in several organs ACh increases cGMP content, and that the intensity of the cholinergic response is correlated to the magnitude of such an increase (George *et al.,* 1970; Lee *et al.,* 1972; Eichhorn *et al.,* 1974). Since the increase in cGMP is prevented by atropine, it has been proposed that this nucleotide acts as a "second messenger" of cholinergic muscarinic action in the brain as well as in other tissues (Lee *et al.,* 1972). Cyclic GMP has been shown capable of preventing the release of Ca^{2+} from mitochondria evoked by cAMP, and this suggests that the balance between the two nucleotides may be important in the control of the amount of free Ca^{2+} in the cytoplasm and, therefore, of the membrane permeability (Borle, 1974).

Cholinergic neurotransmission can be affected by several drugs interfering with ACh synthesis, release, metabolic degradation, and at receptor level. They are presented in Table XVII.

8. Electrocortical Changes

The effects of ACh, cholinergic and anticholinergic drugs on sleep and arousal, motor activity, food and water intake, emotional behavior, self-stimulation, and punished behavior, learning, and memory have been reviewed (Pradhan and Dutta, 1971; Holman *et al.,* 1975). Previous reviews on central effects of cholinergic and anticholinergic agents include those by Longo (1966), Reeves (1966), and Koelle (1970).

There is evidence that direct and indirect cholinergic agonists given systemically cause desynchronization of the EEG in the neocortex and synchronization in the hippocampus, thalamus, and midbrain reticular formation and increase arousal responses; anticholinergic drugs have the opposite effects (Rinaldi and Himwich, 1955; Domino *et al.,* 1968). ACh injected into a carotid artery induces electrocortical arousal in cat *encéphale isolé* (Bonnet and Bremer, 1937) and curarized *encéphale isolé* (Himwich and Rinaldi, 1957).

Since ACh and anti-AChe drugs are effective in activating EEG in *cerveau isolé* preparations (with transection at the midbrain level), it has been suggested that ACh acts at the level of a diffuse thalamic projection sys-

TABLE XVII

Drugs Interfering with Cholinergic Neurotransmission

Class	Drug	Observations	References to main action
Receptor stimulants	Acetylcholine	Muscarinic and nicotinic receptors	See Dale (1914); Goodman and Gilman (1975)
	Methacholine	Choline ester (longer duration of action in comparison to ACh)	
	Carbachol (carbamylcholine)	Choline ester (longer duration of action in comparison to ACh)	
	Bethanecanol		
	Muscarine	Muscarinic receptors	
	Pilocarpine	Muscarinic receptors	
	Arecoline	Muscarinic receptors	
	Oxotremorine	Nicotine-like CNS activity	Cox and Potkoniak (1969)
Cholinesterase inhibitors	Neostigmine	Reversible inhibition of AChe; quaternary amine (no BBB passage); initial increase in sympathetic and parasympathetic activity followed by ganglioplegic action	Paton (1953); Whittaker (1963)
	Physostigmine (eserine)	Reversible inhibition of AChe; tertiary amine (BBB passage)	
	Edrophonium	Reversible inhibition of AChe	
	Ambenonium	Reversible inhibition of AChe	
	Melathion	Irreversible inhibition of AChe	
	Tetraethylpyrophosphate (TEPP)	Irreversible inhibition of AChe	
	Paraoxon	Irreversible inhibition of AChe	
	DFP (diisopropylfluoro-phosphate)	Irreversible inhibition of AChe	
Ganglion stimulants	Nicotine	Initial stimulation followed by ganglioplegic action	
	DMPP (dimethylphenyl-piperazinium)		
	TMA (tetramethylammonium)		
Anticholinergic drugs	Atropine (*dl*-hyoscyamine)	Antagonism at muscarinic receptors	Bonnet and Bremer (1952); Phillis and Tebecis (1967)
	Scopolamine	Antagonism at muscarinic receptors	
	Benztropine	Antagonism at muscarinic receptors; inhibition of DA uptake in the striatum (Coyle and Snyder, 1969)	

Benapryzine (BRL 1288) (*N*-*n*-propyl-*N*-ethylamino-ethyl benzilate)	Antagonism at muscarinic receptors	Robinson (1973)
Chlorpromazine	Antagonism at muscarinic receptors	
Triethylcholine	Inhibition of choline uptake (Bull and Hemsworth, 1963)	
Hemicholinium	Inhibition of choline uptake (see Hebb, 1972a)	
Halogenated derivatives of ACh	Inhibition of ChAc	Chase and Tubbs (1966); Morris and Grewaal (1969)
Styrylpyridine analogs	Inhibition of ChAc (Smith *et al.*, 1967; Baker and Gibson, 1971)	
Botulinus toxin	Inhibition of ACh release	
Procaine	Inhibition of ACh release	
Polypeptide antibiotics		
Aminoglycosides		
Tetraethylammonium (TEA)	Ganglion blockade; monoquaternary ammonium	Acheson and Moe (1946); Furukawa (1955); Pixner (1967)
Trimethaphan	Ganglion blockade; monoquaternary ammonium	
Hexamethonium	Ganglion blockade; bisquaternary ammonium	
Azamethonium	Ganglion blockade; bisquaternary ammonium	
Chlorisondamine	Ganglion blockade; bisquaternary ammonium	
Pentolinium	Ganglion blockade; bisquaternary ammonium	
Sparteine	Ganglion blockade; bisaminic compound	
Mecamylamine	Ganglion blockade; secondary amine (BBB passage)	
Pempidine	Ganglion blockade; secondary amine (BBB passage)	
Tubocurarine	Blockade of GABA receptors (Hill *et al.*, 1972); competitive neuromuscular blockade	Eccles (1946); Chang (1953); Feldberg *et al.* (1957)
Dimethyltubocurarine	Competitive neuromuscular blockade	
Gallamine	Competitive neuromuscular blockade	Salmoiraghi and Steiner (1963); Curtis *et al.* (1966); Payton and Shand (1966)
Succinylcholine	Neuromuscular blockade by depolarization	
Decamethonium	Neuromuscular blockade by depolarization	Paton and Zaimis (1943, 1952)

tem (Bradley and Key, 1958). Electrocortical desynchronization has been obtained in rabbits after topical application of ACh (Longo, 1966) or pilocarpine (Cuculic *et al.*, 1968) on the cortex. The effects of cholinergic and anticholinergic drugs on EEG and behavior are not correlated since atropine induced an EEG synchronization which was not associated with behavioral sleep and diisopropylfluorophosphate, an anti-AChe agent (Table XVII), produced EEG activation without behavioral alertness (Bradley and Elkes, 1957; Bradley and Key, 1958). As mentioned above, reticular thalamic and thalamocortical fibers are thought to be cholinergically mediated (see Section F,2). Studies by Hernández-Peón (1962) and Hernández-Peón *et al.* (1967), in which crystals of ACh were implanted intracerebrally, indicated the presence of a dual cholinergic system extending through the diencephalon and into the mesencephalon which affects sleep and arousal. However, contradictory results have been obtained after IVT or intrahypothalamic injection of cholinergic drugs. Rage accompanied by piloerection and hissing was observed in cats receiving IVT (McLean, 1957; Borison, 1959; Leslie, 1965) or intrahypothalamic (Myers, 1964a,b) carbachol, whereas in the same animal species, stupor or a motionless state was observed after IVT carbachol (Myers, 1964a,b). Furthermore, in cats, after microinjection of carbachol into the same areas of the brain which induce sleep (Hernández-Peón *et al.*, 1967), other investigators found only "defense reactions" involving growling (MacPhail and Miller, 1968). Also, species differences in the effect of ACh or ACh-like drugs were found. Thus, cortical synchronization and behavioral sleep were observed in cats after injecting ACh directly into various areas of the brainstem reticular formation between the caudal mesencephalon and the medulla (Cordeau *et al.*, 1963), whereas in fowl, IVT or intrahypothalamic infusion of muscarine or carbachol evoked behavioral and electrocortical arousal. A similar pattern was also obtained in fowl after injection of ACh with physostigmine in the same sites (Marley and Seller, 1972, 1974).

From other studies, evidence has been provided for an involvement of the cholinergic system in REM sleep. Carbachol, injected directly into the midbrain of cats close to the cerebral aqueduct, produced REM sleep (Baxter, 1969; George *et al.*, 1964). Also after systemic injection of physostigmine, or other anti-AChe drugs, REM sleep was evoked in intact cats with transection or lesions at various levels of the brainstem (Matsuzaki, 1969; Karczmar *et al.*, 1970; Torii *et al.*, 1973) and in normal human subjects (Sitaram *et al.*, 1976). Conversely, atropine selectively suppressed REM sleep when given to normal and midpontine transected cats (Jouvet, 1962); and in humans, scopolamine decreased REM and increased non-REM sleep (Segales *et al.*, 1969). Collectively, these experiments suggested that cholinoceptive receptors responsible for REM sleep are present within the caudal portion of the pontine region.

9. Iontophoretic Application

Several excellent reviews on this topic can be found in the literature (Krnjević, 1964, 1965; Phillis, 1965; Salmoiraghi *et al.*, 1965; Salmoiraghi and Stefanis, 1967; Bradley, 1968; Curtis and Crawford, 1969; Pradhan and Dutta, 1971).

Using the iontophoretic technique, cholinoceptive neurons have been found in the hypothalamus, brainstem, cerebral cortex, limbic system, basal ganglia, cerebellum and spinal cord. In many of these regions, ACh had both excitatory and inhibitory effects on sensitive neurons; however, ACh was not universally active. In the cerebral cortex, for example, less than 30% of the cells were cholinoceptive (Krnjević and Phillis, 1963c,d); the majority of these were excited by ACh and the inhibitory effects were comparatively rare (Randic *et al.*, 1964). It has been suggested that the cortical cholinoceptive cells are probably radiations from the brainstem reticular formation (Krnjević, 1964). At the level of the brainstem, more than half the neurons in the medulla and pons of the cat were shown to be cholinoceptive (Bradley and Wolstencroft, 1965; Bradley *et al.*, 1966). In a sample of more than 600 neurons from these areas in which almost all were spontaneously active, 35.5% were found to be excited by ACh and 22% inhibited (Bradley *et al.*, 1966).

So far as hypothalamic neurons are concerned, 30% were cholinoceptive and excitatory, and inhibitory effects occurred in a roughly equal number (Bloom *et al.*, 1963). In the rostral hypothalamus, after ACh application, no changes in the discharge rate were detected in 50% of the neurons, whereas the remainder were either excited or inhibited (Dyball *et al.*, 1974). In other areas, such as the caudate nucleus and the hippocampus, most (more than 80%) of the active neurons responded to application of ACh, and excitation was the response most commonly elicited (Bloom *et al.*, 1965; Biscoe and Straughan, 1966).

G. PROSTAGLANDINS

1. Distribution and Release in the CNS

Prostaglandins (PG's), a class of C_{20} unsaturated fatty acids, have been identified as natural constituents of the brain in several species (see Horton, 1969). In the dog brain, four different PG's have been found. These are distributed widely throughout all regions of the brain without any selective accumulation, a fact suggestive of the existence of four types of neurons, each secreting a different prostaglandin (Holmes and Horton, 1967). In the brain, PG's are associated with but not confined to the

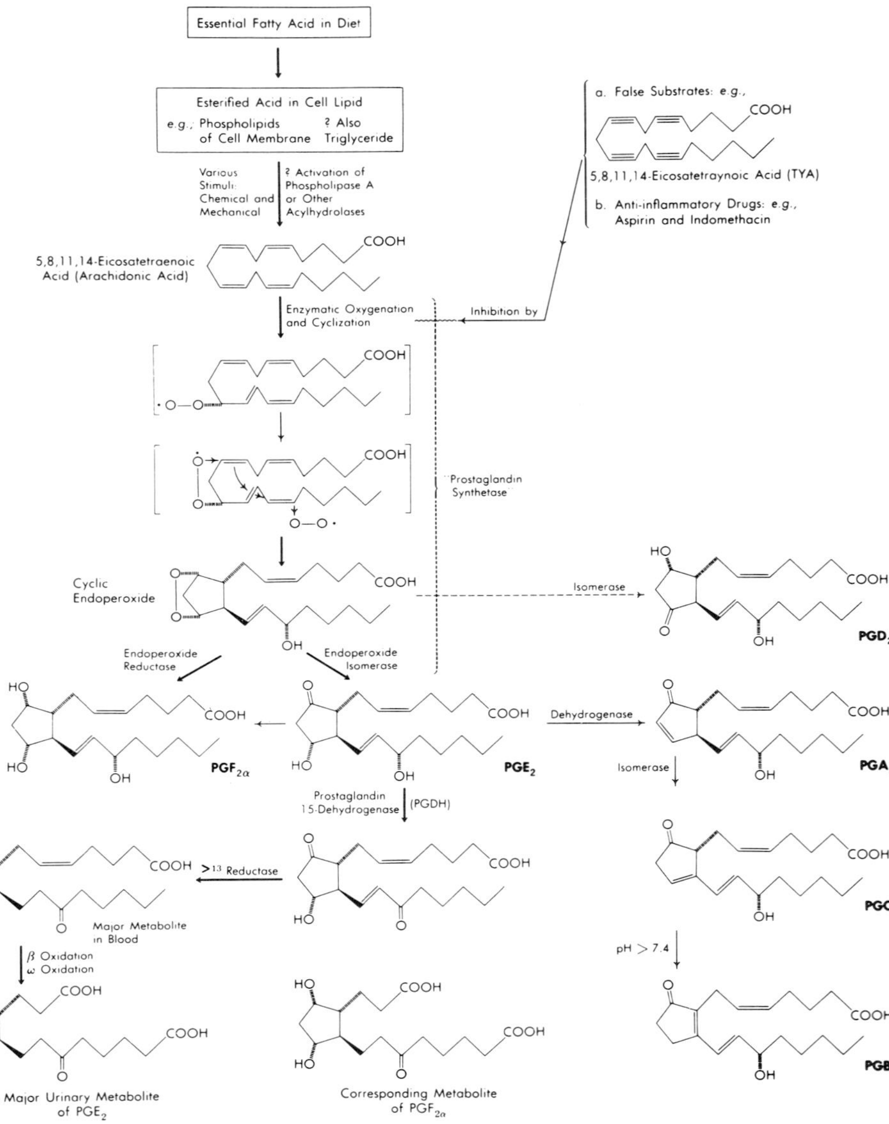

Fig. 29 Synthesis and catabolism of the two principal mammalian prostaglandins, PGE_2 and $PGF_{2\alpha}$. The inhibitors shown act at early stages and prevent formation of the endoperoxide. Enzymatic reduction of PGE to PGF occurs in several tissues. The sequence of breakdown is provided only for PGE_2; $PGF_{2\alpha}$ undergoes an analogous series of events. Some additional metabolites are not shown. (Reproduced, with permission from Goodman and Gilman, 1975.)

nerve-ending fraction (Kataoka *et al.*, 1967), and it is not certain whether in nerve terminals they are present in synaptic vesicles (Wolfe *et al.*, 1967).

PG's are released from the surface of the brain and spinal cord, and are normal constituents of the CSF (see Horton, 1969). Afferent nervous stimulation increases the output of PG's from the superfused spinal cord of the frog and from the somatosensory cortex of the anesthetized cat (Ramwell and Shaw, 1966; Ramwell *et al.*, 1966). Prostaglandins are also released spontaneously from the cerebellar cortex of the cat (Coceani and Wolfe, 1965) and the ventricular system of cats and dogs (Feldberg and Myers, 1966; Holmes, 1970). In dogs, the output from the cerebral ventricles is increased by perfusing 5-HT through the ventricles but is unaffected by CA's, amphetamine, tranylcypromine, or CPZ (see Horton, 1969).

2. Biosynthesis and Metabolic Degradation

Arachidonic acid (α-11-*cis*-5,8,11,14-eicosatetraenoic acid) is converted to PGE_2 or $PGF_{2\alpha}$, whereas PGE_1 and $PGF_{1\alpha}$ are formed from 8,11,14-eicosatrienoic acid (see Horton, 1972; Samuelsson, 1972; Sih and Takeguchi, 1973; Flower, 1974). Figure 29 illustrates the synthetic and catabolic pathways of PGE_2 and $PGF_{2\alpha}$. The multienzyme system which catalyzes the transformation of unsaturated fatty acids to prostaglandins has been named prostaglandin synthetase. This enzyme activity is found in high-speed particulate fractions. The addition of glutathione greatly enhances the formation of PGE's in part at the expenses of PGF's.

The number of enzymes taking part in the synthesis is not known; they appear to be present in every mammalian tissue investigated to date. The initial step in PG biosynthesis is a dioxygenase reaction which leads to a formation of a substrate called (ω-10) hydroperoxide. In the next step ring closure occurs and a cyclic endoperoxide is produced. This intermediate, which is most likely unstable, may be modified by "isomerization" to PGE or PGD or by reductive cleavage to PGF (see Flower, 1974).

PGA and its isomer PGB are dehydration products of PGE. It seems likely that much of the PGA detected in tissue has been formed nonenzymatically during extraction and isolation procedures, although PGA does appear to occur naturally in the seminal fluid. Only the nonesterified acids are substrates for the enzyme, and since the cellular concentration of free fatty acid precursors is generally rather low, it follows that the activity of lipolytic enzymes such as phospholipase A may be important in regulating the supply of substrate to the enzyme (Fig. 30).

Most PG's are rapidly inactivated *in vivo*. Four enzymatic reactions involved in the breakdown of PG's have been found: (1) prostaglandin dehydrogenase (15-hydroxy-PG-dehydrogenase, PGDH) which leads to

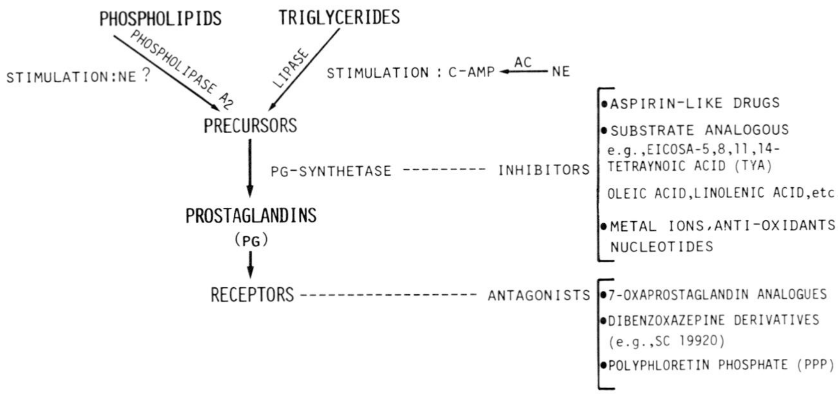

Fig. 30 Schematic illustration of some mechanisms by which several drugs may interfere with prostaglandins (for details see text).

the formation of the 15-dehydro derivative, (2) prostaglandin reductase which catalyzes the formation of a compound saturated at the C-13 to C-14 double bond, and (3) β- or (4) ω-oxidation. The most significant of these appears to be the dehydrogenase reaction; this NAD-dependent enzymatic activity is found in the high-speed cell supernatant. PGE_1 and PGE_2 are the best substrates; PGE_3, PGF, and PGA are less readily oxidized. A histochemical method enabling the observation of PGDH in CNS has recently been devised and has been found useful in studying the precise distribution pattern of PG's in certain areas of the brain (F. E. Bloom, personal communication).

In order to investigate the function of PG's in the CNS, several pharmacological manipulations are possible (Fig. 30). They include administration of drugs (1) able to inhibit PG synthesis, e.g., aspirin-like drugs (Vane, 1971), paracetamol, substrate analogs, or 7-oxaprostaglandin analogs, such as 5,8,11,14-tetraynoic acid (TYA) (see Flower, 1974), (2) able to block PG receptors, e.g., polyphloretin phosphate (PPP) or dibenzoxazepine derivatives (SC-19920) (Eakins and Sanner, 1972); or, (3) able to inhibit PG-dehydrogenase, e.g., the substrate analogs *rac*-prostanoic acid 15-epimers of PGE_1 or several nucleosides and nucleotides, methylxanthines, sulfhydryl inhibitors (see Flower, 1974).

3. Electrical Changes

Direct evidence that PG's act on central neurons has been obtained by microiontophoretic application to cells of the brainstem in rats (Avanzino *et al.*, 1966). The spontaneous firing of a proportion of these neurons was increased by PGE_1, PGE_2, and $PGF_{2\alpha}$; a smaller proportion was inhibited by PGE_1 and $PGF_{2\alpha}$ but not by PGE_2. Sedative effects have been obtained after iv or IVT injection of PGE's in several species (Horton, 1964; Horton and Main, 1965a,b; Nisticò and Marley, 1973, 1976). The site from which behavioral and electrocortical sedation or sleep are elicited appears to be the hypothalamus (Nisticò and Marley, 1973). In various animal species an increase in synthesis and/or release of PG's of the E series within the anterior hypothalamus seems to be the mechanism which stimulates pyrogen fever (Milton and Wendlandt, 1970, 1971a,b; Feldberg and Saxena, 1971a,b; Nisticò and Marley, 1973, 1976).

The role of prostaglandins in the regulation of peripheral adrenergic activity has been previously discussed (see Section B,6); evidence of a similar role in the CNS is still lacking. A clue for a physiological role played by PG's in the regulation of certain effects of NE has been provided by microiontophoretic studies which have shown that topical application of PGE_1 and PGE_2 antagonizes the NE reduction in the discharge rate of cerebellar Purkinje cells (Hoffer *et al.*, 1969). On the other hand, CA's directly or through activation of AC can regulate PG synthesis (Fig. 30).

H. HISTAMINE

1. Regional Distribution in the CNS

Although several biochemical studies indicate that histamine (H) in the CNS may have a transmitter function, neurophysiological data in support of this concept are lacking (White, 1966; Snyder and Taylor, 1972).

Biological, chemical, fluorimetric, and enzymatic–isotopic procedures have been used for estimation of histamine in brain tissue (Table II). In earlier studies (Harris *et al.*, 1952) it was shown by a BA method that in the rat histamine is concentrated chiefly in the hypothalamus, a finding confirmed later (Adam, 1961; Green, 1970). Recently, the regional localization of histamine in the brain and also in individual nuclei of the hypothalamus has been studied in the monkey (Taylor *et al.*, 1972) and in the rat (Brownstein *et al.*, 1974a) by using a very sensitive enzymatic–isotopic assay. In the monkey hypothalamus, the highest concentrations were found in the mammillary bodies, SON, VMN, the ventrolateral nucleus, and the ME. High levels of histamine occurred also in the pineal body, in the raphe nuclei, and in the zona compacta of the substantia nigra. A similar pattern of distribution has been also reported for H in the human brain; the highest levels were found in the hypothalamus, particularly in the mammillary bodies and mid-hypothalamus (Lipinsky *et al.*, 1973).

In the rat, highest concentrations of histamine were found in the ME (Table XVIII). One cannot ignore that at this level, part of it may be located extraneuronally, e.g., in the mast cells (Green, 1970). In general, brain levels of H are about one-tenth the concentration of CA's and 5-HT. Since no mast cells are present in mammalian brain, it is unlikely that H within the brain is present in mast cells except in the ME and pineal body (see Snyder and Taylor, 1972).

A major portion of brain histamine is localized in nerve terminals as shown by subcellular fractionation studies in which brain histamine has been reported to be present in the synaptosomal fraction along with histamine-*N*-methyltransferase (see Snyder and Taylor, 1972). However, until recently, no histaminergic nerve tracts were observed, and attempts to demonstrate the presence of a specific uptake mechanism for terminating the effect of histamine have been unsuccessful (Snyder and Taylor, 1972). Recently, lesion studies indicate that histamine-containing neurons are organized to form an ascending bundle arising from the brainstem, running ipsilaterally with CA and 5-HT neurons in the MFB and projecting diffusely into the whole telencephalon, especially in the cerebral cortex and the hippocampus (Schwartz, 1975). After deafferentation of the MBH

TABLE XVIII

Histamine Content of Hypothalamic Nuclei[a]

Area	Histamine content (ng/mg protein)
Nucleus suprachiasmatis	6.3
Nucleus periventricularis	3.7
Nucleus supraopticus	3.0
Nucleus paraventricularis	2.4
Nucleus hypothalamicus anterior	3.3
Nucleus lateralis anterior	2.7
Nucleus arcuatus	6.2
Nucleus ventromedialis	3.6
Nucleus dorsomedialis	4.0
Nucleus perifornicalis	2.2
Nucleus hypothalamicus posterior	4.6
Nucleus premammillaris dorsalis	3.9
Nucleus premammillaris ventralis	8.1
Corpus mammillare	4.3
Median eminence	17.8

[a] From Brownstein *et al.* (1974a).

in the rat, the level of H did not change significantly from the control value in the ARC n., VMN, DMN, and ME (Brownstein *et al.*, 1976), which suggests that histamine is present in the posterior two-thirds of the hypothalamus in cells which are intrinsic to this area. These findings, however, contradict the earlier data of Dismukes *et al.* (1974), who found that after unilateral isolation of the hypothalamus histamine increases in the hypothalamus on the lesioned side.

Histamine has also been detected in the mammalian pituitary (Adam and Hey, 1966); part of it is mast-cell bound, but there is evidence for the cat pars distalis that histamine is localized in other cells than mast cells. Histamine levels are higher than any other amine in the rat pituitary, and its concentration in the posterior and intermediate lobes is about five times higher than in the anterior lobe (see Table VI).

A-24 hour rhythm for histamine with maximum concentrations during the dark phase of the illumination cycle has been shown in rat midbrain and caudate nucleus (Friedman and Walker, 1968, 1969), whereas in the hypothalamus the maximum concentrations occurred during the early light period and the smallest during the early dark period (Orr and Quay, 1975).

2. Biosynthesis and Metabolic Degradation

Histamine is synthesized from histidine (Fig. 31) by a reaction catalyzed by L-AAAD or by a specific histidine decarboxylase (see Waton, 1956). In

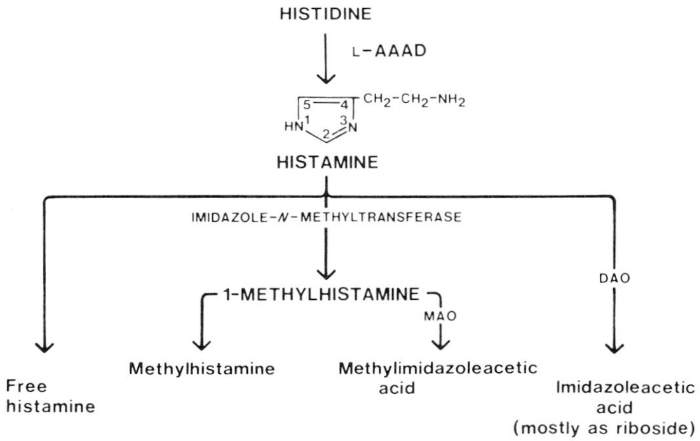

Fig. 31 Enzymatic steps involved in the biosynthesis and metabolism of histamine and common urinary metabolites. DAO, diamine oxidase (for details see text). (Modified from Meyers *et al.*, 1974.)

TABLE XIX

Drugs Interfering with Histamine by Various Mechanisms

Class	Drug	Observations	References to main action
Releasing agents	48/80 (condensation product of formaldehyde + *p*-methoxy-*N*-methylphenylethylamine)		See Goth (1973); see Douglas (1975)
	Polypetides (e.g., bradykinin, kallidin, substance P) and proteins		
	Venoms and toxins		
Synthesis inhibitors (histidine decarboxylase)	Hydroxylamines		Leinweber (1968)
	Hydrazines		Reid and Shepherd (1963)
	Hydrazinoacids		Reid and Shepherd (1963)
	α-Methylhistidine	Formation of α-methylhistamine (inactive)	Kahlson *et al.* (1963)
H_1 receptor antagonists	Diphenhydramine	(All possess local anesthetic activity)	See Schacter (1973); see Douglas (1975)
	Tripelennamine		
	Pyrilamine (mepyramine)		
	Chlorpheniramine		
	Chlorcyclizine		
	Promethazine	Anticholinergic activity	
	Cyproheptadine	Antiserotoninergic and anticholinergic activity	

	Methapyrilene	
H_2 receptor antagonists	Burimamide {*N*-methyl-*N*′-4-[4(5)-imidazolyl]butylthiourea}	Black *et al.* (1972)
	Cimetidine (*N*-cyano-*N*′-methyl-*N*″[2-(5-methyl-4-imidazolylmethylthio)ethyl]-guanidine (SKF 92334)	Brimblecombe *et al.* (1975)
	Metiamide [*N*-methyl-*N*′-2(-5-methyl-imidazole-4-yl)methylthio]ethylthiourea	Black *et al.* (1973)
Histamine-*N*-methyl-transferase inhibitors	Tryptamine	See Creveling and Daly (1971)
	Imidazoles	
	Chlorpromazine	
	Quinacrine	
	Bufotenin	
	Serotonin	
DAO or histaminase inhibitors	Iproniazid	Holmstead and Tham (1952); see Creveling and Daly (1971)
	Semicarbazide	
	Aminoguanidine	

the brain, the action of the latter enzyme, for which specific inhibitors such as α-hydrazine histidine have been discovered, seems more important. The neuronal localization of L-histidine decarboxylase in rat cortex is suggested by its development pattern during ontogeny (Martres *et al.*, 1975) and demonstrated by subcellular fractionation studies which have revealed that a major fraction of the enzyme activity is associated with the isolated nerve ending (Baudry *et al.*, 1973). Substantial synthesis of histamine occurred in the hypothalamus (White, 1959, 1960) with maximum formation about 1 hour after IVT administration of ^{3}H-histidine.

At least a portion (about 40%) of brain histamine is stored in a dynamic state and turns over extremely rapidly ($t_{\frac{1}{2}}$ about 2 minutes). The rapid turnover rate is similar to that of ACh and is more rapid than that of brain NE and 5-HT, the $t_{\frac{1}{2}}$ of which are several hours (see Snyder and Taylor, 1972). Stressful situations such as cold and restraint accelerate histamine turnover and partially deplete its levels.

Histamine metabolism can follow two pathways: the first, catalyzed by imidazole-*N*-methyltransferase, leads to the formation of methylhistamine which is broken down by MAO and then by aldehyde dehydrogenase to form methylimidazoleacetic acid; the second pathway is catalyzed by histaminase or DAO (diamine oxidase) and then by aldehyde dehydrogenase, the terminal produce being imidazoleacetic acid (Fig. 31). A list of drugs interfering with the histamine mechanisms of neurotransmission is presented in Table XIX.

3. Central Effects

That the amine is unable to cross the BBB must be taken into consideration when one interprets the results attributed to central effects of histamine after its systemic administration (Brandon, 1955; Bovet *et al.*, 1958). Intraventricular injection of histamine in cats produced intense sedation accompanied by tachypnea and salivation (Feldberg and Sherwood, 1954), and similar effects have been also obtained in mice (Haley, 1957). However, in the latter, IVT administration of 48/80, a histamine releaser, induced excitation which was not antagonized by mepyramine, a receptor antagonist (Rocha-Silva, 1959). An involvement of H in sleep-arousal control has been recently suggested (Schwartz, 1975). Indeed, several experimental data, in addition to the anatomic distribution of histaminergic neurons, such as the well-known sedative properties of most antihistamines (the H_1-receptor antagonists) or the decreased histamine turnover in brain of barbiturate-treated rats (Pollard *et al.*, 1974), indirectly support this hypothesis.

Histamine given IVT in cats increased blood pressure; such increase was prevented by mepyramine (Trendelenburg, 1957; White, 1965; Sinha *et al.,* 1969). The hypertensive effect and heart rate increase were confirmed, later, in rats after intrahypothalamic (Brezenoff and Jenden, 1969) or IVT (Finch and Hicks, 1975) injections; these effects seem to be mediated by H_1 receptors (Finch and Hicks, 1975). In rats, parenteral administration of antihistamine drugs decreased water consumption, an effect which was antagonized by IVT histamine (Gerald and Maickel, 1972); accordingly, a role of histamine in the hypothalamic control of thirst has been postulated (Leibowitz, 1973). An involvement in the thermoregulatory mechanisms has been suggested by experiments in which hypothermia followed microinjections of histamine or its precursors directly into the rostral hypothalamus of the rat (Green *et al.,* 1975). Such an effect, which has also been reported in the cat (Clark and Cumby, 1975), can be blocked by systemic administration of H_1-receptor antagonists (Green *et al.,* 1976).

Small amounts of histamine (1–4 μg) inhibited trans-callosally evoked potentials in the cat cortex, suggesting an inhibitory role (Gilfoil *et al.,* 1960). After microiontophoretic application of histamine, no action or weak depression was found in the spinal cord (Curtis *et al.,* 1961; Phillis *et al.,* 1968), medullary reticular formation (Bradley, 1968; Haas *et al.,* 1973), or lateral geniculate body of the cat (Curtis and Davis, 1962). Applied iontophoretically to neurons of the lateral geniculate body and spinal cord, histamine had little or no activity; however, on cortical neurons it had an inhibitory effect, although less marked than that of GABA (Cooper *et al.,* 1974). Recently, it has been found that in the rat hypothalamus locally applied histamine induced excitation in the majority of neurons (Haas, 1974). The similar electrophysiological responses of certain neurons to histamine and cAMP is suggestive of the involvement of the nucleotide as a second messenger in histaminergic neurotransmission. Histamine is in fact one of the most potent stimulants of cAMP synthesis, an action which appears to be mediated both by H_1 and H_2 receptors (Baudry *et al.,* 1975).

I. γ-AMINOBUTYRIC ACID

1. Regional Distribution in the CNS

In mammals, γ-aminobutyric acid (GABA) has been detected in the brain and spinal cord but not in the peripheral nervous system (Cooper *et al.,* 1974). Like the monoamines, GABA is unevenly distributed in the

CNS where the highest concentrations were found in the corpora quadrigemina, the substantia nigra, the hypothalamus, and the globus pallidus (Cooper *et al.,* 1974; Fahn and Coté, 1968; De Feudis *et al.,* 1970). Lower concentrations of GABA are present in the cerebral hemispheres, the pons, and the medulla. In various areas of the brain, the concentrations of GABA correlate well with those of glutamate decarboxylase (GAD-I), the enzyme responsible for its synthesis from L-glutamate. No correlation exists between the concentration of GABA and that of the degradative enzyme GABA transaminase.

There is ample experimental evidence to show that GABA is an inhibitory transmitter of synapses in the crustacean and mammalian nervous systems (see Iversen, 1972b; Meldrum, 1975). Several data favor a transmitter role for GABA at numerous CNS sites including (1) the demonstration of high GABA content and GAD activity in synaptosome preparations; (2) evidence for the selective release of GABA after activation of inhibitory systems in the cerebral cortex or cerebellum; (3) evidence from intracellular recordings that the changes in membrane potential and in specific ion conductances produced by physiological inhibitory inputs are similar to those produced by iontophoretically applied GABA; (4) evidence that picrotoxin and bicuculline can block both natural inhibition and inhibition produced by iontophoretic application of GABA; (5) presence of a high affinity uptake system selective for GABA.

The concentrations of GABA are comparatively higher in regions such as the cerebellar cortex in which strychnine-insensitive and picrotoxin–bicuculline-sensitive inhibition predominates and lower in areas such as the spinal cord in which inhibition is mainly strychnine-sensitive (Baxter, 1970; Fahn and Coté, 1968; Shank and Aprison, 1970). More detailed studies on the distribution of GABA in the layers of the cerebellar cortex (Kuriyama *et al.,* 1966) have shown that GABA is present in highest concentrations in the Purkinje cell layer. Similar studies in retina indicate that GABA may be associated with inhibitory interneurons (Kuriyama *et al.,* 1968) and with inhibitory neurons in the hippocampus (Fonnum and Storm-Mathison, quoted by Iversen, 1972). Fibers originating in the striatum and projecting terminals within substantia nigra, where the presence of GABA has been recently confirmed by mass-fragmentographic assay (Cattabeni *et al.,* 1975), have been shown to contain GABA as transmitter (Kim *et al.,* 1971a; Fonnum *et al.,* 1974) and to be inhibitory in nature (Precht and Yoshida, 1971 (see also Section K). In a recent study, the highest GABA content within the rat hypothalamus has been found in the LHA, and fluctuations in GABA levels have been observed in both the LHA and the VMN during hypo- and hyperglycemic conditions (Kimura and Kuriyama, 1975). This suggests that GABA may play a role in controlling the functional state of feeding and satiety centers.

2. Biosynthesis, Release, and Metabolic Degradation

GABA is formed in the nervous system by the decarboxylation of L-glutamate, catalyzed by the enzyme GAD-I (see Iversen, 1972; Meldrum, 1975). Both invertebrate and mammalian GAD-I require pyridoxal phosphate as cofactor and are inhibited by several pyridoxal antagonists such as hydrazine compounds (INI, thiosemicarbazide, MAO, etc.) and aminooxyacetic acid. In brain homogenates, 75–80% of GAD-I is associated with synaptosomes, the remaining enzyme is recovered in the supernatant fraction. A second glutamate decarboxylase (GAD-II) has been found, in contrast to GAD-I, in various nonneural tissues (kidney, blood vessels, glial cells). GAD-II is distinguishable from GAD-I because it is not inhibited by chloride ions and is stimulated rather than inhibited by aminooxyacetic acid and the carbonyl-trapping agents. It seems unlikely that GAD-II participates in the biosynthesis of GABA in the mammalian inhibitory nerve terminals. How GABA synthesis is regulated in mammalian CNS is not yet established; by extrapolating information from arthropod neurons it would appear that GABA concentration is dependent on the ratio between GAD-I and GABA transaminase (GABA-T) activity. In lobsters, but not in mammalian CNS, an end-product negative-feedback mechanism occurs since high concentrations of GABA are able to inhibit GAD. In mammalian CNS, the marked inhibitory effect of chloride ion concentration is probably involved in the control of GABA synthesis. It has been suggested that GABA may act as a repressor for GAD synthesis since high concentrations of GABA inhibit the formation of GAD in developing chick and mouse brain (see Iversen, 1972). As noted above, the subcellular location of GAD-I is in the synaptosome fraction, which indicates that GABA is synthesized mainly in the brain within the nerve endings where GAD-I appears to be free within the axoplasm.

The catabolism of GABA involves a pyridoxal-dependent transamination catalyzed by GABA-T. The deaminated product formed, succinic semialdehyde (SSA), is then oxidized to succinate by a specific NAD-dependent dehydrogenase (SSA-dehydrogenase). Subcellular studies indicate that GABA-T and SSA-dehydrogenase are mitochondrial enzymes. A variety of minor alternative pathways for GABA metabolism have been postulated (see Iversen, 1972b).

Several carbonyl reagents have been found capable of inhibiting GABA-T and thus increasing brain GABA concentration (Table XX). They include hydroxylamine, hydrazines, cycloserine, aminooxyacetic acid; some of these compounds inhibit GAD-I *in vitro,* but *in vivo* they inhibit, preferentially, GABA-T, possibly due to the fact that pyridoxal-inhibitor complexes are bound with higher affinity by GABA-T than by GAD-I (see Table XX).

Little information is available on the possible mechanism(s) involved in the storage of GABA within the nerve terminals before its release. Most of the GABA contained in the synaptosome fraction is released into the supernatant, but it has also been reported that small amounts of GABA are associated with synaptic vesicle fractions. Moreover, EM studies have shown that inhibitory nerve terminals in the crustacean and mammalian nervous system contain a large number of synaptic vesicles which could well represent the major storage site for GABA. As suggested by Iversen (1972), it is not necessary to provide for GABA vesicle storage mechanisms since it is possible that GABA is stored in synaptic vesicles at storage sites which have only a low affinity for the amino acid, so that the vesicle-bound GABA is in equilibrium with a large fraction of free axoplasmic amino acid.

As for the other neurotransmitters, GABA is released spontaneously or upon stimulation from the nerve terminals in which it is stored. An increased release of GABA has been demonstrated in response to stimulation of inhibitory synaptic transmission in crustacean and mammalian systems. As for NE and ACh, the release of GABA stimulated by depolarization of nerve impulse is Ca^{2+}-dependent, whereas the spontaneous release of the transmitter during rest is not (Rubin, 1970). In the mammalian CNS, release of GABA has been shown from the perfused fourth ventricle of the cat in response to stimulation of Purkinje fibers (Obata and Takeda, 1969). A similar release from the surface of the cerebral cortex of the cat occurs in response to electrical stimulation of either the surface of the cortex or the lateral geniculate nucleus (Iversen *et al.*, 1971a).

GABA release from the cerebral cortex was shown to be highest under conditions in which inhibitory synaptic activity is greatest, viz., low-frequency, large amplitude EEG activity (Jasper *et al.*, 1965; Crowshaw *et al.*, 1967; Jasper and Koyama, 1969). For NE terminals, the presence of α-presynaptic receptors regulating the release of NE by autoinhibition has been postulated (see also Section B,6). For GABA release, the presence of presynaptic receptors has been hypothesized on the basis of an increase in the release of ^{3}H-GABA from stimulated brain slices in the presence of the GABA receptor antagonist bicuculline (Johnston and Mitchell, 1971).

The putative neurotransmitter amino acids, GABA, glycine, glutamate, taurine, and aspartate, all have specialized high-affinity uptake systems in mammalian brain and spinal cord (Iversen, 1972). In this respect they are unique among amino acids, all of which are transported by a relatively unspecific low-affinity transport system. The existence of a specific uptake mechanism for these amino acids suggests also that this could be important in terminating the action of these compounds after their release. In slices and homogenates of rat brain, ^{3}H-GABA is taken up only by certain nerve terminals, probably those which normally store GABA as a transmitter. In rat cerebral cortex, among the GABA analogs, only 3-hydroxy-GABA, 2-fluoro-GABA, L-2,4-diaminobutyric acid, and 2-guanidinopropionic acid were shown to be competitive inhibitors of GABA uptake. The most potent inhibitors proved to be *p*-chloromercuriphenylsulfonate, CPZ, and imipramine. None of these are, however, specific in their actions on GABA uptake.

Figure 32 depicts schematically the main events involved in GABAergic neurotransmission, and Table XX lists some drugs interfering with GABA synthesis and metabolic degradation at receptor level.

3. Electrical Changes

The classic studies of inhibitory neuromuscular transmission in crustaceans by Fatt and Katz (1953) and later by Dudel and Kuffler (1961) established some fundamental principles for understanding the neurophysiology of the inhibitory synaptic transmission. The inhibitory transmitter may either stabilize the resting membrane potential of the postsynaptic cell or be released on presynaptic excitatory nerve terminals

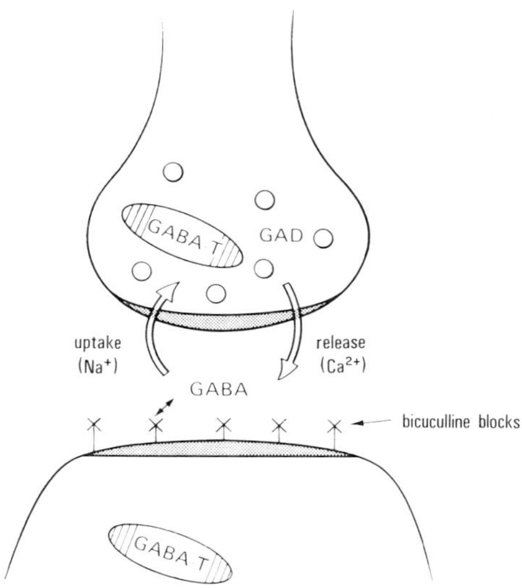

Fig. 32 Synaptic steps involved in GABAergic neurotransmission. After being synthesized from glutamic acid through GAD, GABA is stored in granules from which is released and acts on postsynaptic receptors (blocked by bicuculline). Termination of GABA effects are due to enzymatic degradation by GABA-transaminase (GABA-T) and physical recapture (uptake) into the nerve ending (for details see text). (Reproduced, with permission, from Iversen, 1972.)

TABLE XX

Drugs Interfering with GABAergic Neurotransmission[a]

Class	Drug	Observations	References to main action
Glutamate decarboxylase inhibitors	Hydrazides	Increase in cGMP content in the striatum and cerebellum by isoniazid (Mao *et al.*, 1974a; Biggio and Guidotti, 1976)	Killam and Bain (1957)
	Cycloserine		Roberts *et al.* (1964)
	Hydroxylamine		Roberts *et al.* (1964)
	Aminooxyacetic acid		Dengler (1962); Roberts *et al.* (1964)
	Aminooxypropionic acid		
	Penicillamine		Jaffe (1972)
	Pyridoxine analogs (Toxopyrimidine, 4-deoxypyridoxine, 3-methoxypyridoxine)		Holtz and Palm (1964); Horton and Meldrum (1973)
	3-Mercaptopropionic acid		Lamar (1970); Horton and Meldrum (1973)
	4-Mercaptobutyric acid		
	Hydrazinopropionic acid	More powerful inhibition of GABA-T (Van Gelder, 1968)	
	Allylglycine (2-amino-4-pentenoic acid)		McFarland and Weiner (1965); Horton and Meldrum (1973)
	Estrogens	Exacerbation of epilepsy (Logothetis *et al.*, 1959); epilepsy after topical application on cortex (Marcus *et al.*, 1966; Hardy, 1970)	Baxter (1970)
GABA transaminase inhibitors	Hydroxylamine	More marked depression of GABA-T than GAD	Baxter and Roberts (1961)
	Aminooxyacetic acid	More marked depression of GABA-T than GAD (Perry *et al.*, 1974)	Wallach (1961); Da Vanzo *et al.* (1961)

	Cycloserine	More marked depression of GABA-T than GAD	Dann and Carter (1964); Lehmann (1964)
	Di-*n*-propylacetate		Patry and Naquet (1971)
	Hydrazinopropionic acid		Van Gelder (1968, 1969)
	Ethanolamine-*O*-sulfate		Fowler and John (1972)
	5-Ethyl-5-phenyl-2-pyrrolidinone (EEP)	Little activity *in vivo* (converted to α-ethyl-α-phenyl-GABA)	Carvajal *et al.* (1964)
GABA analogs or GABA-mimetic compounds	2-Pyrrolidinone	Easier passage to BBB than GABA	Hawkins and Saret (1957)
	2-Pyrrolidinone acetamide (Piracetam)		Moyersoons *et al.* (1969)
	1-Hydroxy-3-aminoderivative of 2-pyrrolidinone (HA 966)		Bonta *et al.* (1969, 1971)
	Pyrazolidone derivatives		Motovilov and Kozhevnikov (1968)
	Muscimol (3-hydroxy-5-aminomethylisoxazole)	GABA-like activity	Johnston *et al.* (1968b); Curtis *et al.* (1970a)
	Cycloserine	Inhibition of GABA-T and GABA uptake	
	β-Phenyl-GABA	Facilitated passage across the BBB	Khaunina (1968)
	β-(*p*-Chlorophenyl)-GABA (Baclofen)	Facilitated passage across the BBB; action not antagonized by bicuculline (Curtis *et al.*, 1974)	Keberle and Faigle (1972)
	γ-Amino-β-hydroxybutyric		Hayashi (1959); Crawford and Curtis (1964)
	2-Hydroxy,2-chloro, 2- and 4-methyl-GABA		Beart and Johnston (1973); Buu and Van Gelder (1974)
GABA uptake inhibitors	L-2,4-Diaminobutyric acid (L-DABA)	Convulsive action	Beart and Johnston (1973); Harris *et al.* (1973)
	Nipecotic acid (piperidine-3-carboxylic acid)		Krogsgaard-Larsen and Johnston (1975)
	Muscimol		Krogsgaard-Larsen and Johnston (1975)

TABLE XX *(Continued)*

Class	Drug	Observations	References to main action
	Piperidine-4-carboxylic acid		Beart *et al.* (1972a,b)
	γ-Aminocrotonic acid		Beart *et al.* (1972a,b)
	4,5,6,7-Tetrahydroisoxazolo-[4,5,*c*]pyridin-3-ol	More potent inhibitor of GABA uptake than muscimol	Krogsgaard-Larsen and Johnston (1975)
	Guvacine (1,2,5,6-tetrahydro-pyridine-3-carboxylic acid)	As potent as nipecotic acid	Johnston *et al.* (1975)
	Arecaidine [*N*-methyl-(1,2,5,6-tetrahydropyridine-3-carboxylic acid)]		Johnston *et al.* (1975)
GABA receptor antagonists	Bicuculline	Active at postsynaptic receptors	Curtis *et al.* (1971a,b); Straughan *et al.* (1971); Hill and Simmonds (1973)
	Picrotoxin	Active at presynaptic receptors; decrease in cGMP of cerebellum (Mao *et al.*, 1974b)	Ito *et al.* (1968); Obata and Highstein (1970); Galindo (1969)
	Benzylpenicillin		Curtis *et al.* (1972); Hill *et al.* (1973)
	Tubocurarine		Hill *et al.* (1973)

[a] See Iversen (1972); Meldrum (1975).

to reduce the quanta of excitatory transmitter released (postsynaptic and presynaptic inhibition). Studies of GABA on the membrane potential of crayfish muscle cells showed that the actions of GABA mimic those of the naturally occurring inhibitory neuromuscular transmitters. Both compounds act by causing a selective increase in the permeability of the muscle membrane to chloride ions, thus tending to stabilize the membrane potential (see Iversen, 1972).

GABA seems likely to be an important transmitter in mammalian CNS areas, e.g., cerebellum and cerebral cortex (Krnjević, 1970; Hebb, 1970; Iversen, 1972). Microiontophoretic application of GABA causes hyperpolarization of Deiters' neurons (Obata *et al.,* 1967) and hyperpolarization in the cerebral cortex (Krnjevic and Schwartz, 1966) and in rat caudate neurons (Bernardi *et al.,* 1975). Inhibitory synaptic transmission in the cerebral cortex is strychnine-insensitive (Krnjevic and Schwartz, 1966) and, by intracellular recording techniques, it has been shown that the actions of the naturally occurring cortical inhibitory transmitter are closely similar to those produced by electrophoretically applied GABA (Krnjević and Schwartz, 1967). The alkaloid bicuculline acts as a specific antagonist at the action sites of GABA in mammalian brain (Curtis *et al.,* 1970a). Bicuculline antagonizes the naturally occurring inhibitory transmitter and GABA applied to strychnine-insensitive neurons in the cerebral cortex, cerebellum, and hippocampus (Curtis *et al.,* 1970b). A Na^+-independent binding of GABA to synaptic membranes of mammalian brain has been identified; it appears to represent an interaction with GABA-synaptic receptors (Enna *et al.,* 1975). Application of GABA on dopaminergic and nondopaminergic neurons of the substantia nigra produces inhibition, whereas glutamate produces excitation (Aghajanian and Bunney, 1973). Similar responses to glutamate and GABA were also observed in rat identified VMN neurons (Renaud, 1975) (see also Section B,14).

J. GLYCINE, TAURINE, GLUTAMATE, ASPARTATE

Although some amino acids in the brain have been investigated for their involvement in intermediary metabolism, more recent research has pointed out their possible role as synaptic transmitters (Curtis and Johnston, 1974b; Johnston, 1975; De Feudis, 1975). In addition to GABA, transmitters with synaptic inhibitory roles include glycine and taurine, whereas excitant transmitters include glutamate and aspartate.

Glycine and taurine are neutral whereas glutamic and aspartic acid are acidic amino acids. Glycine was found in high concentrations in the ventral gray matter of the spinal cord, a fact which was suggestive of a transmitter role at this level (Aprison and Werman, 1965). Taurine is present throughout the CNS, where the highest concentration is found in the lateral geniculate body (Guidotti *et al.,* 1972), and also in other tissues such as the heart (Awapara, 1956; Gaitonde, 1970). Glycine is present in higher concentration than in other brain areas in the cervical and lumbar enlargements of the spinal cord and in the medulla (Aprison *et al.,* 1969). Elevated concentrations of glycine and low levels of taurine and glutamate were reported to be present in epileptogenic human cerebral cortex (Van Gelder *et al.,* 1972) and in the cortex of cats made epileptic by topical application of cobalt (Koyama, 1972).

Glutamate is the most abundant amino acid in the CNS, with highest concentrations in the cerebral cortex and caudate nucleus (Johnson and Aprison, 1971). Glutamate seems to be associated with primary afferent fibers (Graham *et al.,* 1967; Roberts, 1973), with the hippocampal mossy fibers (Crawford and Connor, 1973), and with granule cells in the cerebellum (Young *et al.,* 1974). Aspartate is distributed throughout the CNS, with the highest levels in the gray matter of the spinal cord (Graham *et al.,* 1967).

Studies of subcellular distribution in brain homogenates show that these amino acids are not associated with any particular type of subcellular particle (Ryall, 1964; Mangan and Whittaker, 1966; Agrawal *et al.,* 1971). Instead they appear to be cytoplasmic constituents which show redistribution with subcellular fractionation techniques (Whittaker, 1965; Neal and Iversen, 1969). Only a relatively small bound pool of amino acids is present in synaptic vesicles prepared from rat brain homogenates (De Belleroche and Bradford, 1973).

The metabolism of glycine in the CNS is still poorly understood. *In vivo* experiments using radiolabeled precursors have shown that glycine metabolism is associated with glucose and ribose metabolism and that glycine is in equilibrium with carbohydrate intermediates via more than one route (Shank *et al.,* 1973). The rate-limiting step in glycine biosynthesis may be the conversion of D-glycerate to hydroxypyruvate by a "nonphosphorylated" dehydrogenase, since the end product inhibits this enzymatic activity (Uhr and Sneddon, 1972). Hydroxypyruvate is then transaminated into serine, but the transaminating enzyme seems to be lacking in the brain (Walsh and Sallach, 1966) so that the route from hydroxypyruvate to glycine may be via glycolaldehyde, glycolate, and glyoxalate rather than via serine (see Johnston, 1975). There is also evidence that serine is synthesized in the brain via the phosphorylated pathway (Bridgers, 1969). Serine and glycine are interconvertible by serine hydroxymethyltransferase (Broderick *et al.,* 1972).

Taurine in the CNS is synthesized from the precursor cysteinesulfinic acid which is decarboxylated to hypotaurine, and this in turn is oxidized to taurine. The enzymes involved in the synthesis and degradation of taurine to isethionic acid have been found in extracts of rat brain (see Johnston, 1975). The distribution of brain taurine resembles that of GABA, although unexplained variations between species and studies have been reported (Perry, 1971; Guidotti *et al.*, 1972), and there is some evidence that possibly taurine is situated in the same nerve ending fractions as GABA (Agrawal *et al.*, 1971).

In cat spinal and cortical neurons taurine possesses a depressant action on microiontophoretic application (Curtis and Watkins, 1961). This nonessential amino acid is released from the cat cerebral cortex, and in greater concentration during arousal than during sleep (Jasper and Koyama, 1969). Although initial studies indicated that taurine could be an inhibitory transmitter or modulator in the CNS (Davison and Kaczamarek, 1971), there is now evidence suggesting that taurine must be regarded as a stabilizer of membrane excitability (see Barbeau and Donaldson, 1974).

The metabolism of glutamate and aspartate is linked directly by transamination and indirectly by the tricarboxylic acid cycle (TCA) as shown schematically below:

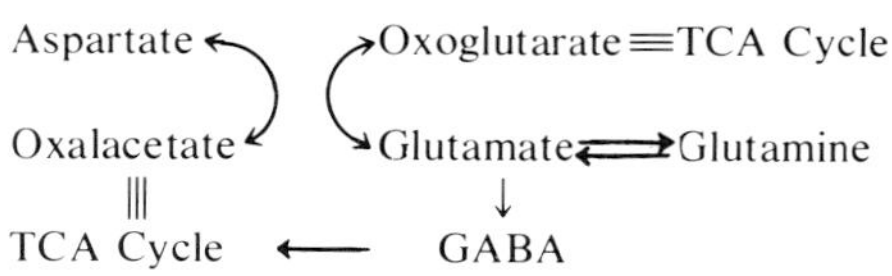

A large pool of these amino acids seems to represent the transmitter amino acids, and a small pool, which may be associated with glial cells, may represent inactivated amino acids taken up from the synaptic cleft. The following enzymes are involved in glutamate metabolism in the CNS: transaminases (glutamate ⇄ oxoglutarate), glutamine synthetase (glutamate + NH_3 = glutamine), glutamate dehydrogenase (2-oxoglutarate + NH_3 = glutamate), glutaminase (glutamine → glutamate), glutamate decarboxylase (glutamate → GABA), γ-glutamyl enzymes (important in the transport of amino acids in the brain and in other tissues) (see Johnston, 1975). There is now evidence for the presence of a specific, carrier-mediated activity uptake and release of the putative amino acid transmitters (Curtis and Johnston, 1974b; De Feudis, 1975). Whereas no proof exists that amino acids are inactivated enzymatically at the extracellular level, it seems likely that an uptake mechanism operates to terminate their effects (Iversen and Neal, 1968; Curtis *et al.*, 1970c). A blood–brain barrier is effective in preventing free diffusion of amino acids from the plasma to extracellular fluids of the brain, and it may involve carrier-mediated transport systems in the choroid plexus (Cohen and Lajtha, 1972; Tate *et al.*, 1973)

The release of amino acids from the CNS has been studied in resting conditions ("spontaneous release") as well as with *in vitro* and *in vivo* stimulation ("evoked release") (see Johnston, 1975). Neurophysiological studies indicate that glycine is an inhibitory transmitter at the spinal level (Werman and Aprison, 1968) and that the spinal convulsant drug, strychnine, acts through selective antagonism of the glycinergic receptors (Curtis *et al.*, 1967, 1971c). Tetanus toxin appears to inhibit the release of glycine and probably also of GABA from nerve terminals (Curtis and De Groot, 1968; Curtis *et al.*, 1973). There is also evidence that some minor tranquilizers, e.g., benzodiazepines, act as stimulants for the central glycinergic receptors and enhance GABAergic inhibitory mechanisms, and it has been suggested that the antianxiety effects of these drugs may be correlated to these actions (see Costa *et al.*, 1975). Glycine appears to be a more potent depressant of neurons in bulbar reticular formation than β-alanine, GABA, or taurine (Hösli and Tebécis, 1970; Haas and Hösli, 1973). The effects of glycine and taurine are blocked by picrotoxin, but on some neurons glycine action is only slightly blocked by picrotoxin. The topographical distribution of amino acid receptors and their antagonists in the CNS has been reviewed (Ryall, 1975; Kelly and Beart, 1975; McLennan, 1975; Snyder and Bennet, 1976).

K. SUBSTANCE P

The isolation of substance P (SP) from the brain and intestine dates back many years (von Euler and Gaddum, 1931; Gaddum and Schild, 1934). The name P refers to the use of "powder" extracts of equine brain and gut by researchers. The hypotensive and smooth muscle stimulating effects of SP were not blocked by atropine, and differed from those evoked by histamine or other substances. It was only a few years ago that the chemical identity of SP had been determined, and now the entire amino acid sequence of this undecapeptide is known (Chang *et al.*, 1971; Tregear *et al.*, 1971). Since the availability of H-Arg-Pro-Lys-Pro-Glu-Glu-Phe-Phe-Gly-Leu-Met-NH_2 has permitted the development of sensitive and specific radioimmunoassays (Powell *et al.*, 1973), a large number of experiments are currently being carried out to determine the physiological role of SP in the CNS (Nicoll, 1975). By solid-phase techniques substance P has been synthesized, and its biological activities were found to be in reasonable agreement with the activities of the natural preparations (Fisher *et al.*, 1974).

Substance P is localized mainly in the alimentary tract and the CNS. Earlier studies had indicated that SP was localized in the dorsal roots rather than in the ventral roots, suggesting that it may be the transmitter in the primary sensory neurons (see Nicoll, 1975). More recent immunohistochemical studies with antibodies for substance P have demonstrated the occurrence of immunoreactivity in the rat and cat spinal and trigeminal ganglia and in many peripheral areas, e.g., skin, sweat glands of both species (G. Nilsson *et al.*, 1974; Hökfelt *et al.*, 1975e,f, 1976a,b), again suggesting that the substance may act as a transmitter or modulator released from sensory nerve terminals (Otsuka *et al.*, 1975). The presence of a Ca^{2+}-dependent release of substance P-like immunoreactivity from the isolated rat spinal cord in response to electrical stimulation of dorsal roots (Otsuka and Konishi, 1976) supports the view that substance P may be an excitatory transmitter at this level (see also below).

As shown by biochemical and radioimmunological studies, the distribution of this undecapeptide in the brain is uneven (Powell *et al.*, 1973; Zetler, 1970), and confined to neurons and not glial cells (Grabner and Lembeck, 1960). An axonal flow for substance P has been demonstrated by axon cutting experiments (Holton, 1959, 1960), and in brain homogenates it is concentrated in the synaptosomal fraction (Kataoka, 1962; Ryall, 1964).

In the rat brain, substance P immunoreactivity occurs in fiber systems of the substantia gelatinosa trigemini, part of the n. tractus spinalis nervi trigemini, the periaqueductal central gray, part of the interpeduncular nucleus, and the medial preoptic area. As for cortical areas, SP-positive nerve endings have been observed only in some limbic areas such as the amygdaloid nuclei, the entorhinal cortex, and the basal hippocampus (Hökfelt *et al.*, 1975e,f). Substance P-positive cell bodies have been found so far only in one rat CNS area, e.g., the medial habenula (Hökfelt *et al.*, 1975e).

The highest concentrations of SP are present in the nerve ending particles of the substantia nigra, whereas the cerebellum is devoid of SP (Duffy *et al.*, 1975; Davies and Dray, 1976). By radioimmunoassay techniques, the highest concentrations in the human brain were also found in the substantia nigra (Powell *et al.*, 1973). This peptide, as well as others, is released from the surface of cat cortex (Ramwell and Shaw, 1963a,b; Shaw and Ramwell, 1968); for SP the release is increased by analeptic drugs. Recently, additional evidence has reinforced the hypothesis that substance P serves a neurotransmitter role in the CNS (see also below) with the demonstration that it can be released from rat hypothalamic tissue in response to a depolarizing stimulus by a Ca^{2+}-dependent mechanism. Apparently, tissue metabolism rather than a reuptake mechanism is the important pathway for terminating the action of the released peptide (Iversen *et al.*, 1976). Although there is no direct evidence that SP is released during neuronal activation, it is of interest that retinal substance P increases in animals maintained in darkness and decreases after exposure to light, implying that its levels may be controlled by impulse activity (Cooper *et al.,* 1974).

Difficulties encountered in the purification of substance P have contributed to the controversy regarding its effects in the CNS. Reportedly, SP is able to cross the BBB (Stern and Hadzovic, 1973). SP administered by intraventricular injection did not exhibit central effects in some studies (Haefely *et al.,* 1962), but in others an effect on the electrical activity of the hypothalamus was reported (Baile and Meinardi, 1967). Similar to another peptide, physalaemin, substance P proved to be a potent neuronal excitant on isolated spinal motor neurons (Otsuka *et al.,* 1972; Konishi and Otsuka, 1974; Henry *et al.,* 1974) and was from 200 to 900 times more potent than L-glutamate in depolarizing motor neurons in the isolated frog and rat spinal cords (Konishi and Otsuka, 1974; Saito *et al.,* 1975).

The cuneate neurons which receive monosynaptic excitation from primary afferents did not change their firing after microiontophoretic application of extracts of substance P (Galindo *et al.,* 1967), whereas the application of pure SP onto these neurons produced excitation (Krnjević and Morris, 1973, 1974), thus confirming Otsuka and co-workers' belief that substance P excites neurons which receive monosynaptic excitatory inputs from primary afferents.

Unlike glutamate, the excitatory action of substance P is very weak and long-lasting (up to many minutes after application). Actually, when glutamate was applied during the action of substance P, its excitatory effect was considerably reduced and this blocking action could be separated from the depolarizing action of SP, thus suggesting that the peptide may be a modulator of endogenous excitatory transmitters (Krnjević and Morris, 1973).

Studies in which substance P and other peptides (physalaemin, bradykinin, bombesin) have been microiontophoretically applied to cortical Betz cells have shown that these substances are excitant, the most potent being substance P (Phillis and Limacher, 1974). In the rat, iontophoretic application of substance P to neurons of the substantia nigra, where in view of its elevated concentration (Duffy *et al.,* 1975) a transmitter or modulator rôle has been suggested, produced an increase in neuronal activity; a similar pattern was also evoked by a similar undecapeptide, e.g., eledoisin (Davies and Dray, 1976). It has been claimed that β-(p-chlorophenyl)-GABA (Baclofen) is a specific antagonist of substance P and that the depression of monosynaptic reflexes by Baclofen therefore constitutes further evidence in support of the hypothesis that SP is the transmitter released by primary afferent fibers (Saito *et al.,* 1975). However, no evidence was obtained that Baclofen is able to selectively antagonize SP-induced excitation of cerebral cortex neurons since ACh excitation was always affected to a comparable extent (Phillis, 1976).

It appears from a recent report that substance P can affect central monoaminergic mechanisms since after IVT injection it significantly increased the concentration of dopa in many regions of the brain (limbic, striatum, hemispheres, diencephalon and lower brain stem), whereas no effects were recorded on 5-HTP concentration (Magnusson *et al.,* 1976). In addition, after inhibition of monoamine synthesis, substance P accelerated DA, NE, and 5-HT turnover. Reportedly, the injection of substance P into the right lateral ventricle produced in most cases rotation of the animal to the left side (dopaminergic stimulation), which lasted several minutes, and behavioral excitation, which persisted even longer. These data suggest that substance P has a stimulatory action on monoaminergic neurons in the brain and that it is used as an excitatory transmitter, e.g., in the substantia nigra where it occurs in high concentration, possibly in nerve terminals impinging on DA cell bodies or dendrites and in other areas in which 5-HT and NE terminals occur.

Another role recently attributed to SP concerns the control of pain (Stewart *et al.*, 1976). Conflicting and misleading results were obtained initially using impure SP which was shown capable of antagonizing the analgesic effects of morphine (Zetler, 1956). More recently, using synthetic SP, it was found that it could be effectively substituted for morphine in mice chronically treated with the latter (Stern and Hadzovic, 1973) and possessed potent analgesic activity (Stewart *et al.,* 1976). Recent reports on the characterization and activity of enkephalin, an endogenous peptide with opioid agonist activity, suggest that this compound may act as an analog of SP (Hughes *et al.,* 1975). Although collectively these studies suggest for SP some role in excitatory synaptic transmission, its functional significance in the CNS has still to be elucidated.

L. FUNCTIONAL INTERACTIONS BETWEEN MONOAMINERGIC, CHOLINERGIC, AND OTHER NEUROTRANSMITTER SYSTEMS IN THE CNS

The complexity of the neuronal network in the CNS where, in some areas, terminals of different, real or putative, transmitter systems intermingle with one another and the functional relations between CA's, 5-HT, ACh, GABA, etc., make it difficult to interpret the significance of experiments carried out with drugs or techniques planned to interfere exclusively with one particular system. With no pretense at being exhaustive, some of these functional interactions are described below.

As already mentioned (see Table X), large doses of L-dopa, in addition to increasing brain DA and NE levels, produce a marked decrease of brain 5-HT, probably due to a displacement of 5-HT from its binding sites by DA formed from the exogenously administered precursor. Competition between L-dopa and precursors of 5-HT at the level of synthesizing enzymes or for entry into the brain and the neurons can be also a contributing factor, particularly in long-term treatment. Conversely, 5-HTP can produce an increase in brain 5-HT accompanied by a reduction of brain DA and NE levels, probably by similar mechanism(s) (see Table XV). Proof that this interaction may be of some importance at the behavioral level is indicated by the fact that some behavioral aspects observed after L-dopa administration can be antagonized by 5-HT neurotransmission inhibitors (Jacobs, 1974) and, reciprocally, behavioral effects of Trp are counteracted by DA inhibitors (Green and Grahame-Smith, 1974). Although these results are probably due to a pharmacological interaction between the substances used, they reflect the presence of a functional link between CA and 5-HT systems and must be considered when evaluating the effects of neuropharmacological experiments on neuroendocrine function (see Chapter IV).

Another interaction to be considered is that of several neuronal systems (DA, ACh, 5-HT, GABA) in regulating extrapyramidal functions. A balance between ACh-DA and 5-HT-H had been suggested for some time (Barbeau, 1962). In recent years, experimental and clinical data have accumulated which strongly support the interdependence of central DA and ACh systems (Andén and Bédard, 1971; Klavans and Rubovits, 1972; Bartholini *et al.,* 1973; Stadler *et al.,* 1973). Dopaminergic terminals in the striatum exert an inhibitory action on small interneurons containing ACh (Albe-Fessard *et al.,* 1967; Connor, 1968; McGeer *et al.,* 1971; Butcher and Butcher, 1974). Drugs known to interfere with the DA system, e.g., DA receptor antagonists, increased striatal ACh release (Stadler *et al.,* 1973) and turnover (Trabucchi *et al.,* 1974), whereas dopaminergic drugs increased rat striatal ACh (Consolo *et al.,* 1974; Ladinsky *et al.,* 1974) and inhibited striatal ACh turnover (Trabucchi *et al.,* 1975).

The presence also of cholinergic neurons innervating the substantia nigra and thus involved in the regulation of DA transmission has been postulated. Nigral applications of cholinergic or anticholinergic drugs were found capable, in fact, of inducing immediate changes in the metabolism of striatal DA (Javoy *et al.,* 1973). On the other hand, atropine and other anticholinergic drugs have been found capable of decreasing DA turnover in the striatum (Bowers and Roth, 1972; Andén, 1972) but not in the dopaminergic limbic system (Andén, 1972). The opposite effect has been described for cholinergic drugs (Corrodi *et al.,* 1967a).

A possible role of GABA in the regulation of cholinergic (Kim *et al.,*

1971a) and dopaminergic (Andén, 1974) functions has also been suggested. Considerable anatomic, biochemical, and physiological evidence exists for a descending pathway in mammalian brain between the caudate nucleus-putamen and substantia nigra (striatonigral pathway) which is mediated by GABA and exerts an inhibitory effect on the substantia nigra (see Section I, and Tarsy *et al.*, 1975). Striatal ablation or transection of striatonigral pathways produce a reduction in the concentration of GABA in the substantia nigra of several animal species (Kim *et al.*, 1971a; Fonnum *et al.*, 1974; Kataoka *et al.*, 1974) and a decrease in the uptake of GABA in the substantia nigra (Storm-Mathisen, 1975), and a degeneration of nigral terminals which, on the basis of autoradiographic evidence, are believed to be the GABA terminals (Bak *et al.*, 1975). It has been postulated that the GAD-containing neurons are under the continuous "trophic" influence of the dopaminergic system. Degeneration of the latter might then result in a biochemical "atrophy" of the GAD neurons (Lloyd and Hornykiewicz, 1974). The existence of this GABA-DA interaction is supported by the rotational behavior occurring in rats after blocking GABA receptors (disinhibition of nigrostriatal activity) (Tarsy *et al.*, 1975) and the inhibition of the striatal DA turnover after administration of aminooxyacetic acid which increases GABA concentrations by inhibiting GABA-T (see Table XX).

In addition to those among dopaminergic, cholinergic, and GABA systems, interactions have also been demonstrated between DA and 5-HT in regulating extrapyramidal function (Cools, 1974; Chase, 1974). Indeed, 5-HT has been shown to activate the release of DA from the striatum (Besson *et al.*, 1969), while a DA receptor stimulation increases central 5-HT turnover (Grabowska *et al.*, 1973). Recent experiments suggest that rotational behavior due to striatal DA receptor stimulation is depressed or enhanced by an elevation or a reduction, respectively, in brain 5-HT content (Nilson and Pycock, 1976). In addition, the possibility that 5-HT may interact with the cholinergic system, initially proposed by Aprison (1962), has gained support with the finding that *in vitro* 5-HT was able to inhibit Che (Barbeau, 1972), However, stimulation of the n. linearis intermedius raphe, while increasing 5-HT release into the lateral ventricles, did not influence ACh (Ashkenazi *et al.*, 1972). Moreover, pCPA decreased brain 5-HT levels but did not affect brain ACh, although in pCPA-treated rats, 5-HTP produced a marked increase in brain ACh output (Pepeu *et al.*, 1974).

III

Hypothalamic Releasing and Inhibiting Hormones

Since the first demonstration of hypophysiotropic action of hypothalamic stalk ME extracts, intense efforts have been made in many laboratories to identify the hypothalamic substances regulating each of the tropic hormones of the AP gland. Such hypophysiotropic "factors" have been isolated, some of them structurally identified, synthesized, and evaluated for diagnostic and therapeutic use. Since the classic criteria for designation as hormones were satisfied, this term was proposed for those factors whose chemical identity was established (Schally *et al.*, 1968a).* The existence of at least nine hypothalamic regulators of the pituitary gland is now reasonably well established. For at least three pituitary hormones there is a dual system of hypothalamic control, one system being inhibitory and one stimulatory (Table XXI).

This chapter is not intended to be an exhaustive survey of all the literature on the hypothalamic neurohormones, but rather a general overview of the neural control of the secretion of the different AP hormones; the chemical isolation, identification, and synthesis of hypothalamic peptides; the methods and techniques available for their assay; the CNS areas thought to be involved in their secretion; the mechanisms of action, and, finally, the recently discovered extraendocrine actions.

A. NEUROHORMONAL CONTROL OF ANTERIOR PITUITARY HORMONES

1. Historical Considerations

The concept of neurohormonal control of the anterior pituitary had initially been derived from studies on the anatomy and physiology of the

* The terms "factor" and "hormone" are used interchangeably.

TABLE XXI

Hypothalamic Releasing and Inhibiting Hormones

Hypothalamic hormone (or factor)	Abbreviation	Structure[a,b]
Corticotropin (ACTH)-releasing factor	CRF	Ac Ser Tyr Cys Phe His [Asp NH_2 Glu NH_2] Cys Pro Val Lys Gly NH_2*
Thyrotropin (TSH)-releasing hormone	TRH	pGlu His Pro NH_2
Luteinizing hormone (LH)-releasing hormone	LH-RH	pGlu His Trp Ser Tyr Gly Leu Arg Pro Gly NH_2
Follicle-stimulating hormone (FSH)-releasing hormone	FSH-RH	pGlu His Trp Ser Tyr Gly Leu Arg Pro Gly NH_2
Growth hormone (GH)-releasing factor	GRF	H Val His Leu Ser Ala Glu Glu Lys Glu Ala OH*
Growth hormone (GH)-release-inhibiting hormone	GH-RIH[c]	H Ala Gly Cys Lys Asn Phe Phe Trp Lys Thr Phe Thr Ser Cys OH
Prolactin-inhibiting factor	PIF	....
Prolactin-releasing factor	PRF	....
Melanocyte-stimulating hormone (MSH) release-inhibiting hormone	MR-IH	H Pro Leu Gly NH_2
Melanocyte-stimulating hormone (MSH)-releasing hormone	MRH	H Cys Tyr Ile Gln Asn OH

[a] The three-letter symbols for amino acids are those recommended in the *Biochem. J.* **102,** 23 (1967), except for pGlu, pyroglutamic acid.

[b] Asterisk indicates proposed structure.

[c] Or somatostatin.

vascular connections between the anterior pituitary and the hypothalamus. In the decade 1930–1940, it was concluded from studies on the rat by Popa and Fielding (1930, 1933) and on the toad by Houssay *et al.* (1935) that, contrary to common belief, the flow of blood was from the hypothalamus toward the pituitary, a finding which was confirmed by Wislocki and King (1936) and later by Wislocki (1937). In 1949, Green and Harris were able to show that, in the rat, blood in the stalk portal vessels flowed toward the AP, an observation later confirmed in other animal species (Torök, 1954; Worthington, 1955).

The recognition of a vascular route connecting the hypothalamus with the pituitary overcame the problem of explaining how in mammals information is relayed from the CNS to the AP lobe of the pituitary—whose secretory elements lack direct innervation—and initiated the idea of neurohumoral involvement in the regulation of the AP. Harris in 1937, in discussing the results of an experiment in which ovulation could be induced in the rabbit by electrical stimulation of the brain, proposed neurohumoral control of the AP but was still inclined to think that direct innervation of the gland was responsible for the ovulation induced by stimulation of the hypothalamus. The neurohumoral hypothesis, essentially the release from hypothalamic nerve fibers of a neurohumoral substance(s) that is carried to the sinusoids of the AP on entering the blood capillaries of the primary plexus of the median eminence, must be attributed to Hinsey (1937) (Fig. 33). Additional support for neurohumoral control was derived from the classic experiments of Harris and Jacobsohn (1952). They showed that female rats bearing an AP transplanted next to the ME were able to mate and to deliver young, whereas rats with transplants under the temporal lobe become anestrous and had attrophied reproductive organs. Since then other anatomic, physiological, and pharmacological data have accumulated, bringing the neurohumoral concept of regulation of the AP into the realm of practical significance (see reviews by McCann and Porter, 1969; McCann *et al.,* 1974; Schally *et al.,* 1973a; Blackwell and Guillemin, 1973; Reichlin *et al.,* 1976).

2. Adrenocorticotropic Hormone

It is well established that the CNS, the hypothalamus in particular, mediates responses to stimuli by liberating a specific corticotropin-releasing factor (CRF) (Burgus and Guillemin, 1970). In turn, CRF stimulates the release of ACTH from the pituitary, which is carried through the general circulation to the adrenal gland where it increases steroidogenic activity.

Although vasopressin was for a while considered to be the major corticotropin-releasing factor (McCann and Brobeck, 1954; McCann and

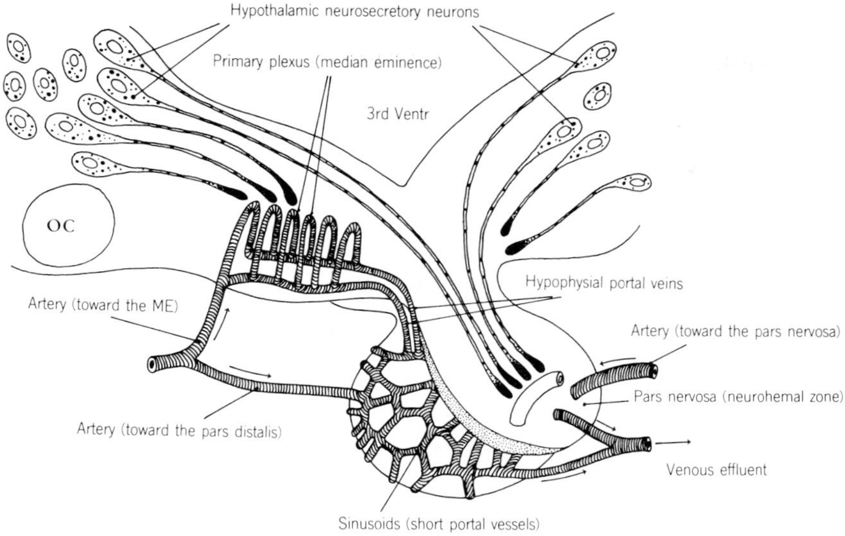

Fig. 33 Hypothalamic pituitary unit. Depicted is the anatomic relation between the base of the hypothalamus and the pituitary gland. Neurosecretory neurons localized in some hypothalamic nuclei send axons which either terminate in relation to blood vessels of the pars nervosa (supraoptic–hypophyseal tract) or in close proximity to the blood vessels of the median eminence. There are two groups of hypophyseal portal veins; the long portal vessels which originate from recurrent capillary loops in the median eminence and pass down the pituitary stalk to end in the sinusoids of the AP and the short portal vessels which take origin from capillaries in the pars nervosa and pass across the pars intermedia to join the AP sinusoidal network. OC, optic chiasm.

Fruit, 1957; Casentini *et al.*, 1959; Martini and Pecile, 1962); see Guillemin, 1964, for a more exhaustive review), many reports clearly demonstrated the existence of a CRF different from vasopressin in the neurohypophysis and the hypothalamus (see Guillemin and Schally, 1963; de Wied *et al.*, 1964).

The first purification of CRF from the hypothalamus was reported independently by two groups of workers (Rumsfeld and Porter, 1959; Royce and Sayers, 1960), and subsequently other purifications were reported (Schally *et al.*, 1962; Dhariwal *et al.*, 1965, 1966). Due to limitation in the availability of hypothalamic tissue, considerable effort was also expended in the isolation and identification in the posterior pituitary of substances possessing ACTH-releasing activity (Guillemin, 1964). A partial amino acid sequence of a CRF purified from powdered porcine neurohypophysis was reported by Schally and Bowers (1964):

Ac-Ser-Tyr-Cys-Phe-His-(Asp-NH_2,Glu-NH_2)-Cys-(Pro,Val)-Lys-Gly-NH_2

The relationship of the CRF's of pituitary origin to hypothalamic CRF is unknown, although the fact that hypothalamic CRF is destroyed by certain proteolytic enzymes (Schally *et al.*, 1968a) would indicate its polypeptide nature. Although CRF was the first hypothalamic neurohormone to be observed (Saffran and Schally, 1955; Saffran *et al.*, 1955; Schally *et al.*, 1958), only crude preparations have been reported so far (see Burgus and Guillemin, 1970; Guillemin, 1971a). The unreliability of the *in vivo* assays used for assessing CRF activity (see below and also Section B,1) and the reported instability of the molecule (Chan *et al.*, 1970) have hindered the isolation of sufficient amounts for the determination of its structure.

A prerequisite for an assay system for CRF is the prevention of endogenous release of CRF in the recipient animal, thus permitting the evaluation of the ACTH-releasing effect on the administration of an exogenous CRF. Since stress-induced rapid ACTH release is blocked after hypothalamic lesions, rats bearing ventral hypothalamic lesions were again suggested as recipients for CRF assay. In the study by Witorsch and Brodish (1972a), ether stress failed to induce plasma B increases in 95% of the lesioned rats. However, unless rigid time parameters are followed in the assay, nonspecific increases in plasma B levels are evoked by nonhypothalamic extracts. Similarly, in the animal preparation of Ondo and Kitay (1972)—dexamethasone-treated rats bearing diencephalic islands—ACTH release could be induced by either stress or vasopressin and histamine administration.

Pituitary incubation methods for the assay of CRF are advantageous in that any *in vitro* effect must be exerted directly on the pituitary gland, thus excluding the possibility of endogenous release of CRF as a contributing factor to the response obtained. The combination of reduced toxicity of the Krebs–Ringer bicarbonate medium and lower pH (4.0–5.0) has been reported to result in an increase in both the range and sensitivity of the CRF assay (Seiden and Brodish, 1971). A log dose–response relationship was obtained between the amount of hypothalamic extract added to the incubation media (0.05–0.2 hypothalamic equivalents) and the plasma corticosterone levels of hypophysectomized assay rats following injection of the incubation medium.

More recently, a semiautomated *in vitro* bioassay has been developed in which the ACTH released by the pituitary tissue into the medium, following exposure to test materials, is determined by automated superinfusion of rat adrenal tissue (Pearlmutter *et al.*, 1974). Using this system, in which most of the ring or tail fragments of vasopressin and oxytocin do not stimulate ACTH release, two peaks were obtained by gel filtration of rat ME extracts, both with little ACTH-releasing activity (Pearlmutter *et al.*, 1975). Full activity of the extract was regained when both peaks were assayed together, which suggested that the CRF activity of hypothalamic extracts is due to the presence of two (or more) factors that interact.

The development of highly specific radioimmunoassays has permitted the direct measurement of plasma ACTH levels (Matsukara *et al.,* 1971). Using a specific RIA method for ACTH and the monolayer pituitary cell culture described by Vale *et al.* (1972a), J. Rivier *et al.* (unpublished data, 1974) were able to demonstrate the presence in sheep hypothalami of a factor that differs from vasopressin, is devoid of intrinsic ACTH activity, is soluble in acid solvents, and stimulates ACTH secretion. This observation has been also made *in vivo* (Blackwell and Guillemin, 1973).

Recently, Jones *et al.* (1976a) reported the isolation of CRF from rat hypothalamus *in vitro* after stimulation with 5-HT (see Chapter IV, Section A,1,b). The separation on Sephadex G-25 of the 5-HT stimulated medium gave two peaks of CRF activity, CRF/A and CRF/B, which were then subjected to further purification either by high-voltage electrophoresis or ion-exchange chromatography on CM cellulose. It appears that CRF/A has a MW of about 2400, and CRF/B, a basic peptide, has a MW of about 1200, and is active at nanogram doses in *in vitro* and *in vivo* assay systems.

Despite the elusiveness of research on the isolation and characterization of hypothalamic CRF, new information has been acquired on CNS control of the CRF–ACTH system. A daily circadian periodicity in the hypothalamic content of CRF with a close temporal relationship with plasma B level has been observed in normal male rats (Hiroshige and Sakakura, 1971). The periodicity apparently manifests a marked sex difference, the circadian rhythm in the female rat (higher CRF content in the morning than in the afternoon) being the reverse of that found in the male rat (Hiroshige *et al.,* 1973). However, in both sexes, the pattern of change in hypothalamic CRF content for a 24-hr period is a mirror image of that of plasma corticosterone. These findings however, have been recently negated by Yasuda and Greer (1976). Despite an unequivocal nycthemeral periodicity of plasma B in intact rats, they were unable to find a significant periodicity of hypothalamic CRF content.

In addition to hypothalamic CRF, evidence from studies on the peripheral blood of rats bearing extensive hypothalamic lesions (Witorsch and Brodish, 1972a,b) demonstrated the existence of an extrahypothalamic "tissue CRF" that is released from peripheral site(s) in response to stress (Lymangrover and Brodish, 1973). It has been proposed that tissue CRF, which similar to hypothalamic CRF is physiologically regulated by pituitary–adrenal activity, may play a major role in conditions of severe stress when the needs of the organism are not supplied by hypothalamic mechanisms(s) (Lymangrover and Brodish, 1973–1974).

Circadian CRF activity rhythms are essentially similar in adrenalectomized rats (Hiroshige and Sakakura, 1971), thus stressing its neural origin and independence of a negative feedback mechanism. Similarly, a circadian periodicity of plasma immunoreactive ACTH levels in the presence of virtually constant plasma F (plasma cortisol) levels has been found in patients with Addison's disease (Besser *et al.*, 1971; Krieger and Gewirtz, 1974), again stressing the lack of involvement of feedback in the genesis of CRF–ACTH periodicity. It appears that ACTH and, therefore, CRF in the human are secreted for episodes of very short duration, the longest observed spanning less than 10 min (Gallagher *et al.*, 1973). The episodes often follow in close temporal juxtaposition; at other times the intervals between bursts are sufficiently long to allow the plasma ACTH to reach zero concentration. Cortisol secretion begins within 5–10 min after the plasma content of ACTH begins to rise, and when the ACTH episodes are separated by very short intervals, the adrenals respond as if they are being continually stimulated. The episodic secretion of either ACTH and cortisol throughout the 24-hr sleep–wake cycle and the fact that a "steady-state" or "basal level" of F is never present for any extended time during the 24-hr cycle (Weitzman *et al.*, 1971) support the concept that the usual closed feedback loop for F secretion in man is inoperative. Thus, a high concentration of plasma F will not necessarily terminate pituitary secretory activity and a low concentration will not initiate the secretory process.

The above findings on the functioning of the pituitary–adrenal relationship in humans were obtained with blood samples taken at 5-min intervals; in fact, to define a rhythm, the frequency of sampling must exceed the frequency of the rhythm (Gallagher *et al.*, 1973). It seems, however, that in humans the hormonal feedback mechanism(s) is involved in the regulation of the CRF–ACTH circadian pattern. Much evidence has been collected on the variation in the inhibiting effect on adrenal steroid secretion of corticoids administered for a 24-hr period; greater inhibition has been demonstrated with corticoids given just prior to the endogenous circadian rise as compared with corticoids given at other times (Demos *et al.*, 1964; Grant *et al.*, 1965). The study of 17-OHCS urinary flow curves plotted at various times for a 24-hr period, during and after constant intravenous infusion of small doses of corticoids revealed two secretory phases in the circadian periodicity of the pituitary–adrenal axis (Angeli, 1974). The first phase (so-called "basal" activity) persists for the entire 24-hour period, whereas the second phase (so-called "impulsive" activity) is superimposed over the first one for a relatively short time during the early morning hours. The two phases exhibit different sensitivity to corticoid inhibition, the "impulsive" phase being highly sensitive and the "basal" phase poorly sensitive, and behave independently as far as recovery after a short-period of total suppression (Ceresa *et al.*, 1970). Similarly, it has been shown that with metyrapone there is a greater ACTH release when the drug is infused just preceding the circadian or "impulsive" rise than when it is infused in the late morning or early afternoon (Martin and Hellman, 1964).

Collectively, these data support the view that in humans the neuroendocrine control mechanism(s) underlying the circadian rise of ACTH secretion is not continuous throughout the 24-hr period but is confined to the nocturnal early morning hours and is qualitatively different from those responsible for the ACTH activity of the rest of the day.

3. Growth Hormone

The study of the regulatory physiology and neuroendocrinology of growth hormone (GH) has advanced significantly in the last 10 years, undoubtedly because of the development of specific and sensitive RIA techniques for the detection of the hormone in the plasma of humans and laboratory animals. New developments include the demonstration that stereotaxic lesioning or stimulation of selective hypothalamic or extrahypothalamic areas induced a marked decrease or a striking elevation of plasma GH, respectively; the recognition of the relevance of brain MA systems in the control of the release of GH (see Chapter IV, Section A,2); observations on cyclic GH secretions as related to EEG patterns; and, finally, the isolation, purification, and synthesis of a growth hormone-inhibiting factor (GH-IF, GIF, or SR-IF) or somatostatin (see Müller, 1973; Martin, 1973a, 1976).

a. GH-Releasing Factor and the Stimulatory Role of the CNS

Although the first demonstration by Deuben and Meites *in vitro* (1964) and Müller and Pecile *in vivo* (1965) of the presence of a growth hormone-releasing factor (GRF) in rodent hypothalamic extracts traces back more than 10 years and that since then many studies provided evidence for the physiological role of GRF (Müller, 1973; McCann *et al.*, 1974), the work on GRF has been hindered by a discrepancy among results obtained by the "tibia test" and those obtained by radioimmunoassay. Most of the conditions such as cold exposure, insulin hypoglycemia, or prolonged fasting found to be associated with reduced levels of pituitary GH when measured by BA displayed no effect on pituitary GH concentration or the circulating GH levels (see Müller, 1973) by RIA developed for measurement of GH in the plasma and pituitary of the rat (Schalch and Reichlin, 1966). The same discrepancies between BA and RIA data affected conclusions on the isolation and purification of a GRF molecule. On the basis of

GH bioassay, material endowed with GRF activity has been purified and isolated from porcine hypothalamic extracts (Schally *et al.*, 1969a). The isolated material appeared to be a decapeptide with an N-terminal valine and C-terminal alanine (Schally *et al.*, 1971a), and was synthesized by Veber *et al.* (1971), who noted a striking similarity between the structure of GRF and that of the β-chain amino terminal sequence of porcine hemoglobin. The depletion of pituitary GH after administration of pure porcine GRF was found to be associated with a simultaneous increase in plasma GH-like activity (Müller *et al.*, 1971a). However, the natural and synthetic materials failed to promote the release of RIA-GH either when injected into rats, sheep, pigs, and monkeys (A. Arimura, personal communication), or when incubated with rat AP halves (Schally *et al.*, 1972a). Moreover, in man no convincing increase in plasma GH levels was found after administration of natural porcine, human, or synthetic material (Kastin *et al.*, 1972b). More recently, however, release of RIA-GH was reported after the infusion into hypophyseal portal vessels of the rat of Sephadex-separated fractions of porcine hypothalamic extracts (Takahara *et al.*, 1975) or synthetic GRF (Wilbur *et al.*, 1975).

Several hypotheses have been advanced to explain the discrepancies between RIA and BA methods in measuring pituitary and plasma GH levels (Müller *et al.*, 1972a; Nicoll, 1972). The statistical validity of the BA method in *in vivo* experiments has been questioned (Roger *et al.*, 1969) and supported (Parkhie and Johnson, 1971). The inability to detect the expected increases in plasma immunoreactive GH has suggested the possibility that plasma GH in the rat may exist in a form not detectable by current radioimmunoassays (Ellis and Grindeland, 1973). Determination of the ratio of the resting levels of BA (tibia test) to RIA-GH has shown a 200-fold excess of BA activity in human plasma and a 50-fold excess in rat plasma. Consistent with previous "depletion" data of pituitary GH concentration (see Müller, 1973), in rats the plasma concentrations of BA hormone could be elevated 2- to 3-fold by cold exposure, insulin hypoglycemia, or prolonged fasting, suggesting that the ratio of the two types of growth hormone (BA-GH and RIA-GH) in plasma can be altered by appropriate stimuli (Ellis *et al.*, 1976).

The existence of GH with varying BA and RIA activities has also been implied by the studies of Vodian and Nicoll (1975) on rat pituitaries incubated with and without synthetic porcine GRF. These authors have also observed that secreted GH *in vivo* can no longer be detected by RIA, which supports Ellis and Grindeland' hypothesis.

Although current opinion is that the isolated GRF is probably not the true stimulatory neurohormone for GH secretion and that the sequence reported may indicate a GH-releasing artifact (Schally *et al.*, 1973a;

McCann *et al.,* 1974), the response of purified hypothalamic fractions or the decapeptide *in vitro* with the sulfation factor and "tibia test" assay (Schally *et al.,* 1968b), the ability to induce the extrusion of GH-containing granules from pituitary somatotropes (Coates *et al.,* 1968; Couch *et al.,* 1969; Wilbur *et al.,* 1974, 1975), or the increase in pituitary cAMP content (Borgeat *et al.,* 1973) cannot be denied. It is also interesting that crude hypothalamic extracts or semipurified GRF preparations, isolated and identified on the basis of positive BA results, were capable of inducing rises in the level of plasma RIA-GH when given systematically (Malacara and Reichlin, 1972), by direct intrapituitary injection (Frohman *et al.,* 1971; Müller *et al.,* 1972a), or by infusion into a hypophyseal portal vessel (Sandow *et al.,* 1972; Takahara *et al.,* 1975).

Partial purification of GH-releasing material from pig or ovine hypothalami that stimulated release of RIA-GH from rat hemipituitaries incubated *in vitro* has been reported by Wilber *et al.* (1971), Stachura *et al.* (1972), and Wilson *et al.* (1974). In these studies the GRF material was present in Sephadex column fractions eluted after the GH-depleting substance reported by Schally *et al.* (1971a). The material purified by Wilson *et al.* (1974) induced a dose-dependent increase in pituitary cGMP, which itself is a potent stimulant for GH release (Peake *et al.,* 1972) (see Section D), but did not affect pituitary levels of cAMP. More recently Nair *et al.* (1976) reported the isolation from bovine hypothalamic extracts of two polypeptide fractions capable of inducing a striking elevation in plasma RIA-GH when injected into adrenalectomized rats. Purification of small portions of the two fractions revealed the presence of a new peptide with GRF activity whose N-terminus seems to be occupied by a pyroglutamyl moiety. Somatotropes from the AP of rats infused into the hypophyseal portal vessels with this material exhibited massive emiocytic activity in the immediate postinfusion periods (Wilbur and Nair, 1976). Further validation is needed to ascertain whether this peptide represents a new GH-RH molecule.

Despite the problems associated with the characterization of a GRF, a number of established physiological and pathological facts have made it appear almost certain that the CNS exerts a stimulatory action on GH release (for a more exhaustive review of this topic, see Müller, 1973; Martin, 1973a; McCann *et al.,* 1974; Reichlin, 1974a). Early observations in man and in the laboratory animal have linked spontaneous diseases of the hypothalamus or hypothalamic damage of the pituitary stalk section, to retardation of growth; only when GH was administered, were normal growth rates restored. Irrefutable evidence that in the human GH secretion was under CNS control was given for the first time by Roth *et al.* (1964), who demonstrated that pituitary stalk section blocked hypoglycemia-induced GH discharge, an observation that since has been confirmed and expanded by more complete studies (Powell *et al.*, 1966; Antony *et al.*, 1969). Conditions which impair reflex GH discharge include tumors of many kinds, granulomatous disease, and several types of metabolic disturbances.

Precise identification of neural pathways involved in GH secretion has not yet been possible in the human but considerable progress has been made in animals with studies utilizing brain lesioning and with stimulation experiments. In the rhesus monkey, electrolytic lesions of the anterior or posterior ME or the inferior portion of the mammillary bodies will inhibit reflex GH responses to hypoglycemia (Abrams *et al.*, 1965) or emotional stress (Brown *et al.*, 1971). In the rat, electrical lesions confined to the VMN, sparing the ME, induce reduction of both pituitary and plasma GH; with increasing lesion size in and around the VMN, a progressive decrease in pituitary weight, pituitary and plasma GH, and linear growth were apparent (Frohman and Bernardis, 1968; Bernardis and Frohman, 1970). Bilateral electrolytic VMN lesions completely abolished the pulsatile secretion of GH or reduced the frequency and amplitude of GH pulses (Martin *et al.*, 1975a).

The opposite picture emerges from electrical stimulation experiments. Electrical stimulation of the VMN and MBH resulted in increased plasma GH levels (Frohman *et al.*, 1968), while stimulation elsewhere in the hypothalamus was ineffective (Martin, 1973b), except for the medial preoptic area whose stimulation caused GH to fall (Martin *et al.*, 1975a). That these regions are involved in GH secretion is also supported by the results of stimulation experiments in the rabbit (McIntyre and Odell, 1974) and dog (Lovinger *et al.*, 1975). It is worth noting that the GH response to VMN stimulation in the rat was clearly a rebound release occurring after termination of the stimulus, whereas during stimulation there was a significant fall in plasma GH levels (Martin, 1973b). Such a biphasic effect on GH release was also observed in sheep following electrical stimulation of the ME and was attributed to an initial release of GH-IF into the portal circulation followed by an intrapituitary rebound from the GH-IF induced inhibition (Malven, 1975a).

Positive, but immediate, GH response also occurred in the rat when the electrical stimulus was applied to brainstem areas, such as the interpeduncular n. and the locus coeruleus, or to "visceral brain" areas, such as the dorsal hippocampus and the basolateral amygdala (Martin, 1973b). An involvement of the hippocampus in the control of GH secretion is consistent with the positive responses obtained in the monkey following stimulation of this area (Smith and Root, 1971). In humans, elevations in GH levels were observed following electrical stimulation of the posterior hypothalamus (Sano *et al.*, 1970), amygdala and cingulum (Luedecke *et al.*, 1976). Stimulation of the corticomedial amygdala elicited lowering of plasma GH levels in the rat (Martin, 1972).

A number of phasic changes in GH secretion, including stress-induced ones, and sleep-associated GH release are probably mediated by the higher brain centers (see Müller, 1973; Martin, 1973a). Since limbic structures have different connections with the VMN (Murphy *et al.*, 1968), it appears that both excitatory and inhibitory inputs to the MBH from the limbic system may act to modulate GH secretion (see also Chapter IV, Sections A,2 and C,2).

b. CNS Glucoreceptors and Control of Growth Hormone Secretion

In view of the reported evidence, the MBH can be viewed as "the final common pathway" to which neural influences for the control of GH secretion are conveyed and, in conjunction with the adjacent LHA, as a center for the integrated homeostatic regulation of energy balance and food intake. 2-Deoxy-D-glucose (2-DG), a glucose analog which causes a state of glucoprivation (Brown, 1962) when injected into the midportion of the lateral hypothalamus, induced in the monkey a sharp increase in plasma GH (Himsworth *et al.*, 1972). Earlier studies had shown that local infusion of glucose into this region prevented the usual GH response to hypoglycemia (Blanco *et al.*, 1966). These data are compatible with the existence within the hypothalamus of glucosensitive elements, which in response to a lack of metabolizable glucose initiate a sustained secretion of GH. Shortage of carbohydrate substrate, which characterizes insulin- or 2-DG-induced glucoprivation, most likely acts to stimulate GRF-secreting structures (or to inhibit GIF structures) through the hypothalamic glucoreceptors. Within the same part of the hypothalamus, stimulation of the chemoreceptors by 2-DG glucoprivation also evokes a complex reflex response which affects eating behavior (Balagura and Kanner, 1971), induces hyperglycemia (Müller *et al.*, 1971b) dissociated from insulin release (Frohman *et al.*, 1973), increases epinephrine secretion (Himsworth, 1970) and gastric acid output (Colin Jones and Himsworth, 1970), and induces hypothermia (Freinkel *et al.*, 1972; Müller *et al.*, 1973a) (Fig. 34). These metabolic, hormonal, and behavioral responses which also include the pituitary release of GH can be viewed as reflex efforts to satisfy the "glucoprivic" receptors. It would appear from neuropharmacological and lesioning experiments (Müller *et al.*, 1972b, 1973b, 1974a) that different populations of glucoreceptor cells are involved in the mediation of glucoprivic responses and that monitoring of glucose utilization by the brain is not exclusively confined to distinct regions such as VMN and LHA (Müller *et al.*, 1974a; Shirsishi and Mager, 1974). In this connection, it is worth

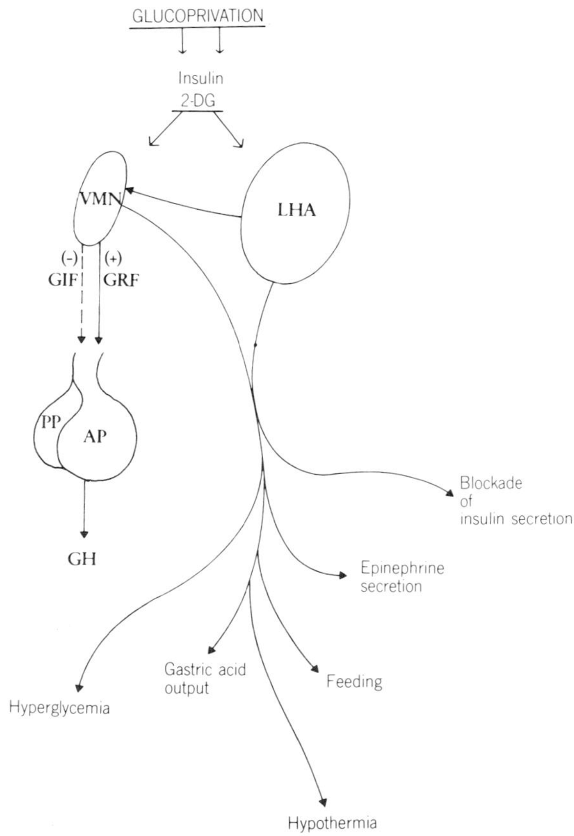

Fig. 34 Schematic representation of the hypothalamic glucoreceptor system involved in the neuroendocrine control of GH secretion. It appears that the increased GH secretion is but one of many neurally related metabolic, hormonal and behavioral responses to the state of glucoprivation. VMN, ventromedial nucleus; LHA, lateral hypothalamic area; GRF, GH-releasing factor neuron; GIF, GH-inhibiting factor neuron; AP, anterior pituitary; PP, posterior pituitary; GH, growth hormone; (+), stimulation; (−), inhibition.

noting that possible glucosensitive areas may be the basal ganglia, e.g., the globus pallidus and the striatum. Electrolytic lesions of the pallidus and striatum eliminated normal feeding responses during 2-DG glucoprivation (Neill and Linn, 1975), and it has been proposed that deficits in consummatory responses to regulatory challenges after electrolytic lesioning of the LHA (the LHA syndrome) (Teitelbaum and Epstein, 1962) are actually due to transection of the ascending nigrostriatal DA system (see Ungerstedt,

1971b). For interactions between glucoreceptors and MA systems see also Chapter IV, Section C,2).

c. *Somatostatin*

In the course of routine screening of fractions from Sephadex columns for GRF, Krulich *et al.* (1967, 1968) observed that certain fractions strongly inhibited the basal release of BA-GH and prevented GRF-stimulated release from rat pituitaries incubated *in vitro*. The inhibitory activity, called GIF, was separable from other RF's and could inhibit the release of GRF both *in vitro* and *in vivo* (Dhariwal *et al.*, 1969a). These results obtained originally by bioassay have also been confirmed by RIA data by Krulich *et al.* (1972) who, in a study on the regional distribution of GIF activity, showed that this activity was mainly contained in the ME; GRF activity was located in the medial hypothalamus in a region slightly dorsolateral to the VMN. Confirmation for the existence of GIF was given by Stachura *et al.* (1972b). They showed that ovine hypothalamic fractions purified on Sephadex G-25 almost completely inhibited GH synthesis and the release of newly synthesized GH (rat pituitary *in vitro* system).

Indirect proof for the existence of inhibitory CNS control and for its physiological significance has been derived from observations in subprimate and primate species. In the rat, surgical isolation of the MBH (Mitchell *et al.*, 1972; Collu *et al.*, 1973; Chihara *et al.*, 1975) and removal of the PO region (Rice and Critchlow, 1976) were associated, respectively, with increased linear growth and elevation in plasma GH levels and block of the stress-induced inhibition of GH secretion. In monkeys, anterior hypothalamic lesions extending to involve the optic chiasm led to exaggerated GH responses to stress (Brown *et al.*, 1971). In humans, abnormally elevated plasma GH levels were reported in a CNS disturbance, the diencephalic syndrome of infancy, which is usually associated with tumors of the optic chiasm and anterior hypothalamus (Bain *et al.*, 1966; Fishman and Peake, 1970). Conversely, electrical stimulation of the PO region decreased GH levels in man (Luedecke *et al.*, 1976).

The validity of the concept that CNS inhibitory influences may play an important role in the physiological regulation of GH secretion has been best confirmed by the isolation from sheep hypothalamus of a tetradecapeptide,

H-Ala-Gly-Cys-Lys-Asn-Phe-Phe-Trp-Lys-Thr-Phe-Thr-Ser-Cys-OH

[GH-RIH or SR-IF (somatotropin release-inhibiting factor or somatostatin)], of which two forms, one linear and the other cyclic (the natural form), have been reported (Brazeau *et al.*, 1973; Blackwell and Guillemin,

1973). After determination of its primary structure (Burgus *et al.*, 1973) and total synthesis (Coy *et al.*, 1973a; Rivier, 1974), this neurohormone was shown not only to suppress the release of immunoreactive GH from dispersed human and rat pituitary cells *in vitro* (Brazeau *et al.*, 1973) but pentobarbital- (Brazeau *et al.*, 1974a), morphine- (Martin *et al.*, 1975b) or suckling- (H. J. Chen *et al.*, 1974) stimulated release of GH from rat *in vivo* as well. Somatostatin, when administered in a suitable long-acting form, also inhibited the pulsatile GH release from the unanesthetized mature rat (Martin *et al.*, 1975a).

The inhibitory effect of GH-RIH on GH release has also been documented in other animal species. In dogs, somatostatin blocked GH release induced by L-dopa (Lovinger *et al.*, 1974a) and insulin-induced hypoglycemia (E. E. Müller, unpublished results). In conscious baboons, iv administration of the peptide suppressed the rise in serum GH normally resulting from insulin-induced hypoglycemia, from systemic iv infusion of L-dopa, and from microinjection of NE into the VMN of the hypothalamus (Ruch *et al.*, 1973). In normal humans, arginine, L-dopa, exercise- or hypoglycemia-induced GH release were inhibited by somatostatin (Siler *et al.*, 1973; Prange-Hansen *et al.*, 1973). It also effectively lowered GH levels in acromegalic subjects (Hall *et al.*, 1973).

Studies on the disappearance, distribution, and excretion of a somatostatin analog, with tyrosine in place of alanine at the N-terminus, have shown that the molecule has a half-life of 2–4 min in mice, rats, and man; that in rats it is concentrated mainly in the pineal and posterior pituitary and that its concentration in the anterior pituitary and hypothalamus is almost nil; and that after 24 hours 70% of it is excreted in the urine (Redding and Coy, 1974).

The existence of somatostatin forces a revision of the concept of hypothalamic control of GH secretion to one involving dual control. However, the physiological role of the inhibitory control of GH secretion, in view of the multiple effects of somatostatin (see below), is uncertain, and accumulated physiological data do not seem to necessarily involve such a control system. A careful review of GH secretion in most, if not all, species, reveals the presence of rather low baseline GH levels, interrupted intermittently by episodic increases in GH release. Such GH release could be due to intermittent GRF secretion without necessarily involving somatostatin action, whose function, at least in rodents, would be that of mediating stress-induced GH inhibition (Martin, 1974; Arimura *et al.*, 1976a; Terry *et al.*, 1976). However, in rats, recent experiments with somatostatin antiserum suggest that although SR-IF does not intervene in the pulsatile release of GH, it regulates the levels at which GH pulses occur (Ferland *et al.*, 1976a). Interestingly, in this study administration of the antiserum also led to an increase in basal TSH levels and potentiated the cold-induced TSH secretion, and in another study, also the TRH-induced TSH secretion (Arimura *et al.*, 1976b) (see below).

The inhibitory action of SR-IF is not confined to GH release; TRH-stimulated TSH secretion has also been shown to be inhibited (Hall *et al.*, 1973; Yen *et al.*, 1974a; Ruch *et al.*, 1973) and, under some circumstances, the TRH-induced PRL secretion (Vale *et al.*, 1973a; Besser *et al.*, 1974). SR-IF does not affect LH-RH stimulation of LH or FSH (Brazeau *et al.*, 1973; Hall *et al.*, 1973). More recently it has been shown that SR-IF is capable of inhibiting the high night levels of serum TSH in normal humans (Weeke *et al.*, 1975) and the increased plasma ACTH levels in patients with Nelson's syndrome (Tyrrel *et al.*, 1975), in patients with Cushing's disease after adrenalectomy without pituitary tumors, and in patients with Addison's disease (Fehm *et al.*, 1977). SR-IF also shows inhibitory action on the release of insulin, glucagon (Alberti *et al.*, 1973; Alford *et al.*, 1974; Mortimer *et al.*, 1974), gastrin, pepsin, and gastric acid (Bloom *et al.*, 1974), and renin (Rosenthal *et al.*, 1976) (Table XXII).

Consistent with these findings, which collectively point to a widespread and rather nonspecific action of the peptide, is the recognition of its ubiquitous distribution in the CNS and in the gastrointestinal tract of the rat (see Section C) and of the existence of a series of different molecular-weight forms of SR-IF in pig hypothalamus, all possessing part of the amino acid sequence of SR-IF (Schally *et al.*, 1975). The widespread distribution of somatostatin in the brain and its interaction with CNS depressant and stimulant drugs (Brown and Vale, 1975) suggest that similar to TRH (Brownstein *et al.*, 1974b) somatostatin may function both as a neurotransmitter in areas of the brain remote from the ME and as a release inhibitory hormone when it is secreted into the pituitary portal circulation (see also Chapter I, Section B,2 and this chapter, Section C).

In the pancreas of many species (man, sheep, ox, pig, rat, chicken) somatostatin seems to be present in particular cell elements, the α_1- or D-cells which differ from the α_2-cells, which store and secrete glucagon, and from the β-cells, which store and secrete insulin (Efendić *et al.*, 1976; Dubois, 1975). The abundance of somatostatin in special cells of the endocrine pancreas suggests, in view of the very short biological half-life of the exogenously administered peptide (Redding and Coy, 1974) and of its well-authenticated effect of acting as an inhibitor of the secretion of both glucagon and insulin by direct action on the endocrine pancreas (Gerich *et al.*, 1974a; Efendić *et al.*, 1976), an *in situ* synthesis of the neurohormone and not a nonspecific adsorption from the circulation (Arimura *et al.*, 1975a). Clarification of this point awaits the availability of sensitive assays for measuring somatostatin in peripheral blood (Kronheim *et al.*, 1976;

TABLE XXII

Some Biological Effects of Somatostatin on Endocrine or Endocrine-Like Cells

Target system	Species	Experimental or clinical situation	Effect	Authors
Pituitary	Rat	*In vivo*	Blocks spontaneous release[a]	Brazeau *et al.* (1973)
		In vivo	Blocks stimulated release	Brazeau *et al.* (1974a); Martin (1973b); Kato *et al.* (1975); Szabo and Frohman (1975)
		Enzymatically dispersed cells	Blocks release	Brazeau *et al.* (1973)
		Enzymatically dispersed cells	Blocks cAMP-induced release	Vale *et al.* (1972b)
		Enzymatically dispersed cells	Blocks PRL release	Vale *et al.* (1973a)
	Dog	*In vivo*	Blocks L-dopa-induced release	Lovinger *et al.* (1974a)
	Monkey[b]	*In vivo*	Blocks insulin-induced release	Ruch *et al.* (1973)
		In vivo	Blocks L-dopa-induced release	Ruch *et al.* (1973)
		In vivo	Blocks NE-induced release[c]	Ruch *et al.* (1973)
	Human	Normal	Blocks insulin-induced release	Hall *et al.* (1973)
		Normal	Blocks arginine-induced release	Siler *et al.* (1973)
		Normal	Blocks L-dopa-induced release	Siler *et al.* (1973)
		Normal	Blocks exercise-induced release	Prange-Hansen *et al.* (1973)
		Normal	Blocks sleep-induced release	Parker and Rossman (1973)
Pituitary		Normal	Blocks TRH-induced TSH	Hall *et al.* (1973)
		Normal	Blocks TRH-induced FSH	Hall *et al.* (1973)
		Normal	Inhibits high TSH levels	Weeke *et al.* (1975)
		Primary hypothyroidism	Inhibits high TSH levels	Lucke *et al.* (1975)
		Addison's disease	Inhibits high ACTH levels	Fehm *et al.* (1977)
		Nelson syndrome	Inhibits high ACTH levels	Tyrrel *et al.* (1975)
		Acromegaly	Inhibits GH hypersecretion	Hall *et al.* (1973)
Pancreas	Rat	*In vivo* perfusion	Blocks stimulated insulin and glucagon release	Gerich *et al.* (1975)
		In vitro	Interacts with secretin receptors	Robberecht *et al.* (1975)
	Dog	*In vitro* perfusion	Blocks stimulated insulin release	Alberti *et al.* (1973)
	Monkey[b]	*In vivo*	Lowers insulin and glucagon levels	Koerker *et al.* (1974)
	Human	Normal	Lowers basal insulin levels	Alberti *et al.* (1973)
		Normal	Blocks stimulated insulin release	Alberti *et al.* (1973)
	Human	Normal	Blocks glucagon secretion	Alford *et al.* (1974)
		Diabetic	Blocks glucagon secretion	Gerich *et al.* (1974b)
		Insulinomas	Inhibits insulin secretion	Christensen *et al.* (1975)
Gastrointestinal tract	Rat	*In vitro*	Blocks arginine-stimulated gastrin release	Hayes *et al.* (1975)
	Dog	*In vivo*	Blocks food, pentagastrin- and histamine induced gastric acid output	Barros D'Sa *et al.* (1975)
	Human	Normal	Inhibits gastrin response to feeding	Bloom *et al.* (1974)
		Anemia perniciosa	Lowers elevated gastrin levels	Bloom *et al.* (1974)
		Duodenal ulcer	Reduces pentagastrin induced acid output	Arnold and Creutzfeldt (1975)
Kidney	Human	Normal	Blocks frusemide-induced renin release	Rosenthal *et al.* (1976)

[a] Unless otherwise specified, the term "blocks" refers to GH release.

[b] Baboon.

[c] NE microinjected into VMN.

Ogawa *et al.*, 1976). If the embryological origin of the α_1-cells were the neural crest anlage (Weichert, 1970), it would explain how pancreatic cells secrete a peptide observed originally in the CNS.

What the mechanism(s) is that regulates the secretion of "pancreatic somatostatin" and its role in the physiological mechanism involved in the secretion of glucagon and/or insulin is presently unknown. The answer to this might be of significance for the understanding of the physiopathology of diabetes. Recently, by immunofluorescence techniques, the distribution of insulin-, glucagon- and somatostatin-containing cells has been evaluated in the islets of Langerhans in many animal species (for details, see Orci and Unger, 1975). The most striking feature in all species examined was the parallel distribution of α_2- and D-cells. Usually the D-cells formed a sparser layer in close proximity to the outer α_2-cell sheath, which occupied the most peripheral area of the islets, and peripheral to the great mass of the centrally located β-cells. Accordingly, it has been proposed that if islet somatostatin functioned as a local inhibitor of the neighboring α_2- and β-cells, it would have a greater influence on glucagon secretion than on insulin secretion (Orci and Unger, 1975). The finding that glucagon secretion is 20 times more sensitive to somatostatin than is insulin secretion supports this view (Gerich *et al.*, 1975).

i. Therapeutic Application of Somatostatin. As noted above, somatostatin has been used successfully in the treatment of acromegaly; however, its short duration of action requires constant infusion of the peptide, which is orally inactive. Discontinuation of infusion is followed by a rebound rise in plasma GH concentration (Hall *et al.*, 1973), the rapidity of which is in sharp contrast to the gradual onset of GH-inhibitory effects of somatostatin. It has been postulated that this "off signal" may result from the accumulation of a stimulant substance during the period of somatostatin administration, which is free to act when the neurohormone is quickly eluted (Carlson *et al.*, 1974). Until GH-RIH is available in a suitable long-acting form, its therapeutic application in acromegaly will be negligible. A prolonged inhibition of GH secretion (24–72 hr) in the rat has been reported with the use of acylated des-Ala^1-Gly^2 somatostatin analogs (Brazeau *et al.*, 1974b), but not confirmed by Ferland *et al.* (1976b). This form of therapy for acromegaly, however, is no longer as important because of the recent introduction of dopaminergic compounds that, in a high percentage of acromegalic patients, induce a marked lowering of GH levels (see Chapter V, Section A,2).

Because of its ability to lower plasma glucose and inhibit GH and glucagon secretion, and in view of the essential role of the latter in the pathogenesis of diabetes mellitus and in the development of diabetic ketoacidosis (Unger and Orci, 1975), it has been proposed that somatostatin may prove useful in the treatment of diabetes mellitus in man (Gerich *et al.,* 1974a,b). In juvenile diabetics somatostatin administration ameliorated the marked lability of blood glucose, and the insulin requirement was reduced by 30–50% (Gerich *et al.,* 1974b; Thum *et al.,* 1975). Moreover, because GH hypersecretion may contribute to the development of microangiopathy, the suppressive effect of somatostatin on GH secretion may benefit patients with the vascular complications of diabetes. It may not benefit equally patients managed on diet and/or oral hypoglycemic agents, since the peptide also inhibits insulin responses to both hyperglycemia and to tolbutamide (Gerich *et al.,* 1974a). This would limit its use to those patients requiring exogenous insulin. Development of somatostatin analogs possessing dissociated biological activities, e.g., capable of suppressing selectively pancreatic insulin or glucagon secretion (Efendić *et al.,* 1975; M. Brown *et al.,* 1976), would be helpful in this context.

Further diagnostic and therapeutic applications of somatostatin include patients with glucagon-producing pancreatic tumors in whom hormone levels are dramatically reduced and in whom blood glucose levels fall (Mortimer *et al.,* 1974) and patients with insulinomas or with pain and bleeding associated with gastric or duodenal ulcers (Bloom *et al.,* 1974). Also for use in these cases, analogs less quickly destroyed than somatostatin but with the same high affinity for the cellular receptor have been sought for long-term therapy.

d. Short Feedback Mechanism

The lack of a specific target gland for GH action suggested the likelihood that, as for prolactin, GH could act on pituitary or hypothalamic structures to regulate its own secretion. Proof for this has accumulated rapidly, especially from animal studies. In the rat, pretreatment with GH, transplantation of a GH-secreting tumor, and implantation of bovine GH or hGH into CNS sites were associated with blockage of the insulin-induced GH release or lowering of pituitary GH content and weight (see Müller, 1973). A similar effect has been observed in primates. In the monkey, iv infusion of GH for 2 hr inhibited GH response to insulin-induced hypoglycemia or Pitressin administration (Sakuma and Knobil, 1970). In the human, exogenous administration of GH for 6 days prevented hypoglycemia-induced elevations in GH (Abrams *et al.,* 1971). Rises in GH levels induced either by a stimulus or by infusion of hGH at a constant rate lessened or prevented the expected hypothalamopituitary response to exercise, arginine, or insulin-induced hypoglycemia (Hagen *et al.,* 1972a). From these studies it is assumed that elevated titers of circulating GH exert an "autofeedback" at the level of the hypothalamic GRF and/or GIF center and/or the pituitary level.

The basic nature of the GH autofeedback mechanism remains obscure, and the question as to whether the observed effects are due to GH itself or to a metabolically mediated response remains unanswered. The possibility of a response mediated by the GH-dependent factor somatomedin (Daughaday *et al.,* 1972), with high levels of the peptide inhibiting and, conversely, low levels increasing GH secretion, although unproved is, nevertheless, highly appealing. Though Abrams *et al.* (1971) did not actually imply that somatomedin is involved in the feedback mechanism, the time relations in their study would fit very well with the reported half life or $t_{1/2}$ of the peptide (Daughaday *et al.,* 1968). The ability of their subjects to secrete GH in response to insulin-induced hypoglycemia began to return about 12 hr after the exogenous hGH was discontinued, and was back to normal in about 48 hr. Similarly, in the study of Hagen *et al.* (1972a), arginine or insulin testings were performed after a time lapse sufficient to allow a GH-dependent increase of somatomedin in the plasma.

The existence of a somatomedin feedback system is consistent with the high concentration of plasma hGH in children who are unable to produce somatomedin (Laron *et al.,* 1966) or in patients severely affected by anorexia nervosa with low plasma somatomedin levels (Frankel and Jenkins, 1975), and with the decreased pituitary GH concentration found in rats infested with spargana of *Spirometra mansonoides,* a worm capable of inducing somatomedin generation and skeletal growth (Daughaday and Garland, 1972). The elevated hGH levels of the somatomedin-deficient children are reminiscent of similar elevated levels of pituitary tropic hormones seen in thyroidectomized and adrenalectomized patients, and raise the possibility of a long loop feedback of GH acting through somatomedin to inhibit GH secretion (Fig. 35). The somatomedin feedback system may alter the sensitivity of the hypothalamic neurotransmitter and/or neurosecretory neurons and thus serve to modulate their response to a variety of short-term metabolic and neurogenic stimuli (see Chapter IV, Section B,2). Another possibility which is not incompatible with the former stems from the chance observation that GH treatment of children led to a suppression of normal TRH-induced TSH response (Root *et al.,* 1973). It may be proposed (I. M. D. Jackson, quoted by Reichlin, 1974a) that GH treatment brings about the secretion of somatostatin, which inhibits GH secretion. Impaired TSH response to TRH would thus be explained on the basis of the established somatostatin inhibition of TSH secretion (see Section A,3,c). Whatever the mechanism is that is responsible for the autofeedback of GH secretion, it appears scarcely operative in acromegaly, since elevated levels of the hormone would be expected to inhibit further release of GH in response to provocative stimuli, a fact contradicted by the persistence of some form of regulatory control in this disease state (see Chapter V, Section A,2).

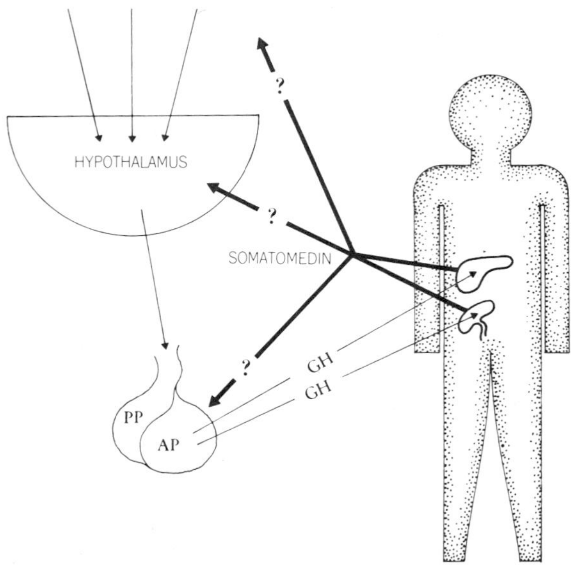

Fig. 35 Hypothetical long loop feedback for the autoregulation of GH secretion. Large arrows indicate the hypothetical sites of action of somatomedin generated by the liver and kidney. AP, anterior pituitary; PP, posterior pituitary; GH, growth hormone.

4. Follicle-Stimulating and Luteinizing Hormone

Two years after the discovery of TRH, another gap between hypothalamus and pituitary was bridged with the isolation of the peptide hormone that controls the secretion of pituitary gonadotropins, LH-releasing hormone/FSH-releasing hormone (LH-RH/FSH-RH). In 1971 Schally *et al.* (1971c) reported for the first time the isolation, amino acid composition, and chemical and biological properties of a nonapeptide, His^1-Arg^1-Ser^1-Glu^1-Pro^1-Gly^2-Leu^1-Tyr^1, which had both LH-RH and FSH-RH activity. In a subsequent paper, Matsuo *et al.* (1971a) reported the definitive structure of the neurohormone, a decapeptide containing, in addition to the nine amino acid residues mentioned above, one residue of tryptophan.

With the knowledge that both the N-terminal and C-terminal carboxyl groups were blocked, Matsuo *et al.* (1971a), using the Edman-dansyl procedure combined with the selective C-terminal titration method of Matsuo *et al.* (1966), sequenced digestion products of LH-RH following chymotrypsin and thermolysin, without separation of the fragments. On the basis of their data they concluded that the structure of porcine LH-RH is (pyro)Glu-His-Trp-Ser-Tyr-Gly-Leu-Arg-Pro-Gly-NH_2. This structure for porcine LH-RH was subsequently confirmed by Schally *et al.* (1971b) with larger quantities of porcine LH-RH. A similar structure has been reported for ovine LH-RH by Burgus *et al.* (1971, 1972), subsequent to the original report by Matsuo *et al.* (1971a). That this structure is correct was confirmed by the synthesis of this material by the solid-phase peptide synthesis of Merrifield (Monahan *et al.*, 1971; Matsuo *et al.*, 1971b), by the classic methods (Geiger *et al.*, 1971), and by a combination of both (Sievertsson *et al.*, 1971).

It soon appeared that the LH-RH decapeptide had a widespread distribution in the animal kingdom. Jeffcoate *et al.* (1974a), using a sensitive and specific RIA for LH-RH (Jeffcoate *et al.*, 1974b) combined with chromatographic procedures, showed that the naturally occurring peptides from rat, rabbits, and chicken hypothalami and synthetic LH-RH are immunologically and chromatographically indistinguishable and that hypothalamic extracts from these species contain only one peptide that is detected in the LH-RH RIA.

The material isolated from porcine hypothalamus stimulated the release of both LH and FHS *in vivo* as well as *in vitro* (Schally *et al.*, 1971e). Synthetic LH-RH/FSH-RH was shown to possess the same properties as the naturally occurring material (Schally *et al.*, 1971c; Burgus *et al.*, 1971). The amount of hormone secreted in response to LH-RH, however, was considerably higher for LH than for FSH (Blackwell *et al.*, 1973). Attempts to dissociate FSH-RH from LH-RH activity were unsuccessful; enzymatic and chemical inactivation of LH-RH activity was accompanied by parallel loss of FSH-RH (Schally *et al.*, 1971d,e,f). These results supported the hypothesis formulated by Schally and collaborators (White, 1970a; Schally *et al.*, 1971e) that one molecule is responsible for the release of both gonadotropins, a view in keeping with the finding that homogeneous decapeptide preparations stimulate the release of both FSH and LH *in vitro* and *in vivo* (Schally *et al.*, 1971e) and with the failure to detect an immunologically distinct FSH-RH in rat SME extracts (Shamanesh and Jeffcoate, 1976). Moreover, discharge of FSH, in addition to LH, is observed after administration of microgram doses of LH-RH in many mammalian species (White, 1970b; Reeves *et al.*, 1972; Arimura *et al.*, 1971), including the human (Kastin *et al.*, 1972a; Zarate *et al.*, 1972a).

Although the possibility cannot be excluded that another hormone, which releases only or predominantly FSH, FSH-RH, is present in hypothalamic tissue (for more details, see Schally *et al.*, 1973a; Blackwell and Guillemin, 1973), to explain the numerous dissociated secretions of the two hormones reported in the literature—i.e., increased LH, but not

FSH secretion during REM sleep (Rubin *et al.*, 1971); presence of 2 or 3 FSH secretory peaks during the normal menstrual cycle in contrast to the sole peak of LH at midcycle (Stevens, 1969); or temporal dissociation in women during the menstrual cycle between plasma LH and FSH peaks (Goebelsmann *et al.*, 1969)—it is not necessary to postulate the existence of a separate neurohormone for FSH release. Using intermittent intravenous infusions of LH-RH in immature male rats, it has been possible to dissociate LH and FSH increases 1 hr after the end of the infusion, since serum LH levels fell reaching control values, but FSH levels remained higher (Vilchez-Martinez *et al.*, 1975). The increase of FSH can be explained by the considerably longer biological half-life of FSH (Bogdanove and Gay, 1969). That a situation similar to that used in these experiments may possibly exist under physiological conditions was demonstrated by the secretion of LH which is pulsatile and shows peaks of immunoreactive material which rapidly disappear (Yen *et al.*, 1972; Vande Wiele and Dyrenfurth, 1973; Foster *et al.*, 1974) and reflects pulsatile release of LH-RH (Carmel *et al.*, 1976). Moreover, the preferential release of LH or FSH from the pituitary gland perhaps could be controlled by other factors, such as sex steroids (Schally *et al.*, 1971f). Thus, a physiological event such as puberty leads to a significant change in the gonadotropic response to LH-RH. Prior to puberty, both in boys and girls, LH-RH induces a small but significant release of LH and a larger response of FSH; in contrast, when the child reaches puberty and as pubertal development continues, LH response increases progressively while the FSH response does not differ significantly from that seen before puberty (Fig. 36) (see Franchimont and Legros, 1975).

Estrogens appear to play an important role in the differential feedback control of LH and FSH secretion. In *in vitro* experiments, Apfelbaum and Taleisnik (1976) observed that while the release of FSH induced by LH-RH was depressed by addition of estrogen to the incubated pituitary that of LH was not affected, a fact which stresses the importance of the hormonal milieu in determining the LH : FSH ratio of response. Tsai and Yen (1971) reported that ethinylestradiol given to normal women abolished the early follicular rise of gonadotropin levels observed in control cycles. The negative feedback effect of estrogens was greater for FSH than for LH levels which showed a significant rebound about 36 hours after cessation of treatment. The differential feedback action of estradiol on FSH and LH release was also evident from experiments in hypogonadal subjects treated with chronic administration of small doses of ethinylestradiol (Yen *et al.*, 1974b). In these subjects the pituitary response to LH-RH had a biphasic course of LH release: an initial enhancement (first 2 weeks) was followed by a diminution (third and fouth weeks). In contrast, FSH re-

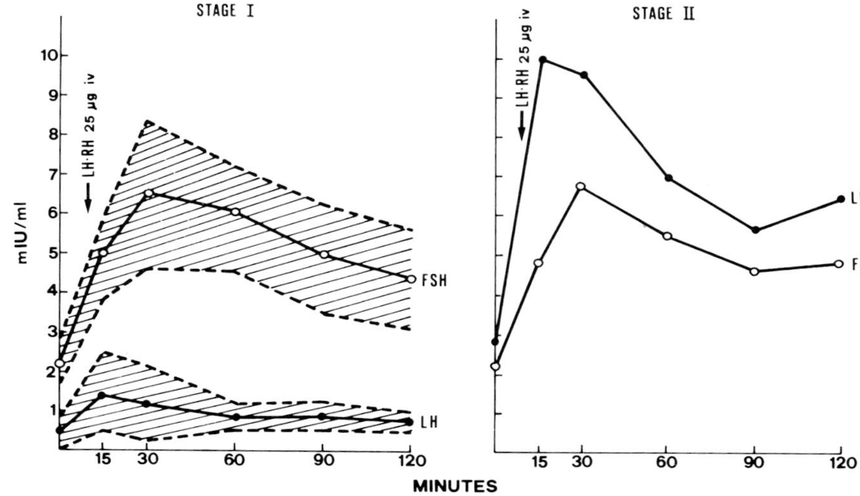

Fig. 36 Mean FSH and LH response to 25 μg of LH-RH in girls at Stages I (13 subjects) and II (9 subjects) of puberty. (Reproduced with permission from Franchimont and Legros, 1975.)

lease showed a preferential inhibition, with a progressive diminution of LH-RH responsiveness. Estrogens in small doses are also capable in female rats of augmenting the pituitary response of LH to LH-RH (Arimura and Schally, 1971), a finding which correlates well with events in the human menstrual cycle and the estrous cycle of animals. The rise in the concentration of plasma estrogen, which precedes the ovulatory peak of LH in rodents and primates, may increase the pituitary response to LH-RH Schally *et al.,* 1972b; Debeljuk *et al.,* 1972a). In contrast to estrogen, large doses of progesterone in cyclic rats (Arimura and Schally, 1970) and in rabbits (Hilliard *et al.,* 1971) or treatment of male rats with testosterone (Debeljuk *et al.,* 1972b) suppressed or increased the release of LH or LH and FSH, respectively. Combination of estrogen and progesterone suppressed the release of LH in female rats and anestrous ewes (Debeljuk *et al.,* 1972a). Similarly, during the menstrual event, the huge amounts of estrogen and progesterone secreted after ovulation may decrease pituitary response to LH-RH (Schally *et al.,* 1972b; Debeljuk *et al.,* 1972a).

Collectively, these and other data suggest that sex hormones acting at pituitary and/or hypothalamic (Mowles *et al.,* 1971) or extrahypothalamic (Stumpf and Sar, 1973) levels affect gonadotropin secretion. Thus, it is not inconceivable to postulate that, ultimately, the differential release of LH and FSH in response to LH-RH may result from the complex interactions of the steroids at the multiple sites discussed above.

Pure, naturally occurring and synthetic LH-RH/FSH-RH significantly influenced both the release and synthesis of LH and FSH in rat AP cultures and increased the incorporation of tritiated glucosamine into LH (Redding *et al.*, 1972; Liu *et al.*, 1976). Similar results have been previously reported for a purified LH-RH/FSH-RH preparation (Mittler *et al.*, 1970). Electron microscopy studies demonstrated that synthetic LH-RH given intracarotidly releases granules from LH cells of rats in persistent estrus (Shiino *et al.*, 1972). More recent studies have shown that the neurohormone can also affect extrusion of LH granules from intact, hemicastrated and castrated rats of either sex as also indicated by an immunocytochemical method using peroxidase-labeled antibody (Mendoza *et al.*, 1973).

With regard to the mechanism by which LH-RH/FSH-RH induces activation of pituitary gonadotropin secretion, evidence has been presented for an increased cAMP formation and AC activity in incubates of rat anterior pituitary exposed to the synthetic hormone ($1 \times 10^{-10} - 1 \times 10^{-9}$ *M*) (Borgeat *et al.*, 1973; Makino, 1973). A close correlation was also observed between rates of LH and FSH release and changes of intracellular cAMP concentrations. This close parallelism between changes of intracellular cAMP concentrations and hormone release holds true also for LH-RH analogs (Borgeat *et al.*, 1974), adding strong support to the evidence for the role of cyclic nucleotide as mediator of LH-RH action (see also Section D).

The biological potency of LH-RH/FSH-RH in various animal species is consistent with the widespread distribution of the neurohormone in the animal kingdom. In the hamster, subcutaneous injections of minute amounts of the hormone induced ovulation and increased the concentration of LH in the plasma (Arimura *et al.*, 1971). Similar effects were obtained following its administration to estrous rabbits (Hilliard *et al.*, 1971). Increased plasma LH and FSH concentrations have also been reported in sheep, following LH-RH/FSH-RH administration by different routes (Reeves *et al.*, 1970, 1971), in Hereford bulls, and prepubertal female pigs (Chakraborty *et al.*, 1973; Baker *et al.*, 1973). Also in anestrous mares a single dose of the neurohormone resulted in increased plasma LH concentrations (Ginther and Wentworth, 1974). LH release was also induced in cockerels given LH-RH (Furr *et al.*, 1973) and premature ovulation in domestic fowl (Van Thienhoven and Schally, 1972). Different from other species, in monkeys LH-RH caused only a small LH increase (Arimura *et al.*, 1973a). Table XXIII reports the effects of natural and synthetic LH-RH/FSH-RH in animals and man.

The stimulation of LH and FSH release in human beings after the administration of highly purified preparations of porcine LH-RH was first

TABLE XXIII

Endocrine Effects of LH-RH/FSH-RH in Animals and Man

Animal species	Sex[a]		Effects	Authors
Frog		♀	Release of LH; ovulation	Thornton and Geschwind (1974)
Chicken		♀	Release of LH; premature ovulation	Furr *et al.* (1973); Van Thienhoven and Schally (1972)
Rat	♂	♀	Release of LH and FSH *in vivo* and *in vitro*	Schally *et al.* (1968a)
	♂	♀	Stimulation of LH and FSH synthesis *in vitro*	Mittler *et al.* (1970); Liu *et al.* (1976); Redding *et al.* (1972)
	♂̸	♀ ♀̸	Extrusion of LH-containing pituitary granules *in vivo*	Shiino *et al.* (1972); Mendoza *et al.* (1973)
	♂		Stimulation of pituitary cAMP *in vitro*	Borgeat *et al.* (1973)
	♂		Stimulation of spermatogenesis	Debeljuk *et al.* (1973)
		♀	Stimulation of follicular maturation	Schally *et al.* (1973c)
		♀	Ovulation	Schally *et al.* (1973c)
Hamster		♀	Release of LH; ovulation	Arimura *et al.* (1971)
Rabbit		♀	Release of LH; ovulation	Hilliard *et al.* (1971); Amoss *et al.* (1972)
Nutria (*Myocastor coypus*)	♂̸	♀̸	Release of LH	Schally *et al.* (1969b)
Sheep	♂	♀	Release of LH	Reeves *et al.* (1970); Arimura *et al.* (1972a)
Pig		♀	Release of LH; ovulation	Chakraborty *et al.* (1973); Baker *et al.* (1973)
Horse		♀	Release of LH	Ginther and Wentworth (1974)
Cattle	♂		Release of LH	Golter *et al.* (1972)
Monkey	♂		Slight LH release	Arimura *et al.* (1973a)
Man	♂	♀	Release of LH and FSH	Kastin *et al.* (1969, 1972c)
	♂		Increased testosterone secretion	Judd *et al.* (1974)
	♂		Stimulation of spermatogenesis	Zarate *et al.* (1973c); Mortimer *et al.* (1974)
		♀	Ovulation	Zarate *et al.* (1972a)
(acromegaly)	♂	♀	Increased GH secretion	Rubin *et al.* (1973); Faglia *et al.* (1973)

[a] ♂̸, Castrated male; castrated female = ♀̸

reported in 1969 (Kastin *et al.*, 1969). Since then an impressive series of clinical studies have demonstrated that injection of purified porcine or synthetic LH-RH increased plasma levels of LH and FSH in children and in men and women under a variety of conditions, induced ovulation confirmed by pregnancy in hypothalamic amenorrhea, and stimulated androgen secretion and spermatogenesis in male infertility (for more details, see Besser and Mortimer, 1976). These data indicate that the neurohormone may be useful in the treatment of sterility in man. Since administration of LH-RH might duplicate the normal events which occur in a woman's body, it is hoped that careful standardization of the dose would avoid superovulation and the resultant multiple births often occurring after administration of pituitary preparations or clomiphene.

Extraction and fractionation on CMC-cellulose columns of serum samples from various species containing immunoreactive LH-RH followed by RIA of the fractions showed that four immunoreactive components of circulating LH-RH could be identified (Jeffcoate and Holland, 1974b). One of these had a mobility identical with that of synthetic LH-RH; of the others, two were eluted before and one after synthetic LH-RH. The nature, site of formation, and possible significance of the extra components are not presently clear.

Calculation of the disappearance time of synthetic ^{3}H-labeled LH-RH in the human gave a half-life of 4 min (Redding *et al.*, 1973a), which agrees with that found by Arimura *et al.*, using a specific RIA (Arimura *et al.*, 1973b). The volume distribution expressed as a percentage of body weight was 8.7%. Extraction of the urine followed by chromatography on CMC and thin-layer electrophoresis revealed two major radioactive metabolites, i.e., pyroglutamic acid (5–10%) and the dipeptide, pyroglutamylhistidine (80–90%). None of the radioactive products found had a mobility identical with that of the intact molecule. Since urine does readily inactivate LH-RH (Jeffcoate *et al.*, 1973), these data suggest that inactivation of LH-RH (probably) occurs before the metabolites are excreted in the urine. Estimation of the production rate of LH-RH in man on the basis of either blood or urine clearances and using radioimmunoassay and specific rat antiserum (Jeffcoate *et al.*, 1974b) gave values of 1.5 μg/day or between 0.7 and 4.0 μg/day, respectively (Jeffcoate and Holland, 1974a).

The very short biological half-life of LH-RH indicates that a simple iv bolus injection of the releasing factor is not sufficient to cause elevation of the LH concentration for long enough to ensure ovulation.

Changes in the C-terminus of the peptide have resulted in derivatives with longer biological half-lives. Fujino *et al.* (1972) synthesized an alkylamine derivative [$Gly^{10}(Pro^{9})$]LH-RH ethylamide which was found to be 2–3 times as potent as the parent hormone in inducing ovulation. Rippel *et al.* (1973) showed that the enhancement of activity of this analog may be due to its increased effective half-life. An even more active analog appears to be [D-Ala^6-Des-Gly-NH_2^{10}]LH-RH ethylamide, whose LH-RH activity corresponds to a multiple of the activities of [D-Ala^6]LH-RH (7 times) and [Des-Gly-NH_2^{10}]LH-RH ethylamide (5 times) (Arimura *et al.*, 1974). Sustained LH release for more than 5 hr has been recently reported in men after oral administration of another superactive analog [D-Leu^6-Des-Gly-NH_2^{10}]LH-RH (Gonzales-Barcena *et al.*, 1975). Another approach to prolonging the duration of action of LH-RH is to attach LH-RH or biologically active LH-RH analogs to soluble polymers. In this context, Amoss *et al.* (1974) reported the extended biological activity as resulting from the coupling of [D-Lys^6]LH-RH to a water-soluble polypeptide, polyglutamic acid. Three hours after injection of the polymer-coupled LH-RH analog, plasma LH and FSH increases were still present, while the activity of a biologically equivalent amount of LH-RH had disappeared within 1 hr and that of the unconjugated [D-Lys^6]LH-RH had disappeared within 3 hr. The possibility has also been envisaged of coupling in a similar manner LH-RH antagonists (see below).

The establishment of the structure of LH-RH/FSH-RH and its synthesis and the availability of its structural analogs or other derivatives should greatly facilitate new approaches to the control of fertility in humans. One major approach is based on the development of a specific antiserum capable of neutralizing endogenous LH-RH. In animals (rabbits, guinea pigs, and ewes), LH-RH antibodies have been produced (Arimura *et al.*, 1973b; Kerdelhué *et al.*, 1973; Jeffcoate *et al.*, 1974b; de la Cruz *et al.*, 1976) (see Section B,2). In rabbits, repeated injections of synthetic LH-RH, adsorbed on polyvinylpyrrolidone, resulted in the production of a specific antisera to LH-RH. The animals producing the antisera showed reduction of pituitary LH and marked atrophy of the testes, suggesting that the antisera interfered with the action of endogenous LH-RH/FSH-RH. The same antisera were subsequently found to be capable of preventing the expected rise of serum LH and FSH in the afternoon of the day of proestrus and of blocking ovulation when administered to cyclic female rats (Arimura and Schally, 1975). In hamsters, administration of a sheep anti-LH-RH serum was effective in completely blocking ovulation when administered at any stage of the estrous cycle (de la Cruz *et al.*, 1976). If, similarly, antisera capable of neutralizing LH-RH could be produced directly in humans, women could be immunized against their own neurohormone with an ensuing decrease in FSH and LH secretion and blockade of ovulation. However, the clinical safety of these materials has yet to be proved.

Another promising method is based on the synthesis of chemical analogs of LH-RH/FSH-RH, i.e., compounds closely related structurally but not identical to it. A suitable analog should be (a) devoid of intrinsic LH-RH activity but capable of blocking the pituitary receptors by competing with the binding of endogenous LH-RH, and (b) resistant to *in vivo* degradation. Several analogs of the decapeptide with either a substitution [Gly^2]LH-RH or a deletion [Des-His^2]LH-RH showed antagonistic properties to LH-RH tested on cultures of dispersed rat anterior pituitary cells (Vale *et al.*, 1972a). The antagonistic activity of [Gly^2]LH-RH is obscured, however, by the fact that it is also a partial agonist and [Des-His^2]LH-RH does not inhibit the LH-RH induced stimulation of LH-release *in vivo* in rats or ovulation in rabbits (Schally *et al.*, 1973b). It was then found that intracarotid infusion of [Des-His^2-des-Gly^{10}]LH-RH ethylamide into estrogen-progesterone ovariectomized rats effectively blocked responses to subsequent intravenous injections of LH-RH. The octapeptide itself possessed only minimal LH-releasing activity (Coy *et al.*, 1973b). Subsequent studies have shown that for a given antagonist, inhibition is enhanced by incorporating either the D-amino acid in position 6 or C-terminal ethylamide, although if both these changes are made in the same analog increased inhibition is often masked by an increase in inherent gonadotropin-releasing activity (Coy *et al.*, 1975). The search for effective inhibitors of LH-RH is now being actively pursued (Ferland *et al.*, 1976c; Labrie *et al.*, 1976a), and chances for the development of a new contraceptive method appear promising.

5. Prolactin

a. Inhibitory Control

It is a widespread notion among students of neuroendocrinology that unlike all other pituitary hormones, but like MSH, the secretion of prolactin in mammals is mainly under a tonic inhibitory influence of the CNS. In the 1950's Desclin (1950) and Everett (1954) observed that in the rat removal of the pituitary from its normal site and transplantation to the kidney capsule caused prolonged maintenance of the corpora lutea (indicative of PRL action) and mammary growth stimulation. It was then shown (see Meites *et al.*, 1963) that placement of lesions in the ventral hypothalamus, disruption of the hypothalamic–pituitary portal circulation, culture or incubation of quartered rat pituitaries was followed by an increase in peripheral prolactin, whereas addition of crude acid extracts of rat hypothalamus to pituitary fragments induced a decrease in PRL secretion when compared to controls. The action of hypothalamic extracts was not duplicated by cerebral cortex extracts (CE). PRL release-inhibiting (PIF) activity was also demonstrated *in vitro* in hypothalamic extracts from the sheep, cow, pig (Pasteels, 1961; Schally *et al.*, 1965; Meites,

1966). Injections of rat hypothalamic extracts into rats were also found capable of preventing the depletion of pituitary PRL induced by suckling (Grosvenor *et al.,* 1965) or cervical stimulation (Kuroshima *et al.,* 1966).

Most of these data on the secretion of PRL were obtained with the systemic pigeon crop sac assay of Riddle *et al.* (1933) or the intracutaneous modification of this method as originally described by Lyons and Page (1935). Evaluation of the specificity of the crop sac bioassay and its correlation with a specific RIA (Raud and Odell, 1971) revealed that when purified PRL preparations were measured, potency estimates by BA were in agreement with those by RIA, but in all cases of a series of crude bovine pituitary extracts, the BA gave higher PRL in potency than the immunoassay, a fact attributable to contamination by ACTH.

Many of the conclusions that had been reached using as end point the crop sac bioassay were subsequently confirmed by specific RIA methods for prolactin. Chen *et al.* (1970) showed that transplantation of a single AP under the kidney capsule of a hypophysectomized female rat resulted in elevated serum PRL concentrations, as measured by RIA, a finding recently confirmed by Panerai *et al.* (1977a). Transplantation of two or four pituitaries resulted in correspondingly higher levels of serum PRL. Similarly, bilateral lesions in the ME or anterior or posterior hypothalamus of ovariectomized rats produced significant elevations in serum PRL as compared to sham-lesioned controls (Chen *et al.,* 1970; Welsch *et al.,* 1971).

Less consistent results were obtained when trying to demonstrate specific activity in extracts from animals of different species either *in vitro* or *in vivo* using the RIA. Arimura (1970) observed that infusion over 30 min iv into the abdominal aorta of freshly prepared acid extracts of rat SME into rats with pituitary grafts under the kidney capsule did not change serum PRL levels significantly. Similarly, Blackwell and Guillemin (1973) were unable to lower plasma PRL levels in rats bearing pituitary transplants following the injection of up to 15 fragment equivalents of ovine hypothalamic extracts. In a further study, Arimura *et al.* (1972b) showed that in Nembutal-anesthetized rats, whose baseline PRL levels are increased, infusion of either 3 SME or CE for 30 min markedly decreased serum PRL levels. In contrast, in animals blocked pharmacologically by perphenazine or reserpine or after surgical ablation of the medial hypothalamus, neither the infusion of CE or SME resulted in a significant reduction of serum PRL. More positive results have been reported by Kamberi *et al.* (1971c), who infused rat hypothalamic extracts directly into the hypophyseal portal vessels of rats and found significant increases in serum LH and FSH levels with concomitant decrease in serum PRL. Also, Amenomori and Meites (1970) reported that injection of SME but not CE reduced serum PRL levels in ether-anesthetized rats. No valid explanation is available to explain the divergent results of these authors; the differences in their findings perhaps reflect differences in routes of extract administration, type of anesthesia, concomitant presence in the extract of a prolactin-releasing factor (PRF). More recently, Takahara and his colleagues (1974a) were able to show that a purified porcine hypothalamic preparation with high PIF activity as tested by an *in vitro* assay, when infused into the hypophyseal portal vessels of urethane-anesthetized male rats, decreased serum PRL levels significantly, as compared to the respective preinfusion levels. However, the PIF activity of the extract could be wholly, or at least in part, accounted for by the CA content of the preparation (see Chapter IV, Section A,4,a,i).

Chemical characterization of PIF has not yet been reported. A substance that inhibits the suckling-induced reduction in pituitary PRL concentration has been recovered in retarded fractions from Sephadex G-25 (Dhariwal *et al.,* 1968). In this study a clear-cut separation of PIF from the LH-RH zone was not achieved. Partial separation of this material from LH-RH was reported by Schally *et al.* (1968a). More recently, chromatographic fractionation of an acetic acid extract of bovine pituitary-SME revealed material with inhibitory activity in the following order of decreasing sensitivity: PRL > GH > TSH; a more highly purified fraction, at the doses tested, inhibited PRL only (Dular *et al.,* 1974). The most highly purified preparations of PIF contained peptide material, as indicated by the reaction with Lowry reagent and by the formation of additional ninhydrin-positive components after acid hydrolysis. The results obtained could not be accounted for by the presence of MA's in the extract (see Chapter IV, Section A,4,a,i) since, in the *in vitro* incubation system used, MA's had no effect (Dular *et al.,* 1974). However, in very recent purification studies Schally *et al.* (1976) were unable to dissociate the PIF activity of porcine hypothalamic extracts from their CA content. The same authors, however (Schally *et al.,* 1977), recently purified and isolated from pig hypothalami a neutral-acidic fraction chromatographically distinct from CA's which possessed PIF activity. The PIF-active substance was identified as GABA (see also Chapter IV, Section 4,C). Evidence in favor of the existence of a CA-free prolactin-inhibiting factor in rat and pig hypothalamic extracts has also been given by Enjalbert *et al.* (1977) and by Dupont and Redding (1975).

Many physiological stimuli or drugs affect prolactin release, and some of them act by changing the hypothalamic synthesis and/or release of PIF. Thus suckling, administration of estrogen, progesterone, testosterone, cortisol, reserpine, CPZ, haloperidol, α-MD, α-MpT, α-MmT, and amphetamine reduced hypothalamic PIF activity and increased PRL release (Meites *et al.,* 1972; Meites and Clemens, 1972). Conversely, L-dopa, iproniazid, pargyline (Lu and Meites, 1971), ergot alkaloids (Wuttke *et al.*, 1972), and PRL itself (Clemens *et al.*, 1969; Voogt and Meites, 1971) increased in rats hypothalamic PIF activity and decreased serum PRL concentration. It has been reported that the rate of secretion of PIF into stalk blood of male rats was increased markedly by IVT-administered DA (Kamberi *et al.*, 1970a), while serum PRL levels decreased (Kamberi *et al.*, 1971a). Similarly, injection of L-dopa into intact or hypophysectomized rats with an ectopic pituitary was found to decrease serum PRL, to increase hypothalamic PIF and induce its release into general circulation (Lu and Meites, 1972) (see also Chapter IV, Section A,4,a). Table XXIV lists the effects of some physiological or pathological states on PRL release in man and animals.

b. Prolactin-Releasing Factor

Although undoubtedly the vast physiological evidence reported above supports the view that PRL secretion by the mammalian AP is under tonic

TABLE XXIV

Serum Prolactin Levels in Animals and Man in Various Physiological or Pathological States

Physiological or pathological states	Animals	Man
Fetus	—	Very high
Newborn	Low	Very high
Puberty	Increase in rats	Increase in females
Mature female vs males	High in females	Slightly higher in females
Aging	High in rats	Lower in females Higher in males
Estrus or menstrual cycle	Increase at proestrus	Increase at midcycle and luteal phase
Pregnancy	Increase	Gradual increase
Postpartum lactation	Increase during suckling	Increase during suckling
Circadian periodicity	Circadian changes	Sleep-related changes
Insulin hypoglycemia	No change	High
Stresses	Increase	Increase
Estrogen administration	Increase	Increase at high doses
Spontaneous galactorrhea	High	High in most cases
PRL-secreting tumors	High	Absence of sleep-related changes
Primary hypothyroidism	No change	Most often within normal range
CNS-AP disconnection	High	High
Acromegaly	—	High
Mental depression	—	High or normal
Renal failure	—	High
Cushing's disease	—	Absence of sleep-related changes

inhibitory hypothalamic control, several findings have suggested that there may also be a stimulatory component in the hypothalamic regulation of PRL secretion. Meites *et al.* (1960) observed that injections of neutralized acid extracts of rat hypothalamus into estrogen-primed rats induced mammary secretion, an effect which was also duplicated by the hypothalami of milking cows (Weller, 1963). Mishkinsky *et al.* (1968) found that the mammogenic activity of hypothalami from postpartum rats was higher than that from virgin rats, and attributed their findings to the presence of a hypothalamic prolactin-releasing (PRF or PRA) activity. However, initiation of lactation in rats does not appear to be an exclusive property of PRL alone, since adrenal corticoids are also essential. Moreover, many specific stimuli can induce mammotropic–galactopoietic responses in estrogen-primed rats (Meites *et al.*, 1963). Subsequent studies of Nicoll *et al.* (1970) provided more direct evidence of a PRF. They showed that incubation of rat pituitary with hypothalamic extracts, after an initial period of inhibition, stimulated PRL secretion. These results, however, could not be confirmed, since Meites (1970) under slightly different *in vitro* conditions, observed only inhibition of PRL release. Further evidence for the existence of a PRA in the hypothalamus was given by Arimura *et al.* (1969), who found that intracarotid infusion of pig SME in sheeps and goats caused a significant increase in RIA-PRL levels.

Also physiological studies on acute regulation of PRL secretion were amenable to interpretation either by an acute release of a PRF or by the acute inhibition of PIF secretion. Thus in the goat (Bryant *et al.*, 1970) and cow (Johke, 1970), milking, oxytocin, and emotional stress, and in the rat, suckling (Amenomori *et al.*, 1970), ether anesthesia (Wuttke and Meites, 1970), hemorrhage, and laparotomy (Neill, 1970) induced a rise in plasma PRL levels. The observation that ether stress which produces PRL release in normal rats was equally effective in stimulating PRL release in reserpine-treated animals, in which a loss of tonic PIF control should have been present, led Valverde *et al.* (1973) to postulate that the stress-induced release of PRL was not due to inhibition of PIF secretion but, most likely, to a prolactin-releasing substance. Consistent also with the existence of a PRF is the finding that in sheep electrical stimulation of the anterior ME led to an immediate rise in plasma PRL levels (Malven, 1975a) and that similar results were obtained by stimulating electrically the POA (Malven, 1975b), which is known to project to the anterior region of the ME.

Significant contribution on behalf of a PRF was also given by Valverde *et al.* (1972), who showed that whole extracts of porcine and rat hypothalamic tissue induced a significant rise in RIA-PRL levels in estrogen–progesterone-sensitized male rats. Chromatographic separation of PRA by Sephadex G-10 demonstrated that it was concentrated in two distinct, separate fractions and could be separated from other recognized hormones of the hypothalamus endowed with prolactin-releasing activity (oxytocin, vasopressin, and TRH). In these studies attempts to demonstrate hypothalamic PIF were unsuccessful. Consistent with the findings of Valverde *et al.* (1972), are those, noted above, of Dular *et al.* (1974). In addition to a PIF, they showed that further purification on Sephadex of fractions from acetic acid extracts of bovine SME was able to achieve separation of inhibiting and releasing material. One fraction distinct from TRH enhanced release of PRL from ovine AP *in vitro* and had no effect on the release of other hormones. Experiments with drugs have also suggested the possibility of a positive control for PRL release in the mammalian hypothalamus (See Chapter IV, Section A,4,a,ii). In addition, preliminary evidence has been given for the existence of circulating PRF in human plasma (Hagen *et al.,* 1976).

In contrast to mammals, in which the existence of a hypothalamic stimulation of PRL release in still a matter of debate (Meites, 1973), there is general agreement that the avian hypothalamus possesses only PRA. This has been demonstrated in hypothalamic extracts from the pigeon, tricolored blackbird, Japanese quail, chicken, duck, and turkey (Meites and Nicoll, 1966; Meites *et al.,* 1972). No satisfactory explanation can be offered for the difference in hypothalamic regulation of PRL secretion in mammals and birds. The very scarcity of DA in the outer layer of SME of avian hypothalamus (Calas, 1977) may be the reason for the lack of PIF activity in birds (see also Chapter IV, Section A,4,a,i). Apparently, agents capable of stimulating PRL release when injected into mammals (e.g., estrogens, androgens or reserpine) have no effect on PRL release in pigeons.

c. *TRH and Prolactin Release*

Although unequivocal evidence for the existence of a distinct hypothalamic PRH (prolactin-releasing hormone) still requires its isolation and chemical characterization, there is no doubt that another hypothalamic neurohormone, TRH, is endowed with prolactin-releasing activity. Tashjian *et al.* (1971) observed that TRH stimulates PRL production *in vitro* in clonal strains of rat pituitary cells removed from pituitary tumors. Indeed pGlu-His-ProNH_2 stimulates the release of PRL and TSH in man, monkeys, sheep, and cattle (see also Section A,6,a). The effect is rapid after intravenous injection of TRH, with peak PRL concentrations occurring at 15 min. Prolactin responses, like those of TSH, are significantly greater in normal women than in men; administration of TRH to women, but not to men, with hypothyroidism elicits a greater than normal PRL as well as TSH response (Bowers *et al.,* 1971a), while in hyperthyroid patients the response is reduced (Bowers *et al.,* 1971a, 1973).

Also, in subprimate species, TRH-induced release of prolactin could be documented. Release of prolactin after TRH has been observed in the pigeon and the domestic fowl in *in vivo* and *in vitro* studies (Hall *et al.*, 1975). In the rat, TRH has little influence on the PRL secretion of normal AP cells in culture (Vale *et al.*, 1973b) or in untreated normal male rats (Mueller *et al.*, 1973). TRH stimulates pituitary cells of rats rendered chronically hypothyroid (Vale *et al.*, 1973b) and *in vivo* increases PRL secretion in the estrogen–progesterone pretreated (EP) male rat (Rivier and Vale, 1974) and in the proestrous female rat (Mueller *et al.*, 1973). In the EP rat, a series of structural analogs of TRH were found to have comparable ability to stimulate both PRL and TSH (Rivier and Vale, 1974).

The above and similar observations on TRH-induced TSH and PRL release have raised the fundamental question of whether TRH plays a physiological role in the regulation of PRL secretion and whether there is a PRF other than TRH in the hypothalamus. Indirect evidence in favor of the view that TRH is the primary regulator of PRL secretion is the observation that in the human the TRH-TSH and PRL dose-response ranges parallel one another; the minimum effective dose (about 10 μg) found to stimulate TSH and PRL release in normal men and women is essentially the same (Bowers *et al.*, 1973). Administration of TRH antibodies suppressed plasma PRL levels in rats (Koch *et al.*, 1977).

Moreover, although the question of whether there is a PRF in the hypothalamus that is not TRH cannot be conclusively resolved with crude or partially purified extracts, it was suggested that the PRL-releasing activity of hypothalamic extracts could be explained by their TRH content (Rivier and Vale, 1974). This suggestion, however, is inconsistent with reports of chromatographic separation of PRL-releasing activity from TRH (see Section A,5,b) and, more recently, of retention of hypothalamic PRL-releasing activity under conditions which destroy or remove the biological activity of TRH (Machlin *et al.*, 1974; Boyd and Jackson, 1975; Szabo and Frohman, 1976). Furthermore, it appears that under a number of physiological or pathophysiological conditions, TSH and PRL can be secreted independently and that brain MA's affect differently the secretion of each hormone (see Chapter IV, Sections A,4,a,b and A,5). Serum PRL, but not TSH, increases in animals and man after section of the pituitary stalk (Lister *et al.*, 1974) or after administration of CPZ (Turkington, 1972a). Stress has been reported to inhibit TSH–thyroid function (Reichlin, 1966), but to stimulate PRL release (Meites *et al.*, 1972). During breast feeding, mammals have high serum levels of PRL but not TSH (Meites *et al.*, 1972). Collectively, the latter observations suggest that most of the stimuli known to release PRL do not release TRH–TSH–thyroid hormones.

A view which is now emerging to account for the dissociated secretion of PRL and TSH under a variety of circumstances postulates that, in addition to the reasons noted before, it may also reflect the action of agents that influence differently at the pituitary level TSH or PRL release. Thus, DA has been shown capable of suppressing *in vitro* PRL but not TSH release (Meites *et al.*, 1972) and of inhibiting TRH-induced increases in the release rate of PRL but not TSH (Chen and Meites, 1975), although recently an infusion of DA was reported to inhibit in the human both the TSH and the PRL response to TRH (Besses *et al.*, 1975; Burrow *et al.*, 1976). Conversely, somatostatin suppresses the TRH-induced TSH release without impairing the PRL response to TRH (Vale *et al.*, 1973a; Szabo and Frohman, 1976) (see also Section A,3,c and Chapter IV, Section A,4,a,i).

The hypothalamus does not appear to be the unique source of PIF and PRF activities. Recently, it has been shown by Blask *et al.* (1976) that the pineal may serve as a supplemental source; crude acid extracts of pineal tissue from three different species, including man, were capable of directly stimulating the secretion of pituitary PRL. Conversely, two bovine pineal peptide fractions directly inhibited PRL release from AP glands. The pineal PRF activity is not due to either TRH or melatonin (see also Chapter IV, Section A,4,b) but is likely due to arginine vasotocin.

d. Circadian periodicity

A circadian periodicity in pituitary and serum concentration of PRL was first observed in the rat (Clark and Baker, 1964; Dunn *et al.*, 1972). In adult male rats two peaks levels of PRL secretion occurred at 5 pm and 11 pm, respectively (Dunn *et al.*, 1972). Nocturnal increases in PRL levels in the rat were maintained during pseudopregnancy (Freeman and Neill, 1972) and in the first days of gestation (Butcher *et al.*, 1972). In the human, circadian periodicity with nocturnal maxima was reported by Nokin *et al.* (1972), who employed a heterologous radioassay and 4-hr blood sample intervals. This finding was subsequently confirmed by many studies which used shorter time intervals for blood samplings (Parker *et al.*, 1973; Sassin *et al.*, 1973). The pattern of PRL release in mature young adults appears to be that of a sleep-related, steadily rising, multiple-peaked release reaching a nycthemeral maximum in late sleep. This pattern is abolished during the last trimester of pregnancy (Nokin *et al.*, 1972; Robyn *et al.*, 1973) and is absent in patients with PRL-secreting tumors (Jacobs and Daughaday, 1973) or Cushing's disease (Krieger *et al.*, 1976). A clearly significant elevation of serum PRL levels is also observed in female rats on the afternoon of proestrus (Niswender *et al.*, 1969; Kwa and Verhofstad, 1967; Neill and Reichert, 1971) which is equivalent to the human rise in serum PRL at midcycle (Vekemans *et al.*, 1972; Robyn *et al.*, 1973), attributable to the increase in circulating estrogens occurring in the same period.

e. Short Feedback Mechanism

As for GH, for the control of PRL secretion the existence of a short or "autofeedback" mechanism has been postulated. Evidence for PRL inhibiting its own secretion accumulated rapidly (see Meites and Clemens, 1972). In rats, transplantation of PRL-secreting pituitary tumors or normal pituitary tissue and injection of PRL or its implantation into the ME can decrease PRL content in the pituitary and blood and inhibit mammary development and lactation. The feedback action of PRL on PRL secretion apparently takes place via the hypothalamus; prolactin, but neither FSH nor LH, markedly increases turnover in the TIDA neurons of hypophysectomized rats (Hökfelt and Fuxe, 1972), a finding consistent with the presence of PRL-sensitive neurons in the same area (Yamada, 1975). This finding was interpreted to suggest that the inhibitory effect of PRL on its own secretion may be partly mediated via release of DA which enhances PIF secretion. Similar conclusions have been reached by other authors based on IVT CA injections (Kamberi *et al.*, 1970a) or treatment with monoamine drugs (Meites and Clemens, 1972) (see also Chapter IV, Section A,4,a and b). Whether the inhibitory feedback action of PRL on its own secretion may be of physiological import is presently unknown. It has been postulated by Meites (1973) that "this mechanism may serve to prevent rapid depletion of anterior pituitary stores by acute stimuli to prolactin release, such as stresses or the suckling stimulus." Lacking the feedback mechanism, the pituitary could be completely and rapidly depleted of prolactin, since stresses of several kinds (i.e., major surgery in the human) or suckling can result in a tenfold or greater increase in plasma prolactin, documented both by BA and RIA (see Frantz *et al.*, 1972).

6. Thyrotropin-Releasing Hormone

Although the existence of a specific factor (TRF or TRH) which controls the pituitary secretion of TSH was discovered some years later than that of CRF (Schreiber *et al.*, 1961, 1962), chemical isolation, identification, and finally synthesis of this hypothalamic tripeptide has been accomplished in only a decade (Schally *et al.*, 1969c; Folkers *et al.*, 1969; Burgus *et al.*, 1970; Gillessen *et al.*, 1970), undoubtedly speeded up by the availability of valid and reliable methods to assay this factor (Redding *et al.*, 1966a,b). The primary structure of TRF, pGlu-His-Pro-NH_2, has been found to be identical in ovine (Burgus *et al.*, 1970) and porcine (Nair *et al.*, 1970) hypothalami, and the same chromatographic mobility has been shown in TRF of human and bovine origin (Bowers *et al.,* 1971c). That native TRH in various species of mammals and also in birds corresponds to synthetic TRH was supported by the observation of its capacity to stimulate the release of thyrotropin in all mammalian and avian (Breneman and Rathkamp, 1973; Scanes, 1974; Bolton *et al.,* 1974) species studied so far. The possibility that the structure of TRF is also pGlu-His-Pro-NH_2 in species as the rat, guinea pig, and frog (Mitnick and Reichlin, 1971; Guillemin, 1971b; Reichlin *et al.,* 1972; Reichlin and Mitnick, 1973) is also inferentially supported from studies on the biosynthesis of TRF. The availability of synthetic TRF in quantity made it possible for the first time to use a chemical marker to follow the incorporation of precursor amino acid in newly formed TRF (Reichlin and Mitnick, 1973a). It was then possible to show that of twenty amino acids tested, only ^{3}H-Pro, ^{14}C-Glu, ^{14}C-His, and ^{14}C-Pro were incorporated when incubated with rat hypothalamic fragments into a substance which, in three chromatographic systems, had the same mobility of the TRF marker (Mitnick and Reichlin, 1971). In successive studies with organ cultures of the guinea pig hypothalamus, it has been found that the entire hypothalamus-ME unit was capable of ^{3}H-TRH labeling from ^{3}H-proline for at least 10 days. In the median eminence, TRH was not synthesized by ependymal elements, but intact nerve terminals were essential for TRH synthesis (McKelvy and Grimm-Jørgensen, 1975). Using short-term incubations of newt *(Triturus viridescens)* brain tissue it has been found that the ability to synthesize ^{3}H-TRH is not confined to the hypothalamus, but is also present in the forebrain, and that ^{3}H-TRH synthesis by hypothalamic fragments is not inhibited by addition of puromycin, cycloheximide, or actinomycin D (McKelvy and Grimm-Jørgensen, 1975), thus leading to the conclusion that the biosynthesis of the tripeptide in the hypothalamus is not ribosomally directed (Reichlin and Mitnick, 1973a; Vale *et al.,* 1973c). The fact that ^{3}H-TRH synthesis was abolished following preincubation with RNase suggested that RNA is required for TRH biosynthesis, although the nature of this RNA involvement has yet to be clarified (McKelvy and Grimm-Jørgensen, 1975). However, subsequent to these studies, there has been serious criticism of the original experiments. It has been found that hypothalamic extracts, similar to those used for the biosynthesis experiments, have active enzymatic TRH-degrading systems and that an "exchange reaction" in the degrading system may account for the incorporation of amino acids into TRH. In addition, negative results were obtained in attempts to confirm the original observations in experiments using porcine or rat hypothalamic extracts and immunoassay for TRH and in experiments on the incorporation of ^{3}H-Pro into immunoprecipitable TRH, in which care was taken to prevent the breakdown of TRH in hypothalamic incubates (for an exhaustive discussion, see Reichlin *et al.*, 1976). Thus, in the light of these conflicting findings a final judgment as to the mechanism of TRH biosynthesis cannot be made at this time.

The main difficulty in demonstrating TRF presence in the blood during the preradioimmunoassay era for RF's was that natural or synthetic TRF is rapidly destroyed when incubated *in vitro* with rat blood, plasma, or lyophilized serum (Redding and Schally, 1969a; Vale *et al.*, 1971), but not with CSF (Kendall *et al.*, 1971). Pretreatment of the plasma with 80% ethanol or heating (56°C for 30 min), almost completely prevented plasma inactivation of TRH.

When plasma fractions of porcine, bovine and human origin were incubated with TRH for 30 min, α-, β-, and γ-globulin fractions caused an 80–90% inactivation of added TRF. The biological activity of the tripeptide was also abolished following incubation with slices of rat liver, kidney, brain, and skeletal muscle (Redding and Schally, 1969b). Degradation of the immunological properties of TRH by blood was delayed at 1°C and inhibited by freezing and thawing the blood or by addition of benzamidine or BAL (Eskay *et al.*, 1976).

Collectively, these results are consistent with the presence of an enzyme in the plasma and in various tissues, including hypothalamic extracts (see above), which is capable of inactivating TRH. Successive studies suggested that during the inactivation of the hormone either *in vitro* or *in vivo* there was no major breakdown of TRH into its constituent amino acids, but only a deamination at the prolyl end giving rise to the free acid, (pyro)Glu-His-Pro-OH, possessing no hormonal activity (Redding and Schally, 1971; Nair *et al.*, 1971a). This conclusion has been questioned by Vale *et al.* (1973c) who, recalling that both pGlu-His-Pro-OH and pGlu-His-Pro-Me, two synthetic tripeptides with TRF activity, are also inactivated by plasma (Vale *et al.*, 1971), suggest that the C-terminal deamination is not the exclusive mechanism of inactivation. Recent studies of Knigge and Schock (1975) have shown that in normal rat plasma, acid TRH (see above) is normally a minor reaction product and that TRH yields Pro and Glu-His as major reaction products. Gradual degradation of TRH to its constituent amino acids has been described after binding to specific receptors of GH_3 strain of pituitary cells in culture (Hinkle and Tashjian, 1975). (For more details on this topic see Griffiths, 1976).

Studies on the distribution and half-life of TRF after injection of the ^{14}C-labeled compound have revealed a very short $t_{1/2}$ (4.16 min), thus indicating a quick removal from the circulation (Redding and Schally, 1972). Explanations for its rapid disappearance involve, in addition to inactivation of TRH in the plasma (see above), a large distribution volume (18.5% of the body weight) and a very rapid excretion. Within 60 min, in fact, approximately 25% of the radioactivity injected is found in the urine as the intact molecule and the free acid of TRH (Redding and Schally, 1972). Since excretion of TRH from plasma into urine is so rapid, the physiological significance of the inactivating plasma enzyme is unclear. Whatever the mechanism of TRH inactivation in plasma and its underlying significance, this effect is a major obstacle which must be overcome if sensitive assay methods for TRH in biological fluids are to be achieved (see Section B).

Immunoreactive (IR) TRH has been found in hypothalamic extracts of man, pig, rat, hamster, chicken, frog, snake, salmon, in the whole brain of the larval form of the primitive cyclostome lamprey, and in the head region of the protochordate amphioxus (Jackson and Reichlin, 1974). TRH is not confined to the hypothalamus but exhibits in mammals as well as in submammalian chordates a widespread extrahypothalamic distribution (Jackson and Reichlin, 1974; Vinokur and Utiger, 1974; Oliver *et al.*, 1974a). In the rat, roughly one-half of the brain IR-TRH is located outside the hypothalamus (e.g., the forebrain section contains only a slightly lower percentage of the total brain TRH), and quantitatively in the lower species (snake, frog, tadpole, salmon) much more IR-TRH lies outside this area (Jackson and Reichlin, 1974) (Table XXV). Interestingly, TRH is also found in species such as the lamprey, the amphioxus (Jackson and Reichlin, 1974), and the snail (Grimm-Jørgensen *et al.*, 1975) which lack TSH or the pituitary. This has led to the suggestion that the TSH-regulating function may be a late evolutionary development and that, in a sense, the pituitary has "co-opted" TRH as a regulatory hormone (Reichlin *et al.*, 1976). TRH is also present in the rat spinal cord (R. Utiger, quoted by Reichlin *et al.*, 1976), in ovine, bovine, and porcine pineal glands (White *et al.*, 1974), and in the neurohypophyses of many species (Jackson and Reichlin, 1974).

The extrahypothalamic distribution of TRH and its presence in areas of the brain believed to be most involved with behavior and emotional state suggest an important role in brain function unrelated to pituitary–thyroid regulation (see Section E).

An impressive series of studies has demonstrated that natural or synthetic TRH when administered intravenously, intraperitoneally, or orally increases the concentration of TSH in the plasma (see Gual, 1973). Synthetic TRH was active when infused (1 ng/min) for 20 min directly into the hypophyseal portal circulation (Porter *et al.*, 1971) and induced increases in the concentration of TSH in plasma even when administered to hypophysectomized rats bearing AP transplants (Bowers *et al.*, 1970; Sirett and Purves, 1974).

A possible role of cerebrospinal fluid in transporting inhibitory or

TABLE XXV

Distribution of TRH in Brain Areas of Several Vertebrates[a]

Species	Brainstem	Cerebellum	Olfactory lobe	Cerebral cortex (forebrain)	Hypothalamus	Pituitary complex
						(Neurohypophysis)[d]
Rat	5[b]	2	6	2	280[c]	155
(Rattus norvegicus)	(4–5)[e]	(1–3)	(5–8)	(1–3)	(240–300)	(150–160)
						(Neurohypophysis)
Chicken	9	1	17	2	41	168
(Gallus domesticus)	(8–10)	(1–2)	(16–18)	(1–3)	(34–49)	(95–240)
Snake	283	135	757	338	564	865
(Thamnophis sirtalis)	(129–359)	(85–186)	(750–764)	(264–381)	(393–731)	(410–1340)
Frog	55	520	326	111	2.270	>5000
(Rana pipiens)	(40–85)	(368–724)	(220–444)	(71–150)	(1520–3620)	(>5000)
Tadpole	303	Not examined	209	447[f]	947	764
(Rana pipiens)	(290–310)		(59–461)		(568–1225)	(508–995)
Salmon	13[f]	3	165	37	235	150
(Salmo sebago)		(2–3)	(154–181)	(22–52)	(188–264)	(31–366)

[a] From Jackson and Reichlin (1974).

[b] Pg/mg tissue wet weight.

[c] The mean TRH concentrations of other mammalian hypothalami studied were 480 (hamster) and 500 (pig); human stalk median eminence (SME) contained up to 300 pg/mg tissue wet weight.

[d] Values for rat and chicken pituitary complex for neurohypophysis alone.

[e] Range of values.

[f] Only 1 fragment extracted.

stimulatory hormones to the AP has been postulated by many investigators (Lofgren, 1960; Knowles and Kumar, 1969; Scott and Knigge, 1970), and proved for TRH (see Knigge *et al.*, 1973). Morphological and functional data suggest the existence of a ventricular route of delivery of TRH to the anterior pituitary. Exogenous TRH administered IVT quickly reaches the AP (Knigge *et al.*, 1971) where it causes TSH release (Oliver *et al.*, 1975), and can be observed by autoradiography in the apical portion of ependyma which lines the recess of the third ventricle (Joseph *et al.*, 1973). Endogenous TRH is present in high concentrations in the rat CSF of the third ventricle (Knigge and Joseph, 1974) and human lumbar CSF (Oliver *et al.*, 1974b).

Although a description of the entire mechanism by which TRH is transported from CSF to pituitary portal blood is not yet possible, available studies strongly suggest that a TRF receptor complex at the ventricular surface of ME may be an initial step (Knigge *et al.*, 1973). The *in vitro* binding properties of ^{3}H-TRH to some elements in ME parallel strongly the behavior of ^{3}H-TRH with a specific receptor in adenohypophyseal plasma membranes (Grant *et al.*, 1972; Labrie *et al.*, 1972) (see also below). Physiologically relevant substances, such as thyroxine and CA's, appear to regulate TRF-transport process at the ME level (Knigge *et al.*, 1973) (see Section A,6,c).

The uptake and transport mechanisms of ME do not seem to be unique for TRH among RF's, since they have also been recently described for LH-RH (Vaala and Knigge, 1974). It would appear, however, that CSF does not serve as a vehicle for transport of LH-RH to the ME under physiological conditions (Cramer and Barraclough, 1975).

In man, iv doses of as little as 15 μg TRH are followed by significant increases in plasma TSH in a dose–response relationship (Haigler *et al.*, 1971). In an extensive study, Snyder and Utiger (1972) noted in normal humans some of the variables affecting the serum TSH response to the iv injection of synthetic TRH. They showed the dose-dependency of the TSH response to a concentration of 400 μg and that the TSH response was not sex- but age-related, a marked fall being present in subjects of increasing age when compared to younger individuals of the same body weight and sex. A higher TSH response to TRH in females than in males has been reported by Lemarchand-Béraud *et al.* (1974) and Hall and Besser (1975).

Administration of TRH to normal subjects causes not only a prompt rise in plasma TSH concentration but also a significant increase in circulating plasma T_3 which occurs 20 min after the elevation in TSH. In contrast, to the clear-cut elevation in plasma T_3, the increase in plasma T_4 is quite small (Hollander *et al.*, 1972), thus explaining the conflicting reports of effect or lack of effect of TRH on serum T_4 (Hall *et al.*, 1970; Hersham and Pittman,

1971; Ormston *et al.*, 1971). Serial measurements of TSH and T_3 after administration of TRH may prove of clinical value, therefore, in simultaneously assessing the ability of the pituitary and thyroid glands to respond to their tropic hormones.

Biological activity has been tested not only in normal man but also in patients with thyroid disorders. Patients with primary hypothyroidism show elevated basal plasma TSH levels and exaggerated TSH responses to iv administration of TRH, while in patients with Grave's disease TRH does not elicit any significant pituitary TSH response (see Gual, 1973). It has also been shown that TRH elicits an exaggerated TSH response in many of Grave's disease patients made euthyroid by drug therapy (von zur Mühlen *et al.*, 1970). In all subjects administration of large doses of T_4 or T_3 decreases or inhibits the TSH response to TRH (Bowers *et al.*, 1971a; Fleischer *et al.*, 1971; Ormston *et al.*, 1971) (see also below). In euthyroid subjects with pituitary tumors the TSH response to TRH is not uniform, ranging from normal to diminished, and appears to depend on several factors which include the clinical course of the disease and the type of therapy used (see Gual, 1973).

a. "Specificity" of TRH Action

TRH with somatostatin represents a striking exception to the concept of neurohormonal specificity. Tashjian *et al.* (1971) were the first to demonstrate that TRH affects PRL release in cultured rat pituitary tumor cells, a finding also confirmed with pituitary cells of rats rendered chronically hypothyroid (Vale *et al.*, 1973b) and in female proestrous rats *in vivo* (Mueller *et al.*, 1973). These studies have been extended to other species, and the PRL-releasing activity of TRH has been evidenced *in vivo* in humans (Jacobs *et al.*, 1971; Bowers *et al.*, 1971b, 1973; Friesen *et al.*, 1972), sheep (Fell *et al.*, 1973), and cattle (Convey *et al.*, 1973) (see also Section A,5,c).

TRH has also proved to be an effective GH releaser, although, in the intact rat (Udeschini *et al.*, 1976) and in normal human subjects (Noel *et al.*, 1974) it produced only slight or no increase, respectively, in plasma GH. An increased GH release following TRH has also been reported in sheep (Takahara *et al.*, 1974b) or bovine (Machlin *et al.*, 1974) pituitaries *in vitro* and in acromegalic and uremic human subjects (Irie and Tsushima, 1972; Schalch *et al.*, 1972; Faglia *et al.*, 1973; Gonzales-Barcena *et al.*, 1973). More recently, TRH-induced GH release has also been described in patients with mental depression (Maeda *et al.*, 1975), anorexia nervosa (Maeda *et al.*, 1976), and liver cirrhosis (Panerai *et al.*, 1977b).

In the rat, positive GH responses to the tripeptide involve large doses of TRH infused directly into a hypophyseal portal vessel (Takahara *et al.*, 1974c), perfused on hemipituitaries *in vitro* (Carlson *et al.*, 1974), administered *in vivo* to hypophysectomized rats bearing an ectopic pituitary (Udeschini *et al.*, 1976), or into ME-lesioned rats (Müller *et al.*, 1976a). Since most of the pathological conditions of the human in which a nonspecific GH-releasing effect of TRH is present frequently show an impaired GH response to CNS-acting stimuli (see Chapter V, Section A,1 and 2), it has been proposed that in these cases TRH releases GH due to the existence of a functional CNS–AP disconnection, which removes the specific neurohormonal influences of GH secretion (Müller *et al.*, 1976a).

As mentioned previously, the distribution of TRH in the brain of mammals as well as in submammalian chordates suggested that this neurohormone may have other functions in addition to that of regulating the release of TSH and possibly PRL. In this context, studies performed in the laboratory animal and in the human have revealed that TRH has significant effects on behavior and on mood state (see Section E).

b. Mechanism of Action of TRH

The availability of the first synthetic hypothalamic neurohormone in pure form has opened new avenues for the study of the mechanism of action of this peptide in the AP gland.

TRH probably acts on the plasma membrane of the thyrotrope, inducing the release of TSH. Studies which have shown that the addition of actinomycin D to pituitary tissue *in vitro* did not affect the release of TSH induced by TRH (Schally and Redding, 1967) suggest that TRH acts mainly by stimulating release rather than by synthesis of TSH and that *de novo* synthesis of TSH is not required for TRH to induce its releasing effect. TRH can also induce TSH synthesis, since its addition to incubation media significantly increases incorporation of ^{14}C-amino acids into pituitary tissue (Schally *et al.*, 1968a). The effect of TRH on TSH release can be inhibited by T_3 and T_4 in a dose-related fashion and the blockade can be overcome by increasing the dose of TRH (Bowers *et al.*, 1967). Similar conclusions were reached by Vale *et al.* (1967). It is apparent from *in vivo* and *in vitro* studies that TRH and thyroid hormones do not directly interact and that the inhibitory effect of T_3 on TRH-induced TSH release requires normal protein and RNA synthesis (Schally *et al.*, 1968a). It is postulated that thyroid hormones acting inside the cell induce the formation of mRNA which controls the synthesis of a regulator protein; the latter inhibits the sequence of events stimulated by the binding of TRH to the plasma membrane of the thyrotrope.

Further support for the concept that thyroid hormone-mediated inhibition of TSH secretion is a consequence of the blocking of TRH secretion at the pituitary level was derived from experiments which showed that a single dose of parenteral T_3 completely abolished cold-induced TSH rise without affecting elevation in blood TRH (Montoya *et al.*, 1975) and that

TRH production rate is actually increased in hyperthyroid rats (Jackson *et al.,* 1974) (see also Section A,6,c).

The availability of a pure synthetic TRH labeled at high specific activity in its proline residue furthered studies on the binding of TRH to pituitary receptor sites. Working with membranes of mouse thyrotropic tumor or intact pituitary cells, Grant *et al.* (1972) showed that labeled TRH specifically binds to the cell plasma membrane, that ^{3}H-TRH competes with unlabeled TRH, and that the inactive TRH analog pGlu-His-OMe and synthetic LH-RH do not compete for binding even at 200-fold concentration. Similar results were reached by Labrie *et al.* (1972) studying the properties of TRH binding to plasma membranes isolated from bovine AP gland. These authors emphasized the specificity of the site of binding of TRH by showing absence of competition when two oligopeptides of hypothalamic origin, i.e., LVP and MIF or various peptide hormones (porcine ACTH, bovine LH, bovine GH, bovine PRL, bovine insulin, and porcine glucagon), were added at concentrations well above the physiological plasma concentration. The ^{3}H-TRH binding data show that pituitary cells possess two TRH receptors with close affinity constants. Vale *et al.* (1973c) calculated that the apparent binding constant of the TRH-receptor interaction as measured using biological activity (release of TSH by enzymatically dispersed rat pituitary cells) is at least 10 times lower than the binding constants determined directly in equilibrium and kinetic-binding experiments. They have suggested the existence of a class of high affinity, low capacity receptors obscured by higher capacity, lower affinity receptors. Alternatively, they propose the concept of spare receptors in which only a fraction of each cell's receptors need to be saturated to yield a maximal biological response.

The establishment of a competitive binding assay for TRH has permitted not only thorough structure-function studies using various analogs of TRH but also insight into the number of factors that can determine the biological activity of an analog (e.g., access to the binding site, biological half-life, and binding affinity). It would appear that the biological activity (TSH release *in vivo* and *in vitro*) correlates better with the binding affinity, thus indicating that for TRH analogs studied to date the limiting factor in their biological activity is their relative affinity for the TRH receptors (for greater detail, see Vale *et al.,* 1973c).

The concentration of ^{3}H-TRH binding sites in rat adenohypophyseal tissue is increased 3- to 4-fold by treatment with estrogen and decreased by thyroid hormone administration (Labrie *et al.,* 1975).

c. Feedback Regulation of TSH Secretion

As mentioned previously, thyroid hormones appear to exert their main effect on TSH secretion by acting directly on the AP. However, in the last few years, a series of *in vitro* studies on the transport capacity of ME has suggested the possibility of thyroid hormone action also exerted on the CNS. In an initial series of studies, Silverman *et al.* (1972) demonstrated that ME possesses an active Na^+-dependent transport for the neutral nonutilizable amino acid, α-aminoisobutyric acid (AIB). These studies were extended to describe qualitatively and quantitatively the mechanism of T_4 accumulation in ME (Silverman and Knigge, 1972). This tissue was also shown capable of accumulating large quantities of T_4 by an energy-dependent mechanism; several metabolic inhibitors (ouabain, iodoacetate, sodium fluoride, etc.) markedly inhibited the active transport. The accumulation of T_4 was three times as great as for AIB, and the carrier responsible for this active transport differed from that for neutral amino acids. The bound hormone remains as T_4 in a "readily releasable" pool; the addition of TSH influences T_4 transport by increasing its uptake and inhibiting its efflux. The failure of LH to alter the course of T_4 efflux was indicative of a degree of specificity of the observed interaction.

The presence of a large, mobile pool of "free" thyroxine at the ME level can serve as a potential regulatory influence on TRH-TSH interactions. An increase of thyroxine in this compartment diminishes the rate of delivery of TRH to portal blood (Knigge *et al.*, 1973). Thus TSH, which decreases the net amount of free thyroxine reaching the pituitary and increases the thyroxine level in ME, through its influence on thyroxine transport exerts indirectly a negative short loop feedback on TRH delivery to the AP. In addition to thyroxine, catecholamines, which have a significant extraneuronal uptake in the ME, negatively influence the rate of transport of TRH to portal capillaries (Knigge *et al.*, 1973). Figure 37 depicts schematically the thyroxine and TRF transport system of ME.

Additional studies are needed before a physiological role can be attributed to this highly effective transport system for T_4 in the ME. A CNS site of action for thyroid hormones is also implied by the findings of Reichlin and Mitnick (1973a), who have shown that T_4 stimulates TRH synthetase formation in rat and frog hypothalami. More recently, however, a negative feedback action of T_3 on TSH secretion acting at a hypothalamic level has been demonstrated in hypothyroid rhesus monkeys (Belchetz *et al.*, 1976). In these animals, T_3 injected bilaterally into chronically implanted hypothalamic cannulae induced a consistent fall in the increased plasma TSH levels; the speed of response observed was consistent with direct inhibition of TRH secretion.

7. Melanocyte-Stimulating Hormone

It is now generally accepted that the hypothalamus exerts an inhibitory control on the secretion of melanocyte-stimulating hormone from the inter-

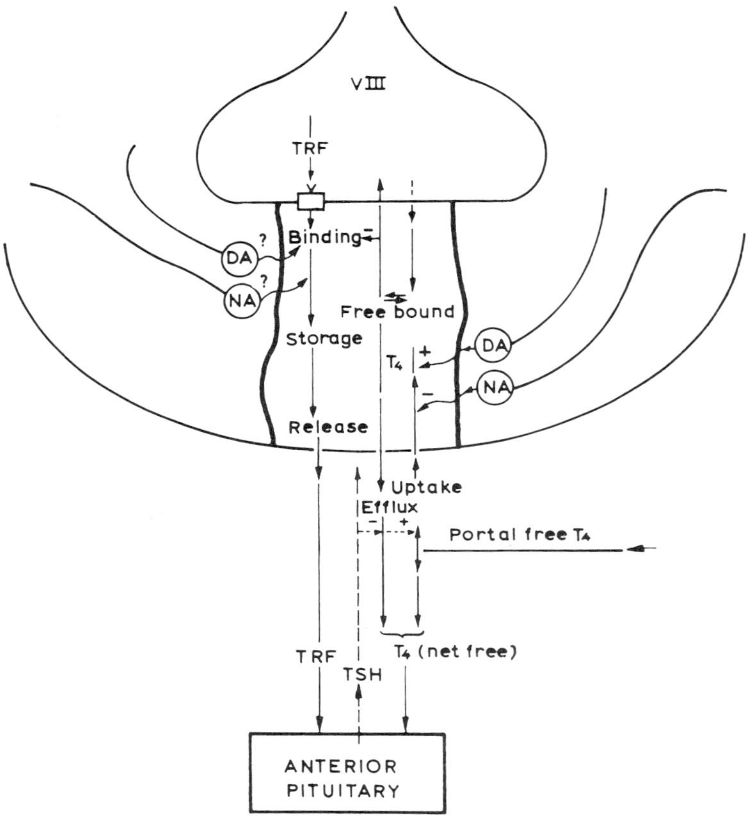

Fig. 37 Schema of thyroxine and TRF transport by median eminence (see text for explanation). VIII, third ventricle; T_4, thyroxine; DA, dopamine; NA, noradrenaline. (Reproduced, with permission, from Knigge *et al.*, 1973.)

mediate lobe of the pituitary (Kastin and Ross, 1964; Taleisnik *et al.*, 1967). An MSH-release inhibiting factor (MIF) activity has been found in hypothalamic tissue from 7 animal species including man (Kastin and Schally, 1971). Nair *et al.* (1971b) isolated MIF from beef hypothalami and determined its structure to be Pro-Leu-Gly-NH_2 (MIF-I) (PLG), which is the carboxyl terminal tripeptide side chain of oxytocin. This structure corresponded to that proposed earlier for rat MIF by Celis *et al.* (1971a) and would be obtained by cleavage from oxytocin by a microsomal enzyme system present in the rat SME. Another peptide, H-Pro-His-Phe-Arg-Gly-NH_2 (MIF-II), a thousandth as potent as MIF-I in inhibiting MSH secretion in the BA based on skin lightening of frogs previously darkened by hypothalamic lesions, has been characterized in extracts of bovine hypothalamus by Nair *et al.* (1972). However, the existence of species variability for compounds with MIF activity has been suggested by the findings of Bower *et al.* (1971), who could not confirm the MIF activity of PLG in an *in vitro* system using bisected rat pituitaries or *in vivo* using frog pituitary. These authors reported, instead, that the ring structure of oxytocin (tocinoic acid), Cys-Tyr-Ile-Gln-Asn-Cys-OH, exhibited powerful MIF activity on rat pituitary tissue but not on frog pituitaries. The ability of plasma to inactivate PLG might perhaps explain some of the differences in response reported. In fact, when large doses (10 mg) of PLG were injected into the dorsal lymph sac of the frog, or even directly into the aortic trunk, no lightening occurred (Kastin *et al.*, 1971). Similarly, in the human, iv injection of PLG did not change plasma concentration of MSH (Kastin *et al.*, 1972c; Thody *et al.*, 1974). Marked lightening of the frog did occur, however, when smaller doses (10 ng) of PLG were applied directly to the surface of the AP (Kastin *et al.*, 1971).

Distribution studies performed in the rat with PLG labeled with ^{14}C-leucine or ^{3}H-proline showed neurohormone accumulation in the pineal, pituitary, kidney, liver, and adrenals (Redding *et al.*, 1973b). The calculated half-life was approximately 9 min and the volume of distribution greater than that of the plasma volume (11.3 ml); both figures are almost twice those described for TRH (Redding and Schally, 1972). The marked accumulation found of radioactive PLG in the pineal is consistent with the view that a feedback system may be operative between the hypothalamus and the pineal, and adds new complexity to the already intricate relationships described for the pineal–hypothalamic–pituitary axis (Kastin *et al.*, 1972d).

Delayed disappearance of labeled PLG was found in hypophysectomized rats when compared to rats with an intact pituitary, while pinealectomy and adrenalectomy did not change the half-time disappearance (Kastin *et al.*, 1974). This may indicate that normally the pituitary gland directly binds and inactivates MIF, while a similar binding, but with lesser inactivation, occurs in the pineal.

Recently, the distribution and fate of MIF-I have been determined in the human (Redding *et al.*, 1974). The half-time disappearance of the neurohormone was about 2 min and its distribution volume greater than that of the plasma volume. MIF-I was slowly eliminated from the body during the first hour (17% of the dose) by urinary excretion of the intact molecule, while a smaller amount was probably metabolized in the tissue. The long retention time of MIF-I in the body, as indicated by its large distribution volume and relatively slow urinary excretion, might account for some of the extraendocrine activity of the peptide (see Section E).

In addition to the existence of hypothalamic inhibitory influences for the secretion of MSH, proof exists for the presence in freshly prepared extracts of rat ME of MSH-releasing activity (Taleisnik and Orias, 1965; Taleisnik and Tomatis, 1967a). These findings have been more recently substantiated with the isolation of a pentapeptide fragment of oxytocin, H-Cys-Tyr-Ile-Asn-OH, endowed at nanogram concentrations with MSH-releasing activity (Celis *et al.,* 1971b). This pentapeptide MRF can also be obtained by cleavage of oxytocin, as suggested by incubation experiments of a mitochondrial fraction prepared from extracts of paraventricular nucleus of male rats (Celis and Taleisnik, 1971). It would thus appear that oxytocin can be degraded by hypothalamic peptidases to produce either MRF or MIF. According to an interesting concept formulated by Taleisnik and Celis (see Howe, 1973), both MRF and MIF are derived from a common precursor, oxytocin, by competing peptidases. According to these authors, the physiological state of the animal influences the rate of synthesis, and, hence, predominance of the respective peptidase, and in this way the levels of hypothalamic MIF or MRF activity. The mitochondrial preparation containing the enzymatic system yielding MRF upon incubation with oxytocin (Celis and Taleisnik, 1971) was found, similar to that yielding MIF (Celis and Taleisnik, 1974), to be estrogen-dependent, since it was present in the ME of female rats at proestrus and estrus and absent in other stages of the cycle (Celis, 1975). Accordingly, plasma MSH values were higher on the morning of proestrus and decreased to undetectable values in other stages of the cycle (Celis, 1975).

The precise hypothalamic area(s) concerned with MSH control has yet to be defined. The existence of a functional relationship between the hypothalamic–neurohypophyseal system and the pars intermedia has been suggested by Legait and Legait (1962). Changes in the histological appearance of the pars intermedia (Duchen, 1962; Roux, 1962; Howe and Thody, 1970) and in its MSH content (Howe and Thody, 1970) paralleled activation of the neurohypophysis on the administration of hypertonic saline or dehydration. Suckling was reported to affect the MSH content of the pituitary (Taleisnik and Tomatis, 1968). Collectively, these data supported the idea of an involvement of the PVN and SON in the control of MSH secretion. In this regard, Howe and Thody (1969) reported that in the rat electrical lesions of the PVN produce a large increase in the MSH content of the neurointermediate lobe, accompanied by rising levels of MSH in the plasma (Thody, 1974). Increase in plasma MSH levels following lesioning of the PVN has also been observed by Taleisnik and Celis (1974). These authors, however (Taleisnik *et al.,* 1966), in studies on the topographical distribution of MRF, had previously reported that this neurohormone is produced by neurons in the PVN with axon endings in the ME and neural lobe. Their conclusions were supported by the results of lesion experiments, from which it appeared that bilateral destruction of the PVN was followed by disappearance of MRF activity in the ME and neural lobe (Taleisnik *et al.,* 1967).

The view that PV nuclei are CNS sites for the inhibitory control of MSH release fits in well with the reported derivation of MIF-I from oxytocin which is manufactured in these areas (Celis *et al.,* 1971a). Destruction of the PVN would deplete oxytocin, prevent MIF production and finally result in an increased MSH secretion.

Also, in assuming PVN as the site of this dual inhibitory and excitatory system, all details of the transport and release of both MIF and MRF have yet to be established. It is unlikely that the released neurohormone may reach the target intermediate lobe through a functional neurovascular link in view of the near avascularity of this structure in most mammalian species (Howe, 1973). A possibility is that these RF's may be liberated at the end of nerve fibers directly innervating the cells of the intermediate lobe, thus functioning as neurohumoral transmitters (see Chapter I, Section A). In this context, however, the possibility has to be considered of a functional interaction of these neurosecretory peptides with the numerous aminergic neurons impinging on the PI (see Chapter II, Section B,5 and Chapter IV, Section A,6).

B. ASSAY METHODS

1. Bioassay

A number of biological assay systems have been used in the past for the identification of releasing and inhibiting factor activity and the successive establishment of the chemical nature of the isolated and purified neurohormone. Table XXVI summarizes the general types of assay.

Among the available methods, *in vivo* assays are generally preferred to the *in vitro* techniques since they more closely mimic the events occurring in the physiological condition; they are, however, rather complex and measure as the indicators of activity the effects on pituitary tropic hormone induced by the releasing factor. Thus, as an end point they determine (1) the increased or decreased level of the pituitary hormone in the circulation, (2) the change induced in the hormone content within the pituitary gland, or (3) the change in the weight or functional activity of the peripheral gland, secondary to increased pituitary secretion, e.g., adrenal cortex (ACTH), liver (GH)-increased somatomedin generation.

TABLE XXVI

Methods of Assay of Releasing and Inhibiting Hormones

In vivo
- A. *Indirect Methods*
  1. Increased or decreased blood level of pituitary hormone or change in target organ secondary to altered pituitary secretion and/or
  2. Alteration in pituitary hormone content in
     - a. Intact animals
     - b. Hypophysectomized animals bearing an ectopic pituitary
     - c. Animals with pharmacological blockade by target gland hormones or CNS-active drugs
     - d. Animals lesioned in the median eminence
     - e. Animals bearing a cannula into hypophyseal portal vessels
     - f. Animals bearing an intrapituitary cannula
- B. *Direct Methods*
  1. Radioimmunoassay (microassay) in different CNS areas
  2. Radioimmunoassay in different biological fluids (plasma, CSF, urine)

In vitro
- A. *Indirect Methods*
  1. Hormone released from pituitary tissue and/or
  2. Hormone content in pituitary in
     - a. Short-term incubation
     - b. Long-term tissue culture of intact cells, cells enzymatically dissociated, tumoral cells
  3. Continuous flow superinfusion
- B. *Direct Methods*
  1. Radioimmunoassay in medium from explants of CNS areas

Bioassays based on the change of a pituitary tropic hormone concentration have, in the past, been subjected to criticism since it would appear that in certain instances a rapid "depletion" of an adenohypophyseal hormone occurs without, or temporarily dissociated, from release of RIA detectable material (see Section A,3,a). Also, they appear to offer a low degree of sensitivity for monitoring the increased level of the released pituitary hormone in the circulation (point A,2,a, Table XXVI), unless the circulating hormone is determined by RIA.

Animals treated with a variety of pharmacological agents appear better suited for RF assay, being more resistent to nonspecific manipulations; however, they may have the disadvantage of exhibiting a low pituitary response for an action of the pharmacological blockade at this level. This certainly is the case for the dexamethasone- or thyroxine-treated rat used for CRF or TRF assay, respectively (Russell *et al.*, 1969; Bowers *et al.*, 1968). Animals with lesions in the ME, and thus devoid of intrinsic neural influences over the gland, are quite good subjects for *in vivo* characterization of RF's provided that incompleteness of the lesion and/or damage of portal vessels are excluded. The *in vivo* assay based on the infusion of material directly into the AP (Hedge *et al.,* 1966; Dhariwal *et al.,* 1969b) has been more recently replaced by the portal infusion system, i.e., administration of the test substance directly into hypophyseal portal vessels, thus imitating the physiological route for releasing factors (Porter *et al.,* 1970a). Although time-consuming and requiring skillful technique, this assay, no doubt, is strikingly more sensitive than the intrapituitary assay (Porter *et al.,* 1970b; Sandow *et al.,* 1972; Takahara *et al.,* 1974c), and allows a better distribution of samples with no entrance into the interstices of the gland and ensuing nonspecific effects.

Assays referring to points A,2,b–f of Table XXVI, in view of possible extrapituitary effects for RF's, have the advantage over that indicated in point A,2,a (intact animals) of better defining the (pituitary) site at which RF's act. Intact rats are also unsuitable when the output of the hormone to be studied is influenced by manipulation of the animal or by mere injection of the material (McCann *et al.,* 1968).

Releasing and inhibiting hormones can also be assayed in either short-term or long-term incubation experiments in which the test extract is added to a cultured pituitary gland. The activity present is monitored as a change in the amount of either (1) tropic hormone released from the pituitary tissue and detectable in the medium or (2) tropic hormone stored in the pituitary. One major advantage is that effects *in vitro* must be exerted directly on the gland. The continuous flow superinfusion or perinfusion incubation system (Tait *et al.,* 1967; Saffran and Rowell, 1969) offers advantages over static incubation conditions of the AP. It makes possible studies of the isolated AP *in vitro* under conditions which resemble more closely the conditions *in vivo* and monitors changes in the secretory activity of the gland over a period of time. Continuous turnover of the medium should minimize the effects of accumulation of secretory or metabolic products as well as of the breakdown of substances secreted by the gland or administered to evoke a response.

An elegant technique for studying the effect of RF's on the incubated pituitary is that which involves the sequential use of an amino acid labeled with different isotopes, thereby permitting the simultaneous examination of releasing factor influences upon pituitary hormone synthesis and release. Thus, for instance, by this technique it has been possible to show that a partial purified extract of hypothalamus increased release of newly synthesized GH, while not affecting release of older, stored GH (Stachura *et al.,* 1972a), and to confirm the existence of GIF in ovine hypothalamic fractions (Stachura *et al.,* 1972b).

2. Radioimmunoassay

Since RIA methods have proved to be highly sensitive and specific, and relatively simple for measuring many substances (Odell and Daughaday, 1971), this methodology has been applied for obtaining measurements of releasing and inhibiting hormones. As these molecules are small peptides and as such likely to be weakly, if at all, immunogenic, preparation of releasing hormones bound to large carrier proteins was undertaken. Synthetic TRH coupled to bovine serum albumin with bis diazotized benzidine (BDB) was found to be highly immunogenic in rabbits. Using ^{125}I-TRH, resulting from iodination of the histidine residue of TRH, a double antibody RIA assay for TRH was developed in which as little as 0.1 ng unlabeled TRH could be detected. A variety of TRH analogs tested in this immunoassay showed only little reactivity, and no reactivity was exhibited by the amino acids constituting TRH, thus indicating that the anti-TRH antibody was highly specific (Bassiri and Utiger, 1972).

The availability of TRH radioimmunoassay has also allowed the determination of this neurohormone in the CNS (see Section C). In addition to serum (Redding and Schally, 1969a), homogenates of brain tissue contain considerable amounts of TRH inactivator(s) (see Section A,6); however, TRH can be readily extracted from hypothalamic homogenates with methanol and its concentration precisely determined (Bassiri and Utiger, 1974). Mean hypothalamic TRH content ranged from 7.1 to 8.4 ng in normal male rats of varying weights, was decreased following chronic, but not acute, daily T_4 treatment or hypophysectomy and little influenced by thyroidectomy or cold exposure. Recently, a RIA for TRH has been described in which the anti-TRH serum used was obtained from rabbits into which TRH coupled to bovine thyroglobulin by means of the BDB reaction was injected (Jackson and Reichlin, 1974). The assay appears to have marked sensitivity (below 7 pg) and considerable precision.

First determinations of circulating LH-RH were reported in women at midcycle by the use of a BA, the end point of which was the increase in RIA-LH in plasma of ovariectomized, estrogen–progesterone–thyroxine-treated rats; however only 6 out of 36 positive results were obtained and 40 ml of serum had to be extracted (Malacara *et al.*, 1972). The availability of synthetic LH-RH made possible the production of antibodies against the neurohormone. Production of specific antisera to LH-RH in rabbits was reported by Arimura *et al.* (1973b), following repeated injections of synthetic LH-RH decapeptide, adsorbed on polyvinylpyrrolidone and emulsified with complete Freund's adjuvant. This allowed the development of two types of RIA's for LH-RH. The first was an ordinary (competitive) type in which a small amount (0.37 ng/assay tube) of unlabeled LH-RH was added to the labeled peptide. This saturated the antiserum's capacity for paradoxical binding (see below) so that further addition of LH-RH inhibited binding of labeled LH-RH. The second assay was based on paradoxical binding and omitted the premixing of labeled and unlabeled LH-RH. In this assay addition of very small amounts (0.5 to 310 pg) of unlabeled LH-RH to the assay tubes increased the amount of label bound to antiserum and allowed construction of a parabolic curve of positive slope when B/T was plotted against arithmetic dose (Fig. 38). With this assay, whose sensitivity is less than 1 pg and is highly specific for LH-RH, plasma LH-RH levels in cycling rats on the day of proestrus were in the range of 15–18 pg/ml and usually undetectable. According to de la Cruz and Arimura (1975), and in keeping with previous findings of Jeffcoate and Holland (1974b) (see Section A,4), most of the IR-LH-RH in plasma determined by ordinary (competitive) RIA is not related to LH-RH, and only assays based on paradoxical binding are capable of detecting true LH-RH activity. This necessitates elimination of nonspecific inhibiting substances for valid estimation of plasma LH-RH by ordinary RIA and explains, perhaps, the failure to report a midcycle surge of RIA-LH-RH in humans (Nett *et al.*, 1973; L. E. Clemens *et al.*, 1974; Jonas *et al.*, 1975).

RIA for LH-RH was also developed by Kerdelhué *et al.* (1973) using

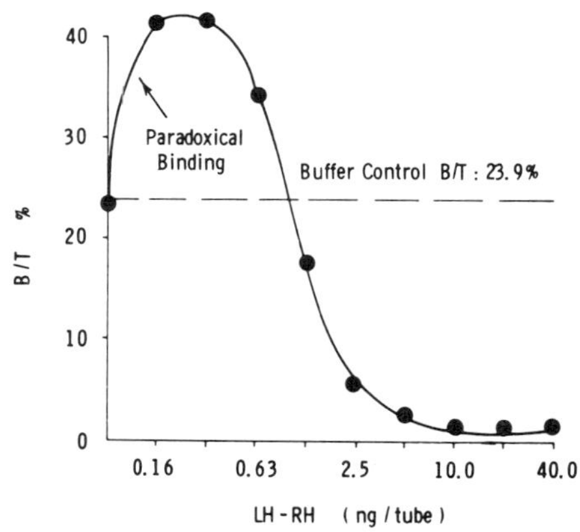

Fig. 38 Effect of increasing doses of cold LH-RH on [^{125}I]-LH-RH antiserum binding. To 0.4 ml 1% egg albumin phosphate buffer was added 0.1 ml labeled hormone, 0.1 ml cold LH-RH, and 0.1 ml antiserum, and incubated overnight at 4°C. Bound and free hormone were separated by dextran-coated charcoal. On the ordinate, percent bound radioactivity over the total radioactivity (B/T%); on the abscissa, increasing doses of LH-RH. (Reproduced, with permission, from Arimura *et al.*, 1973b.)

antisera against synthetic LH-RH generated in guinea pig by injecting the peptide associated only with complete Freund's adjuvant. This antiserum allowed detection of as little as 50 pg of LH-RH. With this method the content of LH-RH in tissue extracts of SME was 30 ng/hypothalamus; plasma levels in a normal female sheep at estrus were 6 ng/ml and in hypophysectomized male and female rats about 4 ng/ml. Jeffcoate *et al.* (1974b) produced antisera against the decapeptide conjugated to albumin in rabbits and against the 3–10 octapeptide conjugated to albumin in rats. They reported that extraction with methanol of serum samples before RIA has the advantage of removing potential interfering substances such as enzymes, proteins or other hormones (see also above) and of concentrating the extract, thus increasing the sensitivity of the assay when applied to serum. The RIA using the rat antiserum in their experiments was 30 times more sensitive than with rabbit antiserum and could detect LH-RH in peripheral serum in concentrations ranging from undetectable levels (less than 0.5 pg/ml) to 3.0 pg/ml (see also Section A,4).

The recent availability of somatostatin (Brazeau *et al.*, 1973) has permitted the development of a sensitive and specific RIA for neurohormone determination in tissue and blood (Arimura *et al.*, 1975b) (see Section A,3,c). The tetradecapeptide synthesized by a solid phase method, conjugated with human serum globulin and emulsified with complete Freund's adjuvant, when injected in rabbits was capable of producing specific antisera. One of the antisera, used at 1 : 7000 dilution, was capable of binding ^{125}I-Tyr^{1}-GH-RIH, a binding which could be inhibited by adding cold GH-RIH to the reaction mixture. A linear dose–response relationship was present between logit of B/T and the log dose of GIF in the range of 2–512 pg. This RIA system failed to show any cross-reaction with TRH, LH-RH, and various pituitary hormones from different species. However, linear GH-RIH and the ring portion of the GH-RIH tetradecapeptide cross-reacted in the RIA system by 17 and 67%, respectively. Dose-related inhibition of ^{125}I-Tyr^{1}-GH-RIH antibody binding by GH-RIH-free plasma and the elimination of inhibition by deproteinization suggested either the presence of substance(s) immunologically indistinguishable from GH-RIH in plasma or nonspecific interference of the RIA system with plasma proteins and the need for deproteinizing plasma before assaying for GH-RIH. A similar RIA for somatostatin has also been developed by Kronheim *et al.* (1976), and, recently, a radioreceptor assay has been introduced (Ogawa *et al.*, 1976). Estimating quantitatively the releasing and inhibiting hormones in selective CNS areas has recently been made possible by the introduction of a microdissection technique used in conjunction with microassay (see Section C). For receptor-binding radiostereoassay for TRH see Section A,6,b.

C. LOCALIZATION

Techniques used for studying neural control of AP hormones also provided information on CNS areas concerned with RF production and/or storage. Examination of the effect of brain lesions on the function of the thyroid led Greer (1951) to delineate a "thyrotropic" area of the brain in the midline between the PV nuclei and the ME. Studies in which AP tissue was transplanted into various regions of the hypothalamus showed that normal structure and function of the anterior lobe are maintained only in those animals in which graft is situated in the so called "hypophysiotropic area" (Flament-Durand, 1965; Halàsz *et al.*, 1968). This area which consists of the ARC nuclei, the ventral pars of the anterior periventricular nuclei, the medial part of the retrochiasmatic area, and the ME was identified as the site of secretion and release of RF's.

A more direct approach for the assessment of the topographic distribution was to bioassay various brain regions for their RF content, a procedure extensively used by McCann and his co-workers (Krulich *et al.*, 1971). In studies in which freshly frozen hypothalami were sectioned in-three planes to determine the localization of the LH-RH by an *in vitro* assay, activity was found in a wide band of tissue extending from the medial POA caudally to the ME-arcuate region (Crighton *et al.*, 1970; Krulich *et al.*, 1971). However, these studies could not determine the localization of the cell bodies of the neurons that presumably secrete LH-RH. In female rats, lesions in the suprachiasmatic-preoptic area resulting in a decrease in the LH-RH stored in the ME-ARC region and the appearance of a state of constant estrus led (Schneider *et al.* (1969) to postulate the existence of LH-RH neurons with cell bodies located rostrally in the hypothalamus and involved in the ovulatory surge of LH. This localization was also supported by the findings of Quijada *et al.* (1971). The remaining LH-RH activity originated presumably from neurons located caudally in the mediobasal tuberal region and was responsible for the release of LH (and possibly FSH) other than that required for ovulation.

More recently, the availability of specific antibodies against LH-RH protein complexes has permitted the use of immunofluorescence and immunohistochemical techniques to test for LH-RH immunoreactivity in hypothalamoinfundibular axons from various mammalian species. Barry and Dubois (1973) and Leonardelli *et al.* (1973), using the immunofluorescence technique, found in guinea pigs and rats LH-RH positive nerve fibers around the capillary loops of the hypophyseal portal vessels in the ME, and LH-RH immunopositive cell bodies in the periventricular gray surrounding the third ventricle (Barry *et al.*, 1973a) as well as in the PO and septal areas (Barry *et al.*, 1973b). In addition to the PO-infundibular pathway, some "extrahypophyseal immunoreactive pathways" were observed in the guinea pig (e.g., running toward the mesencephalon or the hypothalamus or the basal rhinencephalon), possibly involved in the control of sexual behavior (Barry *et al.*, 1974) (see also Section E). Additional and more direct evidence that ARC neurons contain LH-RH immunoreactivity was given by Zimmerman *et al.* (1974) using the immunoperoxidase method. In the hypothalamus of male and female mice, this method localized LH-RH in the perikarya of some ARC neurons and throughout numerous tanycytes of the ME, a finding which might imply that the neurohormone synthesized in ARC neurons and secreted in the third ventricle reaches the portal capillaries via CSF flow to tanycytes (see Knigge *et al.*, 1973, and Section A,6). In this study LH-RH was also found in the organum vasculosum of the lamina terminalis.

None of the other studies on LH-RH pathways performed in the rat were capable of establishing LH-RH immunoreactivity in cell bodies of hypothalamic nuclei (Baker *et al.*, 1974; Ramirez and Kordon, 1975) even when the axoplasmic transport of LH-RH in the LH-RH synthesizing neurons, under enhanced demand caused by castration, was blocked by colchicine (Sétàlo *et al.*, 1975). Use of nonspecific antibodies generated against the LH-RH decapeptide lacking the reactive free amino acid group necessary for conjugation to carrier protein and their cross-reactivity with an immunoreactive substance present in the perikarya, which is not LH-RH itself, could explain the inconsistent results. Similarly, nonspecific binding of IgG, present in the rabbit antiserum used, to rat ependymal cells could be responsible for detection of LH-RH in tanycytes and the pineal gland (see Sétàlo *et al.*, 1975; Ramirez and Kordon, 1975). However, recent detection of both LH-RH and TRH in significant quantities in all circumventricular organs in addition to the ME and the organum vasculosum (Kizer *et al.*, 1976a) supports the function of specialized ependymal cells in RF's transport.

Recently, Silverman (1976), reexamining the distribution of LH-RH in the guinea pig hypothalamus, observed that in this species immunoreactive deposits of the neurohormone are actually present in the cell bodies of hypothalamic neurons in the medial preoptic, anterior hypothalamic, suprachiasmatic, and ARC nuclei. The significance of species differences between the mouse and the guinea pig versus the rat, in which the morphological studies did not show any immunoreactive perikarya, remains to be clarified. Combined use of an immunohistochemical technique and EM (Pelletier *et al.*, 1974a) revealed recently that LH-RH is contained in the secretory granules (750–950 Å) of some nerve endings (10–20%) in the lateral external layer of the ME, thus confirming the hypothesis that hypothalamic RF's are stored in the small secretory granules of the nerve fibers located at the proximity of the primary capillaries of the pituitary portal plexus (Kobayashi *et al.*, 1970; Scott and Knigge, 1970), and supporting the theory that LH-RH synthesized in cell bodies of the ARC nucleus is transported to terminals in the ME by axoplasmic flow and then liberated into the vascular channel. Although distribution of LH-RH nerve terminals corresponds with the distribution of the TIDA neurons (see Chapters I and IV, Section C,3), coexistence of MA's and RF's in the same neurons seems to be negated by the finding of unchanged LH-RH levels in the ME and ARC nuclei of rats whose central catecholaminergic neurons have been chemically destroyed by 6-OHDA (Kizer *et al.*, 1975a), and the identification of distinct LH–RH- and catecholamine-containing neurons in primary cultures of dispersed cells of the basal hypothalamus (Knigge *et al.*, 1977). Studies of subcellular fractionation by both differential and sucrose gradient centrifugation showed most of the LH-RH activity to be located in the synaptosomal band of the P_2 primary (crude mitochondria and synaptosomes) fraction (Ramirez and Kordon, 1975).

Techniques combining microassay and microdissection methods have proved to be useful adjuncts to histochemical methods for the mapping of neurotransmitters and RF's in the CNS of the rat. The hypothalamic distribution of LH-RH in the rat has been studied by Palkovits *et al.* (1974b) by a method in which tiny cylinders, 300–500 μm in diameter, are punched out, under visual control of a dissecting microscope, from frozen sections of hypothalamus (Palkovits, 1973). Pooled cylinders representing separate nuclei are then assayed for LH-RH by RIA (Arimura *et al.*, 1973b). As expected, LH-RH was found in high concentration in the ME and ARC n., and in very small amounts in the suprachiasmatic and SON (Table XXVII). Contrary to the results of Palkovits *et al.* (1974b), who failed to detect LH-RH in extracts of MPOA, Wheaton *et al.* (1975), found LH-RH present in this area in relatively uniform amounts, thus confirming results previously reported using BA and immunofluorescence studies. In addition, some LH-RH was also found in an area medial to the MPOA and immediately rostral to the chiasm, the anterior periventricular area (Kawakami *et al.*, 1975). LH-RH has been found by RIA also in the ME of humans, sheep, pigs, and monkeys, and in relatively large amounts in the pineal gland of porcines and ovines (see Ramirez and Kordon, 1975).

Until the recent use of RIA for releasing hormones, few attempts were made to determine the distribution of TRH activity in the hypothalamus by direct measurement. Equal TRF activity, measured separately, in the ME and in the rest of the hypothalamus was reported in sheep by Guillemin *et al.* (1965). Averill and Kennedy (1967), working with the same animal species but using a different technique for determination of TRF activity, concluded that TRF activity in the ME is approximately twice that in the hypothalamus, although the latter contained 7 times as much

TABLE XXVII

LH-RH Content of the Hypothalamic Nuclei[a]

Area	LH-RH content (pg/μg protein)
Nucleus preopticus medialis	<0.05
Nucleus periventricularis	<0.05
Nucleus suprachiasmatis	Trace (<0.1)
Nucleus supraopticus	Trace (<0.1)
Nucleus hypothalamicus anterior	<0.05
Area hypothalamica lateralis anterior	<0.05
Nucleus paraventricularis	<0.05
Median eminence	22.4 ± 2.2
Nucleus arcuatus	2.9 ± 0.8
Nucleus ventromedialis, pars medialis	Trace (<0.1)
Nucleus ventromedialis, pars lateralis	0.6 ± 0.5
Nucleus dorsomedialis	<0.05
Nucleus perifornicalis	<0.05
Area hypothalamica lateralis posterior	<0.05
Nucleus hypothalamicus posterior	Trace (<0.1)
Nucleus premammillaris dorsalis	<0.05
Nucleus premammillaris ventralis	<0.05

[a] From Palkovits *et al.* (1974b).

TRF. The distribution of TRH in the hypothalamus was studied by Krulich *et al.* (1974) using serial sections from fresh, frozen rat hypothalami for assaying TRF activity (release of RIA-TSH from rat pituitaries *in vitro*). Their results indicated that TRF activity is not diffusely dispersed throughout the hypothalamus, but appears, on the contrary, to be concentrated in the preoptic region. The assumption that ME and dorsal hypothalamus are production sites of TRF was in agreement with the work of Reichlin and Mitnick (1973a), who found active synthesis of TRF from amino acid precursors in hypothalamic tissue from these areas (see Section A,6).

As for LH-RH, the development of sensitive and specific TRH radioimmunoassays has allowed more precise TRH studies (see also Section B). The localization of TRH in hypothalamic and extrahypothalamic brain nuclei has been determined with the aid of the microdissection technique (Palkovits, 1973) (see above). The greatest concentration of TRH was found in the ME (38 ng/mg) which contained 1.04 ng of TRH (Brownstein *et al.*, 1974b); this amount represents about 25% of the TRH found in a 30-mg fragment of hypothalamus (Vinokur and Utiger, 1974). High concentrations were also found in some hypothalamic nuclei (n. periventricularis, ARC n., n. ventromedialis pars medialis, n. dorsomedialis). Outside the hypothalamus, relatively large amounts of TRH were found in the septal and preoptic areas. (Brownstein *et al.*, 1974b) (Fig. 39).

The findings of radioimmunological determinations of TRH in the brain have been confirmed and broadened by the results of immunohistochemical studies (Hökfelt *et al.*, 1975b, 1976c). TRH-containing nerve terminals were found in the dorsomedial nucleus and the perifornical area, in extrahypothalamic nuclei such as n. accumbens, the lateral septal nucleus and in several motor nuclei of the brainstem and spinal cord. Interestingly, the distribution of TRH within the ME (Fig. 40) differs distinctly from that of LH-RH and somatostatin (Fig. 41). Whereas TRH is concentrated in the

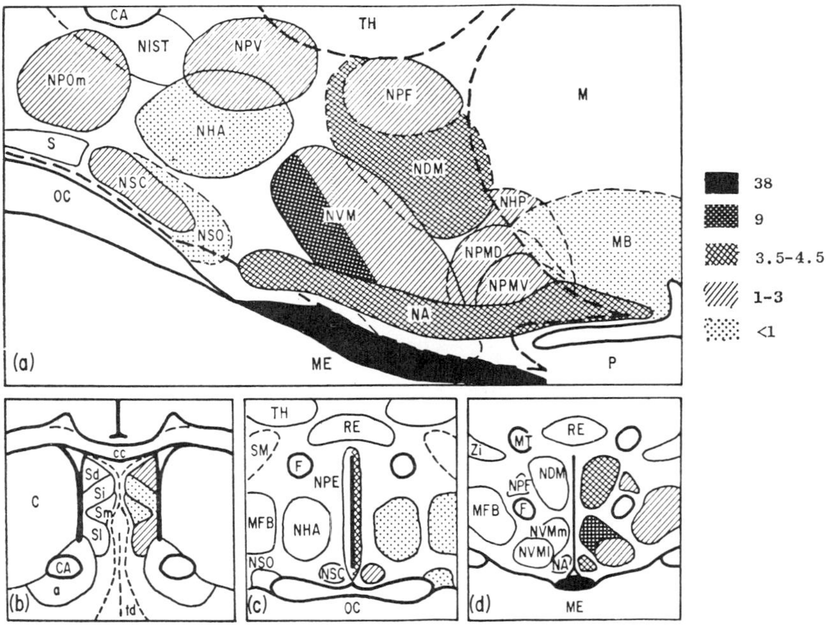

Fig. 39 Localization of TRH in the hypothalamus, septum, and preoptic area. Drawing (a) is of a parasagittal section through the rat hypothalamus and (b–d) are of frontal sections. Drawing (b) depicts the septal region, (c) the anterior hypothalamus, and (d) the tuberal region. C, nucleus caudatus; CA, commissura anterior; CC, corpus callosum; F, fornix; M, mesencephalon; MB, mammillary body; MT, tractus mammillothalamicus; NA, nucleus arcuatus; NHA, nucleus anterior; NIST, nucleus interstitialis striae terminalis; NPO_m, nucleus preopticus medialis; NPV, nucleus paraventricularis; NSC, nucleus suprachiasmaticus; OC, chiasma opticum; P, pituitary; RE, nucleus reuniens thalami; S, nucleus preopticus suprachiasmaticus; SM, stria medullaris; TH, thalamus; ZI, zona incerta; a, nucleus accumbens; td, nucleus tractus diagonalis; key, nanograms of TRH per milligram of protein. (Reproduced with permission from Brownstein *et al.*, 1974b. Copyright 1974 by the American Association for the Advancement of Science.)

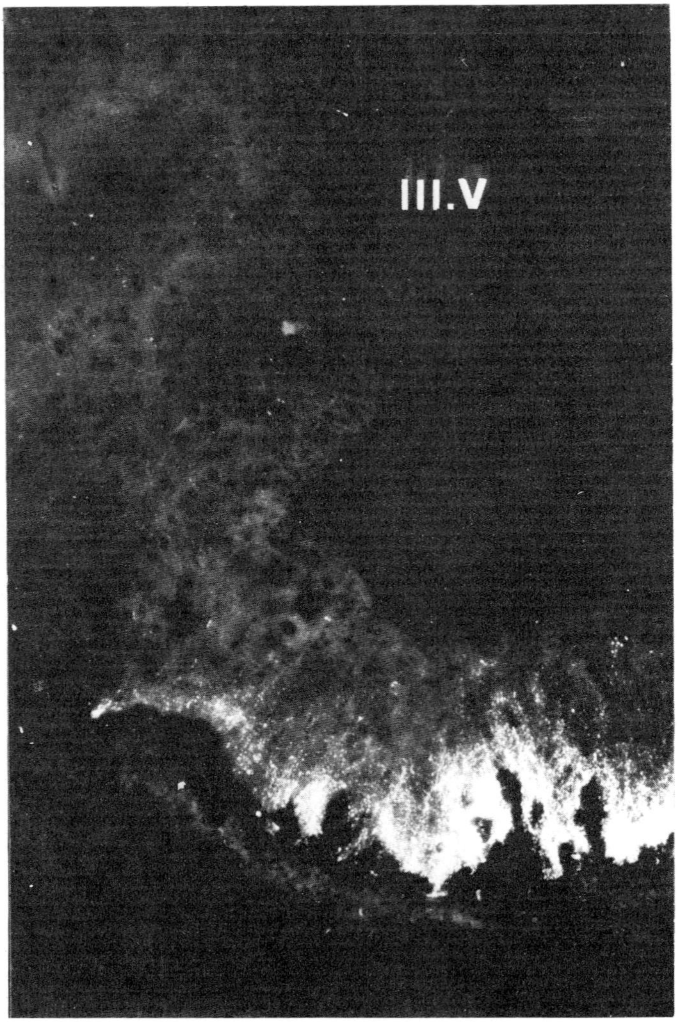

Fig. 40 Immunofluorescence micrograph of the basal hypothalamus including the median eminence. The highest density of TRH-positive fibers is found in the medial part of the external layer close to the capillary loops of the portal vessels, whereas the lateral part contains comparatively low numbers. III.V, third ventricle. (×160). (Reproduced, with permission, from Hökfelt *et al.*, 1975b.)

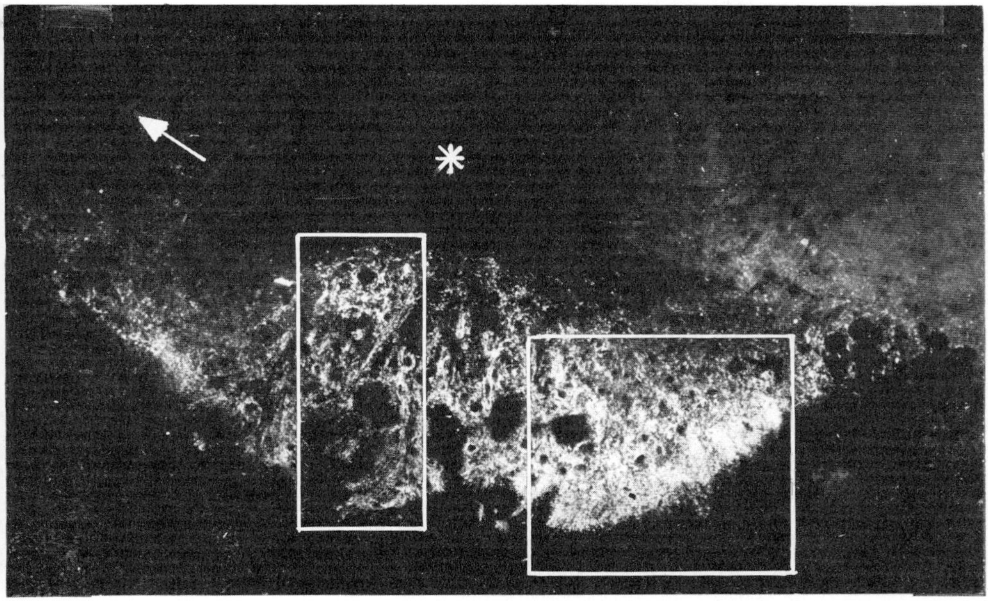

Fig. 41 Immunofluorescence micrograph of the basal hypothalamus after incubation with somatostatin. A dense plexus of fluorescent fibers is observed in the median eminence, and a weak plexus also in the ARC nucleus (arrow). Asterisk indicates third ventricle. (×140). (Reproduced, with permission, from Hökfelt *et al.*, 1975d.)

external medial part, LH-RH is present in the external lateral parts and somatostatin both in the external lateral and external and internal medial parts (see Hökfelt *et al.*, 1976c, and Fig. 54, B and C). While it was highly conceivable that TRH found in the ME is present in axons and nerve endings (Bennett *et al.*, 1975) which terminate on or near the portal vessels, until recently the true neuronal localization of the TRH measured elsewhere (cell bodies, axons, or nerve terminals?) was not clear. Studies in which the *in vitro* uptake of TRH by brain tissue of the rat was evaluated in parallel with the uptake of DA revealed that a temperature-dependent, saturable uptake mechanism is present for both substances in the synaptosome-rich fraction of the cerebral cortex and hypothalamus (Parker and Barnea, 1976).

By coupling somatostatin to human γ-globulin, antibodies could be produced in rabbits against the neurohormone (Arimura *et al.*, 1975b) and used for the immunohistochemical tracing of cell systems containing "somatostatin-like immunoreactivity" (SLI). These studies did not exclude the possibility that the immunoreactivity demonstrated was actually due to two or even more closely related peptides (see Hökfelt *et al.*, 1975d). The structures containing SLI are of two types and seem to represent either neurons or endocrine-like cells (Dubois *et al.*, 1974; Hökfelt *et al.*, 1974b; Pelletier *et al.*, 1974b; Luft *et al.*, 1974; Dubois, 1975; Arimura *et al.*, 1975a).

In the hypothalamus of untreated rats both somatostatin-positive cell bodies and fibers were observed. The few cell bodies were located in the periventricular region of the anterior hypothalamus at the level of the suprachiasmatic nucleus and were the origin of the extensive intrahypothalamic somatostatin-positive fiber plexuses observed in the VMN and the ARC n., the ventral premammillary nucleus, and the ME (Elde *et al.*, 1976). At this level, the distribution of somatostatin nerve endings in the rat (30%) does not correspond to that of LH-RH or TRH (see above). Somatostatin occurs, in fact, both in the lateral and median part of the ME (external and internal layer) (Fig. 41) (Hökfelt *et al.*, 1975d). At this level, nerve endings contain SLI material in granules which are larger (900–1100 Å) than the ones described for LH-RH (Pelletier *et al.*, 1974b). In the stalk, only a few LH-RH positive fibers were seen, while SLI covered the entire surface of the stalk; some somatostatin-stalk fibers also extended into the proximal part of the neurohypophysis (Hökfelt *et al.*, 1975d). With improved immunohistochemical techniques, Johansson *et al.*, quoted by Hökfelt *et al.*, (1975d), were able to locate numerous somatostatin positive cell bodies—approximately 25–50 per section—in the periventricular area of the anterior hypothalamus, and extensive networks of somatostatin-positive nerve endings in the tuberculum olfactorium and the suprachiasmatic nucleus, a finding which agrees better with the results of RIA determinations.

Determination of somatostatin content in the brain by combined microassay (Arimura *et al.*, 1975b) and microdissection (Palkovits, 1973) techniques produced evidence for a more ubiquitous distribution of the neurohormone. In addition to its concentration being determined in the hypothalamus (20–40 ng per hypothalamus), somatostatin was detectable in elevated titers in the septum-preoptic area region (25 ng), the thalamus (17 ng), the midbrain (9.5 ng) and the brainstem (9.8 ng). Within the hypothalamus, the concentration of somatostatin was highest in the ME (309 ng/mg protein) and ARC n. (45 ng/mg protein), although all of the hypothalamic nuclei, especially the VMN (15 ng/mg protein) and the n. periventricular (24 ng/mg protein), contained some of this material (Brownstein *et al.*, 1975b) (Fig. 42). Of particular relevance was the presence of somatostatin in areas of the brain thought to be involved in the regulation of GH (see Sections A,3,a and c): the POA contained considerably more somatostatin than the septum, and the amygdala, notably the

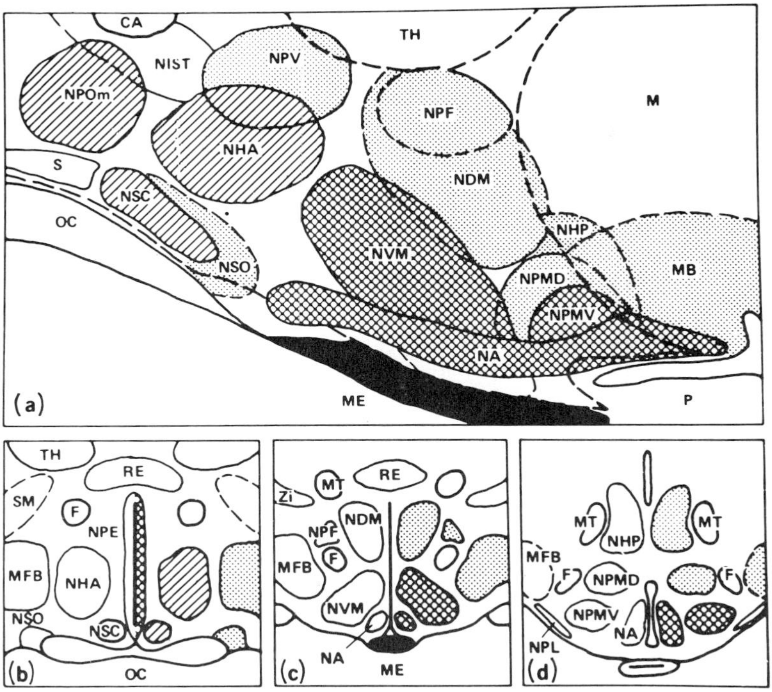

Fig. 42 Localization of somatostatin in nuclei of the hypothalamus. Drawing (a) is of a parasagittal section through this area of the brain. Drawings (b), (c), and (d) are of frontal sections through the anterior hypothalamus, the tuberal region, and the premammillary region, respectively. Relative concentrations of somatostatin are denoted as follows: ■ > ⊠ > ▨ > ⊡ Abbreviations: CA, anterior commissura; F, fornix; M, mesencephalon; MB, mammillary body; MT, mammillothalamic tract; ME, median eminence; MFB, medial forebrain bundle; NA, arcuate nucleus; NDM, dorsomedial nucleus; NHA, anterior hypothalamic nucleus; NHP, posterior hypothalamic nucleus; NIST, nucleus interstitialis striae terminalis; NPE, periventricular nucleus; NPF, perifornical nucleus; NPMD, dorsal premammillary nucleus; NPMV, ventral premammillary nucleus, NPOm, medial preoptic nucleus; NPV, paraventricular nucleus; NSC, suprachiasmatic nucleus; NSO, supraoptic nucleus; NVM, ventromedial nucleus; OC, optic chiasm; P, pituitary; RE, nucleus reuniens thalami; S, preoptic suprachiasmatic nucleus, SM, stria medullaris; TH, thalamus; ZI, zona incerta. (Reproduced with permission from Brownstein *et al.*, 1975b.)

corticomedial portion (Brazeau *et al.*, 1976), was richer in somatostatin than the hippocampus and parietal cortex (Brownstein *et al.*, 1975b).

The distribution of SLI in the hypothalamus might be indicative of an elaborate control system of GH secretion (1) by somatostatin acting neurohormonally at the level of the pituitary gland and (2) modulating synaptically GRF-producing cell bodies at the level of the VMN, a view which obviously requires the support of more experimental data.

In extra-CNS regions, SLI was found mainly in spinal ganglia (10% of the neuronal cell bodies) and the substantia gelatinosa of the dorsal horn of the spinal cord, a finding which may be compared with recent immunohistochemical studies on the localization of substance P in neuronal cell bodies of the rat spinal ganglia and in nerve terminals in the spinal cord (Hökfelt *et al.*, 1975e,f) (see Chapter II, Section K). For substance P, an excitatory transmitter function at the level of primary sensory neurons has been postulated (Otsuka, 1976); by analogy, a similar (inhibitory?) role might be envisaged for somatostatin at the level of a second subpopulation of primary sensory neurons (Hökfelt *et al.*, 1975g).

The extra-CNS localization of somatostatin involves endocrine-like cells of the salivary glands, thyroid, pancreas, stomach, and the epithelial cells in the intestine (Fig. 43) (Luft *et al.*, 1974; Hökfelt *et al.*, 1975a,d; Dubois, 1975; Polak *et al.*, 1975; Molrar *et al.*, 1976). In the thyroid gland,

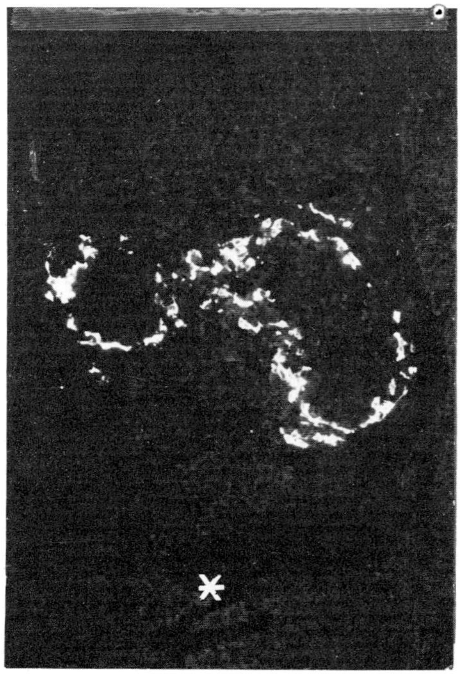

Fig. 43 Immunofluorescence micrograph of pancreas section incubated with somatostatin antiserum pretreated with human α-globulin. Fluorescent cell bodies and processes are observed only in the periphery of the islet. Asterisk indicates blood vessel. (×100). (Reproduced, with permission, from Hökfelt *et al.*, 1975d.)

specific cells exhibiting SLI could be demonstrated, with a localization in the parafollicular–calcitonin-producing cells, although these cells were also fluorescent after pretreatment of the antiserum with calcitonin (Hökfelt *et al.,* 1975d). In the stomach, the presence of SLI in cells with short thick processes, often running parallel to the gastric gland, is consonant with the ability of somatostatin to inhibit the secretion of gastrin, pepsin, and gastric acid (Bloom *et al.,* 1974; Barros D'Sa *et al.,* 1975; Gomez-Pan *et al.,* 1975). In all parts of the intestine, a positive immunofluorescence was shown in structures with a varicose appearance, most likely representing nerve terminals; the fibers were seen chiefly in the basal part of the lamina propria of the mucosa and occasionally penetrated into the muscular layer (Hökfelt *et al.,* 1975f). The role of these somatostatin-containing nerves in the rat gastrointestinal tract has yet to be determined, but it is conceivable that the peptide plays an important part in the regulation of alimentary function. In this respect it may be analogous to other substances such as DA, 5-HT, and H, which, it has been postulated, function as local hormones (see also Chapter I, Section B). It is worth noting also that in the intestinal wall somatostatin distribution is paralleled by the distribution of substance P (Hökfelt *et al.,* 1976a). The sharing of hormonal peptides by hypothalamic and gastrointestinal cells supports the concept of the neuroectodermal origin of the endocrine–peptide-producing cells of the APUD series (Pearse, 1969) (see also Chapter I, Section B,2), although definitive proof of the similarity of CNS and gut somatostatin must still be obtained from further studies. In the pancreas, somatostatin localization in the D cells and its possible significance for islet insulin and glucagon secretion have been previously discussed (Section A,3,c).

From the data available, it appears that somatostatin or somatostatin-like peptides have a widespread distribution in mammals, being localized both in the nerve and in endocrine-like cells. This suggests that somatostatin, in addition to its inhibitory action on GH secretion, may have regulatory functions, some of which were hitherto unknown, at different levels of the organism.

With regard to GRF activity, determined by the ability of frozen brain tissue sections to induce release of GH measured by RIA from *in vitro* pituitaries, this was located in the lateral portion of the VMN (Krulich *et al.,* 1971), which is compatible with the observations that lesions of the VMN cause a decrease in plasma GH levels and that, conversely, electrical stimulation of the same area causes a prompt increase in plasma GH levels (see also Section A,3,a).

The unavailability of natural or synthetic PIF and PRF has hindered more refined studies on their localization in the brain with the use of RIA methods. With the use of several techniques previously used for assaying TRF or GRF activity, Krulich *et al.* (1971) found prolactin-inhibiting activity in the lateral POA and lateral hypothalamus and PRA in the medial suprachiasmatic region. It is worth noting that the latter localization corresponds to an area of distribution of ascending serotoninergic nerve endings (see Chapter II, Section C,2) and for 5-HT a stimulatory role for the release of PRL has been demonstrated (Chapter IV, Section A,4,b).

D. MECHANISM OF ACTION

Hypothalamic neurohormones do not escape from the general rule governing the mode of action of peptide and protein hormones: that of combining with high affinity receptors on the surface of the cell with the resultant generation of one of several intracellular messengers (see also Section A,4). Strong evidence supports for RF's a mechanism of action via the second messenger, cAMP. Addition of crude hypothalamic extracts was reported concurrently by two independent groups to result in a rapid increase in AC activity (Zor *et al.*, 1969a,b) and a dramatic rise in cAMP levels (A. L. Steiner *et al.*, 1969, 1970). There was great specificity in the action of the crude hypothalamic extracts, since no effect on either AC or cAMP resulted when they were incubated with posterior pituitary, adrenal medulla, and thyroid tissue (Zor *et al.*, 1969b; Steiner *et al.*, 1970); however, Chase *et al.* (1969) and Chase (1970) found that acid extracts of hypothalamus and brain cortex produced nonspecific stimulation of AC in many tissues. Increased hormone release following both TRF and LH-RH/FSH-RH administration has been associated with increases in pituitary cAMP, although some contradictory evidence still exists (Guillemin, 1971c). In the case of TRF, Borgeat *et al.* (1972) had shown an increase in cAMP at 15 min and a progressive accumulation to 2 hr of incubation. After LH-RH/FSH-RH administration (1×10^{-10}–1×10^{-9} *M*), Borgeat *et al.* (1973) observed a simultaneous increase in cAMP and in the *in vitro* release of FSH and LH after 120 min of incubation, a finding substantially confirmed by Jutisz *et al.* (1972), Makino (1973), Kaneko *et al.* (1973), and Borgeat *et al.* (1974). Also with LH-RH analogs, a close correlation was found between increased adenohypophyseal cAMP concentration and increased LH and FSH release (Borgeat *et al.*, 1974).

In order to establish that the increased intracellular concentration of cAMP accounts for the effect of RF's on pituitary secretion, the criterion of obtaining release of hormones with cAMP or with its derivative dbcAMP should be satisfied (Rasmussen, 1974). Indeed, the latter was able to induce release of all pituitary hormones *in vitro* and its effects were enhanced by a substimulatory dose of aminophylline or theophylline, two phosphodiesterase inhibitors (Schofield, 1970; Peake, 1973; Wakabayashi *et al.*, 1972). One would also postulate that inhibiting factors would, conversely, lower the cyclic AMP levels in the gland. It was gratifying, therefore, to observe that the newly discovered somatostatin actually decreased pituitary cAMP levels and the GH release stimulated by cAMP, theophylline, and prostaglandins (Vale *et al.*, 1972b; Labrie *et al.*, 1975) and that LH-RH antagonists, with the great ability to compete with LH-RH, always showed close parallelism between inhibition of LH release and cAMP accumulation (Beaulieu *et al.*, 1975; Labrie *et al.*, 1976b).

Cyclic AMP does not appear to be the only cyclic nucleotide involved. Recently, it has been reported that purified ovine GRF preparations increased pituitary concentration of cGMP, a cyclic nucleotide formed from GTP by the catalytic action of guanylate cyclase, and not cAMP (Wilson *et al.*, 1974). In addition, cGMP itself appeared to be a potent stimulant for GH release (Peake *et al.*, 1972). An interesting sex difference was noted in this regard: cGMP was more effective than cAMP in stimulating GH release from female pituitaries, and cAMP was more effective than cGMP on male pituitaries (Cehovic *et al.*, 1972).

As in the case of many other cellular systems in which cAMP has been involved as a second messenger, there is also evidence that other second messengers exist. In particular, in many of these systems Ca^{2+} also serves as second messenger (Hardman, 1974). As an alternative to the hypothesis that RF's act via cAMP, it has been postulated that they induce their dramatic effects on pituitary cells by altering cell permeability. This would be followed by uptake of Ca^{2+}, which in some way activates hormone release. This hypothesis is supported by the observation that increased K^+ in the medium activates the release of AP hormones from pituitaries incubated *in vitro* (Samli and Geschwind, 1968; Wakabayashi *et al.*, 1969). This high concentration of K^+ was associated with depolarization of the pituitary cell membrane (Martin *et al.*, 1973) and an increase in net accumulation of extracellular $^{45}Ca^{2+}$ (Milligan and Kraicer, 1971). Further support for this hypothesis was the fact that removal of Ca^{2+} from the medium blocked secretion from both high K^+ and normal stimuli, although in the latter case it was often necessary to remove residual Ca^{2+} with ethylenediaminetetraacetic acid to block the releasing action completely (Samli and Geschwind, 1968; Wakabayashi *et al.*, 1969). In addition, an increase in intracellular Ca^{2+} has been reported following addition of high K^+ or GRF concentrations to incubated pituitaries (Milligan *et al.*, 1972). Also the digitalis glycoside ouabain has been reported to stimulate *in vitro* release of pituitary ACTH and GH (Fleischer *et al.*, 1972), which is associated with a ouabain-stimulated net uptake of $^{45}Ca^{2+}$ by the pituitary

(Fleischer and Woods, 1973). However, more recent studies by Fleischer and his collaborators (Eto *et al.,* 1974) on pituitary tissue *in vitro,* using verapamil, an agent that inhibits Ca^{2+} entry, have shown that ouabain and K^+ affect the process of "stimulus-secretion coupling" by promoting a specific Ca^{2+} influx, but that hypothalamic hormones and their intermediates (dbcAMP or theophylline) do not. These authors suggested that RF's are a means of initiating hormone release promoting redistribution of cellular Ca^{2+} rather than a critical Ca^{2+} influx. The two actions, cAMP mediation or increased permeability to Ca^{2+}, are not necessarily mutually independent and could be caused by a common mechanism. It is known that cAMP in many tissues activates a class of enzymes, cAMP-dependent protein kinases, which have been proposed as the major if not as the sole site of cAMP action in cells (Rasmussen, 1974). A cAMP-binding protein kinase has been found in the AP (Howell and Montague, 1971). The releasing hormone–receptor interaction might have at least two effects: (1) an activation of AC with resultant increase in intracellular cAMP; and (2) an increase in the Ca^{2+} permeability of the cell membrane due to the activated protein kinase (Labrie *et al.,* 1971) (Fig. 44). It is of note that high K^+ concentrations elevate release of pituitary hormones with no effect on cAMP levels (Zor *et al.,* 1970). This finding suggests inferentially that the ultimate effect of both K^+ and cAMP would be that of altering membrane permeability; the increased Ca^{2+} availability would then promote alone or together with cAMP hormone release. In addition to this late site of Ca^{2+} requirement, an early site has been demonstrated by the observation that increasing concentrations of free Ca^{2+} enhanced LH-RH-induced cAMP accumulation (Labrie *et al.,* 1976b).

There has been rapidly accumulating evidence for a role of microtubules in endocrine gland secretion; they would act in the translocation of hormone-containing granules to the cell surface (Rasmussen, 1970). The microtubular protein (tubulin) might serve as a substrate for cAMP-dependent protein kinases. Contradictory findings have been reported for their role in the secretion processes of AP hormones. While Milligan and Kraicer (1971) reported that the addition of colchicine, which binds to and blocks the contractile microtubular system, can suppress the augmented release of ACTH induced by high K^+ concentrations *in vitro,* Sundberg *et al.* (1973) did not find any blocking activity for colchicine on the basal release of GH, PRL, LH, FSH, and TSH but, actually, an enhancement of the release promoted by high K^+ or by RF's. Similarly, enhancement of basal and cAMP-stimulated GH release from rat pituitaries was observed with colchicine by Lockhart-Ewart and Taylor (1971). In contrast to these findings, more recently blockade of both basal and stimulated (PGE_1, K^+, purified GRF) release of GH was reported following incubation of rat adenohypophysis with vincristine (Labrie *et al.,* 1973). After prolonged

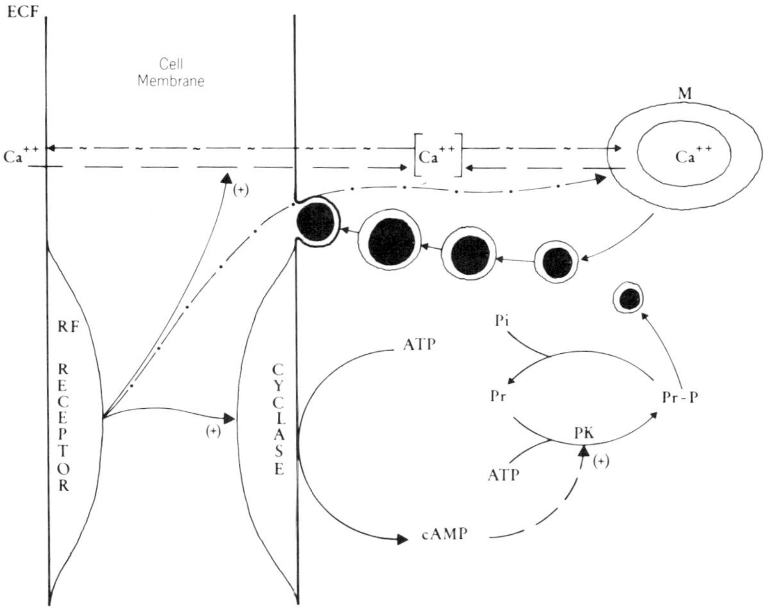

Fig. 44 A model of RF action on the pituitary cell in which RF-receptor interaction leads to: (1) an activation of adenyl cyclase and (2) an increase in the permeability of the cell membrane to Ca^{2+}. The possibility is also envisaged (—·—·—·) that RF's, rather than promoting Ca^{2+} influx, may act to promote redistribution of mitochondrial calcium pool. PK, protein kinase; Pr-P, phosphorylated protein; Pi, pyrophosphate; M, mitochondrion; ECF, extracellular fluid. (Adapted from Rasmussen, 1974.)

incubation in the presence of this periwinkle alkaloid, both somatotropes and mammotropes contained numerous crystalline deposits scattered throughout the cytoplasm which were presumably derived from the microtubules. Thus, on the basis of these divergent results the role of the microtubular system in the pituitary secretory process awaits further clarification.

PG's have also been implicated in pituitary hormone release, and many similarities have been found between PG stimulation of the AP and hypothalamic extract stimulation. Zor *et al.* (1969b, 1970) and MacLeod and Lehemeyer (1970) found increased pituitary AC activity following PGE_1 administration.

PG's are potent stimulators of cAMP accumulation in the pituitary gland, the order of potency of PG's being $E_1 = E_2 > A_1 = A_2 > F_{1\alpha} = F_{2\alpha}$ (Labrie *et al.*, 1976b), and concomitantly releasers of GH and TSH (small rise) (Zor *et al.*, 1970; Hertelendy *et al.*, 1971; Labrie *et al.*, 1976b). 7-Oxa-13-prostanoic acid, a PG antagonist, was found to reduce GH release and to prevent prostaglandin-induced GH and cyclic AMP rises

(Peake, 1973). Apparently, the structural similarity of the antagonist to PGE_1 permitted binding of the antagonist to the receptor without activation of AC.

In addition to direct stimulation of the AP, which is the mechanism responsible for GH and TSH release, PG's are capable of acting at the hypothalamic level to release FSH, LH, and PRL, although, also, for these hormones a direct pituitary site of action cannot be overlooked (Sato *et al.*, 1975a) (see Chapter IV, Section A,3,c). Probably responsible for the effect on neurohormonal secretion elicited at hypothalamic level by PG's is the increase in ME cAMP levels found after *in vivo* administration of PGE_1 (Labrie *et al.*, 1976b). For hypothalamic-inhibiting hormones (e.g., PIF), it has been postulated that, conversely, PG's may act by preventing the activation of the AC system. Ojeda *et al.* (1974b) have shown that PGE_1, injected IVT into ovariectomized, estrogen-treated rats at a dose (1 μg) unable to affect plasma PRL, completely prevented the PRL-inhibiting effect of DA and apomorphine but failed to modify the decrease in plasma PRL-induced by dbcAMP. Although actual changes in hypothalamic cAMP and PGE_1 levels were never measured after drug treatment, the interpretation given by the authors was that PGE_1 may act to elevate PRL by decreasing the activation of PIF-secreting cells by dopaminergic drugs, thus inhibiting the DA-induced cAMP rise in these neurons (see also Chapter IV, Section A,4,a). Collectively, these results suggest that PG's may play a role in the physiological release of AP hormones either by a hypothalamic site of action or directly at the pituitary level.

E. EXTRAENDOCRINE ACTIONS

Localization of hypothalamic hormones both in extrahypothalamic regions of the CNS and in certain extraneuronal regions, such as the pineal gland (see Section C), evidence of high affinity binding sites for TRH in extrahypothalamic tissue (Burt and Snyder, 1974), identification of LH-RH in synaptosome fractions of brain (Ramirez and Kordon, 1975), and localization of LH-RH (Zimmerman *et al.*, 1974; Pelletier *et al.*, 1974a) and somatostatin (Pelletier *et al.*, 1975; Hökfelt *et al.*, 1975d) in nerve terminals suggested a role for these peptides in neuronal function.

1. Behavioral Effects

a. L-*Dopa Potentiation Test*

It was first demonstrated that a hypothalamic hormone may exert effects other than on the AP in 1971 and 1972 by Plotnikoff, Kastin and their co-workers (Plotnikoff *et al.*, 1971, 1972a), who observed that MIF-I enhanced the behavioral effects of L-dopa and inhibited the effects of oxotremorine in both intact and hypophysectomized mice. Not only was MIF-I found to be active in a system considered to be an animal model of Parkinson's disease and mental depression, but the crucial observation was that the presence of the pituitary was not required for the action of the peptide on the brain. The observation in the dopa potentiation test was quickly applied and confirmed for TRH, which was generally more available for animal studies than MIF-I (Plotnikoff *et al.*, 1972b). It was then shown that MIF-I was capable of reducing the sedation effect induced in both monkeys and mice by deserpidine (Plotnikoff *et al.*, 1973), described as an animal model of depression and Parkinson's disease (Sulser and Bass, 1968). Collectively, all these observations led to the theory that both MIF-I and TRH might be of therapeutic value in patients with depression and Parkinson's disease (see below).

On examining the results of the studies of several hypothalamic peptides in many neuropharmacological systems, a different pattern of effects for each hormone tends to emerge. MIF-I is the most potent peptide affecting L-dopa (Kastin *et al.*, 1973a), while TRH is more effective than MIF-I in 5-HT potentiation test, in which 5-HTP is injected instead of L-dopa; LH-RH stimulates L-dopa-induced behavior but not 5-HT behavior (Spirtes *et al.*, 1976). Even somatostatin exerts some activity in the dopa potentiation test, but not as much as MIF-I or even TRH (Spirtes *et al.*, 1976).

b. Pentobarbital Sleeping Time

TRH, MIF-I, and somatostatin have all been tested for their ability to reduce or enhance the sleep and hypothermia produced by pentobarbital and ethanol. TRH administered systemically, but none of the other peptides, was found to antagonize the effects of pentobarbital and ethanol. In contrast to TRH, MIF-I, and LH-RH were devoid of this action and somatostatin enhanced pentobarbital-induced sleep (Breese *et al.*, 1975). Central administration of somatostatin in freely moving rats reduced both REM and SWS (Havlicek *et al.*, 1975), whereas administration of antisomatostatin antiserum lengthened REM sleep (Havlicek *et al.*, 1976). TRH has also been found capable of reversing the lethal effects of pentobarbital (Brown and Vale, 1975) and to antagonize the hypothermia and sedation induced by reserpine and CPZ (Breese *et al.*, 1975).

c. Motor Activity

Small doses of MIF-I induced stereotyped compulsive behavior in unrestrained adult cats, but no change in locomotor activity (North *et al.*, 1973), in agreement with previous findings with rats (Kastin *et al.*, 1973b).

TRH, on the other hand, increased spontaneous activity in the rat whereas somatostatin decreased locomotion (Segal and Mandell, 1974). The increased motility due to intracerebrally injected TRH in mice was enhanced by pretreatment with CPZ, reserpine, and sulpiride (Mora *et al.*, 1976). Like amphetamine, IVT-injected TRH induced in rats pretreated with apomorphine clockwise rotational behavior, suggesting indirect DA release action at the level of the nigrostriatal DA pathway (Cohn *et al.*, 1976).

d. Body Temperature

TRH injected into the lateral ventricle of the cat elicited at nanogram doses a marked hypothermic effect; indeed, when compared to other hypothermic agents, TRH seems to be the most potent one (Metcalf, 1974). Interestingly, TRH effects on body temperature appear to allow a species difference similar to the one revealed for brain MA's. Thus, in rabbits in which IVT NE administration produces hyperthermia (Cooper *et al.*, 1965), TRH also elicits a rise in body temperature (Metcalf, 1974). In view of the reported ability of NE to stimulate TRH release (see Chapter IV, Section A,5), it has been proposed that the hypothermic effect of NE may be mediated by TRH production (Metcalf, 1974). In the mouse, in which TRH at physiological doses does not affect body temperature, it antagonized reserpine and oxotremorine-induced hypothermia (Goujet *et al.*, 1975).

e. Mating Behavior

The fact that the preovulatory rise of gonadotropins precedes the onset of mating behavior and that LH-RH is localized in the brain in the same regions thought to be involved in inducing mating behavior (Reichlin, 1974b), i.e., the POA, led McCann and associates (Moss and McCann, 1973; Moss *et al.*, 1975b) to evaluate the role of LH-RH in the induction of mating behavior. Ovariectomized rats pretreated with estrogens, after systemic injection of LH-RH, displayed a pattern of lordosis behavior very similar to that produced by progesterone. Little or no lordosis behavior was observed in response to LH-RH alone. TRH exhibited no action whatsoever. In addition, LRH-facilitated lordosis occurred in the absence of the adrenals, thus eliminating a possible contribution of adrenal progesterone. The possibility that LH-RH was active via release of gonadotropins was ruled out by showing mating behavior with LH-RH in hypophysectomized, estrogen-primed females (Pfaff, 1973). In addition, estrogen-primed, ovariectomized rats infused with systemically ineffective doses of LH-RH directly in the MPOA or ARC nuclei exhibited increased lordosis behavior when compared with saline-treated rats, suggesting LH-RH acts centrally to induce mating behavior; LH-RH infused into the LHA or cerebral cortex had no significant effect (Moss and Foreman, 1976). In the sexually active male rat as well as in the testosterone propionate-primed, castrated male rat, LH-RH enhanced mating behavior, measured by a significant decrease in the time to the first intromission and to subsequent ejaculation. No mating behavior was observed in the castrated rat treated with LH-RH alone (Moss *et al.*, 1975b).

Collectively, these observations suggest that LH-RH plays a role in mediating sex behavior, possibly by activating neurons in the MPOA, which is considered to be the integrative center for mediating sexual receptivity (Powers and Valenstein, 1972). In this way, in the female, LH-RH, which is responsible for the preovulatory discharge of LH, by activating neurons in the MPOA would ensure synchronization between the ovulatory event and the onset of sexual receptivity.

The direct, central stimulating effect of LH-RH on sexual behavior in rats is in accord with findings reported recently in patients with hypogonadism and psychogenic impotence (Mortimer *et al.*, 1975). In these cases, subcutaneous administration of LH-RH induced a very rapid return of potency, suggesting the prompt elicitation of a mechanism which was not androgen-mediated. The possible therapeutic value of LH-RH in the treatment of impotence is implied by these findings. Some of the effects of hypothalamic hormones on neuropharmacological or behavioral tests are reported in Table XXVIII.

TABLE XXVIII

Effects of Hypothalamic Hormones on Neuropharmacological or Behavioral Tests[a,b]

Neuropharmacological or behavioral tests	TRH	MIF-I	GH-RIH	LH-RH
Pentobarbital sleeping-time	↓	—	↑	—
Ethanol narcosis	↓			
Slow wave and REM sleep			↓[e]	
Strychnine-induced seizure	↑		↓	
Dopa[c] potentiation	↑	↑↑	↑	↑
5-HTP[c] potentiation	↑↑	↑		
Motor activity	↑	Stereotypy	↓	
Body temperature	↑	—		—
Reserpine-induced hypothermia	↓			
Lordosis[d]	—			↑

[a] See text for details.
[b] Key to symbols: —, no effect; ↓, inhibition; ↑, potentiation; ↑↑, marked potentiation.
[c] Associated to pargyline.
[d] Ovariectomized estrogen-primed rats.
[e] Given intraventricularly.

2. Effect on Brain Monoamine Metabolism

Efforts have been made to establish biochemical correlates for the direct CNS behavioral action of hypothalamic hormones; thus, the effect of some of these neurohormones on brain monoamine metabolism has been studied. To be valid, such an effect should be studied in both intact and hypophysectomized animals since hypothalamic hormones have been found to be equally effective in neuropharmacological tests involving both types of animals. The first report on the action of a hypothalamic hormone on brain neurotransmitter metabolism is that of Friedman *et al.* (1973) who showed that MIF-I increased *in vitro* the synthesis of DA in the striatum of intact but not hypophysectomized rats, an effect partially supported by the findings of Mulder and Smelik (1974). Plotnikoff *et al.* (1974) and Kostrzewa *et al.* (1975) were unable, however, to find evidence of an effect of MIF-I on DA in the striatum. Similarly, Spirtes *et al.* (1976), using an *in vivo* system, were unable to show any effect of MIF-I on the striatum of intact rats, while their data were indicative of an increased NE turnover in the frontal cortex. In hypophysectomized rats, MIF-I, as well as α-MSH, seems to reduce *in vivo* 5-HT turnover in the hypothalamus (Spirtes *et al.*, 1976). An involvement of MIF-I in brain MA metabolism is also supported by the results of the IVT injection of the tripeptide, which showed its specific localization in several discrete portions of the rat brain (Pelletier *et al.*, 1977).

For TRH, conflicting results have been reported. Keller *et al.* (1974) demonstrated enchanced NE turnover in the whole brain as well as in various brain regions, whereas other studies found no such effect (Reigle *et al.*, 1974). Similar to the latter findings, it has also been reported (Tuomisto and Männisto, 1973) that TRH is ineffective in inhibiting the *in vitro* uptake of biogenic amines (NE, DA, and 5-HT) by rat synaptosomes or rabbit platelets, and no significant alterations have been found in the metabolism of endogenously administered L-dopa following systemic administration of TRH in rats (Sandler *et al.*, 1973). Collectively, these findings provide little evidence to suggest that alterations in the metabolism of NE or other MA's in brain, such as have been observed with other antidepressant drugs, might be associated with the reported antidepressant effect of TRH (see below), although proposals to the contrary have been formulated (Kulig, 1975).

An interaction of TRH with the cholinergic system of the brain has also been suggested by the reduction of TRH antagonism of pentobarbital and ethanol sleeping time produced by intracisternally injected atropine (Breese *et al.*, 1975). The potentiating effect of TRH on the excitatory action of ACh on cerebral cortical neurons supports this view (Yarbrough, 1976).

Proof for direct action of hypothalamic hormones on the brain has been derived from experiments undertaken to establish whether MIF-I, TRH, somatostatin, and/or LH-RH have an effect on the electrical activity of brain neurons. Microiontophoretic application of TRH and LH-RH inhibited or excited the spontaneous discharge of single unit-action potentials in the PO and anterior areas. The releasing factors did not influence the firing of any of the neurons tested in the cerebral cortex (Dyer and Dyball, 1974). The response of the unit of LH-RH fluctuated during the estrous cycle, with the larger percentage being present in proestrous than in diestrous 1 (Kawakami and Sakuma, 1974). Neurons sensitive to hypothalamic hormones, although more frequent for TRH in the hypothalamus than elsewhere in the brain, were also found for LH-RH and somatostatin (linear or cyclized form) in the cuneate nucleus, the cerebellar cortex, and the cerebral cortex (Renaud *et al.,* 1975), areas in which these peptides have either not been detected or are low in concentration (see Section C). Changes in the electrical activity of the brain after MIF-I administration have also been observed in the rabbit and in man (Plotnikoff and Kastin, quoted by Kastin *et al.,* 1975). MIF-I given in a very huge dose altered the background EEG activity of restrained albino rabbits from a sleeplike pattern to one characteristic of continuous arousal. Changes in the contingent negative variation (CNV) of healthy human beings have also been found after administration of MIF-I (R. Hall *et al.,* unpublished observations). It would appear that the response pattern of peptide-sensitive neurons is not restricted to hypothalamic neurons involved in pituitary regulation. Microiontophoretic studies with substance P indicate a predominantly excitatory action when applied to neurons elsewhere in the brain (Krnjević and Morris, 1974) (see Chapter II, Section K).

3. Therapeutic Applications

a. Parkinson's Disease

Behavioral effects induced in the animal model of Parkinson's disease and mental depression (see above) were the rationale for therapeutic trials with hypothalamic hormones in disorders of the extrapyramidal function and in depressed patients. It appears that MIF-I has a relatively nontoxic, albeit modest, antiparkinsonian effect in some patients (Kastin and Barbeau, 1972; Woods and Chase, 1973). However, it appears to worsen rather than improve preexisting L-dopa dyskinesias (Woods and Chase, 1973; Chase *et al.,* 1974), in spite of its ability to enhance the motor effects of L-dopa in rodents (Plotnikoff *et al.,* 1971). Recently, however, it was reported that MIF, injected intravenously, greatly enhanced the effect of an oral L-dopa dose on motor performance. The amelioration in motility was accompanied by improvement in intellectual functioning (Barbeau, 1975). In Parkinsonian patients TRH was without effect (Chase *et al.*, 1974).

Although at the height of behavioral activation TRH causes a significant increase in cerebral DA levels in mice given a MAOI (Breese *et al.*, 1974), neither tripeptide affects brain DA levels or turnover in *in vivo* experiments (Spirtes *et al.*, 1976; Plotnikoff *et al.*, 1974). For MIF, which does not seem to release CA's in an amphetamine-like manner (Huibodro-Toro *et al.*, 1975), the most likely site of action, if it is synaptic, is the postsynaptic membrane. In favor of this hypothesis is the fact that it greatly enhances apomorphine-induced hyperactivity in intact as well as in hypothalamus-lesioned rats (C. Burnett, E. Strother, R. F. Butterworth, and A. Barbeau, unpublished data). However, many other possibilities must still be considered and tested, and it should be noted that clinical effects of MIF are not necessarily derived by mechanisms mediated by dopaminergic systems (Cox *et al.*, 1976). Conceivably, MIF possesses some direct action on central pathways subserving extrapyramidal function, which would justify long-term clinical trials using higher dose levels of this peptide, alone or in combination with L-dopa, in Parkinson's disease.

b. Mental Depression

A dramatic and rapid improvement in depressed patients given a single injection of TRH in a double-blind controlled trial was first reported by Prange *et al.* (1972). since then several groups have reported that TRH has mood-elevating properties in depressed women (Kastin *et al.*, 1972e; Tiwary *et al.*, 1972; Wilson *et al.*, 1973a). In one study (Wilson *et al.*, 1973b), normal women were also found to respond to TRH administration with a transient elevation in mood. However, in successive studies, the beneficial effects of TRH in depressed patients have not been confirmed (Sørensen *et al.*, 1974; Mountjoy *et al.*, 1974; Coppen *et al.*, 1974), and TRH has been found to be ineffective in other psychotic patients (Drayson, 1974; Dimitrikoudi *et al.*, 1974). In 5 parkinsonian patients, administration of TRH did not produce any measurable motor improvement, but a true elevation of mood in 2 out of 3 patients who were taking L-dopa (McCaul *et al.*, 1974).

Collectively, it appears from these reports that TRH is not so therapeutically useful as it was thought before. Its therapeutic efficacy, however, might be ignored in favor of its use as a potential diagnostic tool in differentiating types of depression (Ehrensing *et al.*, 1974). In this regard, Kastin *et al.* (1972e) were the first to report that in depressed patients, despite the absence of known endocrine disease, the response of plasma TSH levels to TRH was limited. This was confirmed by Prange *et al.* (1972) and McCaul *et al.* (1974) and was expanded to include age-related changes as well as plasma prolactin levels (Ehrensing *et al.*, 1974). Interestingly, "nonspecific" release of GH (Maeda *et al.*, 1975; see also Section A,6,a) and limited PRL response (Legros and Breulet, 1975) have been reported in depressed patients following TRH.

In conclusion, the main point raised by these studies is that the hypothalamic peptides exert action on the CNS. Although the animal models used to represent human diseases are somewhat imprecise and nonspecific, they clearly demonstrate a CNS action by hypothalamic peptides. Concerning the mode for controlling brain neuronal function, two possible routes of access to neurons can be postulated for hypothalamic hormones (see Chapter I, Section B,2). The first is that by which the releasing or inhibiting peptide is secreted into the portal vessels from which it could be carried through the systemic circulation back to the brain. An alternative possibility is that hypothalamic peptides may act as neurotransmitters secreted from branches of peptidergic neurons which synapse locally within the hypothalamus or exit from the hypothalamus to make synaptic contacts in extrahypothalamic areas. This trans-synaptic mode of action is supported by studies which have revealed the existence of some "extrahypophyseal immunoreactive pathways" for LH-RH which do not pass to the ME, but from the hypothalamus pass near the epithalamus and the basal rhinencephalon (see Section C). A single neuron or group of neurons endowed with this double neurotransmitter and neurohormonal action might thus promote both behavioral and anterior pituitary secretion changes. Beyond their specific usefulness in the treatment of Parkinson's disease and mental depression, the evidence is irrefutable that new vistas for investigation have been made possible by the concept of extraendocrine effects of hypothalamic hormones.

IV

Brain Neurotransmitters and the Regulation of Anterior Pituitary Function

The two main components of the neural–endocrine communication system, e.g., the neurohumoral and the neurhormonal components, having been elucidated, we will now consider the roles of specific brain neurotransmitters in controlling secretion of anterior pituitary hormones in both experimental animals and man. Two main strategies have been used in experiments by researchers attempting to establish neurotransmitter–hypothalamic hormone relationships: (1) varying the functional activity of brain neurotransmitter systems by administration of neurotransmitters or their precursors and by neuropharmacological treatment and measuring the effects on endocrine function (Table XXIX); (2) using natural or induced changes in endocrine activity and studying the effect on brain neurotransmitter metabolism (Table XXX). The main advantages, limitations, or drawbacks of the most commonly used approaches or techniques will be discussed in the appropriate subsections. We will attempt to review, without being exhaustive, some of the evidence linking brain neurotransmitters to the secretion of hypothalamic hormones and hence AP hormones (point 1) or to neuroendocrine feedback regulation (point 2). Although a larger number of compounds have been identified or proposed as neurotransmitters (see Chapter II), DA, NE, and 5-HT have so far received the major attention of researchers in this field.

A. MANIPULATIONS OF BRAIN MONOAMINES AND THE SECRETION OF THE ANTERIOR PITUITARY

1. Brain Neurotransmitters in the Control of ACTH Secretion

Failure to isolate CRF from hypothalamic or neurohypophyseal sources in amounts sufficient to characterize its structure did not impede

TABLE XXIX

Approaches Most Commonly Used in the Study of Neurotransmitter–Neurohormone Interactions[a]

Approach	Response measured[b]
In vivo	
1.[c] Central administration (intracisternal, intraventricular, intracerebral) of the neurotransmitter or a specific agonist or antagonist drug, or systemic administration of a neurotransmitter precursor, an agonist, or an antagonist drug (A–D)	A. Increased or decreased blood levels of AP hormones or change in target organ secondary to altered pituitary secretion
	B. Response to stimuli acting at CNS or/and pituitary level
2.[c] Intracerebral implant of the neurotransmitter or a specific agonist or antagonist drug (A–D)	C. Alteration in AP hormone content
3. Electric stimulation of discrete brain areas after pretreatment with drugs which block or stimulate neurotransmitter synthesis (A and C)	D. Increased or decreased RIA-RF content in different CNS areas or/and biological fluids
4. Administration into hypophyseal portal vessels or into the pituitary of the neurotransmitter or a specific agonist or antagonist drug (A–C)	
In vitro	
5. Addition of the neurotransmitter or a specific agonist or antagonist drug to the incubation medium of an AP alone (E and G)[d] or in presence of hypothalamic fragments (E, F) or to hypothalamic fragments alone (F)	E. Alteration in AP hormone content in the pituitary or/and the incubation medium
	F. Increased or decreased RIA-RF content in the hypothalamic fragments or/and the incubation medium
	G. Response to stimuli acting at pituitary level

[a] Advantages or drawbacks of these approaches are discussed in the text.

[b] A–G in parentheses refer to the type of response which has to be measured.

[c] Points 1 and 2 may refer to animal: intact, deafferented, hypophysectomized, or hypophysectomized, bearing an ectopic pituitary.

[d] Points 4 (B) and 5 (E and G) are suitable to study a direct neurotransmitter effect at pituitary level.

an extensive series of studies devoted to the elucidation of the neurohumoral systems which influence central regulation of the hypothalamohypophyseal–adrenal axis (HHAA). Three possible modes of neurotransmitter participation in the neuroendocrine process may be identified: (1) involvement in the elicitation or mediation of the external

TABLE XXX

Techniques Used to Study the Effect of Hormones on Brain Neurotransmitter Metabolism

1. Histochemical or/and biochemical evaluation of neurotransmitter levels or turnover in the whole brain, in discrete CNS areas or in single CNS nuclei and/or
2. Determination of the activity of enzymes involved in neurotransmitter biosynthetic or catabolic pathways in the whole brain, in discrete CNS areas or in single CNS nuclei in
   a. Normal animals
   b. Animals with endocrinological manipulations
   c. Animals undergoing physiological events (different stages of the estrous cycle, pregnancy, sleep, etc.)

inputs conveyed to the hormone-secreting cells of the hypothalamus and representing the open loop component which is capable of surmounting the closed loop negative feedback system; (2) interference with the circadian rhythm of ACTH secretion regulated by a biological *zeitgeber* in the limbic system of the brain; and, finally, (3) an action on the closed loop feedback corticosteroids–CRF–ACTH, that constantly operate, at least in nonprimate species, to stabilize the rate of ACTH secretion.

a. Catecholamines

A considerable body of evidence is available relating CA metabolism to the activity of the HHAA (see Ganong, 1975). The association of sympathoadrenomedullary and pituitary adrenocortical activity in stressful situations suggested very early that CA's might play a role in regulating ACTH secretion. On the basis of Vogt's findings (1954), showing that brain CA's were maximally concentrated in the hypothalamus, the hypothesis was suggested that these neurohumors might be the RF's regulating AP hormone secretion. However, both *in vitro* (Guillemin, 1955; Guillemin and Rosenberg, 1955; Saffran and Schally, 1955) and *in vivo* studies (Briggs and Munson, 1955; Martini *et al.,* 1960; De Wied *et al.,* 1969) disproved the possibility of CA's being the hypothalamic CRF. Of particular import in this regard was the observation that neither systemic E nor NE were able to elicit ACTH secretion in midbrain-transected rats (Martini *et al.,* 1960). Similarly, the hypothesis formulated by Long (1952) that E released from the adrenal medulla was an important factor in stimulating ACTH secretion was not confirmed (Ganong, 1963). On examining the possible regulatory role of central CA's on the activity of the HHAA, the numerous literature contributions on this topic can be divided into evidence (1) denying any relationship, (2) in favor of an excitatory role, or (3) implying the existence of an inhibitory role for CA's.

Indications against an involvement of brain CA's in CRF secretion were provided by Smelik (1967), and were based on the inability of reserpine to alter the steady-state system when administered into the ME. These negative results might merely indicate that the site(s) of the interaction between NE (inhibitory) and peptidergic CRF-producing neurons was not the ME. Preziosi *et al.* (1968) were also inclined to disregard the increase in plasma B found following prenylamine administration, a sympatholytic agent, as proof of specific removal of a central CA inhibitory action. They interpreted the observed activation of the HHAA as due to the marked blood pressure-lowering effect of the drug, disregarding the fact that α-MD, a less potent antiadrenergic agent which similarly provided a marked hypotensive action, failed to stimulate ACTH secretion. De Schaepdryver *et al.* (1969a,b), using CA depletors and trying to prevent the stress-induced rise of plasma B by pretreating the animals with nialamide, were unable to demonstrate any correlation between CA's and ACTH secretion. Similarly, Moore and co-workers (Carr and Moore, 1968; Hirsch and Moore, 1968) found that depletion of brain CA's by a combination of reserpine and α-MpT failed to alter plasma B response to various stressors or to subsequent injection of reserpine; and, no changes in hypothalamic CRF, pituitary ACTH content, and plasma B or in the response to stress were found after α-MpT administration by Bhattacharya and Marks (1970). Repeated injections of small amounts (McKinney *et al.*, 1971; Kaplanski *et al.*, 1972) or oral administration (Uchimura *et al.*, 1972) of α-MpT decreased brain NE content but failed to stimulate permanently the HHAA, leading these authors to conclude that CA's do not likely play a major role in ACTH regulation. The same conclusions were reached by other workers because of a lack of adrenocortical activation during chronic depletion of central NE levels by reserpine (Carr and Moore, 1968; Abe and Hiroshige, 1974) or 6-OHDA (Kaplanski and Smelik, 1973a; Lippa *et al.*, 1973; Ulrich and Yuwiler, 1973; Abe and Hiroshige, 1974; Kumeda *et al.*, 1974; Kaplanski *et al.*, 1973–1974).

In accord with these negative results, L-dopa administration did not alter in man either urinary 17-OHCS excretion when given chronically (von Werder *et al.*, 1970) or plasma F when administered acutely (Eddy *et al.*, 1971). In patients with nontumor Cushing's disease, chronic L-dopa treatment, at doses usually associated with marked clinical improvement in Parkinson's disease, did not ameliorate the clinical or laboratory parameters of the disease (Krieger, 1973).

Evidence that CA's may stimulate ACTH secretion was initially derived from observations of peripheral administration of amines and/or amine precursors. Vogt (1944) first reported that iv injections of E in dogs caused a rapid and sustained rise in plasma F; it was found, subsequently, that systemic E administration stimulated adrenocortical function in rats (Long and Fry, 1945; Recant *et al.*, 1948; Gershberg *et al.*, 1950; Ronzoni and Reichlin, 1950), in dogs (Pashkis *et al.*, 1949), and in man (Godowski, 1948; Hills *et al.*, 1948; Corcoran and Page, 1948; Vernikos-Danellis and Marks, 1962). Epinephrine added to isolated adrenal glands (Hechter, 1949; Vogt, 1951) or injected into hypophysectomized rats (Gershberg *et al.*, 1950) was ineffective. These observations along with the ability of E to stimulate the transplanted pituitary (McDermott *et al.*, 1950) led Long (1952) to formulate the hypothesis that medullary E released by stress may be the humoral regulator of ACTH secretion. Arguments disproving this hypothesis, in addition to the study of Martini *et al.* (1960), include the ineffectiveness of electric stimulation of the dorsomedial hypothalamus, which increases E secretion, to change corticosteroid production (Goldfien and Ganong, 1961) and the reported ability of morphine and pentobarbital, drugs presumably acting on the CNS, to block *in vivo* the ACTH-releasing activity of E (Briggs and Munson, 1955). Therefore, the excitatory effect of systemically injected E, NE, and DA (Eskin and Shchedrina, 1971; King and Thomas, 1968), which do not appreciably cross the BBB, is most likely caused by peripheral neuronal stimulation conveyed to the hypophysiotropic area of the hypothalamus by ascending spinal pathways. An activation of α-adrenergic receptors appears to be involved in the peripheral stimulation by sympathomimetic drugs, since phentolamine prevented the ACTH and cortisol rises induced in humans by administration of the adrenergic derivative methoxamine, which does not cross the BBB (Nakai *et al.*, 1973). Also, sympathomimetic agents such as amphetamine, methylamphetamine, and dexamphetamine, which cross the BBB, were shown capable of stimulating ACTH secretion in man (Besser *et al.*, 1969). Epinephrine, injected directly into the pituitary, produced inconsistent effects on ACTH secretion (Hiroshige *et al.*, 1968; Dhariwal *et al.*, 1969b).

Better evidence for a possible excitatory role of brain CA's was obtained by injecting the amines directly into the brain or the ventricular system. Endröczi *et al.* (1963) were the first to demonstrate that E, NE, or ephedrine injected into the posterior hypothalamus or ventral tegmentum increased adrenal venous corticosteroid output in cats. H. P. Krieger and Krieger (1970) confirmed the possible adrenergic stimulation of ACTH secretion by implanting NE in several diencephalic areas of the cat; the increase of plasma corticosteroids elicited by the implant was not eliminated by prior systemic administration of dexamethasone (D. T. Krieger and Krieger, 1970). In guinea pigs administration of NE into the posterior hypothalamus or rostral midbrain resulted in an increase of 17-OHCS (Naumenko, 1967, 1968); this could be prevented by midbrain transection (Naumenko, 1968). Injection of NE and DA into the lateral ventricle of anesthetized (with Nembutal) (Hiroshige and Abe, 1973) or "gentled" unanesthetized rats (Abe and Hiroshige, 1974) elicited a significant increase in ACTH secretion, but did not alter the stress-induced secretion (Hiroshige and Abe, 1973).

In summary, it is difficult to draw a definitive conclusion on the physiological significance of the excitatory action of CA's in the regulation of the activity of the HHAA. Most of the foregoing data were based on (1) the effect of systemic administration of amines which barely cross the BBB or (2) on amine or drug administration or implantation into various CNS areas. On considering the first series of data, these might be explained by peripheral stimulation of α-adrenergic receptors (Vernikos-Danellis, 1968; Van Loon, 1970, 1973; Nakai *et al.*, 1973; Naumenko *et al.*, 1973; Ganong, 1974); results of the experiments in which drugs were administered or implanted into various CNS areas might instead be due to either nonspecific activation of other neuronal systems (see also Section 3,a) or, taking into account the blockade effect of midbrain transection (Naumenko, 1968; Naumenko *et al.*, 1973), to stimulation of an efferent pathway via the brainstem to the spinal cord and ensuing activation of peripheral mechanisms relaying to the hypothalamus and CRF neurons (Naumenko *et al.*, 1973).

i. Catecholamines and the inhibition of ACTH secretion. Neural mechanisms mediating tonic inhibition of ACTH secretion have been demonstrated. Removal of the whole brain, leaving only an isolated hypothalamus, was associated with elevation of corticosteroid plasma levels in the dog (Egdahl, 1960); in rats, complete deafferentation of the hypothalamus by a stereotaxic knife also led to high levels of plasma B (Halàsz, 1969).

Several lines of evidence link catecholaminergic pathways to the occurrence of a tonic inhibition. In several animal species (rat, mouse, guinea pig, and monkey), stress stimuli increasing ACTH secretion, such as cold exposure (Maynert and Levi, 1964; Gordon *et al.*, 1966); electric shock (Bliss and Zwanziger, 1966; Moore and Larivière, 1964; Paulsen and Hess, 1963; Ordy *et al.*, 1966); prolonged swimming (Barchas and Freedman, 1963); hypoglycemia (Vogt, 1954); hemorrhage (Coleman and Glaviano, 1963); and physical stress (Welch and Welch, 1968; Bliss and Zwanziger, 1966; Corrodi *et al.*, 1968; Elefthériou and Church, 1968), were associated with a decrease in brain NE content or a stimulated brain NE turnover (Gordon *et al.*, 1966; Thierry *et al.*, 1968; Corrodi *et al.*, 1968; Bliss *et al.*, 1968; Gibson *et al.*, 1969), with prevalence of the catabolic phase (Bliss *et al.*, 1968).

The effect of stress on central DA neurons has not been as well demonstrated. No change in brain DA content or turnover could be detected in the rat after different types of stress procedures (Gordon *et al.,* 1966; Streefkerk, 1967; Engel *et al.,* 1968; Corrodi *et al.,* 1968; Bliss *et al.,* 1968; Fuxe and Hökfelt, 1969), the overall conclusion being that stress enhances the functional activity of central NE rather than DA neurons.

A thorough survey of the literature relating CA functional activity to ACTH secretion suggests the possibility of an inverse relationship. Attempts to elucidate the role played by biogenic amines in the central regulation of ACTH secretion involved at first the use of pharmacological agents known to decrease brain MA tone. Reserpine, the most commonly used drug in this respect, was shown capable of stimulating ACTH secretion in the rat (see Van Loon, 1973; Hodges and Vellucci, 1975; Eisenberg, 1975), dog (Egdahl *et al.,* 1956), monkey (Harwood and Mason, 1957; Rosenthal and Mason, 1961), and man (Khazan *et al.,* 1961). Reserpine at doses which failed to decrease brain NE and 5-HT content by greater than 50% did not increase ACTH secretion. Hypophysectomy and adrenalectomy neither altered the effect of reserpine on brain MA content nor the adrenal response to exogenous ACTH, suggesting that the reserpine effect on brain MA's was not mediated through the pituitary or the adrenals. Analogs of reserpine, such as syrosingopine, Ro 4-1284, isoreserpine, and isoraunescine, at doses which failed to deplete brain NE or 5-HT content, did not increase ACTH secretion as measured by plasma B levels, adrenal ascorbic acid depletion, or liver TPO activity (Maickel *et al.,* 1961; Westermann *et al.,* 1962). However, larger doses of syrosingopine or Ro 4-1284, which decreased central NE and 5-HT by more than 50%, did increase ACTH secretion (Westermann *et al.,* 1962). An attempt was made by the same authors to demonstrate which of the two amines was involved in the regulation of ACTH secretion. Administration of α-MmT did not alter plasma B or adrenal ascorbic acid and failed to prevent their response to reserpine administration. Since α-MmT affects only adrenergic metabolism, it was concluded that NE and DA were not necessary for the ACTH response to reserpine, and that 5-HT was the neurotransmitter whose depletion was responsible for the ACTH response to the drug. Subsequent investigations suggested that this conclusion may in fact not be valid, since after α-MmT administration, a CA depletor such as tyramine still retained its effect (Andén and Magnusson, 1964). It was thus apparent that a functional NE pool survived α-MmT action (Kopin, 1968), and that reserpine probably was acting on this NE pool to stimulate ACTH secretion. These initial observations, confirmed by other authors (see Van Loon, 1973), favor the hypothesis that CA's exert a tonic inhibitory control on ACTH secretion and that the decreased release of bioactive CA's from nerve endings due to reserpine eliminates this tonic inhibition.

In some instances and only for small doses, reserpine pretreatment was shown capable of producing an increased response of the adrenal axis to stress (Eisenberg, 1975). However, following larger doses or intracerebral implants of the drug, this effect was not always observed (Smelik, 1967; Carr and Moore, 1968; Eisenberg, 1975), probably because of the overall debilitation of the animal or perhaps because high doses of the drug acted on pathways involving other neurotransmitters (Eisenberg, 1975). The use of guanethidine, a drug with reserpine-like action but unable to cross the BBB (see Chapter II, Section B,7), reinforced the proposal that in the rat the inhibitory catecholaminergic action is central in nature. Guanethidine, systemically administered, failed to increase plasma B (Scapagnini and Preziosi, 1971; Scapagnini *et al.*, 1972; Eisenberg, 1975); however, when injected IVT, it induced, 8 hr later, elevation of plasma B and depletion of hypothalamic NE and DA (Scapagnini and Preziosi, 1971; Scapagnini *et al.*, 1972). Another mode of decreasing central adrenergic tone was by blockade of CA biosynthesis. In this regard, α-MpT, no matter when injected systemically throughout the day, induced in rats an increase in both plasma B (Scapagnini *et al.*, 1970) and ACTH levels (Van Loon *et al.*, 1971a; Van Loon, 1973; Scapagnini *et al.*, 1975). Use of the soluble form (methyl ester) of the compound (Scapagnini *et al.*, 1970) avoided the possibility of interpreting the rise as due to unspecific stress (Vernikos-Danellis, 1968; Krieger and Rizzo, 1969) from local irritation (Carr and Moore, 1968).

Evaluation of the time course of α-MpT action on ACTH and B release revealed prompt stimulation (1 hr) of the adrenal axis (Scapagnini *et al.*, 1975). Additional direct effects of α-MpT on the adrenal gland were excluded in experiments in which it proved to be ineffective in increasing plasma B concentration in acutely hypophysectomized rats (Van Loon, 1973). The fact that systemically ineffective doses of α-MpT increased plasma B levels when given into the third ventricle of the rat brain (Van Loon *et al.*, 1971c) provided additional evidence for a CNS mechanism of action and reinforced the concept of a central inhibitory catecholaminergic action on CRF-ACTH secretion.

The use of inhibitors at various steps of CA biosynthesis clarified the respective roles of DA and NE in the mediation of the tonic CRF-ACTH regulation. FLA-63, an inhibitor of D-β-H, decreased hypothalamic NE with no change in hypothalamic DA content and induced a concomitant clear-cut adrenocortical activation, thus implying that NE and not DA is the amine which exerts a major role. Accordingly, an inverse linear correlation could be demonstrated between plasma B and hypothalamic NE content (Scapagnini *et al.*, 1972; Scapagnini and Preziosi, 1973a). Selective depletion of DA was obtained by first giving α-MpT and then replenishing NE stores by DOPS. In these experiments the adrenocortical activation following α-MpT administration was strongly inhibited by DOPS (Scapagnini and Preziosi, 1971; Cuello *et al.,* 1973–1974) (see, however, Chapter II, Section B,6). The relationship between plasma B and hypothalamic NE content following α-MpT and DOPS administration was inversely hyperbolic, suggesting the presence of a threshold NE concentration below which plasma B is elevated (Ganong, 1974; Cuello *et al.,* 1973–1974).

Consistent with the above findings were those obtained in experiments in which the central adrenergic tone was reduced by means of antagonists at CA receptor sites. Among phenothiazines, CPZ was the drug most extensively investigated. The overall conclusion derived from the many reports on the effects of CPZ on the HHAA in animals and humans was that the drug induces adrenocortical activation (De Wied, 1967), acting via an alteration of brain adrenergic tone (Bradley *et al.,* 1966; Sulser and Dingell, 1968) and independent of changes in body temperature (Bhattacharya and Marks, 1969a).

More recently, depletion of CRF from the ME associated with increased plasma and adrenal B levels and decreased pituitary ACTH content has been reported in rats after treatment with CPZ and reserpine. In this study, the time-course of CRF depletion coincided with that of CA depletion (Bhattacharya and Marks, 1969a,b; Marks *et al.,* 1970). Similarly, another phenothiazine drug, perphenazine, has been shown to produce in the rat a reproducible adrenocortical activation (Smith *et al.,* 1973; Chatterton *et al.,* 1974). The action of phenothiazines on the pituitary–adrenal axis is also shared by butyrophenone derivatives (Debreceni *et al.,* 1969) that, similar to phenothiazines, can block central CA receptors.

Classic α-adrenergic blockers interfere with CA regulation of ACTH secretion. Phentolamine, although ineffective in dogs (Ganong, 1974), induced in rats a short increase of plasma B on systemic administration (Scapagnini and Preziosi, 1973b). However, in these experiments the specificity of phentolamine action could be questioned, since activation of the adrenal axis might have resulted only from the hypotensive effect of the drug. Better evidence for the specificity of the phentolamine effect was derived from experiments in which direct administration of systemically ineffective doses of the drug into the third ventricle of the brain increased plasma B concentration (Scapagnini and Preziosi, 1973b). Phentolamine also reduced the inhibitory action of NE on CRF stimulated release *in vitro* (Hillhouse *et al.,* 1975). Similar to phentolamine, phenoxybenzamine injected systemically (Guillemin, 1955) or implanted chronically into the lateral ventricle of the rat brain (Eisenberg, 1975) removed the tonic inhibition of NE on ACTH secretion.

Another approach to ascertain the existence of NE inhibitory pathways was that of studying the effect of functional activation of the central NE tone on the stress-induced stimulation of the adrenal axis. Intravenous injection of L-dopa was able to inhibit in the dog the adrenocortical activation induced by laparotomy stress, whereas systemically injected DA and NE failed to do so (Van Loon *et al.,* 1971b). Pretreatment with a drug affecting endogenous CA stores did modify the amount of L-dopa necessary to induce inhibition (Van Loon *et al.,* 1971b). Similar to L-dopa, clonidine, which crosses the BBB and stimulates α-adrenergic receptors, inhibited the adrenocortical activation due to laparotomy stress (Ganong *et al.,* 1975). The inhibitory action of both L-dopa and clonidine was counteracted by centrally administered phenoxybenzamine, whereas propranolol was ineffective (Ganong *et al.,* 1975).

Functional activation of the central NE tone was also induced by giving MAOI. In mice, pretreatment with iproniazid was able to prevent the adrenocortical activation produced by reserpine or formalin stress (De Schaepdryver and Preziosi, 1959). However, more intense effects of stress, such as those from laparotomy, could not be prevented by the sole use of iproniazid. The drug, however, enhanced the inhibitory effect of dexamethasone (Scapagnini and Preziosi, 1973b), which was unable per se to prevent completely the increase in plasma B following laparotomy (Dallmann and Yates, 1968). The fact that dexamethasone–iproniazid-induced blockade of adrenocortical activation could be removed by phentolamine supported the view that an increased central α-adrenergic tone was responsible at least in part for the blockade effect (Scapagnini and Preziosi, 1973b). The inhibitory effect of iproniazid was partially confirmed by Hirsch and Moore (1968), who tried to analyze whether its action was dependent on a general hydrazine structure or on MAOI activity. They found that both tranylcypromine, a nonhydrazine compound, and pheniprazine, a hydrazine compound, decreased reserpine-induced stimulation of the HHAA. Another MAOI belonging to the nonhydrazine group, pargyline, inhibited the stress-induced ACTH release in rats (Bhattacharya and Marks, 1969b) and enhanced the L-dopa inhibitory effect in dogs (Van Loon *et al.,* 1971b). Also α-ethyltryptamine, an antidepressant indole compound, with MAOI activity, was found capable of blocking in dogs, even when implanted in specific hypothalamic areas, the corticosteroid discharge which follows laparotomy stress (Lorenzen and Ganong, 1967). However, since the drug also possesses a pronounced peripheral sympathomimetic activity, it was suggested that a rise in blood pressure via the carotid and aortic baroceptors was responsible for the inhibition of ACTH secretion. Inferential evidence, however makes the hypothesis of a pressure-mediated inhibition unlikely. In fact, systemic administration of low doses of phenoxybenzamine eliminated the L-dopa pressor response without affecting the inhibition of stress-induced ACTH secretion, and, conversely, NE and DA did not inhibit ACTH secretion at doses inducing a marked pressor response (Van Loon *et al.*, 1971b). It was then concluded that α-ethyltryptamine affected in an inhibitory manner stress-induced ACTH secretion via its central activity on MAO (Ganong, 1974).

Sympathomimetic drugs have been found useful for increasing the central adrenergic activity. In rats (Hirsch and Moore, 1968; Bhattacharya and Marks, 1969b), dogs (Lorenzen and Ganong, 1967; Van Loon *et al.*, 1971d), and monkeys (Toivola and Gale, 1973), amphetamine, administered before a variety of stressful stimuli, reportedly prevented, at least in part, an increase in ACTH discharge. Recently, an amphetamine-induced increase in plasma GH and inhibition of plasma 17-OHCS or cortisol have been observed in monkeys (Smith *et al.*, 1973; Sachar *et al.*, 1976), which confirms early results by Harwood and Bigelow (1960). In the study by Sachar *et al.*, the adrenal response to *l*- or *d*-amphetamine was identical after pretreating the animals with pimozide to eliminate any dopaminergic effect, which suggested that amphetamine could act through activation of central NE neurons.

For a more direct evaluation of the central vs peripheral component of the CA effect, another approach has been used, e.g., injection into the brain of neurotransmitters unable to cross the BBB. It was found that high doses of CA's and drugs that release or protect CA's from inactivation, upon injection into the third ventricle of the dog brain near the ME, inhibited stress- induced ACTH secretion. A CA metabolite such as VMA and a biologically less active isomer, D-NE, had no inhibitory effect (Van Loon *et al.*, 1971d). Similarly, in reserpine-treated rats, administration of either NE or DA at doses ineffective in the untreated animal caused a depression of reserpine-induced increases in adrenal B level (Hall, 1970; Marks *et al.*, 1970).

The presence of a central NE tonus that chronically inhibits CRF secretion is also supported by *in vitro* studies. In experiments, in which the action of putative neurotransmitters was investigated on the release of CRF from rat hypothalamus *in vitro,* NE was found capable of inhibiting the release of the neurohormone in response to ACh and 5-HT (Hillhouse *et al.*, 1975). The same inhibitory activity for NE was demonstrated on the release of CRF from synaptosomes isolated from rat hypothalami (Edwardson and Bennett, 1974). Collectively, the results of these experiments, dealing with acute manipulations of brain CA levels or/and turnover and relatively short-term evaluation of adrenocortical activity, favored the hypothesis of a central NE mechanism tonically inhibiting CRF-ACTH secretion.

In contrast, the hypothesis of a tonic NE inhibition appeared less tenable in conditions of long-lasting depletion of brain CA's. In fact, after a single dose of reserpine, a dissociation was found between the long-lasting depletion of brain CA's and the short-lived activation of the HHAA (Carr and Moore, 1968; Montanari and Stockham, 1962). The same discrepancy was present after repeated injections of reserpine. In rats, daily ip administration of low doses of reserpine (0.5 mg/kg for 9 days) markedly depleted hypothalamic NE content, whereas plasma B levels, after an initial rise, progressively decreased to reach control values on the fifth day of treatment (Scapagnini, 1974; Scapagnini *et al.*, 1976a; Hodges and Vellucci, 1975).

These data can be best explained by the reappearance after reserpine administration of a small functional NE pool produced as a consequence of increased TH activity due to feedback stimulation (see Chapter II, Section B,6). It has been shown, in fact, that long-term treatment with low doses of reserpine progressively increases TH activity in the brainstem and this was reflected by the rate of HHAA function. In rats pretreated for 9 days with reserpine, injection of α-MpT, at a dose (100 mg/kg) which is unable to stimulate the adrenocortical axis in control rats, restored the elevated levels of plasma B. Conceivably, under these conditions, α-MpT suppressed the small NE pool still functioning after the depleting effect of reserpine (Scapagnini *et al.*, 1976a).

The same type of reasoning can be applied to explain the lack of activation of the CRF-ACTH system found after central injection of 6-OHDA (Scapagnini and Preziosi, 1973a; Kaplanski and Smelik, 1973a; Kaplanski *et al.*, 1973–1974; Kumeda *et al.*, 1974). Evaluation of the effects of centrally injected 6-OHDA in rats, shortly after drug administration (24 hr), revealed a marked depletion of hypothalamic NE content, which was accompanied by adrenocortical activation (Scapagnini and Preziosi, 1973a; Cuello *et al.*, 1974; Ganong, 1974) and enhanced response to stress (Fuxe *et al.* 1973a). In contrast, 3, 15, and 30 days later, despite the sustained depletion of hypothalamic NE, corticosterone levels were not modified (Scapagnini, 1974; Cuello *et al.*, 1974); however, as in the reserpine-pretreated rats, in these animals injection of a low dose of α-MpT induced activation of the HHAA (Scapagnini, 1974). Therefore, even after selective degeneration of CA nerve terminals, a pool of NE responsible for the functional inhibition of the HHAA can still be present. The maintenance of a functionally active pool could be attributed to the hyperactivity of NE-containing neurons surviving the degenerative effects of 6-OHDA (N. J. Uretsky *et al.*, 1971), although other reasons, e.g., the existence of receptor supersensitivity following functional denervation, cannot be disregarded for the interpretation of this phenomenon (see Chapter II, Section D). A triphasic action for 6-OHDA has been described on the basis of rat and dog experiments by Ganong *et al.* (1973). A single injection of 6-OHDA causes NE release initially and can inhibit in rats and dogs the stress-induced adrenocortical activation. ACTH secretion is then increased in rats, and reduction of the inhibitory response to L-dopa is present in dogs presumably due to general disruption of adrenergic neurons. Finally, for the reasons noted above, ACTH secretion returns to normal despite persistent NE depletion.

Less consistent results were obtained in trying to apply to primates findings derived from rat and dog experiments. Neither inhibitory nor stimulatory effects of dopaminergic (apomorphine, L-dopa) or adrenergic (L-dopa, DOPS, and clonidine) agonists on the resting levels of plasma F were observed in unanesthetized rhesus monkeys by Chambers and Brown (1976). Some clinical support for the existence of an inhibitory role of central adrenergic origin on ACTH secretion can be found only in the results of von Werder *et al.* (1970), who showed that chronic treatment with L-dopa decreased plasma 17-OHCS response to insulin-induced hypoglycemia in some Parkinson's patients. However, resting levels of 17-OHCS and 17-ketosteroids were normal in these subjects. In contrast, in tranquilized monkeys iv administration of L-dopa clearly enhanced the response of plasma ACTH to metyrapone (Hsu *et al.,* 1976).

More recently, the ACTH lowering effect of CB-154, a dopaminergic compound, has been reported in 6 patients with active Cushing's disease (Lamberts and Birkenhägen, 1976a), and in one patient with Nelson's syndrome who underwent a bilateral adrenalectomy because of pituitary-dependent Cushing's syndrome (Lamberts and Birkenhägen, 1976b). On the basis of these findings and the partial return of plasma hGH response to hypoglycemia in three patients with Cushing's disease after CB-154 administration, hypothalamic dopaminergic depletion has been proposed as a factor in the pathogenesis of Cushing's disease (Lamberts and Birkenhägen, 1976a). However, on recalling the ability of ergot drugs to decrease ACTH secretion from experimental pituitary tumors (MacLeod and Krieger, 1976), the above results are better explained by a direct inhibitory action of CB-154 on the tumor pituitary. No effect of CB-154 was seen in normal subjects, even after stimulation of ACTH secretion by lysine vasopressin (Lamberts and Birkenhägen, 1976c), which agrees with the ineffectiveness of dopaminergic stimulation on F secretion in monkeys (Chambers and Brown, 1976) reported above.

The effects of drugs altering brain CA neurotransmission on the HHAA are listed in Tables XXXI and XXXII.

b. Serotonin

The role of central 5-HT system(s) in the regulation of ACTH secretion has also been extensively studied. Unlike CA's, serotoninergic neurotransmission does not seem to be involved mainly in acute response to stress but, instead, in the circadian periodicity of adrenal cortex secretion. Brain 5-HT content does not show consistent changes after stress (see Ganong and Lorenzen, 1967), although changes in brain 5-HT turnover have been detected in some stressful situations (Thierry *et al.,* 1968; Rosencrans, 1970). A diurnal fluctuation of the amine has been demonstrated in the whole brain of the mouse (Albrecht *et al.,* 1956), hamster (Matussek and Patschke, 1963), rat (Scheving *et al.,* 1968; Quay, 1968; Dixit and Buckley, 1967; Asano, 1971; Okada, 1971), and in specific brain areas of the cat (Reis *et al.,* 1969; Friedman and Walker, 1968) or rat (Scapagnini *et al.,* 1971; Quay, 1965, 1968, 1969; Héry *et al.,* 1972; Krieger, 1975; Simon and George, 1975). Although some studies were unsuccessful in correlating circadian fluctuations of brain 5-HT with the daily rhythm of plasma corticosteroids (Dixit and Buckley, 1967; Friedman and Walker, 1968; Krieger, 1975), others showed a positive correlation between 5-HT content in the limbic system and plasma B (Scapagnini *et al.,* 1971; Annunziato *et al.,* 1973; Simon and George, 1975), suggesting that 5-HT may play a role in the regulation of diurnal fluctuations of ACTH secretion. This view is in line with experimental evidence suggesting that the biological clock responsible for ACTH rhythmicity is outside the hypothalamus (Ganong, 1963). However, the existence of a simple correlation between brain monoamine and plasma hormone levels does not necessarily imply a causal relationship.

Experiments in which central 5-HT activity was varied by surgical or pharmacological means provided more consistent and reliable results. pCPA, the 5-HT synthesis blocker, was unable to alter baseline levels of plasma B when its effect was evaluated only at one time (morning) of the day (Preziosi *et al.,* 1968; McKinney *et al.,* 1971; Scapagnini and Preziosi, 1972). Also, implantation of the drug into the MBH, in spite of lowering hypothalamic 5-HT content, did not affect compensatory adrenal function (Vermes *et al.,* 1973a). Systemic pCPA administration, however, eliminated the diurnal fluctuations of plasma corticosteroids in birds (Boissin and Assenmacher, 1971), rats (Scapagnini *et al.,* 1971; Vernikos-Danellis *et al.,* 1973; Van Delft *et al.,* 1973), and cats (Krieger and Rizzo, 1969). In the rat treated with pCPA, the plasma B concentration was constantly maintained throughout the day at levels which were higher than normal in the morning but lower than normal in the evening (Scapagnini *et al.,* 1971; Vernikos-Danellis *et al.,* 1973). Interestingly, 48 (Vernikos-Danellis *et al.,* 1973) or 72 hr (Annunziato *et al.,* 1975) after pCPA injection, in spite of permanent depletion of brain 5-HT levels, a normal diurnal rhythmicity of plasma B was restored.

Central reduction of 5-HT activity by means of 5-HT receptor blockers similarly suppressed circadian periodicity in rats (Krieger and Rizzo,

TABLE XXXI

Effect of Drugs Altering Brain Catecholamine Transmission on the Hypothalamohypophyseal–Adrenal Axis[a]

Sympatholytic drugs	Species	Route of administration	Action on CA metabolism	Central adrenergic tone: DA	Central adrenergic tone: NE	HHAA activity	Authors
α-MpT	Rat	ip	Inhibition of DA and NE synthesis	↓	↓	↑	Scapagnini *et al.* (1970)
	Rat	IVT	Inhibition of DA and NE synthesis	↓	↓	↑	Van Loon *et al.* (1971c)
	Dog	iv	Inhibition of DA and NE synthesis	↓	↓	↑	Van Loon (1973)
α-MpT + L-dopa	Rat	ip	Inhibition of DA and NE synthesis + replacement of DA and NE stores	→	→	→	Van Loon *et al.* (1971c)
α-MpT + DOPS	Rat	ip	Inhibition of DA and NE synthesis + replacement of NE stores	↓	→	→	Cuello *et al.* (1973–1974)
FLA-63	Rat	ip	Inhibition of NE synthesis	→(↑)	↓	↑	Scapagnini *et al.* (1972)
Reserpine	Rat	ip	Depletion of DA and NE stores	↓	↓	↑	Brodie *et al.* (1961)
	Dog	im	Depletion of DA and NE stores	↓	↓	↑	Egdahl *et al.* (1956)
	Monkey	im	Depletion of DA and NE stores	↓	↓	↑	Harwood and Mason (1957)
	Man	im	Depletion of DA and NE stores	↓	↓	↑	Khazan *et al.* (1961)
Reserpine[b]	Rat	ip	Depletion of DA and NE stores + feedback stimulation + receptor supersensitivity	→(↑)	→(↑)	→(↓)	Scapagnini *et al.* (1976a)
Guanethidine	Rat	ip	Depletion of DA and NE stores	→	→	→	Scapagnini *et al.* (1972)
	Rat	IVT	Depletion of DA and DA stores	↓	↓	↑	Scapagnini *et al.* (1972)
Prenylamine	Rat	ip	Depletion of DA and NE stores	↓	↓	↑	Preziosi *et al.* (1968)
Phentolamine	Rat	ip	α-Blockade	↓	↓	↑	Scapagnini and Preziosi (1973b)
	Rat	IVT	α-Blockade	↓	↓	↑	Scapagnini and Preziosi (1973b)
Phenoxybenzamine	Rat	ip	Long-lasting α-blockade	↓	↓	↑	Guillemin (1955)
	Rat	Intracerebral	Long-lasting α-blockade	↓	↓	↑	Eisenberg (1975)

Chlorpromazine	Rat	ip	NE and DA receptor blockade	↓	↓	↑	Bhattacharya and Marks (1969a,b)
	Monkey	im	NE and (DA) receptor blockade	↓	↓	↑	See De Wied (1967)
	Man	im	NE and (DA) receptor blockade	↓	↓	↑	See De Wied (1967)
Perphenazine	Rat	ip	NE and (DA) receptor blockade	↓	↓	↑	Chatterton *et al.* (1974)
Dehydrobenzperidol	Rat	sc	DA and (NE) receptor blockade	↓	↓	↑	Debreceni *et al.* (1969)
6-OHDA[c]	Rat	IVT	Degeneration of central NE and (DA) nerve endings	(↓)	↓	↑	Scapagnini and Preziosi (1973a)
	Dog	IVT	Degeneration of central NE and (DA) nerve endings	(↓)	↓	↑	Ganong *et al.* (1973)
6-OHDA[d]	Rat	IVT	Degeneration of central NE and (DA) nerve endings; functional recovery of surviving neurons	(↓)→	(↓)→	→	Scapagnini and Preziosi (1973a)
	Dog	IVT	Degeneration of central NE and (DA) nerve endings; functional recovery of surviving neurons	(↓)→	(↓)→	→	Ganong *et al.* (1973)
6-OHDA[c]	Rat	ip	Degeneration of the ME – NE nerve endings outside the BBB	(↓)	↓	↑	Cuello *et al.* (1974)
6-OHDA[d]	Rat	ip	Recovery of the ME – NE nerve endings outside the BBB	→	→	→	Cuello *et al.* (1974)
6-OH-dopa[c]	Rat	ip	Degeneration of central NE and (DA) nerve endings	↓	↓	↑	Scapagnini *et al.* (1974b)
6-OH-dopa[d]	Rat	ip	Degeneration of central NE and (DA) nerve endings + functional recovery of surviving neurons + receptor supersensitivity	→	(↓)	→(↓)	Scapagnini *et al.* (1974b)

[a] Key to symbols: →, no effect; ↑, stimulation; ↓, inhibition; ( ), slight effect.
[b] Repeated daily injections.
[c] Effects evaluated for a short time after drug administration.
[d] Effect evaluated for a long time (3–30 days) after drug administration.

TABLE XXXII

Effect of Drugs Altering Brain Catecholamine Transmission on the Hypothalamohypophyseal–Adrenal Axis[a]

Sympathomimetic drugs	Species	Route of adminis-tration	Action on CA metabolism	Central adrenergic tone DA	Central adrenergic tone NE	Effect on stress-induced activation of HHAA	Authors
L-NE	Rat	ip	α-Receptor stimulation	→	→	→	Van Loon *et al.* (1971c)
	Dog	iv	α-Receptor stimulation	→	→	→	Van Loon *et al.* (1971b)
	Dog	IVT	α-Receptor stimulation + central NE receptor stimulation	→	↑	↓	Van Loon *et al.* (1971d)
Clonidine	Dog	ip	α-Receptor stimulation + central NE receptor stimulation	→	↑	↓	Ganong *et al.* (1975)
Clonidine + phenoxybenzamine	Dog	IVT	α-Receptor stimulation + central NE receptor stimulation + α-receptor blockade	→	→	→	Ganong *et al.* (1975)
DA	Rat	ip	Slight α-receptor stimulation	→	→	→	King and Thomas (1968)
	Dog	iv	Slight α-receptor stimulation	→	→	→	Van Loon *et al.* (1971c)
	Dog	IVT	Slight α-receptor stimulation + central DA and NE receptor stimulation	↑	↑	↓	Van Loon *et al.* (1971d)
L-Dopa	Dog	iv	Increase in DA and NE synthesis	↑	↑	↓	Van Loon *et al.* (1971b)
	Dog	IVT	Increase in DA and NE synthesis	(↑)	(↑)	(↓)	Van Loon *et al.* (1971d)
	Man	po	Increase in DA and NE synthesis	↑	↑	↓	von Werder *et al.* (1970)

L-Dopa + phenoxybenzamine	Dog	iv IVT	Increase in DA and NE synthesis + α-receptor blockade	→	→	→	Ganong *et al.* (1976)
L-Dopa + pargyline	Dog	iv	Increase in DA and NE synthesis + MAO inhibition	↑↑	↑↑	↓↓	Van Loon *et al.* (1971b)
Amphetamine	Rat	ip	Release of DA and NE	↑	↑	↓	Hirsch and Moore (1968)
	Dog	iv	Release of DA and NE	↑	↑	↓	Lorenzen and Ganong (1967)
	Monkey	im	Release of DA and NE	↑	↑	↓	Toivola and Gale (1973)
α-Ethyltryptamine	Dog	iv	MAO inhibition	↑	↑	↓	Lorenzen and Ganong (1967)
	Dog	Intracerebral	MAO inhibition	↑	↑	↓	Lorenzen and Ganong (1967)
Iproniazid	Rat	ip	MAO inhibition	↑	↑	(↓)	De Schaepdryver and Preziosi (1959)
Pheniprazine	Rat	ip	MAO inhibition	↑	↑	(↓)	Hirsch and Moore (1968)
Tranylcypromine	Rat	ip	MAO inhibition	↑	↑	(↓)	Hirsch and Moore (1968)
Pargyline	Rat	ip	MAO inhibition	↑	↑	↓	Bhattacharya and Marks (1969b)
Iproniazid + dexamethasone	Rat	ip	MAO inhibition + TH stimulation	↑↑	↑↑	↓↓	Dallmann and Yates (1968)

[a] Key to symbols: →, no effect; ↑, stimulation; ↓, inhibition; ( ), slight effect; ↑↑, marked stimulation; ↓↓, marked inhibition.

1969). Intracisternal injection of the neurotoxic tryptamine derivative, 5,6-DHT, into newborn rats did not eliminate plasma B circadian periodicity but induced 4-hr advance in the time of peaking (Krieger, 1975). Enhancement of the morning baseline levels of plasma B was present 1 day but absent 1 week after 5,6-DHT-induced degeneration of the 5-HT mesencephalic-hypothalamic pathway (Fuxe *et al.*, 1973a), a finding in keeping with the reported dampening of corticosterone circadian rhythm detectable 5 but not 21 days after IVT injection of 5,6-DHT (Scapagnini *et al.*, 1974a). The above results favor a 5-HT modulating role on circadian fluctuation of ACTH secretion.

5-HT pathways ascending from the raphe nuclei reach the hypothalamus, the amygdala and the hippocampus via the MFB (see Chapter II, Section C,2). Lesions along these pathways will chronically decrease 5-HT content in the hypothalamus and limbic system. Further support for the view that 5-HT pathways play a modulating role in the phasic regulation of ACTH secretion was derived from the observation that in rats lesions of the 5-HT containing raphe nuclei cell bodies abolished circadian rhythm of plasma B (Scapagnini and Preziosi, 1971; Harrington *et al.*, 1973; Scapagnini *et al.*, 1974a). Similarly, lesions isolating the hypothalamus by complete or frontal deafferentation decreased 5-HT hypothalamic content (Popova *et al.*, 1972; Vermes *et al.*, 1973c; Weiner, 1973), blocked diurnal ACTH cycle (see Halàsz, 1969; Palka *et al.*, 1969), increased the morning levels, and affected ACTH response to stress (Vermes *et al.*, 1973b). Additional evidence that the mechanism(s) for the normal circadian fluctuation of B secretion is located in the limbic area and is probably serotoninergic in nature was provided by the observation that section of the fornix (Mason, 1959; Möberg *et al.*, 1971; Lengvàri and Halàsz, 1973) or septal lesions (Harrington *et al.*, 1973) also resulted in cessation of the usual circadian rhythms with intermediate (fornix) or high (septum) B levels. However, 3 weeks after fornix transection, a significant diurnal variation did reappear (Lengvàri and Halàsz, 1973).

While a role of the 5-HT system in the maintenance of the ACTH-corticosterone balance is suggested by the above findings, participation of the amine in the events underlying tonic or stress-induced activation of the HHAA emerges less clearly and, in this respect, the results reported in the literature have been contradictory. The stress response after 5-HT synthesis inhibition appears to differ according to animal species or experimental conditions; no change was found in the cat (Krieger and Rizzo, 1969), while increased (Rosencrans, 1968; Vernikos-Danellis *et al.*, 1973; Vermes and Telegdy, 1973) or decreased (Van Delft *et al.*, 1973) adrenal responsiveness to ether stress was reported in the rat. Proof for a stimulatory role was provided by experiments in which direct injection of 5-HT into the ventricular system of the rat brain or into different hypothalamic, midbrain, and forebrain sites was accompanied by a pronounced functional activation of the adrenal system (Naumenko, 1965, 1968, 1969; H. P. Krieger and Krieger, 1970; Abe and Hiroshige, 1974). Systemic administration of 5-HTP produced adrenocortical activation in normal (Vernikos-Danellis *et al.*, 1973; Popova *et al.*, 1972; Annunziato *et al.*, 1973) or deafferented rats (Popova *et al.*, 1972). The effect of 5-HTP on the CRF-ACTH system in the rat is stereospecific, L-5-HTP being much more active than D-5-HTP, and can be attributed with certainty to action on 5-HT neurons since it was greatly enhanced by fluoxetine, a specific inhibitor of 5-HT reuptake (see Table XV). Fluoxetine alone induced a short-lived increase in plasma B (Fuller *et al.*, 1976). In chair-adapted, unanesthetized, male rhesus monkeys, infusion of 5-HTP (45-mg/kg) induced behavioral sedation and significant cortisol responses (Chambers and Brown, 1976). A modest rise in plasma ACTH and cortisol levels was also reported in the human after oral 5-HTP administration (Imura *et al.*, 1973). Conversely, in normal human volunteers, a 5-HT blocker, methergoline, inhibited the hypophyseal ACTH release in response to metyrapone (Cavagnini *et al.*, 1975) or insulin induced hypoglycemia (Cavagnini *et al.*, 1976), and cyproheptadine, but not methysergide, reduced hypoglycemia-induced cortisol secretion (Plonk *et al.*, 1974) and HHAA response to metyrapone (Plonk and Feldman, 1976). In patients with carcinoid tumors, cyproheptadine has been reported to decrease urinary 17-OHCS excretion and to inhibit the exaggerated metyrapone response seen in such patients (Plonk and Feldman, 1975). Moreover, it has been observed by Krieger *et al.* (1975) that in 3 patients with Cushing's disease long-term administration of this drug was associated with prompt and sustained clinical and biochemical remission of the disease, although the abnormality in circadian periodicity of plasma F concentrations persisted and abnormal ACTH levels remained in 1 of the 2 patients tested. The findings of Krieger *et al.* (1975) have been more recently corroborated by Middler (1976), but not by Scott *et al.*, (1977), and satisfactory remissions have been reported so far in 60% of the 40 patients with Cushing's disease who underwent chronic cyproheptadine treatment (Krieger, 1976). Positive results with the use of cyproheptadine in patients with Nelson's syndrome were observed by Hartwig *et al.* (1976) and by Krieger (1976), but not by Cassar *et al.* (1976).

In keeping with a stimulatory role of brain 5-HT on the HHAA system is the finding that spontaneous release of CRF from incubated rat hypothalami was stimulated by 5-HT in a dose-dependent manner and that the stimulatory action was antagonized by methysergide (Jones *et al.*, 1976a). It is of note that cholinergic blockade by hexamethonium and atropine reduced the secretion of CRF in response to 5-HT (Jones *et al.*, 1976a). This suggests that the 5-HT acts via a cholinergic interneuron (see below) that contains specific serotoninergic receptors since the effect was blocked by methysergide.

In contrast to these findings, which collectively may lead to the assumption that central 5-HT stimulation is associated with increased secretion of CRF-ACTH, other data have been presented which favor, instead, a reciprocal relationship between central 5-HT and adrenocortical activation. Thus, implantation of 5-HT into the LHA blocked stress-induced activation of the HHAA (Vermes and Telegdy, 1972) and compensatory adrenal hypertrophy and hypersecretion in rats (Vermes *et al.*, 1973a). Pretreatment with L-Trp or 5-HTP reduced the stress response in rats and 17-OHCS excretion in humans (Vernikos-Danellis *et al.*, 1973), and, conversely, pCPA administered to rats facilitated stress responses (Rosencrans, 1968; Vernikos-Danellis *et al.*, 1973; Vermes *et al.*, 1973b). Other evidence relating brain 5-HT neuronal function and ACTH secretion in a reciprocal way includes the decrease response to stress of the adrenal system after MAO inhibition (Tullner and Hertz, 1964; Ganong *et al.*, 1965; Bhattacharya and Marks, 1969b; Van Loon, 1973) and the blockade effect of 5-HT on ACTH secretion stimulated by hypothalamic fragments *in vitro* (Vermes *et al.*, 1972). Support for the view of a reciprocal relationship between central 5-HT function and adrenal activation can be found in the experiments in which the effects of adrenal steroids on brain 5-HT metabolism were studied (see Section B,1).

The actual reasons for the contradictory results and hence the views generated by the above-mentioned studies are far from being clear. Nonspecificity of effects of some of the drugs at the doses used (see Tables XIV and XV), differences in experimental design, mostly concerning times of the day at which alterations of the HHAA were induced or investigated, etc., can be responsible. Figure 45 represents schematically the role of brain MA's in the control of the HHAA during resting conditions, stress, or phasic events.

c. Acetylcholine and Putative Neurotransmitters

i. Acetylcholine. A central cholinergic link may also be involved in the regulation of HHAA activity, since in several species an increase in the cholinergic neurotransmission is often accompanied by adrenocortical activation.

In conscious or anesthetized dogs stimulation of nicotinic or muscarinic receptors by nicotine (Suzuki *et al.*, 1973) or pilocarpine (Suzuki *et al.*, 1975), respectively, increased markedly plasma 17-OHCS secretion. Impairment of the adrenocortical response to pilocarpine was found follow-

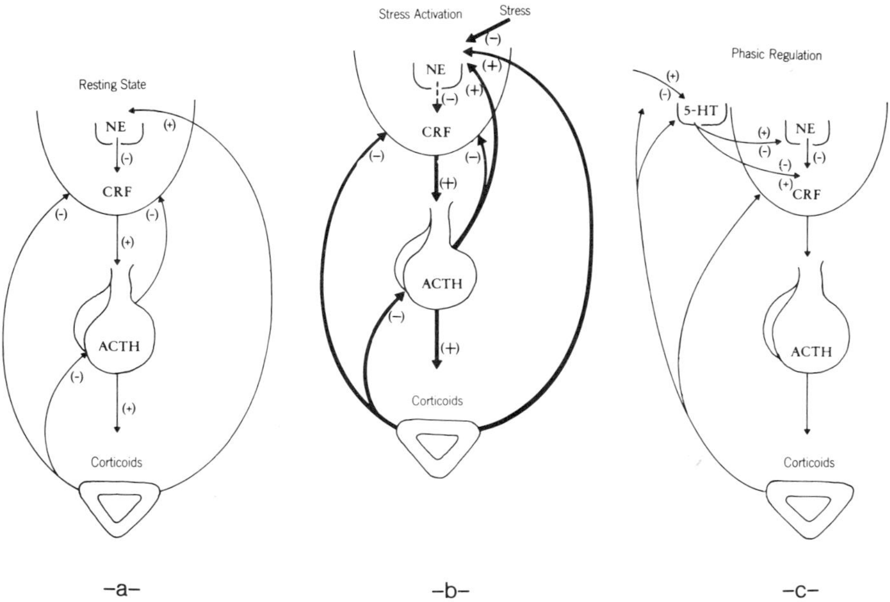

Fig. 45 MA intervention in the control of the hypothalamohypophyseal–adrenal axis during resting conditions (a), stress (b), or phasic regulation (c). (a) CRF which is under a tonic inhibitory norepinephrinergic influence stimulates the release of ACTH by the pituitary; secreted ACTH, in turn, stimulates adrenal corticoid secretion. There is evidence that the negative feedback action of corticoid secretion is directed largely at the pituitary gland but also at the neural control system where corticoids act by increasing NE turnover and so restraining further CRF release (see also Section B,1). (b) The neural component of control is influenced by various nonfeedback neural transients, among them stress. Depicted is the possibility that stress activation may occur through an inhibition of brain NE turnover. (c) Circadian periodicity of ACTH secretion is generated by an intrinsic brain rhythm possibly modulated by 5-HT neurons. Interactions between brain 5-HT and NE pathways are envisaged. NE, norepinephrine; CRF, corticotropin-releasing factor; ACTH, adrenocorticotropin; 5-HT, serotonin; (+), stimulation; (−), inhibition.

ing radiation-induced lesions of the anterior ME, but not of other hypothalamic areas (Suzuki *et al.,* 1975). In rats, when ACTH release was blocked by pentobarbital or morphine and CPZ, carbachol administered systemically acted as a powerful stimulus for the HHAA (De Wied *et al.,* 1969); carbachol was equally effective in stimulating the HHAA when injected directly into the lateral ventricle (Abe and Hiroshige, 1974). In cats, carbachol infused into the basal hypothalamus resulted in ACTH release (H. P. Krieger and Krieger, 1970) that could not be suppressed by pretreatment with dexamethasone (D. T. Krieger and Krieger, 1970).

Conversely, inhibition of central cholinergic activity was found to be associated with inhibition of the stress-induced adrenocortical activation. In rats, implants of atropine in the anterior hypothalamus prevented vasopressin and stress-induced ACTH release (Hedge and Smelik, 1968; Kaplanski and Smelik, 1973b). Atropine implants in nearby regions of the brain or sham-implantation in the same CNS region were ineffective; furthermore, other neurohormones, such as vasopressin, did not appear to be affected by atropine implants, thus implying a specific action of the drug on the neuronal component regulating CRF-ACTH secretion (Hedge and De Wied, 1971). In dogs and cats, however, central (Ganong, 1974) or systemic (Guillemin, 1955; Krieger *et al.,* 1968) administration of atropine failed to inhibit the stress-induced activation of the HHAA.

Other central mechanisms are possibly sensitive to the anticholinergic agents. In cats, the systemic injection of atropine just prior to the time of the expected circadian rise of 17-OHCS blocked this elevation. The fact that atropine was ineffective if administered at other times and the reported inability of the drug to interfere with the acute activation of the system (see above) suggest that there are cholinergic mechanisms and/or structures involved in the regulation of ACTH secretion. In the cat they appear to be related chiefly to cyclic neuroendocrine regulation (Krieger *et al.,* 1968).

Studies *in vitro* provided more cogent evidence relating central cholinergic mechanism(s) to the HHAA. F. A. Steiner *et al.* (1969), using a microelectrophoretic approach, revealed a population of steroid-sensitive hypothalamic neurons (dexamethasone-inhibited) that were activated by iontophoretically applied ACh and inhibited by NE. CRF release from the *in vitro* isolated rat hypothalamus was stimulated in a dose-dependent manner by ACh added to the medium (Bradbury *et al.,* 1974); this effect was antagonized by hexamethonium and, partially, by atropine, suggesting at first that both muscarinic and nicotinic receptors were involved in the ACh stimulation (Hillhouse *et al.,* 1975). That the receptor system may be predominantly nicotinic was suggested by the observation that a pure muscarinic agent, bethanacol, did not significantly alter the release of CRF above basal levels (Jones *et al.,* 1976b). NE also inhibited the ACh-induced release of CRF. ACh was ineffective as a CRF releaser when tested on the ME *in vitro,* which led to the suggestion that it stimulates CRF secretion by acting at the dendritic or soma level of the CRF neuron (Jones *et al.,* 1976b). In an extension of these studies (R. Bock, M. Jones, and E. Hillhouse, unpublished observations) it was shown that ACh has no effect on CRF secretion from hypothalamic blocks containing the VMN and the dorsomedial nucleus, the periventricular nucleus, and the ARC nucleus, and the ME. This suggests that the CRF cell bodies lie outside this area (see also Section C,1).

Finally, ACh was able to stimulate, in a dose-dependent manner, the release of CRF from a suspension of synaptosomes prepared from the sheep hypothalamus; the response was not enhanced by eserine and was blocked by atropine (Edwardson and Bennett, 1974). This suggests that, in sheep at least, there are excitatory synapses with neurosecretory neurons in the ME.

In conclusion, ACh seems to be implicated in the regulation of HHAA activity, possibly playing a stimulatory role whose physiological significance has yet to be defined. From the *in vitro* findings the possibility of an interrelationship with the inhibitory NE regulation can also be suggested.

ii. Prostaglandins. Among other endocrine and metabolic effects, PG's, mostly those of the E series, have been shown to possess stimulatory effects at various levels of the HHAA. PG's directly stimulated adrenal tissue *in vitro* (Falck *et al.,* 1969) and elicited ACTH secretion in several assay systems commonly used to determine CRF activity (De Wied *et al.,* 1969). PG's also act at the CNS level to stimulate ACTH secretion. The PGE_1 stimulatory effect on systemic injection could be blocked by hypophysectomy or pretreatment with cortisol or pentobarbital and morphine, suggesting that the main site of action of the compound was above the adrenals and the pituitary (Peng *et al.,* 1970). Some insight into the CNS site of PG action was derived from observations that in dexamethasone-blocked animals, PG's injected into the ME increased ACTH secretion whereas they were ineffective when injected into nearby regions of the hypothalamus, the AP, or systemically. Higher doses of dexamethasone or morphine also blocked the stimulatory action of PG's at the ME level. $PGF_{1\alpha}$, although less effective than PGE_1, exerted specific stimulatory action on the HHAA (Hedge, 1972). In light of present knowledge, it cannot be determined whether the central stimulatory action of PG's is directly elicited by interaction with CRF-producing neurons or with other regulatory neurotransmitters (i.e., NE).

iii. Melatonin. It appears from *in vitro* experiments that melatonin in concentrations of up to 10 ng/ml does not significantly alter the basal secretion of CRF. However, at doses beyond 10 ng/ml, it significantly reduced the CRF-releasing effect of 5-HT and ACh (Jones *et al.,* 1976b). It has been suggested that melatonin acts as an antagonist on the 5-HT receptor (see Section A,2,d); however, since melatonin also inhibited the action of ACh, its action appears to be distal to the receptor and is probably exerted directly on the CRF neuron. Interaction between the inhibitory actions of NE and melatonin were excluded in *in vitro* experiments in which phentolamine had no effect on the inhibitory action of melatonin on 5-HT stimulated CRF release (Jones *et al.,* 1976b). The finding that melatonin and 5-HT have opposing effects on CRF release is relevant in experiments in which pCPA was used to study the role of 5-HT in the central control of CRF secretion (see Section A,1,b). The results of such experiments must be interpreted with caution, since pretreatment of animals with pCPA interferes with 5-HT as well melatonin biosynthesis.

iv. GABA and Glutamate. In rats, IVT injection of GABA did not modify baseline HHAA activity (Abe and Hiroshige, 1974), although under stress conditions, it appeared to play an inhibitory role in the hypothalamic network regulating CRF secretion. In fact, IVT-infused GABA inhibited stress-induced ACTH release, whereas GABA antagonists, such as picrotoxin and bicuculline, stimulated ACTH release under conditions in which nonspecific stress effects were minimized; the HHAA activation produced by GABA antagonists was prevented by simultaneously administered GABA (Makara and Stork, 1974). The inhibitory role observed *in vivo* is consistent with the findings that *in vitro* the stimulated release of CRF from incubated fragments of the hypophysiotropic area could be inhibited by GABA (Jones *et al.,* 1976b,c). It was found that GABA, while ineffective on the basal release of CRF, reduced the stimulatory action of either ACh or 5-HT (see above), suggesting an action exerted distal to the cholinergic interneuron and possibly directly on the CRF neuron. Since GABA is present in high concentrations in the lateral-PO region of the rat hypothalamus (see Chapter II, Section I,1) and the amount of GABA needed to inhibit CRF release *in vitro* is only a minute fraction of the tissue content, it is suggested that it may play a role in CRF secretion. Its action appears to be mediated via a specific GABA receptor, since picrotoxin significantly reduces the inhibitory action of GABA on 5-HT stimulated CRF release. The possibility that GABA may act via an NE interneuron appears unlikely as the inhibitory action of the amino acid on CRF release is unaltered by phentolamine (Jones *et al.,* 1976b). In keeping with the inhibitory action of GABA on the CRF-ACTH system, in the human administration of baclofen (β-4-chlorophenyl-GABA), a GABA analog which crosses the BBB (see Table XX), reportedly induced a decrease in plasma F values and in plasma F response to insulin-induced hypoglycemia (Invitti *et al.,* 1976).

Glutamate, another putative neurotransmitter of excitatory nature, has been proposed as a participant in the hypothalamic regulation of ACTH release. The infusion of glutamate into the third ventricle of rat brain acted as a potent stimulus for ACTH secretion; however, its effect could be prevented by prior deafferentation of the MBH (Makara and Stork, 1975), suggesting that glutamate stimulates ACTH secretion by interfering with CNS neurotransmission at sites outside the MBH. In keeping with such a conclusion is the reported ineffectiveness of glutamate in stimulating CRF release from rat hypothalami *in vitro* (Jones *et al.,* 1976a).

2. Brain Neurotransmitters in the Control of Growth Hormone Secretion

a. Catecholamines

i. Subprimate species. In early neuropharmacological studies on the rat to determine whether or not brain CA's participate in the reactions elicited by the application of a stimulus that releases GH, the now criticized "tibia test" BA method was used (see Chapter III, Section A,3,a). Among many CNS-active drugs tested as possible inhibitors of GH release in response to insulin hypoglycemia, reserpine and CPZ were the most effective (Müller *et al.*, 1967a). Of the many hypotheses that could be formulated to explain these findings, the possibility that the suppressive action of these drugs might be due to their known antiadrenergic effect was considered most likely. In addition to reserpine, other brain NE depletors suppressed insulin-induced GH release, while peripheral NE depletors did not (Müller *et al.*, 1967b). In rats pretreated with reserpine or CPZ, the response to the GH-depleting effect of GRF was still intact, implying an action by the blocking agents at the level of the CNS and not the AP (Müller *et al.*, 1967a; Sawano *et al.*, 1967). Injection of NE and DA into the lateral ventricle of normal or hypophysectomized rats to circumvent the BBB to CA's was followed by decreased GH levels in the AP (Müller *et al.*, 1968) and by depletion of hypothalamic GRF, which was found in plasma from hypophysectomized but not intact rats (Müller *et al.*, 1970). NE administered IVT induced a depletion of GH pituitary activity at doses at which DA was ineffective, thus suggesting a role for NE as the synaptic transmitter for the release of GRF (Müller *et al.*, 1970).

Subsequent development and use of RIA techniques for the measurement of rat GH in the plasma produced data in conflict with the existing BA data, and thus affected conclusions derived from neuropharmacological studies of GH regulation in the rat (see Chapter III, Section A,3,a). Collu *et al.* (1972) reported reduction of plasma GH levels in the urethane-anesthetized rat after IVT administration of DA and a highly significant GH rise following central injection of 5-HT. Kato *et al.* (1973, 1974a) observed, in the same animal model, a significant increase in plasma GH after systemic administration of either phentolamine or isoproterenol (an α-adrenergic blocker or a β-adrenergic stimulant, respectively) and a dose-related rise after administration of CPZ (Kato *et al.*, 1974a). No increase was detected in response to L-dopa. In an extension of these studies, Chihara *et al.* (1975) observed that complete hypothalamic deafferentation increased plasma GH levels and that in this situation CPZ was still effective in elevating plasma GH, an effect inhibited by pretreatment with α-MT. From these studies they concluded that the effects of the adrenergic agents and inhibitors on GH secretion in the rat are the opposite of those in primates (see below) and that CPZ stimulates GH release by inhibiting the dopaminergic mechanism which is active in the MBH. In keeping with these results are those of Collu *et al.* (1973), who showed that treatment of totally deafferented rats with α-MT prevented GH reduction in response to ether stress. Since DA neurons are preserved in the deafferented rat hypothalamus (Weiner *et al.,* 1972), DA would be responsible for the inhibitory effect of stress on GH release in the rat.

However, reinvestigation by RIA of the role of brain CA's in the control of GH secretion in the infant, unanesthetized rat showed that 6-OHDA (30 or 60 μg) administered IVT into 5-day-old rats induced 5 and 10 days later a clear-cut fall of plasma GH; protection of 6-OHDA-induced depletion of brain NE stores by DMI given before administration of 6-OHDA (see Chapter II, Section D) resulted in a less striking, but still evident GH reduction (Müller, 1976). The latter results, which are in agreement with the previous BA data, suggest that, at least in the postnatal period, both NE and DA exert a stimulatory role on GH secretion in the rat. This view is supported by the observation that in the infant rat functional blockade of NE and DA systems by α-MT significantly lowered GH and that, conversely, systemic administration of L-dopa increased plasma GH levels (Irit Gil-Ad, unpublished results). A stimulatory role of L-dopa on GH release has also been reported by H. J. Chen *et al.* (1974) in the mother rat during suckling.

Although conclusions reached only with the use of intracranial injections of 6-OHDA as the method of damaging central CA-containing neurons may be erroneous because this highly reactive substance could damage other neurons when applied locally in high concentration, combined evaluation of 6-OHDA, α-MT and L-dopa experiments indicates that at least in the infant rat, which when compared to the adult rat has the undoubted advantage of more uniform plasma GH levels (Cocchi *et al.,* 1976), brain CA's exert a stimulatory effect on the neuroendocrine reflex responsible for the tonic secretion of GH. Brain CA's also play a role in mediating the GH lowering effect of stress stimuli: in the infant rat, IVT pretreatment with 6-OHDA reduced or abolished, respectively, the reduction of plasma GH levels induced by cold exposure or insulin administration (Gil-Ad *et al.,* 1976b).

In the adult, unanesthetized rat, administration of FLA-63, a selective blocker of NE biosynthesis (see Chapter II, Section B,6,a) lowered plasma GH, while central sympathectomy by 6-OHDA or systemic administration of α-MT, procedures which affect both the NE and the DA functions, did not change plasma GH levels (Müller *et al.,* 1973c). These results would be consistent with the existence in the adult rat of opposing brain DA inhibitory and NE stimulatory activities. The recent findings of Mueller *et al.* (1976a), who found stimulation of GH release after low but not high doses of specific DA agonists, are not opposed to the proposition of an inhibitory action of DA on GH secretion in rodents. DA agonists at low doses might induce selective stimulation of presynaptic receptors (Carlsson, 1977) and, hence, feedback inhibition of DA neurotransmission (see Chapter II, Section B,11).

Of relevance, but difficult to interpret, is the striking elevation induced in plasma GH levels by systemic administration of morphine (see Martin *et al.*, 1975b), a drug which has a number of diverse effects on brain MA metabolism (Loh *et al.*, 1973; Yarbrough *et al.*, 1972). The GH-releasing effect of morphine, which is also present in the cat (French *et al.*, 1974), is not affected by agents which block synthesis of or target receptors for CA's or the synthesis of 5-HT (Martin *et al.*, 1975b), nor is it blocked by the narcotic antagonist, naloxone (Kokka *et al.*, 1973). It is of note that, similar to morphine, enkephalin, an endogenous peptide with opiate agonistic activity, (Hughes *et al.*, 1975) administered IVT in neonatal rats induced a rise in plasma GH, which, however, as that of morphine, was antagonized by naloxone (Cocchi *et al.*, 1977).

In the mouse, a species which seems to be particularly resistant to GH-releasing stimuli (Müller *et al.*, 1971c), Sinha *et al.* (1972) observed a dramatic decrease in the serum GH levels following systemic administration of either DA or L-dopa, and reduced levels were also present after ip injection of E. Conversely, 6-OHDA administered IVT markedly increased circulating GH levels (E. E. Müller, unpublished results).

Although these experiments suggested a central role of CA's in the modulation of GH secretion in the rat, they gave no indication of the levels at which such control is exerted. This problem has been investigated in elegant studies by Martin (1973b). In these studies both reserpine and α-MT, administered 24 hr before electrical stimulation, had no effect on GH release following VMN stimulation, but completely blocked the GH release following either basolateral amygdala (BLA) or hippocampal stimulation. Hypothalamic responses were also not affected by administration of the α-adrenergic blocking agent phenoxybenzamine or the β-adrenergic blocker propranolol, whereas BLA responses were blocked by phenoxybenzamine but not propranolol. These results indicate that the relay of responses from limbic structures to the MBH may involve CA's, whereas the latter do not intervene in the GH response evoked by direct hypothalamic stimulation. In an extension of these studies, it has also been found that treatment with pCPA, the inhibitor of 5-HT synthesis, blocked BLA-induced GH release without any effect on VMN-induced release.

The role of 5-HT in the relay of neuroendocrine reflexes for GH regulation in the rat based on the pCPA effect must be viewed with caution, however, since pCPA-induced changes in CNS function may also involve the catecholaminergic system (see Table XIV) and melatonin.

Data on the involvement of brain CA's in the regulation of GH secretion in other mammalian species are rather scanty. In the pig, a species in which the response to GH-releasing stimuli is weak (Machlin *et al.*, 1968), no rise in plasma GH has been reported following systemic E injection and peripheral administration of NE proved equally ineffective (Machlin *et al.*, 1968). In the sheep, plasma levels of GH were constantly low during an iv infusion of E, and the plasma GH response to arginine infusion was abolished (Hertelendy *et al.*, 1969). Intracarotid infusion of phenoxybenzamine resulted in a delayed increase in plasma levels of GH in mature cyclic ewes and produced additive effects when combined with an arginine stimulus; infusion of L-dopa by itself had no effect on plasma GH, but suppressed the expected increase of GH due to arginine (Davis and Borger, 1973). In this species, suppression of GH release by E or L-dopa *in vivo* might be explained, at least in part, by the direct effect of CA's on the somatotropes, as indicated by the finding that addition of either NE or DA to the incubation medium abolished TRH-induced GH release by sheep pituitary fragments (Takahara *et al.*, 1974b).

In the pentobarbital-anesthetized dog, iv L-dopa administration stimulated GH secretion (Lovinger *et al.*, 1974b, 1975), an effect apparently mediated by NE and exerted via α-adrenergic receptors since phenoxybenzamine or phentolamine given IVT blocked it. Epinephrine and DA had no effect on plasma GH when given both intravenously and intraventricularly; however, NE, when infused IVT, evoked a significant rise in plasma GH in 2 out of 7 animals tested. Adrenergic stimulation with iv clonidine increased plasma GH, while IVT administration of phenoxybenzamine and phentolamine, at systemically ineffective doses, inhibited GH response to both clonidine and L-dopa. Interestingly, when clonidine was administered directly into the third ventricle, there was no increase in GH secretion (Ganong, 1975). This fact is difficult to interpret, but one interpretation may be that IVT-injected clonidine inhibited NE neurons via activation of adrenergic autoreceptors on or near NE cell bodies (Svensson *et al.*, 1975) (see Chapter II, Section B,6). Like dopamine, the DA-receptor stimulant apomorphine, administered intraventricularly, did not increase plasma GH, and pimozide, a DA receptor blocker, did not abolish the GH response to iv L-dopa (Lovinger *et al.*, 1975).

The effects of GH secretion in subprimate species of drugs affecting brain CA's are listed in Table XXXIII.

ii. Primate species. Results of experiments performed on the rat on the stimulatory effect of brain CA's in the release of GH agreed with those obtained in primates. Single injections of large doses of E initiated a marked and prompt increase in the plasma GH levels of female rhesus monkeys (Meyer and Knobil, 1967). Studies conducted on conscious, fasted baboons showed that α-adrenergic blockade, induced by iv infusion of phentolamine, significantly depressed GH secretion, whereas β-adrenergic blockade with propranolol and ganglionic blockade with trimetaphan was associated with a significant prompt rise in GH. The observed GH changes did not seem to be associated with changes in plasma glucose or NEFA levels, although lowered plasma NEFA and glucose levels were present following propranolol administration (Werrbach *et al.*, 1970). Infusion of systemically ineffective doses of phentolamine directly into the third ventricle or anterior hypothalamus produced a similar fall in the GH level (Toivola *et al.*, 1972).

These data suggested the existence in the CNS of the baboon of α-adrenergic receptors regulating GH secretion. It was interesting to observe that NE microinjected into the VMN consistently elevated GH in conscious baboons (Toivola and Gale, 1972) and that clonidine also triggered very increased GH secretion (Brown, 1973). In contrast to central administration of NE, intrahypothalamic infusion of DA lowered plasma GH (Toivola and Gale, 1970); failure to trigger GH secretion was also observed with apomorphine (Brown, 1973). Recently, acute elevation of plasma GH has been reported in male rhesus monkeys following iv L-dopa; in these studies, apomorphine did not induce GH release (Jacoby *et al.*, 1974; Chambers and Brown, 1976). In other studies, a rise in plasma GH was observed in baboons after both systemic L-dopa administration and infusion of huge doses of DA. Concomitant infusion of phentolamine blocked the DA-induced release of GH, suggesting that is NE and not DA which ultimately acts on the ME-basal hypothalamus to evoke GH release (Illner *et al.*, 1976).

In man the observation by Müller *et al.* (1967a) of an inhibition of GH response to insulin induced hypoglycemia in rats treated with reserpine was confirmed by Cavagnini and Peracchi (1971). Infusion of E in doses sufficient to produce significant hyperglycemia neither induced GH secretion (Roth *et al.*, 1963) nor inhibited arginine-stimulated hGH release (Rabinowitz *et al.*, 1968). The poor penetrability of the BBB by CA's was overcome by administering L-dopa; the results obtained showed that moderate doses of L-dopa caused a significant rise in plasma GH levels in patients with Parkinson's disease (Boyd *et al.*, 1970), as well in normal subjects (Eddy *et al.*, 1971; Mims *et al.*, 1973a).

The stimulant effect on GH secretion by L-dopa was not blocked by

TABLE XXXIII

Effect of Drugs Altering Brain Catecholamine Transmission on GH Secretion in Subprimate Species[a]

Drug	Catecholamin-ergic tone	Species	Dosage and route of administration	Effect on GH secretion	Authors
Epinephrine	→	Rat	30 μg sc	↑[b]	Müller *et al.* (1967a)
	↑	Rat	0.5 μg IVT	↑[b]	Müller *et al.* (1968)
	→	Mouse	1.0 mg/kg ip	↓	Sinha *et al.* (1972)
	→	Pig	1.0 mg sc	→	Machlin *et al.* (1968)
	→	Sheep	0.5 μg/kg/min/150 min iv	↓	Hertelendy *et al.* (1969)
	↑	Dog[c]	iv and IVT	→	Lovinger *et al.* (1975)
Norepinephrine	↑	Rat	0.005–0.5 μg IVT	↑[b]	Müller *et al.* (1970)
	↑	Rat[c]	1.0 μg IVT	→	Collu *et al.* (1972)
	↑	Dog[c]	IVT	↑	Lovinger *et al.* (1975)
Clonidine	↑	Dog[c]	30 μg/kg iv	↑	Ganong (1975)
	↑	Dog[c]	5 μg/kg IVT	→	Ganong (1975)
Dopamine	↑	Rat	0.01–0.5 μg IVT	↑[b]	Müller *et al.* (1970)
	↑	Rat[c]	1.0 μg IVT	↓	Collu *et al.* (1972)
	→	Mouse	1.0 mg/kg ip	↓	Sinha *et al.* (1972)
	↑	Dog[c]	iv and IVT	→	Lovinger *et al.* (1975)
Apomorphine	↑	Rat	5.0 mg/kg ip	↓	Kokka *et al.* (1973)
	↑	Dog	IVT	→	Lovinger *et al.* (1975)
L-Dopa	↑	Rat[d]	100 mg/kg ip	↑	I. Gil-Ad, unpublished results
	↑	Rat	100 mg/kg ip	↓	Müller *et al.* (1973c)
	↑	Rat[c]	10–50 mg/kg iv	→	Kato *et al.* (1973)
	↑	Mouse	1.0 mg/kg ip	↓	Sinha *et al.* (1972)
	↑	Sheep	1 mg/kg intracarotid	→	Davis and Borger (1973)
	↑	Dog[c]	50 mg/kg iv	↑	Lovinger *et al.* (1974b)

FLA-63	↓	Rat	40 + 25 mg/kg ip	↓	Müller *et al.* (1973c)
α-MpT	↓	Rat	200 mg/kg ip	→	Müller *et al.* (1973c)
	↓	Rat[d]	250 mg/kg ip	↓	I. Gil-Ad, unpublished results
Phenoxybenzamine	↓	Rat[c]	50 μg IVT	↓	Collu *et al.* (1972)
	↓	Sheep	0.5 mg/kg intracarotid	↑	Davis and Borger (1973)
	↓	Dog	40 μg/kg IVT	Prevents L-dopa-induced ↑	Ganong (1975)
Propranolol[e]	↓	Dog	10 μg/kg/min/120 min	↑	Altszuler and Hampshire (1976)
Isoproterenol	↑	Rat[c]	100 μg/kg iv	↑	Kato *et al.* (1973)
Chlorpromazine	↓	Rat[c]	500 μg–10 mg/kg iv	↑	Kato *et al.* (1973)
6-OHDA	↓	Rat	200 μg + 100 μg IVT	→	Müller *et al.* (1973c)
	↓	Rat[d]	30–60 μg IVT	↓	Müller (1976)
	↓	Mouse	50 μg IVT	↑	E. E. Müller, unpublished data

[a] Key to symbols: →, no effect; ↑, stimulation; ↓, inhibition.
[b] Determined by bioassay.
[c] Anesthetized.
[d] Infant rat (10–15-day-old).
[e] Combined with E or NE.

either oral or iv glucose given following the ingestion of L-dopa (Boyd *et al.*, 1970), but there was no rise in GH concentration when hyperglycemia was induced and maintained before L-dopa administration (Mims *et al.*, 1973b; Johansen, 1973). Similarly, glucose loading in men significantly decreased the hGH response to apomorphine (Ettigi *et al.*, 1975), suggesting that stimulation of hypothalamic glucoreceptors antagonizes dopaminergic regulation of GH secretion.

L-Dopa-induced GH release was also reduced or activated, respectively, by the concomitant infusion of phentolamine (Kansal *et al.*, 1972) or propranolol (Camanni and Massara, 1974), stressing the involvement of adrenergic α- and β-receptors in the regulation of GH secretion. In line with these findings, previous studies had shown that blockade of α-adrenergic receptors by phentolamine in man prevented insulin-induced GH release; in contrast, blockade of β-adrenergic receptors by propranolol augmented the levels of GH in hypoglycemic subjects, although this was associated with increased hypoglycemia and a decrease in NEFA (Blackard and Heidingsfelder, 1968). However, in this study neither α- nor β-adrenergic blockade had a detectable effect in the absence of hypoglycemia, although a stimulant effect was reported for propranolol alone in Japanese subjects (Yawata and Fukase, 1968). In Caucasian subjects, an increase in circulating GH levels was present only following combined administration of propranolol and E (Massara and Strumia, 1970; Blackard and Hubbel, 1970). In this case, the stimulatory effect of E on plasma hGH in the presence of propranolol must have been the result of unopposed α-receptor activity, since β-stimulation would have been expected to inhibit GH secretion. In keeping with this conclusion, the effect of combined E–propranolol administration was inhibited by concomitant infusion of phentolamine (Massara and Camanni, 1972). The possibility that the permissive action of propranolol on E effect took place through a reduction of E-induced hyperglycemia or/and an increase in NEFA levels was considered but appears unlikely. Neither an exogenously induced hyperglycemia nor a marked increased in NEFA levels due to an infusion of an "intralipid" and heparin was capable of inhibiting the GH response following E-propranolol administration (Massara *et al.*, 1973a). Difficult to interpret is the failure in man of the combined NE–propranolol administration to affect GH levels (Massara and Camanni, 1972) in view of the similar ability of E and NE to diffuse into the hypothalamus after parenteral administration (Weil-Malherbe *et al.*, 1961a). This is not true for dog, in which both L- and D-propranolol administration led to increased GH secretion in response to NE infusion (Altszuler and Hampshire, 1976). Further support for the existence of a noradrenergic regulatory mechanism modulating GH secretion in the human can be derived from the stimulatory effect on GH levels of systemic administration of α-agonist drugs such as methoxamine (Imura *et al.,* 1971) and clonidine (Lal *et al.,* 1975a).

In addition to L-dopa, apomorphine, in subemetic doses, 2-Br-α-ergocryptine (CB 154, Sandoz), or piribedil, all DA-agonist drugs (see Table XI), induced a rise in hGH levels (Lal *et al.,* 1972, 1973; Brown *et al.,* 1973; Camanni *et al.,* 1975a; Silvestrini *et al.,* 1975; Thorner *et al.,* 1976), an effect which is compatible with the existence of a dopaminergic mechanism involved in the release of hGH. Also, 5-butylpicolinic acid, a natural enzyme inhibitor of D-β-H (see Table VII), induced, by itself, a slight elevation in hGH levels (Hidaka *et al.,* 1973) and haloperidol or CPZ, two neuroleptics antagonist to DA, inhibited insulin- or apomorphine-induced hGH rise (Kim *et al.,* 1971b; Sherman *et al.,* 1971; Lal *et al.,* 1973). In the human, L-dopa-induced GH rise was also suppressed by a specific DA blocker, e.g., pimozide (Liuzzi *et al.,* 1976). However, not all drugs enhancing dopaminergic neurotransmission appear effective in increasing GH levels in the human; both amantadine (Cavagnini *et al.,* 1973; Massara *et al.,* 1973b) and an amantadine derivative, 1,3-dimethyl-5-aminoadamantane (F. Camanni and F. Massara, unpublished data), were ineffective, although the former was capable of potentiating L-dopa-induced GH release (Massara *et al.,* 1973b).

Although the GH-releasing effect of L-dopa was first documented in patients with Parkinson's disease (Boyd *et al.,* 1970), contradictory findings have been subsequently reported on the eftect of dopaminergic stimulation in disease states of the extrapyramidal system. In 24 out of 26 patients with Parkinson's disease treated with L-dopa for 1–5 yr, a single dose of the drug induced a consistent GH rise 30–120 min later (Parkes *et al.,* 1976); similar results were obtained by Lebovitz *et al.* (1974), who showed that the magnitude of GH response in these patients given L-dopa did not alter during 1 yr of treatment. In contrast, Malarkey *et al.* (1974) found that only 1 out of 10 patients treated for 1.5–4 yr showed the GH response to L-dopa, and further evidence of disturbed GH regulation was provided by an unexpected suppression of GH following L-dopa administration in 3 patients and a paradoxical increase in GH following glucose administration found in 2. However, with insulin-induced hypoglycemia, each of the patients tested responded with a normal GH secretory peak.

With regard to other DA-stimulant drugs, apomorphine administered acutely was active in only 1 out of 4 Parkinson patients tested (Brown *et al.,* 1973). Modest rises in GH levels were also found in 9 patients for a 24-hr period after a 6-week course of bromocriptine therapy (Shaw *et al.,* 1976), and a slight GH response was observed after a single oral dose of the drug (Debono *et al.,* 1975; Parkes *et al.,* 1976). In another deteriorating disease of the extrapyramidal centers, e.g., Huntington's chorea, it was found that glucose was effective in suppressing GH levels in patients who had previously received L-dopa, whereas it was not so in normal subjects (Podolsky and Leopold, 1973); more recently, a more consistent and prompt GH rise than in normal subjects has been reported in choreic patients following CB 154 administration (Caraceni *et al.*, 1977).

The varied effects of L-dopa and dopaminergic drugs on GH secretion in Parkinson's disease may be due to the altered absorption and/or metabolism of these agents, possibilities that could only be directly examined by measuring the various drug metabolites in the blood and CNS. However, since increased L-dopa treatment has been found to be associated with a more rapid attainment of peak L-dopa plasma concentrations (Muenter and Tyce, 1971) and since the presence of serum PRL suppressive effects following L-dopa in subjects who had shown an altered GH response has been consistently found (Malarkey *et al.*, 1974; Podolsky and Leopold, 1976), an absorptive defect seems unlikely. It would appear, instead, that functional integrity of the nigrostriatal DA neurons is necessary to allow release of GH in response to dopaminergic stimulation, since in Parkinson's disease CA neurons in the hypothalamus are histologically preserved and may function normally. The extrapyramidal centers may thus be proposed as a site of neurotransmitter–neurohormone interaction for the neuroendocrine control of GH secretion (see also Section C,2).

The effects on GH secretion in primates of drugs affecting brain CA's are listed in Table XXXIV. (For the effect of DA stimulant drugs in states of GH hypersecretion see Chapter V, Section A,2.)

b. Indoleamines

While the role of brain CA's in the neuroendocrine mechanism(s) which controls GH secretion has been established more satisfactorily (see above), the involvement of the indoleaminergic system is still a matter of considerable controversy, since conflicting results have been reported in the literature (Table XXXV). Collu *et al.* (1972) have claimed that in the urethane-anesthetized rat, IVT infusion of 5-HT increases GH levels in the plasma, a finding not confirmed by Müller *et al.* (1973c). In "gentled," unanesthetized rats, Smythe and Lazarus (1973a) reported that systemic administration of 5-HTP is also a potent stimulus for the secretion of GH. Combined administration of melatonin or cyproheptadine (Cy) inhibited the GH response to 5-HTP (Smythe and Lazarus, 1973a). The same agents were also capable of inhibiting the GH response which follows pentobarbital anesthesia (Smythe *et al.*, 1975a). The conclusions of these studies were that secretion of GH is mediated through (presumably hypothalamic) 5-HT receptors and that melatonin can act as a general blocker or competitive inhibitor at 5-HT receptor sites.

The possibility that hGH may also be stimulated via 5-HT mechanisms was suggested by the finding that hGH levels are increased following oral administration of 5-HTP (Imura *et al.*, 1973). Furthermore, it was shown that hGH responses to insulin-induced hypoglycemia or physical exercise are inhibited by Cy and methysergide (Smythe and Lazarus, 1974; Bivens *et al.*, 1973a). Because an increase in hypothalamic 5-HT content has been found in the rat during insulin-induced hypoglycemia (Gordon and Meldrum, 1970) and because IVT-injected 5-HT stimulates GH secretion in the rat, Bivens *et al.* (1973a) concluded from their experiments that these two 5-HT antagonists probably interfered with hypothalamic 5-HT stimulation of GH secretion. However, in contrast to these findings, a vast series of experiments show that 5-HT does not play a major role in the GH-releasing mechanism(s).

In the monkey, unlike the positive effects induced by NE, 5-HT microinjected into the VMN was ineffective in elevating plasma GH levels (Toivola and Gale, 1972), although a significant GH response was reported by Chambers and Brown (1976) following intravenous infusion of 5-HTP. Recently, the availability of a specific and sensitive homologous RIA for GH in the dog (Cocola *et al.*, 1976), allowed the study of the participation of 5-HT in the response evoked by insulin hypoglycemia. In 9 unanesthetized dogs, administration of pCPA did not modify baseline canine GH (cGH) and glucose levels, caused a marked decrease of blood 5-HT, and resulted in a magnification of the insulin-induced cGH release (Fig. 46). In 8 dogs, similarly treated with pCPA, administration of 5-HTP 10 min before the hypoglycemic stimulus inhibited the GH secretory response to the latter. In 5 dogs, chronic feeding for up to 40 days of a diet (maize) deficient in Trp (see Chapter II, Section C,4) resulted in a progressive decrease of total and "free" plasma Trp, no significant changes in baseline cGH and blood glucose levels, and a GH response to the insulin stimulus almost identical to that elicited in the same animals under normal conditions. In 3 Trp-deficient dogs, Trp infusion induced a striking reduction of the hypoglycemia-induced GH release (Fig. 47).

Collectively, the above data lend support to the theory that, contrary to the conclusions reached by Bivens *et al.* (1973a) in the human, 5-HT plays in the dog an inhibitory role in hypoglycemia-induced GH release. Failure to duplicate in Trp-deficient dogs the enhanced GH response present after pCPA administration, might perhaps reflect the smaller depletion of brain 5-HT levels induced by the diet regimen, thus allowing persistence of residual 5-HT activity in the serotoninergic system. Similar to the observation in pCPA-5-HTP treated dogs, Trp preloading resulted in a clear-cut reduction of the insulin-induced cGH elevation. These data were interpreted to mean that by replenishing central stores of 5-HT Trp induced

TABLE XXXIV

Effect of Drugs Altering Brain Catecholamine Transmission on GH Secretion in Primates[a]

Drug	Catechol-aminergic tone	Species	Dosage and route of administration	Effect on GH secretion	Authors
Epinephrine	→	Monkey	5–40 μg/kg iv	↑	Meyer and Knobil (1967)
	→	Human	10 μg/kg sc	→	Schalch (1967)
	→	Human	0–14 μg/kg/min iv	→	Massara and Strumia (1970)
Norepinephrine	↑	Monkey	1 μg/min/30 min IVT	→	Toivola and Gale (1970)
	↑	Monkey	10 μg microinjected into VMN	↑	Toivola and Gale (1972)
	→	Human	7–4 μg/min/30 min iv	→	Massara and Camanni (1972)
Clonidine	↑	Human	0.15 mg iv	↑	Lal *et al.* (1975a)
Dopamine	↑	Monkey	15 μg IVT	↓	Toivola and Gale (1970)
	→	Monkey	1.2 mg/kg/30 min iv	↑	Illner *et al.* (1976)
Apomorphine	↑	Human	0.75–1.5 mg sc	↑	Lal *et al.* (1972)
	↑	Monkey	50–75 μg iv	→	Jacoby *et al.* (1974)
Bromocriptine	↑	Human	2.5 mg po	↑	Camanni *et al.* (1975a)
Piribedil	↑	Human	3 mg/2 hr iv	↑	Thorner *et al.* (1976)
Methylamphetamine	↑	Human	15 mg/5 min iv	↑	Rees *et al.* (1970)
Amantadine	↑	Human	200 mg po	→	Cavagnini *et al.* (1973)
	↑	Human	300 mg iv	→	Cavagnini *et al.* (1973)
	↑	Human[b]	300 mg/3 days po	Potentiates L-dopa	Massara *et al.* (1973b)
L-Dopa	↑	Human[b]	500–1000 mg po	↑	Boyd *et al.* (1970)
	↑	Monkey	15–200 mg/kg iv	↑	Jacoby *et al.* (1974)
FA	↓	Human	350 mg po	↑	Hidaka *et al.* (1973)
Reserpine	↓	Human	33 mg/6 days po	Inhibits insulin hypoglycemia	Cavagnini and Peracchi (1971)

Phentolamine	↓	Monkey	0.013 mg/kg/min iv	↓	Meyer and Knobil (1967)
	↓	Monkey	1–5 μg/min/30 min IVT	↓	Toivola *et al.* (1972)
	↓	Human	0.5 mg/min/75 min iv	Inhibits L-dopa	Kansal *et al.* (1972)
	↓	Human	0.5 mg/min/90 min iv	Inhibits insulin hypoglycemia	Blackard and Heidingsfelder (1968)
	↓	Human	0.5 mg/min/90 min iv	Inhibits vasopressin	Heidingsfelder and Blackard (1968)
	↓	Human	3.5 mg/50 min iv	Inhibits propranolol + E	Massara and Camanni (1972)
	↓	Human	0.5 mg/min/120 min iv	No effect on GH sleep peak	Lucke and Glick (1971)
	↓	Human	0.5 mg/min/120 min iv	Inhibits arginine	Buckler *et al.* (1969)
Propranolol	↓	Monkey	0.011 mg/kg/min iv	↑	Werrbach *et al.* (1970)
	↓	Human	0.21 mg/kg/10 min iv	→	Massara and Strumia (1970)
	↓	Human	0.20 mg/kg/120 min iv	↑	Yawata and Fukase (1968)
	↓	Human	0.15 mg/kg/120 few min iv	Potentiates Amphetamine	Rees *et al.* (1970)
	↓	Human	0.15 mg/kg/few min iv	Potentiates insulin hypoglycemia	Blackard and Heidingsfelder (1968)
Propranolol + E	↓	Human	0.21 mg/kg/10 min iv	↑	Massara and Strumia (1970)
	↓	Human	0.07–0.14 μg/kg/min/30 min iv	↑	Massara and Strumia (1970)
Haloperidol	↓	Human	2 mg po	Attenuates insulin hypoglycemia	Kim *et al.*, (1971b)
Chlorpromazine	↓	Human	100 mg po	↓	Sherman *et al.* (1971)
Pimozide	↓	Human	4 mg day/one week po	Inhibits L-dopa	Liuzzi *et al.* (1976)
α-Methyldopa	↓	Human	2–3 weeks treatment	Potentiates insulin hypoglycemia	Steiner *et al.* (1976)
Carbidopa	→	Human	300 mg/day/6 days po	→	G.M. Brown *et al.* (1976b)

[a] Key to symbols: →, No effect; ↑, stimulation; ↓, inhibition.
[b] Normal and parkinsonian patients.

TABLE XXXV

Effects of Drugs Altering Brain Indoleamines on GH Secretion[a]

Drug	Central indole-aminergic tone	Species	Dosage and route of administration	Effect on GH secretion	References
5-HT	↑	Rat	1 μg IVT	↑	Collu *et al.* (1972)
		Rat	1 μg IVT	→	Müller *et al.* (1973c)
		Monkey	10 μg microinjected into VMN	→	Toivola and Gale (1972)
	→	Human[b]		↑	Feldman *et al.* (1975)
Melatonin	↑	Human	500 mg po	Blunts insulin hypoglycemia	Smythe *et al.* (1976)
Trp	↑	Dog	2 gm iv	Slight increase	Müller *et al.* (1976b)
		Monkey	50–100 mg/kg iv	Slight increase	Jacoby *et al.* (1974)
		Human	70 mg/kg po	Slight increase	Müller *et al.* (1974b)
		Human	10 gm iv	→	McIndoe and Turkington (1973)
5-HTP	↑	Rat	10–50 mg/kg ip	↑	Smythe and Lazarus (1973a)
		Dog	10 mg/kg iv	↑	Müller *et al.* (1976b)
		Dog	10 mg/kg iv	Attenuates pCPA-induced potentiation	Müller *et al.* (1976b)
		Monkey	20–60 mg/kg iv	↑	Jacoby *et al.* (1974)
	↑	Human	150 mg po	↑	Imura *et al.* (1973)
		Human	150 mg po	→	Müller *et al.* (1974b)
		Human	150 mg po	→	Benkert *et al.* (1973)
		Human	200 mg po	→	Handwerger *et al.* (1975)

pCPA	↓	Dog	100 mg/kg po/5 days	Potentiates insulin hypoglycemia	Müller *et al.* (1976b)
Cyproheptadine	↓	Dog	48 mg po	Attenuates insulin hypoglycemia	Locatelli *et al.* (1977)
		Human	32 mg po	Blunts insulin hypoglycemia	Bivens *et al.* (1973a)
		Human	5 mg iv	Blocks 5-HTP rise	Nakai *et al.* (1974)
			5 mg iv	Blocks sleep-induced	Chihara *et al.* (1976)
		Human	60 mg po	Blunts exercise-induced	Smythe and Lazarus (1974)
		Human	60 mg po	Attenuates L-dopa	Smythe *et al.* (1976)
Methysergide	↓	Dog	4 mg sc	Attenuates insulin hypoglycemia	Locatelli *et al.* (1977)
		Human	16 mg po	Attenuates insulin hypoglycemia	Bivens *et al.* (1973a)
	↓	Human	16 mg po	Potentiates sleep-induced	Mendelson *et al.* (1975)
Methergoline	↓	Dog	17 mg sc	Potentiates insulin-induced	Locatelli *et al.* (1977)
		Human	24 mg po	Potentiates arginine-induced release and inhibits glucose-induced suppression	Pontiroli *et al.* (1975a)

[a] Key to symbols: →, No effect; ↑, stimulation; ↓, inhibition.
[b] Patients with carcinoid syndrome.

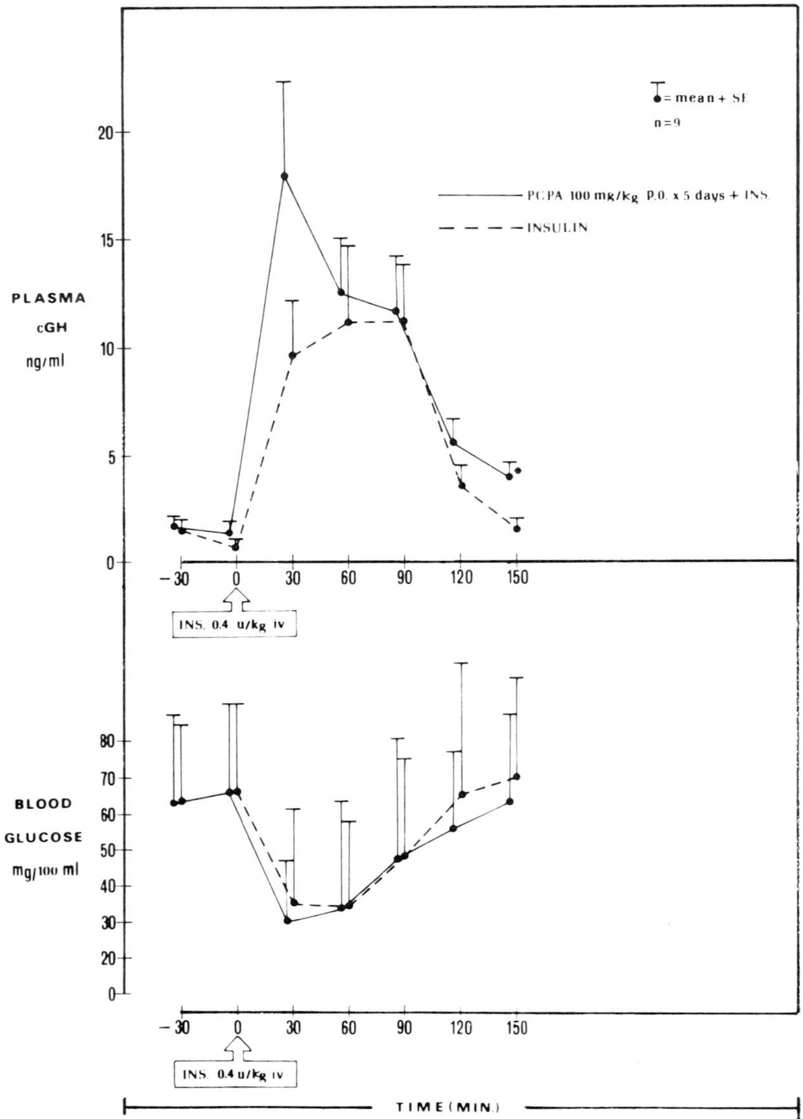

Fig. 46 Effect of insulin-induced hypoglycemia on plasma canine GH (cGH) and blood glucose levels of dogs pretreated or not with parachlorophenylalanine (100 mg/kg po daily for 5 days). (Reproduced, with permission, from Müller *et al.*, 1976b.)

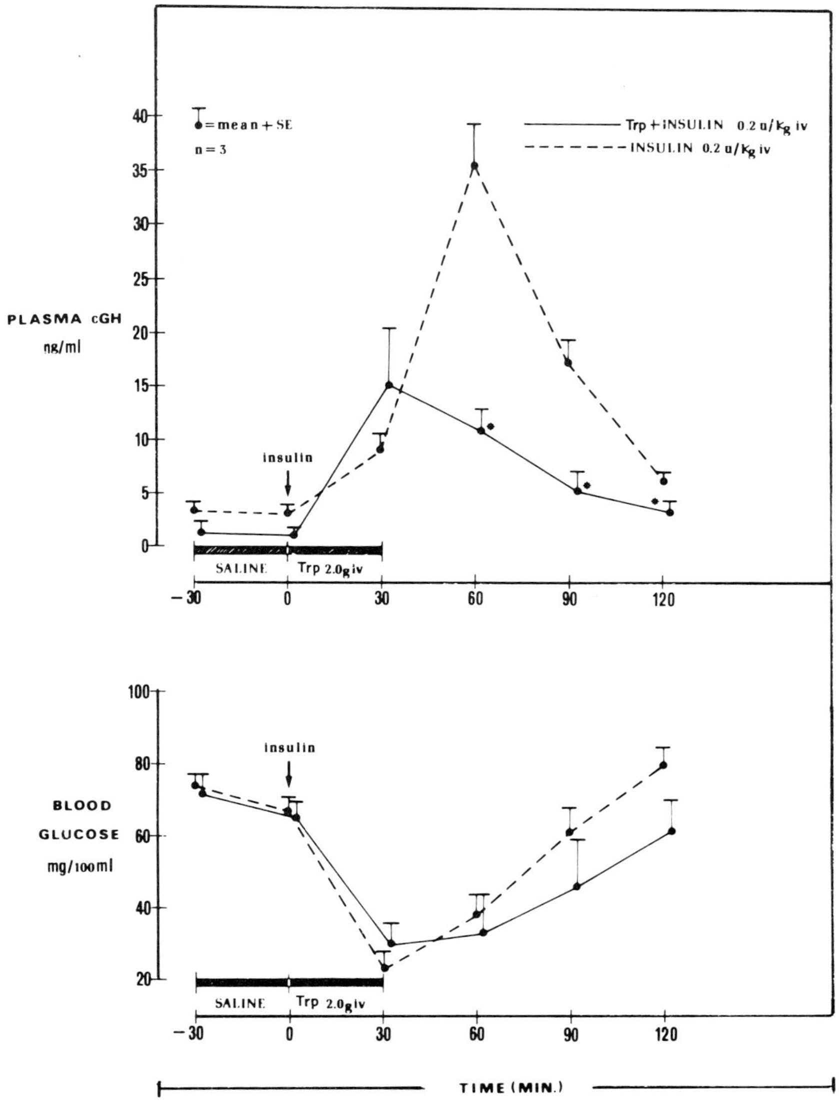

Fig. 47 Effect of tryptophan (Trp) infusion (2.0 gm iv over 30 min) on the GH response to insulin-induced hypoglycemia of dogs fed a Trp-deficient diet. (Reproduced, with permission, from Müller *et al.*, 1976b.)

activation of 5-HT receptor sites, which was responsible for the inhibitory effect on GH release (Müller *et al.*, 1976b).

For the human, data which disprove a major role for the 5-HT system in GH release mechanism(s) have also been presented. In fact, Trp given orally elicited only a slight increase in plasma hGH, and administered with insulin inhibited the GH response which followed administration of insulin alone (Müller *et al.*, 1974b). Moreover, in contrast to the findings of Imura *et al.* (1973), oral administration of 5-HTP did not induce a significant increase in plasma hGH (Müller *et al.*, 1974b), and, similarly, no effect on plasma hGH of 5-HTP given in combination with a peripheral L-AAAD inhibitor was observed by Benkert *et al.* (1973) or, more recently, by Handwerger *et al.* (1975). In addition, methysergide, previously shown capable of decreasing stimulated GH secretion during insulin tolerance testing (Bivens *et al.*, 1973a), enhanced sleep-related GH secretion (Mendelson *et al.*, 1975), although opposite results on sleep-related GH rises were reported after cyproheptadine administration (Chihara *et al.*, 1976). Indirect proof for an inhibitory role of the 5-HT system is also derived from the reported enhancement of the arginine-induced hGH release by methergoline (Pontiroli *et al.*, 1975a), a drug with anti-5-HT activity (see Table XIV), and from the suppression of the exercise-mediated hGH rise following fenfluramine administration (Sulaiman and Johnson, 1973), a drug that activates the brain 5-HT system (Costa *et al.*, 1971).

c. *Acetylcholine and Putative Neurotransmitters*

Reports on the effect of other neurotransmitters on the CNS control of GH are few. One report concerns the possible involvement of the cholinergic system (Blackard and Waddel, 1969). In this study atropine was administered to normal subjects and the increase in GH secretion which follows insulin-induced hypoglycemia was measured. There was no effect of atropine on the GH-stimulated release, thus suggesting that the cholinergic system is not involved in the release of GH.

In the dog, administration of two antihistamine derivatives, promethazine (Fargan, Farmitalia) and diphenhydramine (Benadryl, Parke-Davis), did not modify hypoglycemia-induced GH release, whereas another antihistamine compound, 1-methyl-2-[2-(α-methyl-*p*-chloro-benzhydryloxy)ethyl]pyrrolidone hydrogen fumarate (Tavegil, Sandoz), was effective (Locatelli *et al.*, 1977) and, in the human, significantly reduced the GH release induced by arginine administration (Pontiroli *et al.*, 1975b).

PG's have been shown capable of inducing release of GH in the rat in both *in vitro* and *in vivo* systems (see Chapter III, Section D and Szabo and Frohman, 1975; Udeschini *et al.*, 1976). An intravenous drip of PGE_1 increased GH levels in man, probably mediated by action on the pituitary or hypothalamus and not by stimulation of the sympathetic nervous system (secondary to the fall in blood pressure) (Ito *et al.,* 1971). This effect may be specific for PGE_1, since PGA_1 in man did not modify plasma GH (Wilson *et al.,* 1971). More recently, it was reported that in the human, aspirin, which can block PG synthesis, inhibited insulin-induced GH release (Cavagnini *et al.,* 1977). However, in contrast to these findings, which point to a stimulatory role for PG's on GH secretion, is the observation that another blocker of PG synthesis, e.g., indomethacin, induced a striking increase in baseline GH concentrations in both the rat (Rigler *et al.,* 1976) and man (Cavagnini *et al.,* 1977).

The involvement of melatonin in the GH-release mechanism(s) in both rodents and humans has been discussed in the preceding section. Data in the literature so far are compatible with the idea that some pineal constituent diminishes GH synthesis or secretion and, as a consequence, restricts growth (Sorrentino *et al.,* 1972; for details, see Reiter, 1974).

In the human, administration of the GABA analog, baclofen, would result in a decrease in the hypoglycemia-induced GH release (Invitti *et al.,* 1976). With regard to substance P, stimulation of growth hormone and prolactin release have been reported after its systemic, but not intracerebroventricular, administration in urethane-anesthetized rats, intact or bearing extensive hypothalamic lesions (Kato *et al.,* 1976a; Rivier *et al.,* 1977).

Strictly linked with and dependent on the nature of the neurotransmitter is that of the possible site(s) at which monoamines or monoamine blockers act to regulate hormone secretion, a topic which will be discussed in Section C.

d. Comments

The studies mentioned here are without doubt compatible with the view that brain neurotransmitters intervene in the process of GH secretion; however, some aspects of this neurohumoral control are still unclear and worth investigating.

In the rat much of the conflicting literature on the role of CA's on GH might well be explained by differences in the experimental design used with respect to the site of drug administration, condition of the animal preparation, and the lack of specificity of the neuropharmacological agents used.

Direct intraventricular infusion of unphysiological amounts of neurotransmitter can affect in addition to the specific receptor sites, other neuronal systems; injection by the same route of an amine precursor (e.g., DA) may cause its uptake, metabolism, and release from presynaptic endings in its final form (e.g., NE). Moreover, for 5-HT it has been shown that its IVT administration leads to the formation of artifactual pools which are not present if the amine is synthesized *in vivo* (Gallagher *et al.*, 1975).

Urethane, used as anesthetic, reduces circulating GH levels (Collu *et al.*, 1972), and also affects transmission of signals from limbic structures to the hypothalamus (Cross and Dyer, 1971), thus very likely altering the response of the latter to CNS-acting drugs. In this context it is of note that CA neurons, as already mentioned, are important in the relay of GH responses from extrahypothalamic structures to the hypothalamus (Martin *et al.*, 1973), and that in the conscious rat CPZ administration was found repeatedly inactive in increasing GH levels (E. E. Müller, unpublished results).

When given systemically, neuropharmacological agents can affect hormone secretions, not necessarily by acting on CNS structures (see also Section A,1,a); for example, E, NE, and 5-HT do not appreciably cross the BBB, and clear-cut evidence for phentolamine is lacking. Therefore, it is entirely possible that stimulating or blocking effects of systemically administered drugs may occur outside the CNS. The above considerations and others (see below) indicate the need for caution in interpreting the results of experiments in which neuropharmacological agents are used to alter neuroendocrine function.

From the results obtained in the rat during postnatal development, it seems probable that the two adrenergic neurotransmitters NE and DA exert a stimulatory effect on the baseline secretion of GH. In the adult rat the evidence presented favors the existence of a dual effect for NE and DA on RIA-GH secretion, the former having a stimulatory influence, the latter inhibitory. However, the finding that GH in the adult, unanesthetized rat is secreted in episodic bursts (Martin *et al.*, 1975a) suggests caution in interpreting results from neuropharmacological studies in which the frequency of blood sampling was likely lower than the frequency of the pulsatile secretion. Longitudinal studies for the assessment of the effects of a given experimental procedure on GH regulation in the rat should prove more rewarding. In the mouse the central dopaminergic inhibitory influence(s) seems to be predominant; in fact, concomitant reduction of both NE and DA due to 6-OHDA resulted in increased GH levels in this species.

In the pig and sheep the available evidence is rather scanty, and the results, when available, favor an inhibitory role for NE. The proof presented for an inhibitory role of E in these species is questionable, since it is based mainly on the effect of high doses of systemically administered E. This compound might be capable of inducing the observed effects through the feedback action of increased NEFA levels in plasma. A direct effect of CA's on sheep somatotropes is suggested by *in vitro* experiments.

In the anesthetized dog, the available data point to a stimulatory role of the brain NE system and suggest that brain DA and E receptor sites are not involved. However, the use of anesthesia in the dog experiments makes it necessary to exercise caution in accepting the results produced as definitive proof.

The same dual role for NE and DA thought to be present in the adult rat seems to extend to the monkey, a species in which the central NE system may be vitally important in GH regulation. However, results of more recent experiments which also indicate a stimulatory role for DA do not permit us to draw definite conclusions. In man, the evidence available is compatible with the hypothesis that both NE and DA serve as neurotransmitters controlling GH release. The GH stimulant effect of L-dopa is not unrefutable proof of a stimulant role for DA, since the precursor amino acid could act to increase NE levels in the hypothalamus or limbic system or/and affect the brain 5-HT system (see Chapter II, Section K), and these effects might mediate hGH release. In the human, only 5% of the infused dopa is converted to NE or metabolic products of NE (Goodall and Alton, 1972), so that it seems reasonable to assume that L-dopa elicits GH release by activation of a dopaminergic site. In keeping with this view are also the observations that pyridoxine, which in the form of pyridoxal phosphate acts as a coenzyme of L-dopa decarboxylase (see Chapter II, Section B,6), inhibits L-dopa-induced GH response (Mims *et al.,* 1975), conceivably for a peripheral acceleration of the conversion of L-dopa to DA, and administered alone increases hGH levels (Delitala *et al.,* 1976).

Further evidence for the direct stimulation of DA receptors in the release of GH is derived from the reported stimulant action of apomorphine, CB-154, and piribedil, the elicitation of GH stimulation following blockade of NE synthesis, and the suppression by antidopaminergic drugs of the stimulated release of GH. The likelihood of a dopaminergic mechanism is also inferentially supported by the effectiveness of amphetamine derivatives in stimulating hGH secretion (Rees *et al.,* 1970), bearing in mind that some of the central effects of sympathomimetic amines result from an enhancement of DA receptor activity (see Table XI). It must be pointed out, however, that in monkeys the GH-releasing effect of *l*- and *d*-amphetamines was completely preserved after pimozide pretreatment (Sachar *et al.,* 1976), so that an NE-mediated effect for sympathomimetics cannot be denied. The possibility that the stimulatory effect of L-dopa on GH secretion in the human might result from a final activation of a serotoninergic receptor due to a release of the brain indoleamine from vesicular stores, although appealing, does not appear to be realistic. As reported by Liuzzi *et al.* (1976), methergoline, a 5-HT receptor blocker, did not modify in normal human subjects L-dopa-induced GH rise. Release of brain indoleamines occurs in the laboratory animal only with very high doses of L-dopa (Algeri and Cerletti, 1972). In addition, there is no convincing evidence for an action of 5-HT as a GH releaser; on the contrary, some of the evidence obtained from both subprimates and man supports an inhibitory role for this neurotransmitter, at least in the neuroendocrine activation of GH release which follows insulin-induced hypoglycemia.

Most of the experiments in which a stimulatory role for the serotoninergic system has been claimed were based on the effect of 5-HTP (Smythe and Lazarus, 1973a; Imura *et al.,* 1973; Smythe *et al.,* 1975a). Now it is known that 5-HTP administered systemically increases brain 5-HT, in part owing to a selective accumulation by 5-HT neurons (Corrodi *et al.,* 1967b), but doses such as those used in some rat experiments (10–50 mg/kg) may lead to substantial 5-HT accumulation in cells that do not ordinarily contain this indoleamine and also interfere with CA and their precursors for transport, storage, and metabolism within the CNS (see Chapter II, Section K). Thus, 5-HTP-induced changes in CNS function may largely reflect action on CA or other central neurohumoral systems. It must be recalled that, in addition, the systemic infusion of L-Trp or L-5-HTP in the rat at doses most commonly used in rat or human experiments, far from activating brain 5-HT neurotransmission, were found to depress the firing of brain serotoninergic neurons (Gallagher and Aghajanian, 1976). Thus, the possibility cannot be ruled out that the stimulatory action on GH release observed in the human or the experimental animal after Trp or 5-HTP administration may actually reflect the inhibitory role of 5-HT neurotransmission in GH release.

The neuropharmacological specificity of 5-HT antagonists, cyproheptadine and methysergide, is also speculative. Cyproheptadine was originally developed as an antihistaminic agent, and was subsequently shown to also possess anticholinergic and anti-5-HT activity to a relatively high degree (see Table XIV). Recently, its ability to block L-dopa- (Smythe *et al.,* 1976) and apomorphine-induced hGH release, which might be indicative of an anti-DA action, has been reported (Winkelmann *et al.,* 1975). Moreover, the reported blockade effect of Tavegil on the stimulated release of GH in both the human and the dog may indicate that the role of cyproheptadine in this case is the result of an intrinsic antihistaminic activity.

Methysergide surely possesses 5-HT receptor-blocking activity (Kara *et al.,* 1961; Fiszer and De Robertis, 1969). However, proof has also been given for its inability to counteract central 5-HT activation (Haigler and Aghajanian, 1974) and in favor of 5-HT receptor-stimulating properties (Jalfre *et al.,* 1974).

Melatonin has been proposed as an antagonist of 5-HT at a receptor-site level only on the assumption that all *O*-methylated derivatives of monoamines result in derivatives that inhibit monoamine action at hypothalamic receptor sites (Smythe and Lazarus, 1973b; Smythe *et al.*, 1975b), and because it shares with Cy the ability to block the 5-HTP effect on GH secretion. It cannot be ignored, however, that the action of melatonin on GH secretion, which has been attributed to 5-HT receptor blockade, may instead be due to 5-HT receptor stimulation. It is noteworthy that 5-HT and melatonin share in common the ability to block gonadotropin and to stimulate PRL secretion (see Sections A,3,b and 4,b).

3. Brain Neurotransmitters in the Control of Gonadotropin Secretion

a. Catecholamines

The existence of central neurohumoral mediation in the control of pituitary gonadotropic secretion was proposed some 30 years ago on the basis of ovulation-stimulating and -blocking experiments. In 1947, Sawyer, Markee, and Hollinshead demonstrated that the α-adrenergic blocking agent Dibenamine was able to inhibit ovulation in the rabbit. Subsequent studies (Sawyer, 1952) showed that, conversely, IVT injection of E or NE stimulated the ovulatory process in noncopulatory rabbits. Also, in cyclic, ovulatory mammals such as rats and hamsters treatment with drugs capable of modifying the central CA tone affected ovulation. Brain monoamine depletors (Barraclough and Sawyer, 1957; Coppola *et al.*, 1966; Meyerson and Sawyer, 1968) or specific inhibitors of CA synthesis (Lippman *et al.*, 1967), when administered before the critical period, were active in this sense. This drug effect could be counteracted by treatment with a monoamine precursor (Coppola, 1971) or by reducing amine catabolism with MAOI (Meyerson and Sawyer, 1968).

In the past 10 years interest in this field has been stimulated by sophisticated techniques for demonstrating and localizing monoamines in the brain (see Chapter II, Section A) and for quantitative determinations of their turnover (see Chapter II, Sections B,12 and C,7) and by the development of RIA for the direct determination of gonadotropins in the plasma (Niswender *et al.*, 1968). The introduction of the Falck–Hillarp fluorescence histochemical technique not only allowed demonstration of the topography of brain monoaminergic fibers in the hypophysiotropic area of the hypothalamus but also of the functional changes that catecholaminergic and, chiefly, TIDA neurons undergo during the different phases of the estrous cycle.

Estimation of amine turnover by studying the decline of DA stores in the rat ME following inhibition of synthesis by α-MT and by using the histochemical fluorescence technique for analysis of DA revealed a selective decrease of DA turnover in proestrus and early estrus as compared with other stages of the ovarian cycle. Since these events accompany stimulation of pituitary LH release, an inhibitory action of TIDA neurons on peptidergic neurons in the ME secreting LH-RH was inferred (Fuxe *et al.*, 1969; Fuxe and Hökfelt, 1969; Ahren *et al.*, 1971). This theory was supported by the finding that several conditions involving the endocrine system, such as pregnancy, pseudopregnancy, and lactation, which share in common low circulating LH and FSH and high PRL levels, were associated with functional activation of TIDA neurons (see Fuxe and Hökfelt, 1969). Support for the view that TIDA neurons exert tonic inhibition on LH-RH secreting structures was also derived from experiments in which administration of 17-estradiolbenzoate or contraceptive steroids and clomiphene, at doses inhibiting LH secretion, markedly increased DA turnover in the system and at the same time exerted a feedback inhibition on LH-RH secretion (Fuxe *et al.*, 1974).

As a corollary of these findings, DA-receptor stimulant agents such as ergocornine, 2-Br-α-ergocryptine, apomorphine, piribedil and lergotrile (see Table XI) produced blockade of PMS-induced ovulation in prepuberal rats (Fuxe *et al.*, 1975b) which could be counteracted by pimozide. Also nicotine appeared to lower LH secretion via selective activation of the TIDA neurons (P. Eneroth, unpublished results). In keeping with these results, Uemura and Kobayashi (1971) had found that a mixture of DA and cholesterol implanted into the ME caused prolonged diestrus. All these findings lent support to the view that the lateral tuberoinfundibular DA pathway innervating the lateral zone inhibits the secretion of LH-RH from the ME and challenged the proposal formulated by McCann, Kamberi and their co-workers that DA was the neurotransmitter responsible for the activation of LH-RH-secreting structures.

These authors used *in vitro* techniques and IVT injections of relatively high doses of DA. Enhanced LH-RH release was reported from incubates of rat hypothalamic tissue in the presence of DA, while NE or 5-HT were ineffective and E slightly inhibitory; the response to DA was completely inhibited by phentolamine and unaffected by pronethalol, a β-adrenergic blocker (Schneider and McCann, 1969). Interestingly, the releasing action of DA was blocked by estradiol in this *in vitro* system, and this blockade could be prevented by the addition of either puromycin or cycloheximide, inhibitors of protein synthesis (Schneider and McCann, 1970). This suggested to these authors that the negative feedback action of estrogens to inhibit gonadotropins might be mediated, at least in part, by an action on the RF neurons, with ensuing formation of an inhibitory peptide or protein capable of blocking the action of DA on the RF neuron. This mechanism would be similar to that of thyroid hormones in blocking the action of TRH on the pituitary (see Chapter III, Section A,6,b). The existence of actinomycin D-sensitive steps at the level of MPOA involved in the stimulatory feedback action of estrogen on LH release has been more recently proposed by Kalra (1975) as a result of *in vivo* experimentation.

Intraventricular injections of DA (1.25 μg) into the rat third ventricle resulted in increased release of pituitary LH, while direct perfusion of the amine into the AP was ineffective. For greater quantities of DA, the amount of LH released was inversely related to the dosage. Both E and NE were active at much higher doses (Kamberi *et al.*, 1970b). That the action of DA was not directly on the AP but through the hypothalamus was also implied by detection of increased LH-RH concentrations in pituitary stalk plasma of rats given IVT DA (Kamberi *et al.*, 1969). In accordance with this view, release of IR-LH-RH was reported recently from synaptosome preparations isolated from whole sheep hypothalami or ME's following incubation with 10^{-10} *M* DA (Bennett *et al.*, 1975).

Further evidence for the stimulatory involvement of DA neurons in the control of ovulation was obtained by Kordon and Glowinski (1969). They showed that selective restoration of NE levels with DOPS following ovulation blockade by α-MpT did not restore LH release, whereas restoration of NE and DA levels with L-dopa did. In addition, the infusion of the false transmitter-forming precursor, α-MD, directly into the ARC-ME completely blocked ovulation (Kordon, 1971).

Results obtained with IVT amine injection are conflicting and thus difficult to interpret. Rubinstein and Sawyer (1970) reported that IVT infusions of DA were ineffective, as compared with similar infusions of E, in inducing ovulation in the pentobarbital-blocked proestrous rat; a positive effect for E was also reported by Raziano *et al.* (1971). More recently, Sawyer and his co-workers (1974) showed that in estrogen-primed rabbits, IVT DA was not only completely ineffective in inducing ovulation but that higher doses (50 μg) actually induced a sustained blockage of the ovulatory effect of NE. In experiments in which LH was measured by RIA, DA prevented for some weeks an increase in serum LH following infusion of IVT NE.

Similarly, in the rat, NE injected IVT at high doses induced ovulation in more than 50% of rats made anovulatory by electrolytic-induced lesions in the anterior hypothalamus or by exposure to continuous illumination, while DA was ineffective (Tima and Flerkò, 1974). Ojeda *et al.* (1974b) showed that IVT-injected DA inhibited PRL secretion but failed to influence LH secretion. Brain CA's, namely NE, in addition to being needed for the release of ovulatory hormones at the critical period, seem to be necessary in the cycling rat for the gradual release of gonadotropins (so-called basal secretion) (Teresawa *et al.,* 1975) which ensures estrogen secretion before proestrus (Lawton and Sawyer, 1968).

In the past few years studies of the effects of pharmacologically induced changes in catecholaminergic transmission by peripheral administration of synthesis blockers or amine precursors have failed to confirm the stimulatory role of a dopaminergic synapse in LH-RH release. They have supported, instead, NE as the neural CA which facilitates LH release. In experiments involving electrochemical stimulation of the PO or ME-ARC region of proestrous rats, LH release was blocked by α-MpT; the blockade could be reversed by the administration of either L-dopa or DOPS shortly after treatment. Pretreatment of the animals with DDC, used to lower NE levels selectively, also blocked the response to PO stimulation; this blockade could be partially reversed by the administration of DOPS, whereas L-dopa had little effect (Kalra and McCann, 1973). Similar results were obtained in experiments in which CA synthesis blockers and amine precursors were given to proestrous rats and their effects in blocking the preovulatory elevation in plasma LH and ovulation were determined (Kalra and McCann, 1974).

Further experimentation using Caligaris' model, in which the injection of progesterone into spayed estrogen-treated rats produces discharge of pituitary FSH and LH in the plasma (Caligaris *et al.,* 1968), led to similar results (Kalra *et al.,* 1972). Simultaneous depletion of DA and NE by α-MpT as well as selective NE depletion by DDC blocked gonadotropin peak. Replenishment of the depleted pool of CA's by L-dopa did not restore LH-FSH secretion; this result was obtained by DOPS which restores only NE depletion. In prepuberal castrate male and female rats, administration of α-MpT was almost completely ineffective in counteracting the LH rise that follows gonadectomy, while selective blockade of NE synthesis by FLA-63 completely suppressed the postcastration LH rise (Cocchi *et al.,* 1974).

In male adult rats, IVT injection of 6-OHDA (see Chapter II, Section D) induced a reduction of plasma LH levels only in the immediate postinjection period, while no difference (Kitchen *et al.,* 1974) or increased LH levels (Fenske and Wuttke, 1976) were present in samples taken 2 days or more later when activity of the surviving CA neurons can ensure functional restoration and the NE released stimulates hypersensitive postsynaptic membranes (see Chapter II, Section D). Plasma FSH concentrations in the treated animals showed no significant difference from the control levels at all intervals tested (Kitchen *et al.,* 1974). Similar results were obtained in an extension of these studies, involving direct infusion of 6-OHDA into the POA or the MFB (Kitchen, 1974), two areas leading into the ventral ascending NE pathway which projects to the internal layer of the ME as well as to the septal areas (see Chapter II, Section B,4).

Additional support for the involvement of positive NE inputs in gonadotropin release was the finding that in completely deafferented rats, in which this neurotransmitter is totally depleted within the hypothalamic island while DA is preserved (Weiner *et al.,* 1972), plasma LH and FSH levels were considerably below control values and did not peak at proestrus (Blake *et al.,* 1972).

The pulsatile secretion of LH also appears to be in the rat under the stimulatory activity of the noradrenergic system and the inhibitory activity of the dopaminergic system. In ovariectomized rats, selective inhibition of NE synthesis by FLA-63 or dopaminergic stimulation with apomorphine induced decreased plasma LH levels and disappearance of the pulsations, whereas simultaneous blockade of DA and NE by α-MpT had no effect on plasma LH levels; administration of 1-hydroxy-3-aminopyrrolidone-2 (HA-966), a centrally acting DA-release blocking agent (Bonta *et al.,* 1971), significantly raised plasma LH levels (Gnodde and Schuiling, 1976).

Collectively, the results of these studies indicate the presence in the rat of a stimulatory NE activity and give evidence for the existence of an inhibitory dopaminergic synapse.

Results once considered definitive proof of a positive involvement of DA in the physiological regulation of gonadotropin secretion in rodents can be reinterpreted now, on examining the key experiments which led to this view. When a putative neurotransmitter is infused directly into a brain ventricle, it barely penetrates, in physiological amounts, specific receptor sites; in fact, penetration of the fraction not yet catabolized in the synaptic cleft is more difficult than when the neurotransmitter is released into it from the presynaptic ending. A better physiologic method to use, therefore, for specific receptor stimulation, is the administration of an amine precursor that can be taken up by the presynaptic terminals, metabolized to the active amine, and then released (see also Section A,2,d). The fact that doses of NE 20 times greater than those of DA were needed to evoke the same stimulatory effect (Kamberi *et al.,* 1970b) does not militate against the possibility that only a small fraction of NE, the natural transmitter, reaches the specific receptor site on the peptidergic neurons. On the other hand, stimulatory effects induced by exogenously administered DA could well be attributed to its uptake, metabolism, and release from the presynaptic ending as NE. Support for this view derives from some of the experiments of Kamberi *et al.* (1970b) in which the differential time course of NE and DA stimulatory effects on gonadotropin secretion was followed. The effect of noradrenergic stimulation did appear within few minutes and probably resulted from the small amount of the neurotransmitter that managed to reach the receptor site; in contrast, dopaminergic stimulation needed at least 10–20 min to occur, emphasizing the possibility that the time lapse was required to permit the synthesis of the active transmitter after initial DA uptake. Also, suppression of the DA effect by the α-adrenergic receptor blocker, phentolamine (Schneider and McCann, 1970), is more consistent with the possibility that DA acted after conversion to NE.

In conclusion, from these neuropharmacological and electrical stimulation experiments and from other observations (for references, see McCann *et al.*, 1974), and also in view of the knowledge acquired on the distribution of LH-RH neurons (see Chapter III, Section C), the existence of an NE synapse lying in the preoptic or anterior hypothalamic area which makes contact with LH-RH neurons and which mediates the increased release of LH-RH and LH which occurs during the preovulatory discharge of gonadotropins or in response to removal of steroid negative feedback may be postulated (see also Section C). However, one possibility that should be considered is that the mechanism(s) heretofore attributed to NE may be exerted in part by brain E. The enzyme PNMT, which converts NE to E, has been found to be widely distributed in the brain (Saavedra *et al.*, 1974b), and its presence in selected hypothalamic areas parallels that of TH (Hökfelt *et al.*, 1976c) (see Chapter II, Section B,2). On the other hand, evidence for a direct stimulatory function of brain E in LH release in rodents has been presented by Sawyer and associates (Sawyer, 1952; Rubinstein and Sawyer, 1970) and Raziano *et al.* (1971). A sustained release of E from the adrenal medulla during "stress"-activation of the sympathetic nervous system instead would block spontaneous pituitary LH release and ovulation (Blake, 1976).

Caution should be exercised in trying to apply to primates the knowledge obtained in rodents on the action of CA neurons in the phasic activation of gonadotropin release. In the rat the MBH regulates tonic gonadotropin secretion and more rostral areas initiate the cyclic ovulatory discharge, whereas in the rhesus monkey the major components of the neuroendocrine system that control the tonic mode of gonadotropin secretion as well as LH and FSH surges appear to be present in the MBH-hypophyseal unit (Krey *et al.*, 1975). Thus, monkeys would not require norepinephrinergic connections from the MPOA to the MBH for the transmission of the diurnal neurogenic signal requisite for ovulation in the rat. Accordingly, pharmacological blockers of ovulatory activity in rats, e.g., pentobarbital, reserpine, or phenoxybenzamine, did not interfere with the positive feedback action of estrogen on gonadotropin release in the monkey (Knobil, 1974). However, in more recent studies it has also been observed that in the rhesus monkey bilateral destruction of the ventral POA-AHA blocks spontaneous ovulation and impairs the ability of the hypothalamic–hypophyseal axis to release LH in response to estrogen (Norman *et al.,* 1976). It is clear that the effects of pharmacological manipulation of the adrenergic system on gonadotropin secretion in this species require further investigation. In the human, it has been recently reported that blockade of dopaminergic receptors by pimozide or of NE synthesis by fusaric acid significantly reduced the midcycle surge of LH without significantly affecting the corresponding FSH burst (Leppäluoto *et al.,* 1976) (see also Chapter V, Section A,1). Extension of these neuropharmacological studies is awaited with interest.

Deficiency in hypothalamic CA metabolism have been used to explain age-related alterations in gonadal function of rodents (see also Chapter II, Section B,13). Quadri *et al.* (1973) demonstrated that daily systemic administration of E, L-dopa, and iproniazid, drugs expected to improve brain CA function, could cause the resumption of cycling in aged, constant estrous rats. The efficacy of peripherally administered E was attributed to the higher permeability of the BBB in older than in younger rats, although proof that this is the case is lacking. It was then shown by administering L-dopa to constant estrous rats that the sensitivity of the neuroendocrine control system for gonadotropin secretion is actually altered in aging (Watkins *et al.,* 1975). This is also true for PRL (Watkins *et al.,* 1975) and GH (Sachar *et al.,* 1972) secretion.

b. Indoleamines

The fact that the pineal gland is the primary site of melatonin synthesis (Wurtman *et al.,* 1968b) has led to the concept that the effects of this gland on gonadal development and function may involve melatonin as well its precursor, 5-HT. The existence of a functional link between the pineal gland and the gonads is an old and widespread notion; prolonged absence of the pineal affects gonadal function. After pinealectomy, polyestrous mammals such as hamsters, rats, and mice undergo pubertal changes earlier than do intact representatives of the same species. Furthermore, the gonads of adult rats hypertrophy after pinealectomy, a phenomenon which can be prevented by the administration of extracts of pineal tissue (Kitay and Altschule, 1954; Wurtman *et al.,* 1968b; Reiter and Fraschini, 1969). In experiments designed to test whether the action of 5-HT and melatonin on gonadal function was a direct consequence of their influence on the release of pituitary gonadotropins, it was observed that either substance when administered IVT to male rats suppressed LH (Kamberi *et al.,* 1970b) as well FSH release (Kamberi *et al.,* 1971b). When 5-HT or melatonin was perfused into the AP via a cannulated portal vessel, neither

LH nor FSH release was affected, implying an action of the indoleamines exerted through the hypothalamic–hypophyseal complex. However, surprisingly, administration of pCPA did not modify plasma LH or FSH in either normal or castrated male rats (Donoso *et al.,* 1971). Prevention of ovulation has been observed following IVT injection of 5-HT in immature PMS-progesterone treated rats (Zolovick and Labhsetwar, 1973) or systemic injections of 5-HT on the day before proestrus (O'Steen, 1964; Labhsetwar, 1971), an effect, which, in view of the alleged BBB impermeability to indoleamines (see also Section 4,b) must be attributed to a peripheral action. The immediate precursor of the amine 5-HTP was also reported to have the same effect (Kordon *et al.,* 1968).

A direct action of 5-HT on LH-RH secreting structures was implied by the findings of Leonardelli *et al.* (1974), who showed that 5-HT infused IVT into normal or gonadectomized guinea pigs enabled hypothalamic neurons in the preoptic and suprachiasmatic areas to fluoresce in the presence of an anti-LH-RH serum. When 5-HT was not infused, no fluorescent hypothalamic neurons were evidenced. Similar findings were obtained after systemic injection of melatonin. These results would indicate an inhibitory action of the indoleamines on LH-RH release. In sheep, both intact and with lesions of the AHA, 5-HT and melatonin infused IVT induced prolongation of the estrus cycle, delay of ovulation, and a block of the preovulatory peak of LH (Domanski *et al.,* 1975). The effect of indoleamines in lesioned ewes in which the inhibitory action of the AHA on the transmission of LH-RH was eliminated (Domanski *et al.,* 1972) was indicative of an inhibitory action exerted on LH-RH release occurring at the MBH level. Similarly, suppression of central 5-HT activity by pCPA in constant, estrous anovulatory (CEA) rats induced the appearance of fresh corpora lutea in about 50% of the animals (Trentini *et al.,* 1974). Since the same effect had been observed in CEA rats bearing bilateral suprachiasmatic lesions or frontal hypothalamic deafferentation, following pinealectomy or cervical sympathectomy (Mess *et al.,* 1971; Trentini *et al.,* 1973), the suggestion has been made that pinealectomy might act in inducing ovulation through suppression of a CNS 5-HT mechanism (Trentini *et al.,* 1974). This assumption is supported by the fact that (a) melatonin injection suppresses the effect of pinealectomy in CEA rats (Mess *et al.,* 1971) and increases 5-HT concentration of the hypothalamus and the midbrain (Anton-Tay *et al.,* 1968); (b) 5-HT treatment counteracts the ovulation-inducing effect of pinealectomy in CEA rats (Tima *et al.,* 1973).

Action of melatonin mediated by 5-HT neurons would be consistent with the reported higher effectiveness of melatonin in inhibiting ovulation when given intraocularly than when given subcutaneously or intravenously (Pomerantz and Sorrentino, 1973). The pathway of intraocularly injected melatonin would be through retinohypothalamic fibers which synapse on neurons of the SCN of the anterior hypothalamus (Rieke, 1958), an area of distribution of 5-HT terminals and presumably of formation and release of LH-RH (see Chapter II, Section C,2 and Chapter III, Section C). Presence of HIOMT in the retina (Cardinali and Rosner, 1971) suggests that melatonin formed in the eye might follow this pathway, inhibiting LH-RH synthesis and/or release in the anterior hypothalamus.

While there is no longer any doubt that melatonin can inhibit LH secretion from the AP (Collu *et al.*, 1971; Reiter and Sorrentino, 1971), there is still some disagreement regarding the specificity of melatonin as an anti-FSH substance. According to Fraschini and co-workers (Fraschini *et al.*, 1971; Fraschini and Martini, 1970), melatonin specifically inhibits LH synthesis and secretion but does not influence FSH secretion. These findings are in apparent disagreement with the previously mentioned data of Kamberi *et al.* (1971b) and those of Vaughan *et al.* (1970, 1971) and Sorrentino (1968), in which melatonin was found to block compensatory enlargement of the ovary, a response which at least partially relies on FSH secretion (Benson *et al.*, 1969). Besides melatonin, other natural products of the pineal which inhibit gonadotropin metabolism include 5-methoxytryptophol and 5-hydroxytryptophol (Fraschini, 1969). In the human, serum melatonin measurements during the menstrual cycle showed elevation at the time of menstrual bleeding and lowest concentration at the time of ovulation, thus suggesting that low concentration of the indole at midcycle is a necessary factor for ovulation (Wetterberg *et al.*, 1976). In postmenopausal women, melatonin reportedly exerted an inhibitory effect on baseline LH values (Fideleff *et al.*, 1974); however, in an extension of these studies the same authors (Fideleff *et al.*, 1976) were unable to observe any significant change in basal or poststimulation LH and FSH levels either in man or postmenopausal women after treatment with melatonin. Further studies are needed to clarify this subject.

While the possible mechanism of action of 5-HT, which is present in the hypothalamic–tuberoinfundibular pathway and found in abundance in the region of the SCN, is more easily understood, that of melatonin is not so clear (see also Section A,2,d). Although the administration of melatonin causes biochemical changes in several tissues (Wurtman *et al.*, 1968b), its primary target organ in mammals appears to be the brain. Melatonin is relatively nonpolar, and readily enters the brain from the general circulation (Kopin *et al.*, 1961; Wurtman *et al.*, 1964), becoming concentrated in the midbrain and hypothalamus after its injection into the lateral cerebral ventricles (Anton-Tay and Wurtman, 1969) or the cisterna magna (Cardinali *et al.*, 1973). On the basis of results obtained from stereotaxic implantation studies, Fraschini and co-workers (1968, 1971) postulated the existence of specific melatonin receptors located in the ME and in the mid-brain reticular formation. The midbrain location of melatonin receptor sites coincides with the concentration of the 5-HT neuronal cell bodies and is consistent with the possibility that it may influence the hypothalamopituitary system by altering the metabolism of 5-HT-containing neurons, as suggested by the report that peripherally administered melatonin elevates brain 5-HT levels (Anton-Tay *et al.*, 1968) (see also reasons discussed in previous paragraphs).

Further studies are needed, however, before this model can be accepted as a mechanism of action of melatonin, especially since failure to confirm the influence of melatonin on brain 5-HT exists (W. W. Morgan and R. J. Reiter, unpublished observations). In addition to the pineal, melatonin has been found in minute amounts in the hypothalamus, and its presence is apparently independent of the pineal, since its hypothalamic levels are unaffected by pinealectomy (Koslow, 1974) (see also Chapter II, Section E).

c. *Prostaglandins*

PGE's were shown capable of stimulating LH release both in Nembutal-blocked proestrous (Tsafriri *et al.*, 1973; Spies and Norman, 1973) and ovariectomized rats (Harms *et al.*, 1973, 1974) "*in vivo*." Conversely, inhibitors of prostaglandin synthesis were capable of interfering with the process of ovulation. At high dose levels, systemic administration of indomethacin was reported to reduce hypothalamic prostaglandin content in rats (Behrman *et al.*, 1972), but failed to suppress the proestrous LH release (Tsafriri *et al.*, 1973). However, aspirin, another inhibitor of PG synthetase, blocked ovulation when administered directly into the AHA. Simultaneous injection of $PGF_{2\alpha}$ largely restored both the incidence and rate of ovulation, indicating that aspirin blocked ovulation by inhibiting biosynthesis of PG's in the AHA (Labhsetwar and Zolovick, 1973). Further proof for a hypothalamic site of action was suggested by the finding that infusion of PGE_1 into the third ventricle of ovariectomized rats increased LH but not FSH release while intrapituitary infusion was ineffective (Spies and Norman, 1973; Harms *et al.*, 1973, 1974). However, in estrogen-primed rats, PGE_2 increased LH release also after direct intrapituitary injection (Harms *et al.*, 1974). Suggestions for a direct pituitary site of action also resulted from the experiments of Sato *et al.* (1975a), who showed that plasma LH levels were increased by the iv injection of PG's in rats with an experimentally injured ME, that pretreatment with indomethacin suppressed the response of the rat AP to LH-RH, and, finally, that PGE_1 and PGE_2 injected into the AP increased plasma LH levels. Evidence for the effect of PGE_2 or PGE_1 on AP-LH release *in vitro* is conflicting: incubation with PGE_1 failed to alter basal LH release (Zor *et al.*, 1970), but in a perfusion system, PGE_1 or PGE_2 increased LH release in a dose-dependent manner (Dowd *et al.*, 1973).

Collectively, the results obtained favor the CNS as a major site of action for PG's, without denying an action exerted directly on the AP. The mechanism of PG action on the CNS has yet to be clarified, although PG's presumably act directly or indirectly on neurons which discharge LH-RH into hypophyseal portal vessels. The observed prevention of PGE_2-induced release of LH in immature rats by prior administration of an anti-LH-RH serum (Chobsieng *et al.*, 1975) and the decreased hypothalamic (Sato *et al.*, 1975b) or increased plasma (Ojeda *et al.*, 1975) LH-RH content of rats given systemic or central PGE_2 administration support this theory. The recent observation that in adult male rats the most potent stimulators of LH release after intraventricular infusion were PGE_2, $PGF_{2\alpha}$, and $PGF_{2\beta}$ (Warberg *et al.*, 1976) led to the theory that the potency of PG's in stimulating LH release depends on the presence of both the 5,6-*cis* double bond and the 11-hydroxyl group (see Chapter II, Section G,2). These structural features would be essential for the activation of a PG receptor at the level of the brain, a prerequisite for the release of LH-RH or other factors from the brain.

The action of PGE_2 in LH-RH release appears to take place directly on the LH-RH neuron and is not transsynaptic, since a variety of adrenergic, dopaminergic, serotoninergic, or cholinergic receptor blockers were ineffective in blocking PGE_2 action in inducing LH release (Harms *et al.*, 1976). The data would disprove the mechanism proposed by Labhsetwar and Zolovick (1973), according to which PG's and CA's interact functionally to potentiate hypothalamic adrenergic neurotransmission mediating the ovulatory effect of progesterone.

d. Acetylcholine and Putative Neurotransmitters

The role of a cholinergic synapse in mediating hypothalamic control over gonadotropin release is suggested by some studies, although it is still too early to draw definitive conclusions.

Sawyer (1952), Everett (1964) and their co-workers showed that large doses of systemically administered atropine blocked ovulation, a finding that has been confirmed more recently following central injection of high doses of the drug (Libertun and McCann, 1973). Consistent with previous studies that had shown that implantation of atropine into the rat hypothalamus inhibits the ovarian compensatory hypertrophy which follows unilateral ovariectomy (Monti *et al.*, 1970; Sala *et al.*, 1971) has been the more recent report that ACh can increase FSH (Simonovic *et al.*, 1974) and LH (Fiorindo and Martini, 1975) release from hypothalamopituitary coincubates *in vitro,* an effect completely prevented by atropine, and increase LH release following IVT injection (Justo *et al.,* 1975). The reported effect of ACh would be muscarinic in nature, as implied by the atropine antagonism and by the reported inhibitory effect of stimulation of nicotinic receptors on gonadotropin secretion (Blake, 1974). However, fairly large ACh doses were used to elicit these effects in the *in vitro* studies, and this raises some question as to whether or not ACh or atropine was acting specifically.

Immediate inhibition of gonadotropin secretion followed by a delayed release has been reported in ovariectomized estrogen-primed rats following systemic administration of cholinergic drugs such as pilocarpine or eserine (Libertun and McCann, 1974), and this plus the ineffectiveness of IVT-injected carbachol to stimulate gonadotropin release (C. Libertun and S. M. McCann, unpublished data) cast doubt on the actual role of hypothalamic cholinergic synapses in mediating hypothalamic control over gonadotropin release. Presence of cholinergic receptors in the amygdala whose activation inhibits gonadotropin release has been recently claimed on the basis of implantation studies with cholinomimetic drugs (Piva and Borrell, 1976).

Suggestions for a possible involvement of histamine (see Chapter II, Section H) have been derived from the reported release of LH in ovariectomized, estrogen-primed rats (Libertun and McCann, 1976; Donoso *et al.,* 1976) following IVT administration of huge doses. In male rats IVT-injected histamine did not cause significant LH changes with doses of 2.5–3.0 μg, but induced a decrease when 60 μg were administered (Donoso and Banzan, 1976). Thus, the relationship of histamine to gonadotropin release is presently unclear. With regard to transmitter amino acids, Ondo (1974) has recently reported that very large doses of centrally injected GABA stimulated release of LH in the male rat, and similar results have been obtained following IVT injections of lysine and glutamate (Ondo *et al.,* 1976).

4. Brain Neurotransmitters in the Control of Prolactin Secretion

a. Catecholamines

As already reported in the section on the neural control of PRL secretion (Chapter III, Section A,5), PRL is unique among the AP hormones in that its secretion in mammals is tonically inhibited by the CNS. The existence of a prolactin-inhibiting factor has been proved; however, the nature of PIF still remains elusive and, as will be discussed, the possibility that under physiological conditions CA's, namely DA, may be PIF has not been excluded.

First experimental data linking in a reciprocal way the catecholaminergic tonus in the CNS with PRL secretion go back to the studies of Barraclough and Sawyer (1959) and Coppola *et al.* (1965), which showed that treatment of rats with CPZ or CA synthesis inhibitors resulted in the induction of pseudopregnancy. More recent data (Van Maanen and Smelik, 1968) showed that reserpine implanted in the MBH also caused pseudopregnancy which was prevented by treatment with MAOI. Since then an exhaustive series of studies involving more specific pharmacological agents and direct measurements of circulating PRL levels have strengthened the theory that brain CA's exert an inhibitory action on PRL secretion. Blockade of NE and DA synthesis by α-MpT induced a rapid elevation in serum PRL (Donoso *et al.,* 1971; Meites *et al.,* 1972) which could be reversed by administration of L-dopa (Donoso *et al.,* 1971). Unlike L-dopa, administration of DOPS to reinitiate, selectively, NE synthesis did not decrease but rather induced a small rise in plasma PRL levels. In addition, a variety of pharmacological manipulations of the type already described in the preceding sections led to the conclusion that DA is the inhibitory transmitter which suppresses PRL release and that NE probably exerts a stimulatory role (Donoso *et al.,* 1971; Meites and Clemens, 1972; Meites, 1973) (see also below).

Further information on the catecholaminergic mechanism(s) controlling PRL secretion and on the respective role of the NE and DA neurons involved was provided by studies which employed the technique of surgical isolation of MBH of Halàsz and Pupp (1965). It was demonstrated that total or lateral deafferentation of the MBH produced in female rats, respectively, a complete (Weiner *et al.,* 1972) or partial (Cuello *et al.,* 1973) destruction of NE terminals within the deafferented island, leaving DA-containing neurons relatively unaffected. In both completely (Blake *et al.,* 1972) or partially (Weiner, 1973) deafferented rats, serum PRL levels remained unchanged, while plasma LH and FSH secretions were impaired. Similarly, after complete hypothalamic deafferentation no change was present in serum PRL levels in unstressed male rats (Krulich *et al.,* 1975) or in monkeys (Butler *et al.,* 1975). Collectively, these data suggested that NE neurons were not involved in the tonic inhibition of PRL secretion.

The role of DA-containing neurons present in the hypothalamic island was studied by administering α-MpT (Weiner, 1975) or reserpine (Krulich *et al.,* 1975). Elimination of the dopaminergic inhibition within the fragment following both drugs resulted in a sharp increase in serum PRL, whereas systemic L-dopa administration had the opposite effect (Krulich *et al.,* 1975). These data can be interpreted to mean that DA-containing neurons are capable of inhibiting tonic PRL secretion in the absence of extrahypothalamic inputs. However, they are unable to explain whether the activity of the DA neurons surviving inside of the island and innervating PIF-secreting neurons is responsible for the mechanism which maintains the tonic PRL inhibition, or whether DA itself gaining access to the portal circulation and transported to the AP is. The fact that electrolytic lesions in the ME of deafferented rats caused an elevation in serum PRL (Krulich *et al.*, 1975) can just as well be due to destruction of PIF neurons or of TIDA neurons whose terminals abut on the basement membrane of the pericapillary space of the hypophyseal portal capillaries.

An interesting finding of these studies was that both complete as well as posterolateral deafferentation were equally effective in preventing stress-induced PRL release, whereas anterofrontal deafferentation was only slightly effective (Krulich *et al.*, 1975). The extended anterolateral cut, similar to complete isolation of the hypothalamus, was able instead to eliminate the proestrus release of PRL (Blake *et al.*, 1972) and the induction and maintenance of PRL-mediated pseudopregnancy (Carrer and Taleisnik, 1970) (see also below). On the basis of the knowledge acquired on monoaminergic pathways entering MBH through the posterior or lateral hypothalamus (see Chapter II, Sections B,2 and C,2), it may be hypothesized that interruption of NE or, alternatively, 5-HT nerve fibers were responsible for the suppression of the stress-induced PRL rise; suppression of the phasic activation of PRL secretion by the frontal cut instead would be best explained by interruption of NE fibers deriving from cell bodies in the parafornical region and entering the MBH ventrodorsally (Björklund *et al.*, 1970). The existence of this stimulatory input for PRL secretion is also supported by the increased PRL release induced in proestrous rats (Wuttke and Meites, 1972), pseudopregnant rabbits (Tindal and Knaggs, 1972), or ewes (Malven, 1975b) by electrical stimulation of the POA. These MA inputs to the MBH may act via PRF stimulation, inhibition of PIF neurons, or by acting as PRF itself. Figure 48b is a hypothetical representation of the NE pathways involved in the regulation of PRL secretion as envisaged by the deafferentation studies.

Conclusions derived from these studies are that DA, directly or indirectly through PIF stimulation, is responsible for a tonic inhibition of PRL secretion which can occasionally be overcome by acute (stress) or cyclic events.

i. DA Inhibition of prolactin secretion: a CNS vs a pituitary site of action. In addition to the studies reported above, based mainly on surgical or neurophysiological approaches, a good deal of neuropharmacological data are available on the inhibitory role of DA on tonic PRL secretion. Intraventricular administration of relatively high doses of DA (1.25 μg) or of high doses of E and NE (100 μg) produced a striking reduction in the circulating levels of PRL and increased PIF in the portal circulation. DA, NE, and E, perfused into the AP via a hypophyseal portal vein, had no effect on PRL release (Kamberi *et al.*, 1970a, 1971a). A single ip injection of L-dopa elevated hypothalamic PIF activity and also evoked the presence of PIF (*in vitro* assay) in the systemic blood of intact and hypophysectomized rats bearing a pituitary graft underneath the kidney capsule (Lu and Meites, 1971, 1972). In contrast, a single injection of α-MD in intact rats induced a marked rise in serum PRL levels (Lu and Meites, 1971). Drugs such as apomorphine, ergocornine, CB-154, lisuride, and piribedil, all possessing DA-stimulant activity, suppressed baseline or stimulated PRL secretion in rats (Wuttke *et al.*, 1971; see Floss *et al.*, 1973; Lawson and Gala, 1975; Gräf *et al.*, 1976; MacLeod *et al.*, 1976). The same effect was produced in sheep by direct perfusion of DA (Davis *et al.*, 1975). The opposite effect was observed when α- or β-blocking agents, e.g., phenoxybenzamine, phentolamine, or propranolol, were injected systemically (Lawson and Gala, 1975) or intraventricularly (Gala *et al.*, 1972). Since DA exhibits both α- and β-adrenergic activity at the periphery (Ariëns, 1962), the effect of peripherally administered adrenergic blockers could also be amenable to suppression of DA-receptor activity. A more specific DA-receptor blocker, pimozide, when implanted into the ME, led to a rapid rise in PRL, whereas similar implants in the AP produced only a slight response (Ojeda *et al.*, 1974a).

These observations and others (see Meites, 1973), plus the failure to demonstrate unequivocally that free DA from the brain can reach in significant amounts the AP (Ben-Jonathan *et al.*, 1975), have led some authors to postulate that DA acts primarily via PIF which then inhibits PRL release. However, the possibility that the CA's may also act directly on the lactotropes is supported by abundant experimental evidence. Either DA (Birge *et al.*, 1970), DA precursor (MacLeod and Lehemeyer, 1972), or mimetic drugs (Floss *et al.*, 1973; Smalstig *et al.*, 1974) were shown capable of inhibiting PRL release by AP incubated *in vitro*, an effect which could be prevented by specific DA blockers (MacLeod and Lehemeyer, 1974). Suppression of elevated baseline PRL levels was also induced by DA-mimetic drugs administered to hypophysectomized rats bearing an ectopic pituitary (Gräf *et al.*, 1977) or electrolytic lesions of the ME (Shaar and Clemens, 1976) or when administered to stalk-lesioned rhesus monkeys (Diefenbach *et al.*, 1976) or humans (Woolf *et al.*, 1974). Accurate studies on the binding of DA, DA-like agents, and DA-receptor blockers to the AP suggested the presence of a single affinity binding site for DA to the microsomal fraction of the AP and that DA and apomorphine receptor sites are identical (Kimura and MacLeod, 1975a). The amount of DA needed to saturate the specific hypophyseal receptors, as calculated from inhibition studies of cold DA on ^{3}H-DA binding (Kimura and MacLeod, 1975b), appeared to be lower than that present in hypothalamic extracts provided with PIF activity (0.15–1.5 ng) (Takahara *et al.*, 1974a). It would appear that DA-receptors in the anterior pituitary exhibit denervation supersensitivity to dopaminergic stimulation, such as those of the caudate nucleus (see Chapter II, Section D), after stereotaxic lesioning of DA neurons of the MBH (Cheung and Weiner, 1976). DA injected directly into hypophyseal portal vessels inhibited PRL secretion (Takahara *et al.*, 1974a), in contrast to the earlier, negative findings of Kamberi *et al.* (1971a) (see above), following intraportal administration of DA, possibly attributable to the instability of this amine in the saline solution used. Systemic administration of L-dopa constantly decreased basal as well as stimulated secretion of PRL in rats, sheep, dogs, and primates (see Meites, 1973; Frantz, 1973; De Wied and De Jong, 1974). This effect depends, at least partially, on the direct action of L-dopa on the AP, since it was also present in animals with a functional or anatomical disconnection between the CNS and the AP (Welsch *et al.*, 1971; Shaarr and Clemens, 1972). The specificity of the L-dopa action on PRL secretion was demonstrated by the prevention obtained with DA-receptor blockers (see De Wied and De Jong, 1974). In animals bearing a PRL-secreting pituitary tumor, an ergot derivative, ergocornine, reduced both circulating PRL levels and the size of the tumor (Quadri and Meites, 1973).

In animals, that DA inhibition at the pituitary level takes place at the same receptor sites at which TRH stimulates PRL secretion was shown by the ability of TRH in *in vitro* experiments to overcome the inhibitory effect of DA (Hill-Samli and MacLeod, 1974), CB-154, and apomorphine (Hill-Samli and MacLeod, 1975) on PRL secretion and, conversely, of DA to suppress TRH-induced PRL release (Dibbett *et al.*, 1974). Also, in man, TRH-evoked PRL release was inhibited by L-dopa (Noel *et al.*, 1973) or CB-154 administration (del Pozo *et al.*, 1974) or by infusion of dopamine (Besses *et al.*, 1975).

In addition to an acute effect on PRL release, the inhibitory action of dopaminergic stimulation possibly also involves a regulatory action on PRL synthesis, as suggested by the finding that CB-154 administered before estrogens prevented the estrogen-induced rise in serum PRL by inhibiting DNA synthesis (Davies *et al.*, 1974).

In conclusion, this series of studies and the recent demonstration of DA, but not E or NE, in hypophyseal portal blood of female rats (Ben-Jonathan *et al.*, 1977) support the view that DA can act directly on the AP to inhibit tonic PRL release. Undoubtedly, on considering the amount of DA contained in the ME of adult rats, with a half-life in the hypothalamus of approximately 1.5 hr (see Weiner, 1975), the ME possesses the potential capability of producing DA in amounts sufficient to inhibit PRL secretion in the event of a consistent release into the portal blood. Presently, however, the existence of a hypothalamic PIF distinct from DA cannot be negated on the basis of the reported experimental evidence (see Chapter III, Section A,5,a), and is inferentially supported by the observation that incubation of DA with ventral hypothalamic fragments released an inhibiting factor whose action on the pituitary, unlike that of DA, was not blocked by haloperidol (Quijada *et al.,* 1973–1974).

ii. Role of norepinephrine. The role of NE on tonic PRL release, although extensively investigated, remains to be clarified. A biphasic effect on PRL release *in vitro* after NE and E were administered was found by Koch *et al.* (1970); very low doses of the amines (10–20 ng) stimulated PRL release; on the other hand, high doses (100–1000 ng) inhibited PRL release. Direct infusion of graded doses of NE (6–600 ng) in glucose solution into a hypophyseal portal vessel of male rats decreased serum PRL levels (Takahara *et al.,* 1974a), whereas no inhibition had been previously observed by Kamberi *et al.* (1971a) following infusion of higher NE or E doses in a similar test system. Failure of these compounds to exhibit PIF activity was explained by their destruction in salt solution through oxidation before the infusion (Takahara *et al.,* 1974a) (see also below). No significant effects on plasma PRL titers were reported recently by Vijayan and McCann (1976) after IVT injection of either E or NE (4–20 μg) into ovariectomized, EP-primed or ovariectomized rats. These data as a whole, although conflicting, did not support the proposal formulated by Nicoll (1971) that NE was acting as PRF. On the other hand, the inhibition of PRL release observed following high doses of NE may reflect nonspecific stimulation of DA receptor sites. In primates, e.g., chair-adapted ovariectomized monkeys, administration of α-adrenergic (phentolamine and phenoxybenzamine) or β-adrenergic (propranolol) blocking drugs increased serum PRL levels (Quadri *et al.,* 1976), indicating an inhibitory role for the NE system on the resting secretion of the hormone.

A careful evaluation of a series of experiments leads to the assumption that NE also plays a role in the neuroendocrine reflex responsible for stimulated PRL secretion. In ovariectomized, estrogen-treated rats systemic administration of the central NE receptor agonist clonidine increased PRL levels (Lawson and Gala, 1975). It had previously been reported that restoration of central NE activity by DOPS after α-MpT blockade further increases PRL levels (Donoso *et al.,* 1971). These data are of particular interest in view of the findings of Fuxe and Hökfelt (1970) that specific activation of NE receptors by clonidine decreased the activity of the TIDA neurons. Hence, it is conceivable that the increases in PRL secretion noted following clonidine, NE, and DOPS administration resulted from the depression of the dopaminergic system. Other experimental evidence supporting this type of action stems from the fact that

6-OHDA injected into the third ventricle, which affects mainly the NE system (see Chapter I, Section B,2), inhibited release of PRL (Kizer *et al.*, 1975d). Moreover, after pretreatment with 6-OHDA, systemic administration of haloperidol stimulated an increase higher than normal in PRL levels, suggesting an even greater DA blockade by haloperidol when given after destruction of CA neurons. Since 6-OHDA reduced NE content of various hypothalamic nuclei by nearly 90%, and reduction of DA content was not complete, these data appear consistent with the presence of a stimulatory NE pathway regulating PRL release through the dopaminergic system (Kizer *et al.*, 1975d). Such an NE pathway would intervene in the mechanism(s) underlying the stress-induced PRL release by inhibiting the tonic inhibition of PRL release exerted by TIDA neurons. Support for this view has been gained by more recent neuropharmacological studies (Krulich and Marchlewska-Koj, 1976). It has been shown, in fact, that in male rats selective NE depletion by FLA-63 or β-adrenergic blockade by propranolol depressed the stress-induced PRL release, whereas phenoxybenzamine had no such effect. In contrast, DOPS caused an elevation of serum PRL which could be prevented by propranolol. Collectively, it appears that activation of β-receptor sites intervenes in the mechanism by which the central noradrenergic system exerts an inhibitory effect on TIDA neurons.

A variety of stressful stimuli enhance PRL secretion in several vertebrate species (see Chapter III, Section A,5), while in the rat Nembutal anesthesia (Ajika *et al.*, 1972) or complete or posterolateral hypothalamic deafferentation (Krulich *et al.*, 1975) can suppress the stress-induced PRL release. Equally effective in this context are DA precursors or DA-receptor stimulants (see De Wied and De Jong, 1974). Hence, blockade of the stress-induced PRL release is achieved either by severing proximally neural pathways, at least partially NE in nature, or by acting more distally on DA receptor sites located at the pituitary level.

Another neuroendocrine reflex in which intervention of the stimulatory NE pathway can be envisaged is the PRL release occurring in the cyclic rodent at proestrus (Neill, 1970). As in pseudopregnancy, the proestrus rise in PRL levels can be blocked by anterior or complete hypothalamic deafferentation (Blake *et al.*, 1972) or by administration of DA-stimulant drugs (Kovacs and Fluckiger, 1974; Feierabend and Fernström, 1976) and, conversely, can be stimulated by DA blockers (Dickerman *et al.*, 1974). Since the PRL surge at proestrus is an estrogen-dependent event (Blake *et al.*, 1972), and estrogens may stimulate NE turnover (see Section B,3), a complex hormonal–neurotransmitter–neurohormone interaction might take place before estrus. Estrogen would activate the anterolateral NE pathway, with ensuing inhibition of the activity of TIDA neurons; this

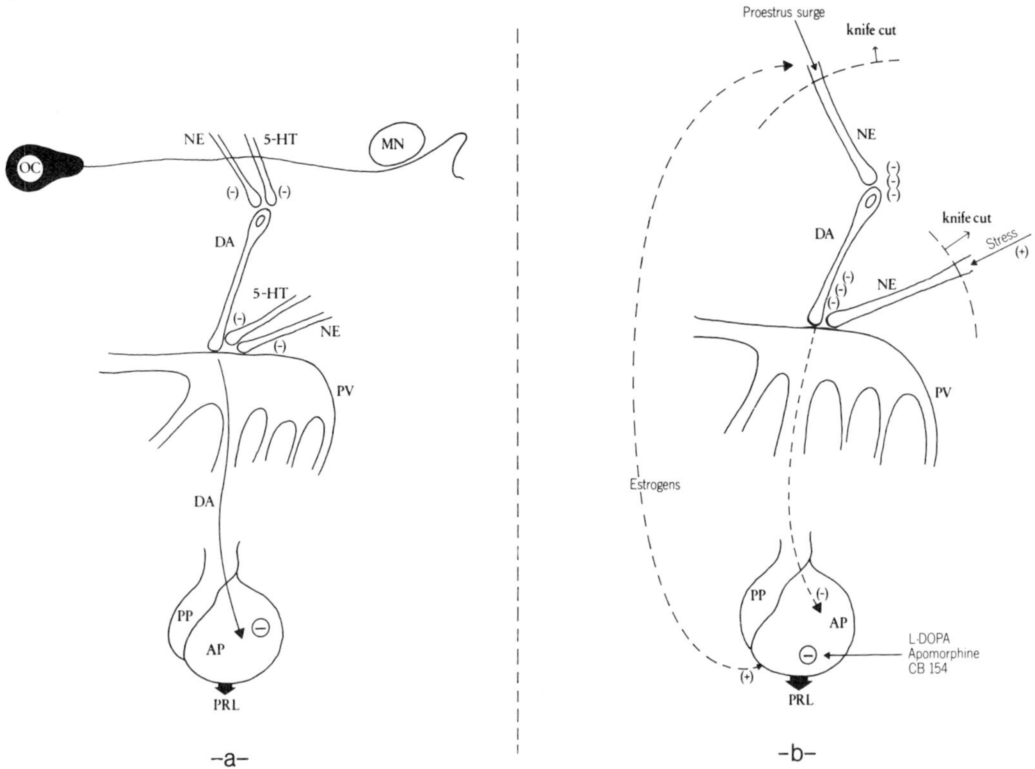

Fig. 48 (a) Schematic representation of central NE and 5-HT pathways exerting an episodic inhibitory influence on DA pathways controlling tonically prolactin secretion by way of a prolactin-inhibiting factor, or directly at the pituitary gland. (b) It is postulated that stress-induced prolactin release or the prolactin surge at proestrus may be mediated by a functional activation of central NE pathways impinging on DA inhibitory pathways. Dopaminergic drugs acting directly at the pituitary gland can block the increased prolactin release. OC, optic chiasm; MN, mammillary nucleus; DA, dopamine neuron; NE, norepinephrine neuron; PV, portal vessels; AP, anterior pituitary; PP, posterior pituitary; PRL, prolactin; (−), inhibition. This same description applies also to Fig. 49.

timely derepression of PRL secretion would be fully evident in the afternoon of proestrus. That a positive input of the adrenergic pathways is present in the afternoon surge of PRL is supported by the finding that the latter was suppressed by α-adrenergic blockers in ovariectomized, estrogen-treated rats (Subramanian and Gala, 1976).

Figure 48 illustrates schematically the postulated stimulatory NE pathways impinging on TIDA neurons, and Fig. 48b the mechanism(s) which would underlie stress-induced PRL release or the PRL release at proestrus.

b. Indoleamines

First experimental evidence for the role of the brain 5-HT system in the control of PRL secretion was provided by studies in which the effect of central administration of the transmitters or systemic administration of precursors was investigated. A threefold increase in plasma PRL levels was reported after IVT injection of 5-HT or melatonin (Kamberi *et al.*, 1971b; Porter *et al.*, 1971–1972; Caligaris and Taleisnik, 1974). The response to melatonin was rapid but short-lived compared to the 5-HT response (Kamberi *et al.*, 1971b); also NAS, an intermediate in the pathway of melatonin formation from 5-HT (Chapter II, Section E), stimulated PRL release on IVT injection (Porter *et al.*, 1971–1972), with a time course similar to that of melatonin. Serotonin metabolites were more active as PRL releasers than melatonin or its metabolites (Kato *et al.*, 1976b), thus arguing against the possibility that the stimulatory action of 5-HT may be totally or in part due to its progressive transformation into the more physiological mediator, melatonin.

Recently, it has also been shown that systemic injection of 5-HT induces a sudden rise in rat plasma PRL levels (Lawson and Gala, 1975), a finding which might be explained by the reported rapid elevation of brain 5-HT levels after its systemic administration (Bulat and Sujek, 1968). However, other authors did not find any effect of peripherally administered 5-HT on plasma PRL levels (Kamberi *et al.*, 1971b; Lu and Meites, 1973), possibly due to the fact that blood samplings in the latter studies were obtained at least 20 min after the injection, whereas in the study of Lawson and Gala the first sampling was at 10 min, and that PRL response is no longer evident after 30 min. It cannot be denied, however, that PRL levels increased as a result of stress effects (Dunn *et al.*, 1972).

As already emphasized, a more correct approach to increase the neuronal function of a given neurotransmitter is by way of systemic administration of physiological precursors. In the rat, a prompt rise in plasma PRL levels was present after ip administration of L-Trp or 5-HTP (Lu and Meites, 1973; Marchlewska-Koj and Krulich, 1975; Mueller *et al.*, 1976b), the immediate 5-HT precursors. Pretreatment of male rats with fluoxetine, a 5-HT reuptake inhibitor (see Table XV), strongly potentiated the PRL-releasing effect of 5-HTP as well as the PRL release induced by either stress or blood withdrawal (Krulich, 1975). Similarly, iv infusion of L-Trp or oral administration of 5-HTP stimulated PRL secretion in man (McIndoe and Turkington, 1973; Kato *et al.*, 1974b), an effect whose specificity rests on the finding that the alleged 5-HT blocker cyproheptadine inhibited 5-HTP-induced PRL secretion (Kato *et al.*, 1974b). 5-HT also stimulated the release of PRL from a clonal strain of rat pituitary cells in culture; even more active than 5-HT was 5-HIAA. In this system Trp, 5-HTP, and 5,6-DHT were inactive (Fang, 1976).

PRL secretion can also be affected by decreasing in several ways central 5-HT activity. Tonic PRL secretion was affected inconsistently by the use of neuropharmacological tools aimed to decrease brain 5-HT levels. Administration of pCPA was found ineffective in changing plasma PRL levels in postpartum female (Kordon *et al.*, 1973-1974), ovariectomized female (Gallo *et al.*, 1975), and castrated male rats (Donoso *et al.*, 1971). However, in normal male rats, acute impairment of central 5-HT activity by pCPA or persistent depletion of brain 5-HT and 5-HIAA levels by 5,7-DHT was accompanied by a marked reduction in baseline plasma PRL levels (Gil-Ad *et al.*, 1976a). Similarly, ovariectomized, estrogen-primed rats responded to pCPA with a decrease in serum PRL (Chen and Meites, 1975). In normal male rats, a TP-deficient diet (maize diet), although resulting within 4 days in a depletion of brain 5-HT content, did not change baseline PRL levels. However, ingestion of the maize diet supplemented with Trp, at a dose roughly corresponding to the daily intake of Trp, resulted in a striking increase of baseline plasma PRL levels (Gil-Ad *et al.*, 1976a). Baseline plasma PRL levels were unaffected in normal male rats following administration of methysergide (Marchlewska-Koj and Krulich, 1975) and, surprisingly, the levels of PRL were even enhanced by methysergide in ovariectomized female rats (Gallo *et al.*, 1975).

In conclusion, the effects induced by a decrease in brain 5-HT activity on unstimulated (baseline) PRL secretion are difficult to evaluate at present. It would appear, however, that 5-HT neurons do not exert appreciable activity on tonic PRL secretion. Their action appears to be more relevant, instead, in conditions of stimulated PRL secretion. In an animal preparation, such as the estrogen-primed, ovariectomized female rat, the rise of plasma PRL levels induced by an acute injection of progesterone (Caligaris *et al.*, 1974) was partially prevented by treatment with pCPA or by injection of methysergide. The blocking effect of pCPA on estrogen–progesterone-induced PRL release was reversed by administration of 5-HTP (Caligaris and Taleisnik, 1974). Since the release of PRL by the injection of progesterone into ovariectomized, estrogen-primed rats depends on an estrogen activated mechanism facilitated by progesterone (Caligaris *et al.,* 1974), the above results indicate that 5-HT-containing neurons mediate the effect of ovarian steroids on prolactin secretion. Apparently the involvement of 5-HT neurons in the mechanism is of an inhibitory nature, as suggested by the finding that both picrotoxin and strychnine, known to block inhibitory synapses in the CNS (see Table XX), prevented the rise in serum PRL present after the ovarian steroid treatment (Caligaris and Taleisnik, 1974). The fact that neither pCPA nor methysergide completely inhibited the increase in serum PRL present after administration of ovarian steroids suggests that in addition to a 5-HT-mediated mechanism, other mechanisms at CNS or AP levels may also be involved. In this connection, the presence of estrogen-sensitive NE pathways impinging on a system tonically inhibiting PRL secretion has to be considered (see Section A,4,a,ii.)

The ability of DA injected into the third ventricle to abolish the serum PRL rise following estradiol benzoate administration confirms the major role exerted by TIDA neurons in maintaining the tonic inhibition of PRL secretion and their resistance to the modulating action of estrogen (Caligaris and Taleisnik, 1974).

Stress-induced PRL release appears to be, at least in part, dependent on central 5-HT activity. Pretreatment of normal male rats with methysergide reportedly blocked the PRL-releasing effect of etherization, bleeding (Marchlewska-Koj and Krulich, 1975), and restraint stress (Meltzer *et al.,* 1976), whereas methysergide did not inhibit the increase of serum PRL induced by either pimozide or α-MpT (Marchlewska-Koj and Krulich, 1975). These results would indicate that a 5-HT regulatory system activated by stress either stimulates a hypothalamic PRF or inhibits a PIF that could as well be the TIDA system.

Changes in the central 5-HT activity may also affect the suckling-induced increase in blood PRL levels. In lactating rats, pCPA completely abolished the suckling-induced increase of plasma PRL, an effect negated by 5-HTP (Kordon *et al.,* 1973-1974); equally effective was the blockade of 5-HT receptors by methysergide (Gallo *et al.,* 1975). L-Dopa (Chen and Meites, 1975) and DA-like drugs (Flückiger and Kovacs, 1974; Smalstig *et al.,* 1974) could also depress the suckling-induced rise, while, conversely, inhibition of CA synthesis by α-MpT had a facilitatory effect (Voogt and Carr, 1975). It appears unlikely that brain CA's are involved directly in the central neuroendocrine reflex responsible for the suckling-induced PRL release; rather the action of L-dopa and DA-like compounds is best explained as occurring directly on hypophyseal receptor sites (see Section A,4,a,i). It is likely that the tonic inhibitory influence of TIDA neurons on PRL release in the absence of suckling can be prevented by other neuronal mechanisms activated by suckling which facilitate PRL release. Based on the evidence presented, it is suggested that this facilitator system is serotoninergic in nature (Kordon *et al.*, 1973-1974).

Two circadian surges of plasma PRL, one diurnal and one nocturnal, are present in the male rat (Dunn *et al.*, 1972) or pseudopregnant female (McLean and Nikitovich-Winer, 1975). In male rats, pretreatment with pCPA in addition to reducing brain 5-HT content 24 hr later, also abolished the diurnal rhythm of plasma PRL at low basal levels. Similarly, destruction of midbrain raphe nuclei by electrolytic lesions caused, along with 5-HT depletion in the forebrain, a significant reduction in plasma PRL measured at times corresponding to the normal two peaks of PRL release (Mulloy and Möberg, 1975). In the human, pretreatment with methysergide dramatically reduced sleep-related PRL secretion (Mendelson *et al.*, 1975). Interestingly, absence of nocturnal elevation of plasma PRL concentrations has been observed in Cushing's disease (Krieger *et al.*, 1976); this might indicate the existence of abnormal, brain 5-HT metabolism in this condition (see Section B,1). Collectively, these observations suggest that both in rats and humans a central 5-HT neuronal system is involved in the regulation of the circadian or sleep-related PRL rises.

In conclusion, it appears that changes in central serotoninergic activity can be responsible for the modulation of the activated PRL secretion. At least for suckling, stress, and circadian variations, the induced release of PRL can be linked to serotoninergic modulation of inhibitory DA inputs (Fig. 49) or directly to PIF or PRF neurons. For the stress-induced PRL rise the intervention of the central NE system must also be considered (see Section A,4,a,ii). The equivocal results obtained by altering 5-HT neuronal function in the unstimulated condition might perhaps be attributable to the different levels of gonadal steroids present in the different experimental situations. As suggested by Taleisnik and Celis (1973) for the secretion of MSH, a hormone that shares with PRL the characteristic of being tonically inhibited by the CNS (Chapter III, Section A,7 and this chapter, Section A,6), it is possible to suggest for PRL secretion the presence of a dual regulatory mechanism: (1) a short-acting mechanism which depends on the abrupt inhibition of TIDA neurons by NE and/or 5-HT inputs, and may account for the rapid adjustment of PRL secretion to neurogenic stimuli and (2) a long-lasting, estrogen-dependent regulatory mechanism responsible for determining, through a hypothalamic or pituitary action, the baseline activity of the system.

c. *Acetylcholine and Putative Neurotransmitters*

Contradictory findings have been reported on the role of ACh on PRL secretion. Central or systemic administration of cholinergic drugs was reported to reduce serum PRL levels in female rats (Grandison *et al.*,

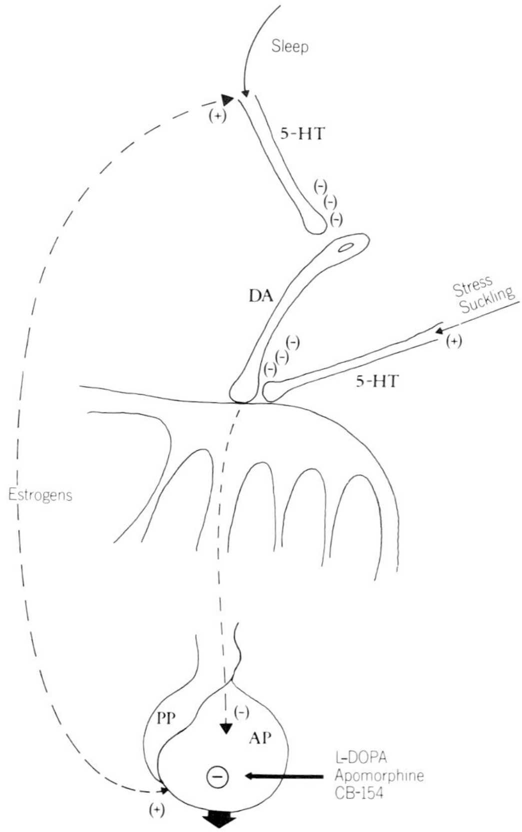

Fig. 49 Schematic representation of central 5-HT pathways exerting a modulatory action on inhibitory DA pathways or (not represented) on PIF or PRF secreting neurons. It is postulated that stress-, sleep-, and suckling-induced PRL rises might be amenable to functional activation of the 5-HT pathways.

1974). The same drugs were also capable of reducing in normal female or estrogen-treated male rats the PRL rises present at proestrus or following 5-HTP administration (Grandison and Meites, 1975), or the daily nocturnal surge of PRL and those ensuing ether or restraint stress (Grandison and Meites, 1976). Moreover, cholinergic agonists, either muscarinic or nicotinic, proved to be effective in inhibiting the afternoon surge of PRL in ovariectomized, estrogen-treated rats (Subramanian and Gala, 1976). On the other hand, evidence has been presented in favor of a stimulatory role of the cholinergic system on PRL release. Libertun and McCann (1973) observed that subcutaneous or IVT injections of the cholinergic blocker, atropine, into male and female rats inhibited the release of PRL, and McLean and Nikitovich-Winer (1975) reported that very high doses of atropine inhibited the nocturnal increase in PRL secretion, an effect which was reversed by pretreatment with eserine. Atropine, again at large doses, also partially inhibited the afternoon PRL surge occurring in ovariectomized, estrogen-treated rats (Subramanian and Gala, 1976). Other studies were unable to show any effect on basal PRL levels either with agonists or antagonists of the cholinergic receptors (Lawson and Gala, 1975).

Histamine reportedly induced PRL release in both ovariectomized, estrogen-primed (Libertun and McCann, 1976) or untreated male (Donoso *et al.*, 1976) rats following IVT administration of huge doses. Pretreatment with antihistaminic drugs prevented the PRL rise due to restraint stress (Libertun and McCann, 1976) or IVT-injected histamine (Donoso and Banzan, 1976). Anticholinergic, antiserotoninergic, or antiadrenergic drugs did not (Donoso and Banzan, 1976).

Inferential evidence for the participation of GABAergic neurons in the central mechanism(s) controlling PRL secretion was derived from the observation that the release of PRL induced by estrogen and/or progesterone in spayed female rats was inhibited by picrotoxin or strychnine. In addition, 5-HT induced PRL release was blocked by picrotoxin, whereas the stimulation induced by TRH or α-MpT was unaffected (Caligaris and Taleisnik, 1974). These data suggest that inhibition of PRL secretion by the GABA antagonist took place at the CNS rather than at the pituitary level. A CNS site of action would be implied more directly by the observation that huge IVT administered doses of GABA induced PRL release in proestrous or ovariectomized female rats (Mioduszewski *et al.*, 1976) or in intact male rats (Ondo and Pass, 1976), but were ineffective when injected systemically into hypophysectomized rats bearing an AP transplant, or added to AP *in vitro* (Mioduszewski *et al.*, 1976). In contrast to these findings which indicate a PRL-releasing activity of the amino acid, Schally *et al.* (1977) reported (see also Chapter III, Section A,5,a) that both natural GABA and synthetic GABA inhibit PRL secretion in many *in vitro* and *in vivo* systems by direct action on the pituitary gland. However, the doses at which the amino acid exerted its PIF activity were so huge that no physiological significance can be attached to the suppressive effect of GABA on PRL secretion.

5. Brain Monoamines in the Control of TSH Secretion

Early studies on the involvement of MA's in the function of the hypothalamohypophyseal–thyroidal axis (HHTA) dealt mainly with effects exerted directly at the level of the thyroid gland (Melander, 1969,

1970; Ericson *et al.,* 1974; Wilber *et al.,* 1969). They will be reviewed briefly in the last part of this section. More recently, most of the studies in this area have focused instead on the functional interaction of brain MA's with TRH-producing elements of the hypothalamus (Grimm and Reichlin, 1973; Kotani *et al.,* 1973; Tuomisto *et al.,* 1973; Annunziato *et al.,* 1974), although the possibility that MA's influence HHTA as neurohormones delivered to the AP with RF's or reach the AP by the general circulation to alter its sensitivity to RF's has not been discarded (K. O. Nilsson *et al.,* 1974, 1975) (see also Section C).

In vivo and *in vitro* experiments using microiontophoretic, pharmacological, and surgical approaches support the notion of central MA participation in the regulation of TRH secretion. The critical observation that isolated fragments of rat hypothalamus (Mitnick and Reichlin, 1971; Reichlin *et al.,* 1972) or guinea pig hypothalamic fragments in tissue culture (McKelvy and Grimm-Jørgensen, 1975) incorporate labeled precursor amino acids into a peptide with the chromatographic properties of TRH (but see also Chapter III, Section A,6) offered an experimental tool by which the regulation of the secretion of TRH could be directly studied *in vitro* by measuring the rate of discharge of pulse-labeled neurohormone from isolated hypothalamic fragments. Using the pulse-labeled mouse hypothalamus as a model, the neuropharmacology of TRH regulation was studied. It was found that both DA and NE brought about the release of TRH (Grimm and Reichlin, 1973), while, conversely, addition of a combination of iproniazid and α-MpT to the fragments depressed spontaneous TRH release (Grimm and Reichlin, 1972). The observation that disulfiram blocked the TRH-releasing effect of DA led the authors to conclude that NE was the active principle stimulating TRH release and that DA was effective after conversion to NE (Grimm and Reichlin, 1973). Impairment of TRH-synthesizing ability was also found in hypothalami from rats pretreated with reserpine (Reichlin *et al.,* 1972; Reichlin and Mitnick, 1973b), a fact which had to be attributed to a decrease in central NE activity only, since 5-HT added to the *in vitro* system did not stimulate, but actually depressed TRH release to the medium (Grimm and Reichlin, 1973). Acetylcholine in the presence of eserine and the ACh analog carbachol failed to alter the release of TRH to the medium as compared to controls.

The studies of Reichlin and co-workers offered several conclusions about the mechanism and physiology of the biosynthesis and release of newly formed TRH; however, the methodology used in these studies did not allow for conclusions on stored TRH. A more sophisticated approach involving the isolation of hypothalamic synaptosomal fractions containing TRH-peptidergic endings allowed the study of MA's influence on total release of RIA-TRH in the medium (Bennett and Edwardson, 1975). The incubation of ME synaptosomes from sheep in the presence of low concentrations of DA stimulated the release of TRH, while on synaptosomes prepared from whole hypothalami DA was ineffective (Bennett *et al.*, 1975). This fact can possibly be attributed to the higher concentration of D-β-H in the ME rich in NE nerve endings when compared to the whole hypothalamus. In the same system, 5-HT inhibited TRH release. Collectively, the findings obtained by *in vitro* experiments would indicate that the regulation of TRH release is under a dual monoaminergic control system, with a positive noradrenergic and a negative serotoninergic component. The stimulatory effect of NE does not seem to depend on its interaction with α- and β-receptors, since neither phenoxybenzamine nor propranolol abolished the effect of NE stimulation (Grimm and Reichlin, 1972).

Experiments performed *in vivo* with the use of the amine precursor, L-dopa, showed it incapable of altering TSH secretion in normal subjects (Burrow *et al.*, 1974), but effective in depressing plasma TSH concentrations in subjects with primary hypothyroidism (Minozzi *et al.*, 1975; Rapaport *et al.*, 1973; Tsukui *et al.*, 1974). A careful analysis of L-dopa effect on plasma TSH levels in the hypothyroid subjects studied by Minozzi *et al.* revealed, however, an early TSH peak preceding the usual inhibitory phase. In the light of the accumulated evidence on the role of the specific MA's on TRH release, the diphasic effect might be due to time-differentiated activity of L-dopa on the brain MA system: first, activation of the central NE followed by activity on the DA or 5-HT systems. Similar to its effect in humans with elevated baseline TSH levels, L-dopa depressed, in several animal species, cold-induced TRH-TSH secretion (Annunziato *et al.*, 1975a; Onaya *et al.*, 1975; Onaya and Hashizume, 1976; Tuomisto *et al.*, 1975).

The use of drugs capable of impairing the central monoaminergic tone by interfering with MA synthesis, storage, or action at specific sites provided more useful information. Systemic administration of α-MpT did not change baseline plasma TSH levels in 1–9 hr in the rat (Annunziato *et al.*, 1975a), thus suggesting the inability of the central adrenergic tone to influence tonic TRH release.

NE, on the other hand, appeared to play a facilitatory or stimulatory role in the activation of the HHTA elicited in rodents by sudden and severe cold exposure, as implied by the finding that α-MpT applied 30–120 min before cold exposure was able to prevent cold-induced TSH release (Annunziato *et al.*, 1974; Onaya *et al.*, 1975; Onaya and Ashizume, 1976). However, if too long an interval elapsed from the α-MpT administration, none or opposite effects were observed (Tuomisto *et al.*, 1973, 1975). Similar to α-MpT, more selective reduction of central NE tone by disulfiram (Tuomisto *et al.*, 1975) or DDC (Tuomisto *et al.*, 1975; Onaya *et al.*,

1975; Onaya and Ashizume, 1976) blocked cold-induced activation of the thyroid axis. Also, blockade of NE receptors by systemic administration of phentolamine (Kotani *et al.,* 1973; Tuomisto *et al.,* 1975) or CPZ (Tuomisto *et al.,* 1975) depressed the cold-induced activation of the HHTA. Central sympathectomy by 6-OHDA inhibited cold-induced TSH release at 1 but not at 3, 15, or 30 days after treatment; administration of α-MpT at a dose (75 mg/kg) unable to alter brain CA's in untreated rats, prevented in rats pretreated 30 days earlier with 6-OHDA the cold-induced TSH secretion (Scapagnini *et al.,* 1976b), thus stressing the necessity of a small functional pool of NE for the restoration of stimulated TRF-TSH secretion in 6-OHDA treated rats (see also Section A,1,a,i).

Experiments to counteract the impairment of central CA tone by α-MpT with the administration of amine precursors gave consistent (Onaya *et al.,* 1975; Onaya and Hashizume, 1976) or negative results (Annunziato *et al.,* 1975a). Finally, in completely deafferented rats in which hypothalamic NE is undetectable (Weiner *et al.,* 1972), cold stress was unable to increase plasma TSH levels, which are unmodified in the resting state (Hefco *et al.,* 1975; Annunziato *et al.,* 1975b). Overall, the above *in vivo* results are consistent with the hypothesis, formulated on the basis of the *in vitro* experiments, that NE plays an excitatory or at least facilitatory role in TSH secretion.

The role of DA is more questionable. Administration of agents capable of elevating the central dopaminergic tone, such as L-dopa (Annunziato *et al.,* 1975a, 1975b; Onaya *et al.,* 1975), L-dopa combined with DDC (Onaya and Hashizume, 1976; Tuomisto *et al.,* 1975), apomorphine (Tuomisto *et al.,* 1975), and CB-154 (Miyai *et al.,* 1974), was associated with inhibition of TSH secretion. However, failure of the specific DA receptor blocker, pimozide, to abolish the inhibitory effect of L-dopa or apomorphine (Tuomisto *et al.,* 1975) argues against the hypothesis of a dopaminergic inhibitory action. Also difficult to interpret is the reported inhibition of the TRH-TSH system by reserpine (Tuomisto *et al.,* 1973; Kotani *et al.,* 1973), which affects both CA and indoleamine stores.

The inhibitory role of 5-HT appears fully established on the basis of results of *in vivo* and *in vitro* animal and human experiments. 5-HTP was capable of decreasing plasma TSH levels in hypothyroid but not in euthyroid individuals (Yoshimura *et al.,* 1973), and it markedly inhibited TSH secretion in response to cold in rats (Onaya *et al.,* 1975). A dramatic reduction of serum TSH levels was also found in rats by systemic administration of Trp (Mueller *et al.,* 1976b). Evidence linking TRH-TSH secretion to blockade of 5-HT biosynthesis was not as clear-cut. The use of too small doses (Tuomisto *et al.,* 1975) or prolonged pCPA treatment (Shenkman *et al.,* 1973) combined with indirect evaluation of HHTA function hampered reaching definitive conclusions.

In contrast to the above findings are the more recent observations of Mannisto and Tuomisto (1976). They reported in an abstract that pCPA was capable of preventing in a dose-dependent manner both the cold-induced TSH response of rats adapted to 30°C and the increased TSH levels of propylthiouracil-treated rats; administration of 5-HTP reversed the action of a single pCPA administration in the hypothyroid rats. In accordance with these findings, which suggest that 5-HT is a stimulatory transmitter in the central regulation of TRH-TSH secretion in the rat, are the observations of an enhanced TSH release in rats given an intraventricular infusion of 5-HT; this effect was blocked by systemic administration of Cy (Jordan *et al.*, 1976).

As a corollary to the existence of functional interactions between brain MA's and HHTA, changes in MA content and/or metabolism should be anticipated by endocrinologic manipulations which alter thyroid function. Morphological and biochemical findings support this theory. In rats, an increase in specific dopaminergic fluorescence intensity in the ARC and periventricular hypothalamic nuclei was induced by cold exposure, an effect which was almost completely prevented by T_4 administered 3 hr previously (Lichtensteiger, 1969). Administration of thiouracil or T_4 without any cold stimulus activation did not change DA fluorescence in the ARC nucleus (Fuxe and Hökfelt, 1969). However, biochemical determination of DA in the anterior or medial hypothalamus revealed a significant drop after T_4 treatment (Brown *et al.*, 1972) and, in experimental hyperthyroidism, hypothalamic NE content was decreased. This was associated with decreased VMA excretion (Fernandez *et al.*, 1973), although one result to the contrary has been reported (Engström *et al.*, 1975). All the above results, although implying an interrelationship between thyroid function and brain MA's, must be evaluated remembering that changes in CA content or histochemical modifications do not necessarily reflect biochemical changes.

Direct measurement of MA turnover and of the activity of MA synthesizing enzymes offered new insights in this subject. Earlier studies by Parker (1972) had shown that alteration of thyroid function was associated with increased brainstem NE turnover. TH activity, however, was significantly increased only in hypothyroid adult rats (Emlen *et al.*, 1972) and CA synthesis was enhanced in normal rats after multiple T_4 injections (Engström *et al.*, 1975). The regulatory influence of thyroid hormones on NE and TH was also investigated in developing rat brains. Experimental hypothyroidism induced by ^{131}I administered at birth interfered with the developmental increase of TH and brain NE levels, whereas treatment of neonatally thyroidectomized rats with T_3, if initiated early enough, restored to normal the observed neurochemical changes in NE metabolism (Rastogi and Singhal, 1974a).

So far the best experimental evidence for the influence of HHTA axis on CA metabolism was produced using the recently developed microdissection technique (Palkovits, 1973) combined with a sensitive enzymatic assay (Coyle and Henry, 1973) to determine TH and D-β-H activity in individual hypothalamic nuclei (Kizer *et al.*, 1974) (see Chapter II, Table II). TH activity was found to be increased in the ME and in the dorsomedial-periventricular and ARC nuclei of hypothyroid rats (Kizer *et al.*, 1974). TH activity of the ME, but not that of other brain areas, decreased below control levels after administration of T_4 to thyroidectomized rats (Kizer *et al.*, 1974). With regard to the nature of the catecholaminergic neurons involved, it would appear that the majority of the endocrine-responsive neurons in the ME are dopaminergic on the basis of the findings that a bilateral ventral bundle lesion destroying most of the NE fibers to the ME had no effect on the increase of TH activity in the ME following thyroidectomy (Kizer *et al.*, 1976b). Therefore, the influence of thyroid ablation and T_4 replacement therapy on the TH activity in the ME and ARC nuclei seems to suggest a stimulatory participation of ARC-ME dopaminergic neurons in TRH secretion, a view which does not fully agree with the results of the aforementioned neuropharmacological studies. Absence of changes in TH activity in the PO nucleus argues against an anterior hypothalamic regulatory site for TRH secretion involving catecholaminergic terminals, although this site of interaction is not excluded for T_4 feedback involving other neurotransmitters (Kizer *et al.*, 1974). 5-HT, for instance, might be implicated since experimental hypothyroidism decreased hypothalamic Trp-H activity and 5-HT content (Rastogi and Singhal, 1974b), and T_3, specifically, reversed this effect (Rastogi and Singhal, 1974b). Also, the sensitivity of central CA receptors can be stimulated by thyroid hormones (Engström *et al.*, 1975); in this instance, a time lag was found between the appearance of increased monoamine synthesis and the enhanced receptor sensitivity, implying that the presynaptic (increased synthesis rate) preceded the postsynaptic event (increased receptor sensitivity).

In summary, most of the existing data discussed permit one to construct a hypothetical model of brain MA–TRH–TSH interaction. Central NE tone would play a stimulatory or at least facilitatory role in the phasic neurohormone secretion; conversely, DA would exert an inhibitory effect. No definitive conclusion can be drawn for 5-HT, although most of the evidence presented favors an inhibitory role. Finally, MA's apparently do not play a major role in the tonic regulation of TSH.

In addition to an established MA–TRH–TSH interaction at the CNS level, some data might be interpreted on the basis of direct MA action at the AP level. L-Dopa has been found capable of blocking TRH-induced TSH release (Refetoff *et al.*, 1974; Tsukui *et al.*, 1974), an effect also induced by systemic DA infusion (Besses *et al.,* 1975) but not by apomorphine (Nilsson *et al.,* 1975). Similar to apomorphine, CB 154 did not affect the TRH-TSH response in normal or hyperprolactinemic women (Hirvonen *et al.,* 1976). The possibility that the L-dopa effect might result from NE-stimulation at the pituitary level, although in keeping with the reported suppression by phentolamine of TRH-induced TSH secretion (K. O. Nilsson *et al.,* 1974), is in conflict with the lack of D-β-H activity at the pituitary level (see Chapter II, Section B,5) and the reported inability of E to affect TSH secretion when infused directly into the AP (von Euler and Holmgren, 1956). With regard to 5-HT, pCPA has been shown capable of impairing TRH stimulation of TSH secretion (Coleoni and Masini, 1974).

Finally, MA's appear capable of influencing thyroid activity by acting directly on the gland. CA's added *in vitro* or given systemically exhibited stimulatory effects on the secretion of the thyroid gland (Wilber *et al.,* 1969; Melander and Sandler, 1972). *In vivo,* systemic administration of E, ISO, or DA also induced release of thyroid hormone in hypophysectomized or T_4-blocked rats (Melander, 1969, 1970), an effect which could be prevented by adrenergic blockers (Melander, 1970).

These findings suggest that biogenic amines directly interact at the level of the thyroid gland with adrenergic receptors; however, their mechanisms(s) of action has yet to be clarified. With regard to other brain neurotransmitters evidence accumulated so far is meager. Marked inhibition of the acute thyroidal endocytotic response to cold has been reported following administration of oxotremorine, a cholinomimetic agent (see Table XVII), or brocresine phosphate, a histamine depletor (Onaya and Hashizume, 1976).

6. Brain Monoamines in the Control of MSH Secretion

The concept that CNS exerts a tonic inhibitory control on the release of MSH has been well documented in mammals and discussed at length previously (see Chapter III, Section A,7). However, the mechanism by which CNS regulates MSH secretion has been a subject of significant controversy, and only in the past few years has some light been shed on this subject. It is well known that in amphibians the hypothalamus exerts an inhibitory control on the secretion of MSH (Etkin, 1967; Jørgensen and Larsen, 1967). Transection of the lower infundibular stem near the ME brings about intense darkening of anurans. In these species, this operation disturbs mainly the nervous pathway to the pituitary, rich in Gomori-positive and CA-containing axons (Enemar *et al.,* 1967). Recognition that inhibitory control on MSH release is not exerted through Gomori-positive axons was provided by the finding that extirpation of the PON from which they are derived did not affect normal color adaptation in frogs (Dierickx, 1965, 1967). This suggested that the inhibitory control might be exerted through the CA-containing nerves present in the PI (Iturriza, 1966; Enemar *et al.*, 1967). In a series of elegant studies, Iturriza (1966, 1969) showed that monoamine-depleting drugs, such as reserpine and α-MpT, administered to normal "bleached" toads appreciably increased the melanophore index (MI), a measure of the grade of melanophore dispersion. Conversely, administration of monoamine precursors, such as L-dopa, or MAOI produced a marked decrease in the MI of normal "darkened" toads or of toads darkened by adrenalectomy, hypothalamic transection, or both. Direct infusion of DA or NE into the anterior chamber of the toad's eye, containing an autotransplanted PI, induced a fall in the MI. CA solvents and hypothalamic extracts heated to destroy CA's, but possibly still containing possible RF's, injected into the eye or iv did not produce any change in MI. These and related observations (Ito, 1974) indicate that in amphibians CA's are mainly responsible for the actual hypothalamic inhibition of MSH release and disprove Taleisnik and Tomatis' theory (1967a,b) of a dual neurohormonal control exerted by an MIF stored in the hypothalamus and an MRF in the neurohypophysis. The chemical nature of this inhibitory communication system appears to be specifically dopaminergic. Recently, Smith (1975) showed that apomorphine and CB 154 implanted directly into the frog pars intermedia caused inhibition of MSH release, and similar findings have been obtained by Morgan and Hadley (1976) on the addition of another ergot drug, ergonovine, to incubated frog pituitary glands.

Microscopic observations of the PI of amphibians (Iturriza, 1964; Dent and Gupta, 1967) are consistent with functional studies, and nerve fibers containing a primary CA, namely DA, have been histochemically visualized among the cells of the PI (see Chapter II, Section B,5). Contacts between aminergic fibers and endocrine elements are of the type categorized under "neurosecretomotor junction" (Bern, 1966) (see Chapter I, Section A).

Also in mammals, the catecholaminergic system intervenes in an inhibitory fashion in MSH secretion, but the evidence available suggests, in addition to a direct action of CA's on the PI (Bower *et al.*, 1974; Morgan and Hadley, 1976), the presence of a functional link between CA's and the neurosecretory neurons that release MIF (Taleisnik *et al.*, 1972). Administration in male rats of reserpine or Dibenamine was followed by a decrease in pituitary MSH content, while no change was produced by propranolol. The action of the two drugs appeared to be exerted on the CNS and not on the pituitary, since neither drug was able to prevent the drop in pituitary MSH content that follows iv injection of SME extracts with

MSH-releasing activity. These data suggested that catecholaminergic fibers were implicated in the inhibitory influence of the CNS on MSH secretion.

An attempt was made to localize in the rat the sites at which CA's interfere in the control of MSH secretion, by implanting NE and DA into the third ventricle and studying their effect in preventing the decrease of pituitary MSH content which follows the administration of hypertonic saline, a stimulus apparently requiring the intactness of NE terminals in the SON (Fuxe and Hökfelt, 1967). Implants into the third ventricle were effective in preventing the MSH release induced by hypertonic saline, while implants either into the lateral ventricle or directly into the AP failed to prevent MSH depletion. These findings plus the observation that the implantation of NE into the third ventricle also resulted in a depletion of hypothalamic content of MIF, probably reflecting increased release from storage cells, were interpreted to mean that neurons related to the production and release of MIF are innervated by catecholaminergic fibers, which, under resting conditions, exert a tonic activating action (Taleisnik *et al.*, 1972). To date no data are available on the role of DA or NE in the inhibitory system or on the use of more specific CA agonist or antagonist drugs.

The above results must be evaluated remembering that the end point on which most of them rely is the "depletion" of pituitary BA-MSH (see Chapter III, Section A,3,a), so that corroboration by concomitant measurements of MSH in the plasma is desirable.

In an extension of these studies, the same authors provided evidence for the involvement of 5-HT pathways in the MSH-secretion mechanism (Taleisnik *et al.*, 1973-1974). In male rats, administration of pCPA or IVT injection of methysergide prevented the drop in pituitary MSH content induced either by hypertonic saline or vaginal stimulation, implying that 5-HT neurons were mediating the effect of these stimuli. Consistent with this view was the depletion of pituitary MSH content after administration of 5-HT into the third ventricle. The fact that picrotoxin blocked MSH depletion induced by hypertonic saline, vaginal stimulation, or 5-HT injection led to the assumption that an inhibitory mechanism was involved in the release of MSH induced by the aforementioned stimuli. Accordingly, administration of the inhibitory transmitter GABA mimicked the effect of different stimuli on the release of MSH and was similarly sensitive to the administration of picrotoxin but not of methysergide. This GABA–picrotoxin interaction would suggest that GABA may act as a transmitter agent of the inhibitory synapse inhibiting the system that tonically suppresses MSH secretion in the intact animal. However, the fact that another inhibitory neurotransmitter, e.g., glycine (see Chapter II, Section J), which induces only postsynaptic inhibition (Curtis and Johnston, 1974a), similarly affected pituitary MSH content may also indicate for GABA an action on the postsynaptic level.

The complex interactions resulting from the described pharmacological manipulations and the previously derived theories on CA involvement have been depicted in an intricate neuronal scheme in Fig. 50, which to a certain extent must be considered speculative. The MIF neurosecretory system would be under the dual influence of an activating system mediated by catecholaminergic neurons which exert a tonic action, thus allowing a permanent release of MIF, and of an inhibitory system responsible for curtailing the rate of afferent impulses in the catecholaminergic neurons by presynaptic inhibition. The latter would be mediated by 5-HT pathways and would have GABA as a transmitter agent in the inhibitory synapse. The proposed monoaminergic mechanism(s) for the control of MSH release is certainly intriguing, but before being definitively accepted requires (1) corroboration with studies using more agonist or antagonist drugs which affect selectively monoamines; (2) evaluation, as an end point, in addition to the BA-MSH levels in the pituitary, of hormone levels in the plasma. In this context, in the human administration of CPZ failed to alter IR-β-MSH

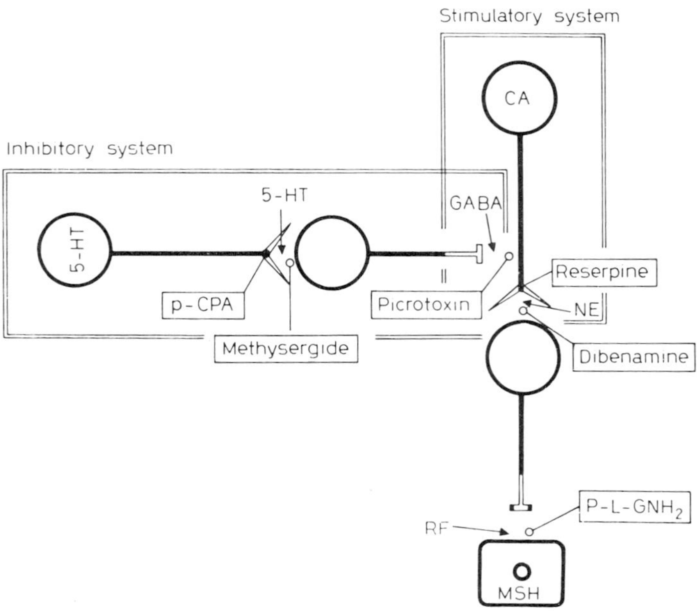

Fig. 50 Postulated neuronal circuit involved in the control of MSH secretion (see text for explanation). CA, catecholamine; NE, norepinephrine; 5-HT, serotonin; p-CPA, *p*-chlorophenylalanine; GABA, γ-aminobutyric acid; P-L-GNH$_2$, prolyl-leucyl-glycinamide; RF, MSH-releasing factor. (Reproduced, with permission, from Taleisnik *et al.*, 1973–1974.)

TABLE XXXVI

Neurotransmitters and Anterior Pituitary Hormone Secretion[a]

Hypothalamic–pituitary axis	Subject[b]	E	NE	DA	5-HT	ACh	PG's	H	GABA
CRF/ACTH	Exp. An.		↓	↓	↑	↑	↑	—	↓
	H		→	→	↑	—	—	—	—
TRH/TSH	Exp. An.		↑[c]	↓[c]	↓[c]↑?	↓?	↑	—	—
	H		—	↓[d]	↓[d]	—	→	—	—
LH-RH/FSH-LH	Exp. An.	↑[e]	↑[e]	↑↓?	↓	↑	↑	↑	↑
	H		→	↑↓?	—	—	—	—	—
GH-RF–GH-IF/GH	Exp. An.	↓	↑	↓	↑↓?	—	↑	↑?	—
	H	↑[f]	↑	↑[g]	↑↓?	→	↑	↑?	↓?
PIF–PRF/PRL	Exp. An.	↑	↑	↓	↑	↑↓?	↑	↑	↑↓?
	H		→	↓	↑	—	—	—	—
MIF–MRF/MSH	Exp. An.		↓	↓	↑	—	—	—	↑
	H		—	—	—	—	—	—	—

[a] Key to symbols: ↑, stimulation; ↓, inhibition; →, no effect; —, action not ascertained; ?, action still questionable.

[b] Exp. An. experimental animal; H, human.

[c] Cold-induced TSH secretion.

[d] Hypothyroid subjects.

[e] Induces ovulation.

[f] In combination with propranolol.

[g] Inhibitory in acromegaly.

levels (Plummer *et al.*, 1975), which might be attributed, however, to the fact that β-MSH does not naturally exist in man (Scott and Lowry, 1974).

In mammals, the existence of CNS influences mainly inhibitory to MSH secretion and the antagonistic actions of the adrenergic and the serotoninergic systems calls for strict analogy with the control of PRL (see Section A,4), a hormone which responds to many of the stimuli active for MSH release (suckling, cervical stimulation, hypertonic saline).

Table XXXVI lists the stimulatory or inhibitory action on AP secretion of the real or putative neurotransmitters, as derived from the experimental evidence reviewed in Section A.

B. CHANGES IN ENDOCRINE FUNCTION AND BRAIN MONOAMINE METABOLISM

1. Pituitary–Adrenal Axis and Brain Monoamine Metabolism

In a previous section (see Section A,1), evidence has been offered for the involvement of central monoamine neurotransmitters in the tonic release of ACTH and/or circadian periodicity of adrenal function. Here consideration will be given to studies done on the opposite aspect of the problem, e.g., the neurochemical transmitters involved in the negative feedback effect of the adrenal corticoids on neural tissue.

The numerous studies investigating the effect on brain MA metabolism of the various components of the HHAA have involved (1) the suppression of peripheral levels of corticoids by adrenalectomy or adrenal steroid synthesis inhibitors followed by replacement therapy; (2) the suppression of ACTH secretion by hypophysectomy and subsequent replacement with corticosteroids and/or ACTH; (3) the administration of exogenous corticosteroids or ACTH in normal animals.

Fluorescence studies of central DA and NE neurons have been extensively performed to evaluate corticoid–brain MA interactions. In the rat, after adrenalectomy, a marked increase in the fluorescence intensity of TIDA neurons of the ME with a narrowing of the fluorescent strip at the ME surface was reported by Akmayev and Donàth (1966). No changes were found in the total ME fluorescence after hydrocortisone administration. Subsequent investigations by the same authors (Akmayev *et al.*, 1967) demonstrated 10–14 days after adrenalectomy an increase in the number and size of dense vesicles in ME nerve endings. These changes were not observed after thyroidectomy or castration. However, far from relating these findings to an increased CA turnover, they attributed the increased fluorescence to an accumulation of CA's. These findings, however, were not substantiated by Fuxe and Hökfelt (1967), who were unable to observe changes in the TIDA neurons of the ME 1, 2, or 14 days after adrenalectomy. Similarly, chronic treatment with high doses of compound F, the stress of injecting NaCl, or insulin-induced hypoglycemia were shown to be ineffective in altering ME fluorescence. Furthermore, neither F treatment nor adrenalectomy altered the depletion of DA in the TIDA neurons after inhibition of CA synthesis (Fuxe and Hökfelt, 1967), whereas corticosteroid treatment slightly increased DA turnover in the nigrostriatal and mesolimbic DA neurons of adrenalectomized rats (Fuxe *et al.*, 1973a). In guinea pigs, inhibition of corticosteroid synthesis by metyrapone did not alter the fluorescence intensity of TIDA neurons (Barry, 1968). All together these results would exclude an influence of corticosteroid levels on TIDA neuron activity.

Changes in brain NE turnover have been described in relation to altered adrenocortical function. An increase in NE turnover in both the hypothalamus and cerebral cortex of the rat was found 6 days following adrenalectomy; conversely, in the adrenalectomized rat, multiple doses of corticosteroids counteracted in a dose-dependent manner the increased turnover of hypothalamic NE. The same corticosteroid regimen was ineffective in intact rats (Fuxe *et al.*, 1973a). Therefore, it appears from these fluorescence studies, that NE but not hypothalamic DA neurons are affected by changes in circulating levels of corticosteroids.

Biochemical measurements of brain CA content or turnover after adrenalectomy and corticosteroid replacement did not allow definitive conclusions. Unlike A. M. Uretsky *et al.* (1971), who were unable to find changes in brain NE levels after adrenalectomy, De Schaepdryver *et al.* (1969a) found reduction in brain DA and NE levels 24 hr but not 3–5 days postoperatively, times at which CA levels were even increased. The turnover of brain NE was reported to be accelerated 6 but not 2 days after adrenalectomy by Javoy *et al.* (1968).

More rewarding results were obtained by the study of the key enzymes involved in CA metabolism. In fact, even though brain MAO activity was not modified by changes in HHAA activity (Garattini *et al.,* 1961), the activity of TH in discrete areas of the rat brain appeared strictly related to the presence of increased or decreased levels of plasma B (Kizer *et al.,* 1974). TH activity in the ME was significantly decreased after adrenalectomy and, conversely, increased following dexamethasone treatment, a result consistent with the postulated inhibitory role of hypothalamic CA's on ACTH secretion. The fact that the changes in TH activity were confined to the ME suggests that only at this level are the steroid-responsive catecholaminergic terminals located which participate in glucocorticoid feedback regulation. Even though reduced TH activity was also found in the ARC and dorsomedial nuclei of adrenalectomized rats as compared to sham-operated controls, no significant difference in TH activity in adrenalectomized and dexamethasone-treated animals was present in these nuclei.

These experiments on TH could not discriminate between changes in enzymatic activity occurring in NE or DA nerve terminals. Nevertheless, the detection of changes in the TH activity of the MFB which paralleled those of the ME (Kizer *et al.,* 1974) suggests that the alterations in TH activity present in the neurohemal zone were likely secondary to an impaired neurotransmission in NE fibers originating in the mesencephalon and conveyed rostrally to the hypothalamus. It must be pointed out, however, that increased DA levels have been found in the ME following dexamethasone and corticosterone treatment (Fekete *et al.,* 1976), so that the involvement at this level of dopaminergic neurons cannot be ruled out.

Collectively, with caution suggested because of existing contradictory data, results of experiments of acute suppression of the adrenal function indicate that in this condition hypothalamic NE content and synthesis are decreased. Such an effect is counteracted by corticosteroid replacement therapy. Therefore, these findings are in keeping with the inhibitory role of NE on the CRF-ACTH system and with the existence of a negative feedback regulation by corticosteroids operating through stimulation of NE synthesis.

The interactions between adrenal function and brain CA metabolism have also been investigated by evaluating the effect of an abrupt suppression of ACTH secretion. A marked decrease in the turnover of NE neurons in several CNS areas was reported in the rat following hypophysectomy. Neostriatal DA neurons were unaffected. Injection of purified ACTH or synthetic 1-24-corticotropin failed to reverse the decreased NE turnover of hypophysectomized rats, but decreased DA turnover in the neostriatum and the limbic forebrain. Only in intact rats was brain NE turnover increased, although slightly by the two peptides (Fuxe *et al.*, 1973a). Although the possibility was envisaged of a negative feedback action of ACTH on CRF release partly mediated by NE pathways, the lack of any effect by ACTH replacement and the fact that after hypophysectomy abrupt modification of the whole hormonal equilibrium occurs argued against the physiological significance of these findings. Different results have been recently reported by Van Loon *et al.* (1976), who tried to dissociate in the rat effects of changes in ACTH secretion on NE neurons from those on DA neurons. They showed that hypophysectomy increased TH activity in the hypothalamus, brainstem, and heart but not in the hippocampus or striatum. Chronic ACTH administration prevented the hypophysectomy-induced increase in TH activity and increased TH activity in the striatum. It would appear from these studies that the effects of changes in ACTH secretion are considerably different in brain regions containing predominantly DA neurons from those with mixed populations or predominantly NE neurons.

Results difficult to interpret and at times inconsistent were obtained in experiments in which the effect on brain CA levels were evaluated by the systemic administration of huge amounts of natural or synthetic glucocorticoids to intact rats (Scapagnini *et al.*, 1969; Shalyapina, 1967). In rats neonatally treated with hydrocortisone acetate both NE and 5-HT levels were elevated significantly in the MBH 30 days later (Ulrich *et al.*, 1975), and an increase in PNMT activity was present both in the medulla and the hypothalamus of newborn rats receiving 6 daily injections of dexamethasone (Moore and Phillipson, 1975b). The latter findings are of relevance when one considers that E is one possible product of PNMT-catalyzed reactions.

The many reports on the effects induced by adrenal steroids on brain 5-HT metabolism have resulted in contradictory views, which reflect, in part, the same divergent opinions noted above for the role (inhibitory or stimulatory, respectively) of the central 5-HT system in controlling the HHAA (see Section A,1,b). On the basis of acute adrenalectomy and corticosteroid replacement therapy experiments, some researchers believe that adrenal steroids stimulate central 5-HT neuronal function. Adrenalectomy in the rat decreased the 5-HT content of the whole brain (De Schaepdryver *et al.*, 1969a), of the hypothalamus (Vermes *et al.*, 1973b), and of several limbic areas (Telegdy and Vermes, 1975), a fact which might be attributed to the reduction of Trp-H activity found in this condition by Azmitia and McEwen (1969), but not confirmed by Lovenberg (1972). Uptake of labeled 5-HT was also lower in the hypothalamus and mesencephalon of adrenalectomized rats (Telegdy and Vermes, 1975). Conversely, administration of corticosteroids to adrenalectomized rats brought back to normal the decreased Trp-H values (Azmitia and McEwen, 1969), as well as brain 5-HT content and uptake (Vermes *et al.,* 1973b; Telegdy and Vermes, 1975). In normal rats, administration of corticosteroids increased 5-HT content in the limbic system within a few minutes, induced brain Trp-H activity in the midbrain and stimulated the conversion of Trp to 5-HT (Millard *et al.,* 1972). More recently, in mice, a single injection of a huge dose of hydrocortisone stimulated within 1 hr the uptake of Trp by nerve terminals and thereby accelerated the rate of synthesis of 5-HT (Nickers and Sze, 1975). The effect was apparently specific for the glucocorticoids since androgenic and progesterone-like steroids were found to be ineffective. These results were expanded further by showing that glucocorticoids may act as regulatory factors for brain Trp-H. In fact, the normal developmental rise of Trp-H levels in neonatal rat brain was blocked by adrenalectomy and repeated injections of cortisone induced a rise in brain Trp-H in neonatal but not in adult brains (Sze *et al.,* 1976). Resistance of the Trp-H activity of individual nuclei of the limbic system, hypothalamus, and midbrain of the adult rat to various endocrine manipulations, including adrenalectomy or dexamethasone treatment, has been recently observed (Kizer *et al.,* 1976c).

Most of the above data along with the finding that some stressful situations are associated with increased brain 5-HT content (see Section A,1,b), lead one to postulate that corticosteroids play an important role in the maintenance of the normal functioning of the serotoninergic system and that the latter is involved in the negative feedback of corticoids at the level of the hypothalamus, even though the neuronal systems mediating feedback may lie distal to the terminal neurons secreting CRF (Vernikos-Danellis *et al.,* 1973) (see Section A,1,b). It is known that the nuclei of CNS neurons are target sites for glucocorticoid action (McEwen *et al.,* 1972a). It appears now that the nerve endings of CNS neurons may also be target sites for glucocorticoid action and that the hormone may also rapidly regulate neurotransmitter metabolism (see also Chapter II, Section B,9).

In apparent contrast to these data are the results of many studies showing that peripheral administration of natural (Curzon and Green, 1968, 1969, 1971; Scapagnini *et al.,* 1969) or synthetic (Nisticò and Preziosi, 1969; Scapagnini *et al.,* 1969) glucocorticoids induces after few hours a sharp decrease of brain 5-HT content. However, the latter and the previous series of results are not necessarily contradictory. The stimulatory effect of corticosteroids occurring within a short interval of time is likely due to an induction of Trp-H and/or an increased 5-HT neuronal uptake, while the decrease in brain 5-HT content evident long after adrenal steroid administration most probably reflects induction of liver TPO with ensuing shift of the available Trp to the liver kynurenine pathway (see Chapter II, Section C,4). In addition, 5-HT levels in whole brain or specific brain sites, such as hypothalamus, following peripheral treatment with steroids reflect the sum of the action of corticoids on extrahypothalamic areas, e.g., the limbic system, and the subsequent alterations in their neuronal output onto a given area. This must be considered when evaluating results of apparently conflicting observations.

2. Growth Hormone and Changes in Brain Serotonin Turnover

Studies on the effect of varying endocrine function on 5-HT metabolism are concerned mainly with the effects of adrenal steroids (see preceding section); fewer studies have been devoted to the investigation of the effect of GH on 5-HT metabolism. In animals, treatment with GH resulted in increased weight and granularity of enterochromaffin cells (Mira, 1960) and increased urinary excretion of 5-HIAA (Laborit *et al.*, 1959), whereas opposite effects on the enterochromaffin cells were noted following hypophysectomy (Ghinghelli, 1961). This suggests that an increase in serum GH may stimulate 5-HT synthesis and turnover in peripheral 5-HT-rich pools. In contrast to this would be the effect of GH on central 5-HT metabolism, as implied by findings obtained in experimental conditions sharing in common a lack of pituitary GH. In hypophysectomized rats, 15 days following surgery, brain concentrations of both Trp and 5-HIAA were markedly higher than in age-matched intact controls, while brain 5-HT levels were comparable in the two groups (Cocchi *et al.*, 1975a). On recalling that the rate of 5-HT formation in the brain is regulated principally by Trp availability (see Chapter II, Section C,4), the presence of elevated brain Trp levels in hypophysectomized rats and the reported inability of hypophysectomy to affect brain Trp-H (Deguchi and Barchas, 1973) suggest that pituitary ablation may be associated with increased brain 5-HT metabolism. The observation of a concomitant increase in brain 5-HIAA concentration further substantiates this hypothesis, since an alteration of the efflux rate of 5-HIAA as a result of hypophysectomy could be excluded (Di Giulio *et al.*, 1975).

Direct proof that a lack of GH could be among the factors responsible for the increased 5-HT metabolism present in hypophysectomized rats was provided by experiments on GH-replacement therapy. In hypophysectomized rats, GH therapy induced a significant reduction in brain

Trp and 5-HIAA concentrations and did not modify brain 5-HT (Cocchi *et al.*, 1975a). Consistent with these findings in the rat were those obtained in genetically Snell-Bag dwarf mice, which are characterized by a selective deficiency in the secretion of GH and PRL from the AP (Gruneberg, 1952). Similar to the hypophysectomized rat, GH-deficient mice had brain Trp and 5-HIAA concentrations markedly higher than the age-matched heterozygote controls. GH treatment decreased Trp and 5-HIAA levels both in dwarf mice and controls, with no change in brain 5-HT concentrations (Table XXXVII), whereas PRL replacement therapy did not change significantly brain Trp levels (Cocchi *et al.*, 1975a).

Collectively, the above data point to a deficiency of GH as a major determinant of the observed alterations but do not explain the possible mode of action of GH in affecting brain 5-HT metabolism. The lipid-mobilizing effect of GH can hardly account for the decreased brain Trp concentrations found after GH administration, since increased plasma NEFA levels would lead to displacement from albumin of the Trp-bound fraction and ensuing increased plasma-"free" and brain Trp concentrations (see Chapter II, Section C,4). Rather, the elevated brain Trp levels of the hypophysectomized rat and the dwarf mouse might be due to a primitive alteration of the availability of circulating amino acids with an increase in the availability of circulating Trp, a hypothesis which fits well with the observation that pituitary ablation delayed the disappearance of ^{3}H-Trp from the plasma (Di Giulio *et al.*, 1975), probably due to an altera-

TABLE XXXVII

Brain Concentrations of Trp, 5-HT, and 5-HIAA in Dwarf Snell-Bag Mice *(dw/dw)* and in Heterozygote Controls *(dw/+)*[a]

	Brain		
Treatment[b]	Trp (μg/gm)	5-HT (μg/gm)	5-HIAA (μg/gm)
(dw/+)	4.20 ± 0.21[c]	1.15 ± 0.055	0.33 ± 0.020
(dw/dw)	6.40 ± 0.33[d]	1.15 ± 0.024	0.65 ± 0.026[e]
(dw/+) + GH	3.87 ± 0.19	1.33 ± 0.093	0.26 ± 0.026[f]
(dw/dw) + GH	5.57 ± 0.20[g]	1.30 ± 0.010	0.57 ± 0.014[h]

[a] From Cocchi *et al.* (1975a).

[b] GH was administered daily (200 μg ip) for 7 days; mice were killed 20 hr after the last injection of GH.

[c] Each value is the mean of at least 6 determinations ±SEM.

[d] $p < 0.001$ vs *(dw/+)*.

[e] $p < 0.001$ vs *(dw/+)*.

[f] $p < 0.02$ vs *(dw/+)*.

[g] $p < 0.05$ vs *(dw/dw)*.

[h] $p < 0.02$ vs *(dw/dw)*.

tion of the hepatic metabolism of the amino acid (see Chapter II, Section C,4). Thus, GH action on brain Trp metabolism could be best explained by its effect, which favors amino acid utilization by tissues (Knobil and Hotchkiss, 1964). Although determinations of "free" and "bound" Trp levels in the plasma of the hypophysectomized rat and the dwarf mouse are needed to definitively prove this theory, the observation that total Trp levels were significantly lower in the plasma of some acromegalic patients in the active stage of the disease is of interest (Cocchi *et al.,* 1975b). However, difficult to reconcile with these findings are those of Essman *et al.* (1973), who showed elevated mean fasting serum 5-HT levels in 5 acromegalic patients and their considerable reduction following transsphenoidal resection of the pituitary tumor.

In addition to these studies, other evidence which links the growth hormone to brain 5-HT metabolism can be derived, although inferentially, from the studies of Drucker-Colin *et al.* (1975). They showed that in the rat exogenously administered GH increases REM sleep, a condition, which at least in the human, is positively correlated with "free" plasma Trp levels (C. N. Chen *et al.,* 1974). Also, these findings, when extrapolated to the human, are difficult to reconcile with some of the previous observations in the acromegalic subjects. In conclusion, although most of these data indicate that GH, directly or indirectly through somatomedin, affects brain 5-HT metabolism, it appears that further investigations are needed for a better understanding of the complex GH–5-HT interaction(s).

With regard to catecholamines it has been reported by Fuxe *et al.* (1974a) that huge doses of GH increased DA turnover in the TIDA neurons of the hypophysectomized rat; consistent with these findings would be the reduced DA turnover present in the brain of the GH-deficient dwarf mouse (A. Groppetti, unpublished results). Recently, modification of the action of some neuroactive drugs has been reported following GH treatment (Tang and Cotzias, 1976). These studies showed that in mice large doses of bovine GH changed the distribution and increased the accumulation of L-dopa, ^{3}H-apomorphine, and ^{3}H-thymidine in the brain. The relevance of these findings for disease states of man, e.g., Parkinson's disease, in which the cumulative excess of GH released during L-dopa or CB-154 treatment may change the therapeutic potential of the drug, or acromegaly, in which GH overproduction is cured by dopaminergic compounds (see Chapter V, Section A,2), requires further experimentation.

3. Pituitary-Gonadal Function and Brain Monoamine Metabolism

An extensive series of studies on the uptake and binding of labeled estrogen in rat brain has revealed a selective neuroanatomic distribution of the hormone. Similar to corticosterone, estradiol binds to brain cell nuclei, but, unlike corticosterone, it binds preferentially to the hypothalamus, the PO area, and the amygdala. The sites of estradiol binding are also the sites where the implanted hormone exerts feedback effects on gonadotropin secretion or facilitates the lordosis reflex (see McEwen *et al.*, 1972b).

Among the possible mechanism(s) underlying these "stimulatory" feedback effects (Stumpf and Sar, 1973), an interaction exerted by estrogen on the classic neurotransmitters, which are predominantly distributed in the same CNS areas, has to be considered. In fact, evidence has been presented that suggests that gonadal steroids can modify catecholaminergic activity in the CNS interfering with different steps of CA metabolism. Manipulation of the hypothalamohypophyseal–ovarian axis induces pronounced effects on CA synthesis. Acceleration of hypothalamic NE turnover in adult male or female rats (20 days after gonadectomy) was first reported by Anton-Tay and Wurtman (1968) after IVT injection of ^{3}H-NE. Since administration of FSH to normal or to hypophysectomized-gonadectomized rats also increased the turnover of brain ^{3}H-NE (Anton-Tay *et al.*, 1968), it was concluded that the effects of gonadectomy were mediated at least in part by the pituitary hormones. Consistent with these data, Bapna *et al.* (1971), using a method for determining the rate of synthesis of NE based on the changes of specific radioactivity of tyrosine and NE with time after IVT injection of ^{3}H-tyrosine (see Chapter II, Section B,12), showed that ovariectomy resulted in an enhanced NE synthesis rate in the anterior hypothalamus, with no change in NE synthesis in the posterior hypothalamus. In these experimental conditions, no changes were found in hypothalamic 5-HT turnover as determined by the rate of conversion of ^{3}H-Trp.

The involvement of gonadal steroids in ovariectomy effects was demonstrated by showing that simultaneous replacement of estrogen and progesterone to ovariectomized rats decreased the formation of ^{3}H-NE from ^{3}H-tyrosine (Anton-Tay and Wurtman, 1970; Bapna *et al.*, 1971) and brought back to normal the increased hypothalamic DA and NE levels (Donoso, 1972). These observations prompted an investigation of the effects of ovariectomy and ovarian steroid replacement on rat hypothalamic TH as a possible mechanism for gonadal steroid influence on NE synthesis (Beattie *et al.*, 1973). It was shown that ovariectomy caused a two- to threefold increase in TH on day 4, which lasted for at least 60 days. In these conditions, estradiol replacement further stimulated the enzymatic activity, whereas progesterone decreased the TH activity in both ovariectomized and ovariectomized, estrogen-treated rats. It was thus concluded that CA synthesis in the rat hypothalamus may be altered by ovarian steroids, with progesterone having an inhibitory effect and estrogen a stimulatory effect. Subsequent *in vitro* studies showed that progesterone's inhibitory action on hypothalamic TH was not dependent on the substrate concentration, but very likely was due to an interference with the enzyme-reduced pteridine cofactor complex (Beattie and Soyka, 1973). A significant rise in median eminence TH activity was also observed in the male rat 10 days following gonadectomy, which was partially reversed by testosterone (Kizer *et al.*, 1974). Interestingly, the finding that destruction of over 90% of the noradrenergic input to the ME failed to block gonadectomy-induced changes in ME-TH activity suggests that the dopaminergic terminals are endocrine-responsive (Kizer *et al.*, 1976b). However, the selective increase of TH activity in the ME of castrated male rats reported by Kizer *et al.* (1974, 1976b) has not been confirmed by the similar studies of Nakahara *et al.* (1976). They found instead that following castration the TH activity increased in the n. periventricularis and in the medial zona incerta and remained unchanged in the other hypothalamic nuclei. Although changes of TH activity in the POA-AHA region were not detected, the increased TH activity found in the n. periventricularis and in the medial zona incerta may influence the POA-AHA region since dopaminergic cell bodies of the incertohypothalamic neuron system (A13 and A14 cell groups) project to the POA-AHA region (Björklund *et al.*, 1975a) (see Chapter II, Section B,4), participating in the control of gonadotropin secretion.

Other evidence for an interaction between sex steroids and CA's derives from studies on CA changes at various stages of the estrus cycle. During vaginal proestrus, when estrogen secretion rate plateaus, CA content rises in the anterior hypothalamus and brain ^{3}H-catechol accumulation is 4 times more rapid than in diestrus (Zschaeck and Wurtman, 1973). For alterations in fluorescence intensity of TIDA neurons, as related to different stages of the estrus cycle, see Section A,3,a.

Also the release of CA's appears to be influenced by sex steroids. In brain synaptosome preparations, estradiol slightly decreased the passive efflux of ^{3}H-NE, while progesterone increased it (Janowsky and Davis, 1970). Many reports in the literature suggest that gonadal steroids can also influence MAO activity (Cavanough and Zeller, 1964; Collins and Southgate, 1970; Grant and Pryse-Davies, 1968). Estrogens are known to have an inhibitory effect on MAO activity *in vitro* (Collins *et al.*, 1970), whereas *in vivo*, Holzbauer and Youdim (1972) found the highest values of MAO activity on the day of proestrus. Kamberi and Kobayashi (1970) measured MAO activity in the entire hypothalamus and in different hypothalamic regions, in the amygdala and cerebral cortex of male and 4-day-cycle female rats. They showed in the entire hypothalamus as well in the ME and the anterior hypothalamic area two peaks of activity on the morning of proestrus and of estrus with a dramatic fall occurring in the afternoon of proestrus, coincidental to gonadotropin discharge. Thus, there appeared to be in the afternoon of proestrus an inverse relationship between increasing ovarian estrogen secretion and MAO activity. The amygdala exhibited less marked cyclic activity and it followed the pattern of the hypothalamus. Much lower levels of activity of MAO were present in other regions of the brain, none of which showed cyclic changes. Similarly, no cyclic changes in MAO were found in the hypothalamus of male rats.

Another biochemical link between estrogens and CA's in the CNS, which might play a role in the gonadotropin control mechanisms, has been envisaged (Fishman and Norton, 1975). As previously reported (see Chapter II, Section B,10), a major pathway for the inactivation of CA's is by *O*-methylation. COMT catalyzes the transfer of a methyl group from *S*-adenosylmethionine to a catechol resulting in the formation of an *O*-methylated derivative. On the other hand, the major route of estrogen metabolism in man as well as in laboratory animals is the formation of catechol estrogen through ring A-2-hydroxylation (Fishman, 1963). Although conversion of estrogen to a catechol derivative produces a loss of estrogenic activity, it has been reported by Kuppen and co-workers (Kuppen *et al.*, 1969; Ball *et al.*, 1972) that the catechol estrogens are excellent competitive inhibitors of the *O*-methylation of catecholamines by COMT. In fact, this enzyme shows a much higher affinity for the catechol estrogens than for the CA's, resulting therefore in a prolongation of neurotransmitter action on receptor sites. This interaction first observed in peripheral adrenergic neurons, may soon be seen also in the CNS, since, recently, Fishman and Norton (1975) provided evidence that rat CNS tissue is capable of metabolizing estrogen at the C-2 position. Interestingly, the capacity for this reaction in the rat is concentrated in the hypothalamus and is minimal in the cerebral cortex; the greatest conversion occurs in tissue from ovariectomized rats. More recently, catechol estrogen formation was demonstrated in human hypothalamus; differing from the rat, however, parietal cortex and the pituitary were also active, which may simply reflect species differences or may be a result of tissue development since the human material was of fetal origin (Fishman *et al.*, 1976).

The role of the *in situ* generated catechol estrogen in the control of gonadotropin secretion has yet to be fully established. The recent finding that 2-hydroxyestrone binds to the hypothalamic cytosol receptor (Davies *et al.*, 1975) and proof that it can elevate serum LH levels in immature male rats (Naftolin *et al.*, 1975) implicate these natural estrogen metabolites in the control of reproductive processes. It may be hypothesized that they represent in the CNS a crucial biochemical link in the interaction of estrogen and catecholamines. Hypothalamic catechol estrogen, in fact, by inhibiting catecholamine metabolism might prolong, at proestrus, when estrogen secretion increases, the action of the catecholaminergic neurotransmitter on peptidergic neurons producing LH-RH/FSH-RH (Fig. 51).

A further estrogen-CA interaction of significance for understanding the feedback regulation by estrogens of the synthesis and/or release of LH-RH/FSH-RH is envisaged in the estrogen-elicited accumulation of cAMP in hypothalami from immature female rats both *in vivo* and *in vitro* (Gunaga and Menon, 1973; Gunaga *et al.*, 1974). This phenomenon is blocked by either α- or β-adrenergic antagonists (Weissman *et al.*, 1975), which suggests that physiologically an enhanced release of CA's after an initial binding of estrogen to cytoplasmic receptors in catecholaminergic cell bodies might lead via cAMP formation to alterations in release and/or synthesis of LH-RH/FSH-RH.

Another possibility whereby sex steroids can interfere with CA's is by inhibition of CA uptake processes (see Chapter II, Section B,9). Estradiol in large doses inhibits NE uptake into adrenergic nerve terminals (Uptake$_1$) without affecting the DA membrane process (Janowsky and Davis, 1970; Nixon *et al.*, 1974; see Iversen 1973). Although the concentrations of the steroid used in these *in vitro* studies were relatively high, one must take into account that the circulating hormones could be highly concentrated in specific areas of the brain (Luttge *et al.*, 1973) (see Chapter II, Section B,9). Finally, estrogen also produces a dose-dependent inhibition of Uptake$_2$, which is the mechanism by which CA's accumulate in the extraneuronal sites and are so inactivated (Fig. 51).

With regard to brain 5-HT metabolism, in keeping with neuropharmacological data showing that activation of 5-HT receptors induced inhibition of the cycle release of LH (see Section A,3,b), it was shown that estrogen increases 5-HT turnover in castrated female rats (Fuxe *et al.*, 1973b). This effect would be related to the inhibitory feedback action of estrogen on LH secretion. Administration of progesterone to the estrogen-primed, ovariectomized rats, a procedure expected to induce peak LH secretion (Caligaris *et al.*, 1968), restored to normal brain 5-HT turnover.

C. SITES OF ACTION OF MONOAMINES IN AFFECTING ANTERIOR PITUITARY FUNCTION

While on the basis of information acquired in the preceding sections the role of brain MA's in the control of AP hormones appears to be well established, the anatomic loci within CNS or elsewhere at which neurohormone and neurotransmitter interact still require precise identifi-

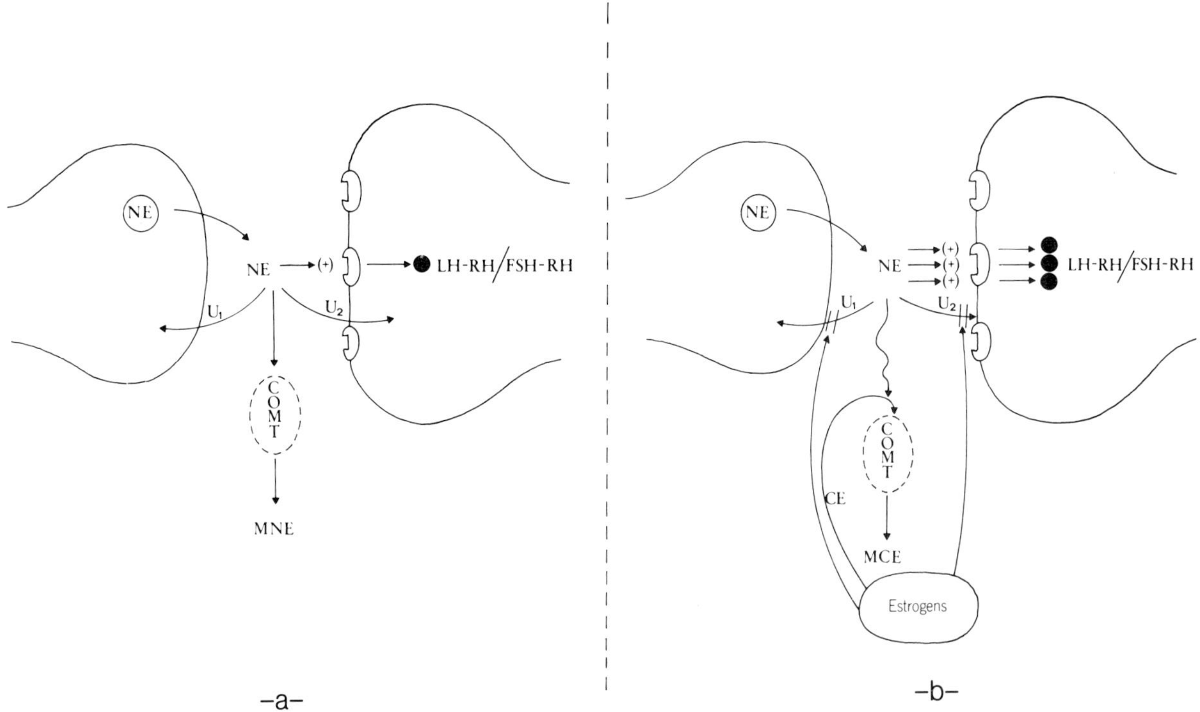

Fig. 51 Diagram illustrating the postulated interaction between catechol estrogens and catabolism of brain CA's. In (a), a synapse between a NE neuron and a LH-RH/FSH-RH neurosecretory neuron is depicted; in the synaptic cleft, NE is catabolized by *O*-methylation. In (b), formation of catechol estrogens is envisaged; they would compete with CA's for *O*-methylation thus leading to preservation of NE and magnification of NE action at post-synaptic level. The inhibitory effect of estrogens of $Uptake_1$ and $Uptake_2$ mechanism(s) is also illustrated. NE, norepinephrine; COMT, catechol-*O*-methyltransferase; CE, catechol estrogens; MCE, *m*-catechol estrogens. U_1, Uptake; U_2, $Uptake_2$; MNE, metanephrine.

cation. The recent development of specific and sensitive assay methods for the detection of hypothalamic hormones will permit a complete mapping of their distribution areas and, in conjunction with the knowledge on the topographical localization of brain monoamine systems, offer a better understanding of neurotransmitter-releasing hormone interrelationships. Further insight will be achieved by the awareness that a specific monoamine is involved in the release or in the inhibition of the release process of a given anterior pituitary hormone.

Many are the possible loci at which AP secretion could be affected by neurotransmitters, some of which have been previously considered by Wurtman (1971). The interaction includes the possibility that there exist (1) cell elements which contain both MA's and neurohormones, with the former affecting the secretion of the latter, (2) direct axodendritic or axosomatic contacts of neuronal terminals which liberate a given neurotransmitter with cells which manufacture releasing or inhibiting hormones, or (3) indirect connections, though the agency of neurons distant to the transducer neuron. Other possibilities are the existence of (4) axoaxonic synapses between peptidergic and neurotransmitter nerve terminals (at ME level?), (5) an interaction by which MA's at the ME level control delivery of RF's to the AP or, finally, (6) neurotransmitter release into the portal circulation to directly stimulate or inhibit pituitary hormone release or alter pituitary sensitivity to RF's. Peripheral MA's might affect AP function once delivered to the general circulation or through direct action on the pituitary or indirectly by affecting peripheral nerve pathways communicating functionally with the CNS (7). These candidate sites for neurotransmitter-neurohormone interactions are schematically represented in Fig. 52.

Present scarcity of knowledge on the function of other identified or proposed neurotransmitters on neuroendocrine regulation and the complexity of their functional interactions permits us to only indicate some of the areas in which brain monoamine fibers and neurohormones probably interact to control the secretion of AP hormones.

1. ACTH

In the dog, evidence has been obtained, using neuropharmacological methods (Section A,1,a,i), that the site at which central adrenergic neurons act to inhibit ACTH secretion is inside the BBB, and, hence, above the basal ME. In fact, CA precursors which readily cross the BBB inhibited stress-induced HHAA activation, while the native transmitters unable to cross did not. Moreover, blockade of peripheral L-AAAD (e.g., by carbidopa, see Table VII), which after amine precursor administration

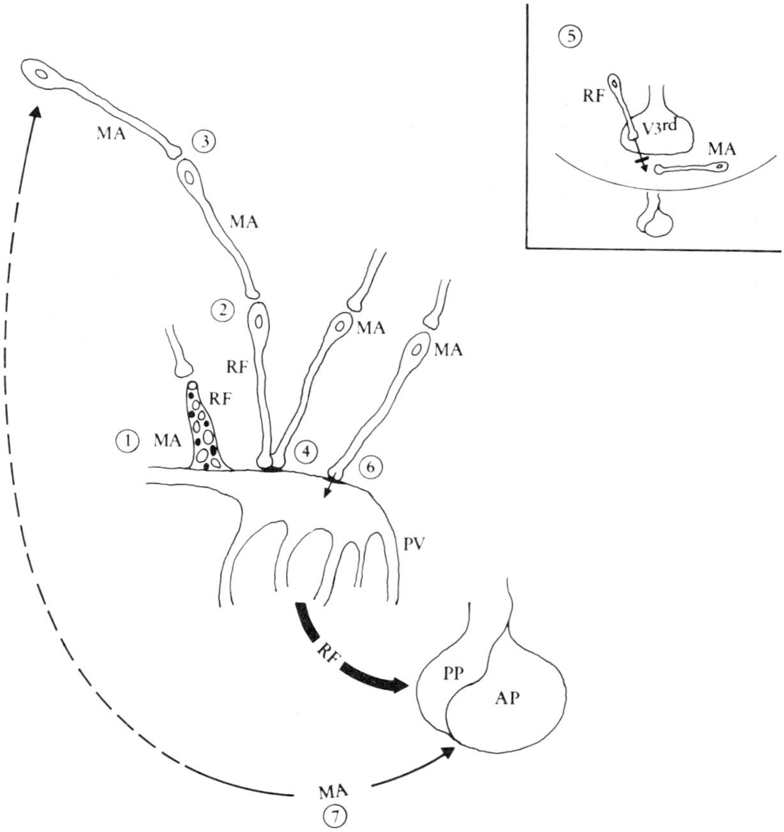

Fig. 52 Possible sites at which MA's might interact with neurosecretory neurons which control anterior pituitary function (see text for explanation). MA, monoaminergic neuron; RF, neurosecretory neuron; PP, posterior pituitary; AP, anterior pituitary. V3rd, third ventricle.

(L-dopa) induced clear-cut peripheral effects (e.g., inversion of L-dopa pressor response), did not alter the central inhibition of ACTH secretion (see Ganong *et al.,* 1976), and carbidopa alone did not alter in man baseline plasma F levels (G. M. Brown *et al.,* 1976b). Evidence that NE inhibition of ACTH secretion takes place "inside the BBB" is supported by earlier studies in which the effect of pretreatment with MAOI, following electrical stimulation of various brain regions, was investigated (Ganong *et al.,* 1965). It was found that stimulation along the dorsal longitudinal fasciculus, mammillary peduncle and amygdala were blocked, while stimulation in the basal hypothalamus was not. These results suggest that the site of inhibition should include an area above the basal hypothalamus, but neither posterior to the mammillary body nor more lateral than the amygdala. Similarly, the evidence which has been accumulated in the rat suggests that in this species, too, the inhibitory effect of adrenergic discharge is at least in part exerted "inside the BBB." Peripheral CA depletors, such as guanethidine, induced adrenocortical activation only when injected directly into the brain ventricles (Scapagnini *et al.*, 1972); NE, the native neurotransmitter responsible for the inhibition, was ineffective when given systemically (see Section A,1,a,i). In addition, 6-OH-dopa, which enters the brain and is converted to 6-OHDA in CNS neurons (see Chapter II, Section D), had a greater effect on plasma B than equal doses of 6-OHDA (Scapagnini *et al.*, 1974a). The finding that 6-OHDA was much more effective when administered directly into the third ventricle than when infused into the lateral ventricle (Scapagnini *et al.*, 1974a) would indicate a site of action beneath the floor of the third ventricle.

Unlike the dog, and probably the human, in the rat the neurotransmitter–neurohormone interaction appears to take place also in that part of the ME which is outside the BBB. Systemic administration of large doses of 6-OHDA produced a small but significant decrease of ME-NE content which was associated with a slight but clear-cut adrenocortical activation (Cuello *et al.*, 1974). This would imply the existence at the ME level of axoaxonic inhibitory contacts, since CRF-cell bodies should be inside the BBB. Such a location is implied also by the reported alterations in TH activity of catecholaminergic terminals of the ME following dexamethasone treatment of adrenalectomized rats (Kizer *et al.*, 1974) (see Section B,1). Thus, it would appear that at least in the rat the long feedback loop of plasma corticosteroids on CRF-ACTH secretion is directed largely at this level. Another candidate area in the feedback action of glucocorticoids might be the SON, as implied by the observation that both dexamethasone and corticosterone treatment increased in the rat the NE content of this area (Fekete *et al.*, 1976).

On the closed-loop, negative-feedback system is superimposed an open-loop, neural component responsible for circadian rhythms, sleep-related changes, and emotional and physical stress rises. As suggested by the observation that immobilization stress in the rat increased selectively TH activity in the ARC nucleus (Palkovits *et al.*, 1975), one is tempted to suggest that the catecholaminergic system at this level may also mediate the action of the neural transient capable of breaking through the corticosteroid feedback loop. For other information on the interactions between corticosteroids, ACTH, and central MA neurons see Section B,1.

2. Growth Hormone

Many of the stimuli which have been shown to induce GH release in the human require intact cerebral CA transmission (Fig. 53). Thus, insulin-

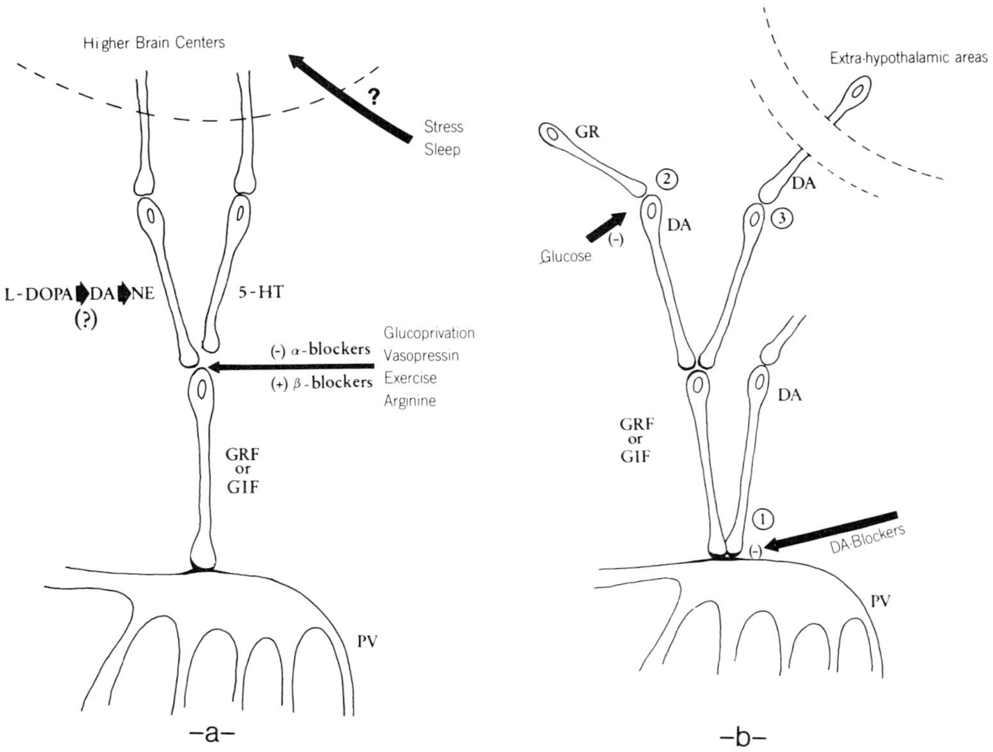

Fig. 53 (a) Possible hypothalamic and extrahypothalamic sites of action of some GH-releasing stimuli and their dependence on NE control. Depicted is also the possibility that L-dopa may act through NE mediation. 5-HT indicates a serotoninergic neuron which like the NE neuron might be involved in relaying impulses from higher brain centers to the hypothalamus. (b) Possible sites of DA-GRF/GIF neuron interactions; 1 indicates an axoaxonic synapse at ME level; 2 indicates a synapse relaying impulses from chemoreceptor to dopaminergic neurons and the inhibitory effect of glucose; 3 indicates an interaction with a DA pathway originating extrahypothalamically (see text for explanation). In view of the dual system of control for GH secretion, positive or negative GH responses might result either by stimulation of GRF neurons or inhibition of GIF neurons or vice versa, respectively. NE, norepinephrinergic neuron; DA, dopaminergic neuron; 5-HT, serotoninergic neuron; GR, glucoreceptor neuron; GRF, GH-releasing factor neuron; GIF, GH-inhibiting factor neuron; PV, portal vessels.

hypoglycemia-, vasopressin-, and exercise-induced GH release are prevented or facilitated, respectively, by α- or β-adrenergic blockade. Activation of the adrenergic system also appears to be present during arginine-induced GH release, as implied by the inhibiting effect of phentolamine (Buckler *et al.*, 1969) (see Section A,2,a,i and ii). In the rat, electrical stimulation of the hippocampus, BLA, interpeduncular n., and locus coeruleus as well as of the VMN or ARC n. induced rises in plasma GH levels, whereas stimulation of the corticomedial amygdala (Martin, 1973b) or the MPOA (Martin *et al.*, 1975a) inhibited GH release. Interruption of MA function by α-MpT or pCPA completely blocked the GH release induced by electrical stimulation of the limbic structures, leaving unaffected the GH rise induced by VMN stimulation (Martin, 1973b). This would suggest that CA's and perhaps 5-HT are involved in the relay of responses from higher CNS centers to the MBH—a view consistent with the presence of monoaminergic limbic-hypothalamic connections—but do not play a major role in the GH response evoked by direct stimulation of the final common pathway.

The plasma GH rise induced in the rat by stimulation of the interpeduncular n., an area of origin of the mesolimbic DA system, appears to be rather puzzling in the light of the postulated inhibitory effect of DA on GH release in rodents (see Section A,2,a,i). If the interpeduncular stimulatory response is mediated over the DA pathway, it would be unlikely that this is directly related to hypothalamic dopaminergic mechanisms. Instead, it might be mediated through higher (amygdaloid or hippocampal) centers. Of relevance is the GH rise noticed following electrical stimulation of the locus coeruleus which is the site of origin of the dorsal NE pathway (see Chapter II, Section B,2).

Evidence for the presence of a hypothalamic α-adrenergic receptor stimulating GH release was obtained by direct infusion of phentolamine into the anterior hypothalamus or third ventricle of conscious baboons. In these studies GH was lowered after central phentolamine infusion (Toivola *et al.*, 1972). For analogy with the monkey, phentolamine blockade of GH release should also take place at the same hypothalamic level in the human. In view of the acknowledged inhibitory role of the anterior hypothalamus on GH secretion and of the preferential GIF localization in the same area, it may be postulated that the action of phentolamine involves direct stimulation of GIF-secreting neurons. However, another possible area of interaction is the internal layer of the ME where a rich distribution of both NE and somatostatinergic nerve endings is present (see Chapter II, Section B,4; Chapter III, Section C; and Figs. 42 and 54). Refractoriness to α- and β-adrenergic modulation of pyrogen- (Lufkin *et al.*, 1971) or sleep-induced GH release (Lucke and Glick, 1971) calls for the existence in the human, as in rodents, of higher brain center mediation. It is of interest that the onset of slow-wave sleep, which is accompanied by GH release in man, may involve hippocampal activation (Jouvet, 1967). 5-HT might be the neurotransmitter involved in the CNS control of sleep-related GH secretion, possibly in an inhibitory way as implied by the enhancement of sleep-related GH secretion noted after methysergide administration (Mendelson *et al.,* 1975).

In the human, a blood-borne signal, such as glucose, which presumably acts directly on glucose-sensitive structures in the VMN-LHA (see Chapter III, Section A,3,b), is capable of blocking the GH response to L-dopa (Mims *et al.,* 1973b) or apomorphine (Ettigi *et al.,* 1975) if given before but not after (Boyd *et al.,* 1970) the dopaminergic stimulus. This suggests that stimulation of (hypothalamic) glucoreceptors (but see also Chapter III, Section A,3,b) antagonizes DA receptor stimulation of GH release and implies the existence of a complex interaction between neurotransmitter, chemoreceptor, and neurosecretory (GRF or GIF) cells (Fig. 53b). Also, the ME can be proposed as a site for the stimulatory action of DA on GH release on recalling that somatostatinergic nerve terminals are present in almost the entire ME (Figs. 42 and 54), even though other brain areas, such as the striatum or the limbic system (see above), have to be considered as possible sites of the interaction (Fig. 53b). In this context, the finding that lesions of the pars compacta of the substantia nigra are capable of decreasing DA in the VMN and ME (Kizer *et al.,* 1976d) suggests the possibility of multiple and functionally related sites of interaction.

The recent discovery of presynaptic receptor sites for dopaminergic neurons in the striatum, the activation of which would result in a feedback inhibition of DA synthesis and release (see Chapter II, Section B,11) and our present scarcity of knowledge on the existence and function of DA receptors in extra-CNS areas further complicates the understanding of the mode of action of DA-like drugs in stimulating or inhibiting pituitary hormone release. Experiments with the use of DA agonist or antagonist drugs acting only or predominantly at pre- or postsynaptic level might help clarify the situation. For the possible site(s) of action of dopaminergic compounds in acromegaly see Chapter V, Section A,2.

Reported evidence for the presence of neurohormonal inhibitory influences on GH release (see Chapter III, Section A,3,c) adds new dimensions to the proper interpretation of the physiology of GH control and to the role in this mechanism(s) of monoaminergic inputs. The possibility must be considered that a given neurotransmitter in inducing a stimulant effect may act through a depression of an inhibitory influence. In line with this reasoning it might be postulated that the phylogenetic difference in the control of GH release between the human and all the other species resides in the different hierarchy of the stimulatory and the inhibitory centers. In the human, the inhibitory center might be the predominant one (lower resting values of GH in the human than in the other species). Thus, a given neurotransmitter might be stimulatory in the human by acting through a depression of an inhibitory center whose tone in resting conditions supercedes the tone of the antagonistic center.

With regard to 5-HT, the known ability of 5-HT agonist or antagonist drugs to interfere with hypoglycemia-induced GH release (Müller *et al.*, 1976b) suggests an interaction of the central 5-HT system with glucose chemoreceptors, located presumably in the hypothalamus. Here a candidate area may be the preoptic-anterior and periventricular hypothalamus where 5-HT nerve terminals are distributed and bipolar somatostatinergic neurons have been detected (see Chapter III, Section C, and Alpert *et al.*, 1976).

3. Gonadotropins

The nature of the involvement of DA neurons in the control of gonadotropin secretion, namely LH, has been a subject of considerable debate, since evidence has been presented in favor of either a stimulatory or an inhibitory role (see Section A,3,a). Whatever the nature of the control, it would appear that the neurotransmitter–neurohormone interaction occurs mainly at the level of the TIDA neurons. Iontophoretic studies suggest that DA and NE present in the MBH modulate the neuronal activity of the ARC nucleus, whose axons project to the ME (see Chapter II, Section B,15). Mapping of hypothalamic neurotransmitters and hypothalamic neurohormones (Chapter II, Section B,4; Chapter III, Section C; and Fig. 54) have shown similar massive distribution of DA (TH) and LH-RH nerve terminals in the external layer of the ME, supporting the view of dopaminergic control of LH-RH secretion at this level. However, in spite of the postulated DA involvement in the control of the release of LH-RH, no clear-cut evidence has been given so far for the existence of direct axoaxonic synapses at the ME level even though their existence is supported by recent correlation studies on the distribution pattern of TH and LH-RH-immunofluorescence in corresponding regions of the rat ME (Fuxe *et al.*, 1976). (See also Chapter I, Section B,2.)

For NE a site of interaction with LH-RH-secreting neurons appears to lie more rostrally at the levels of the MPOA, as also implied by the blockade effect of the LH release which follows electrical stimulation of this area by drugs interfering with NE synthesis (see Section A,3,a). NE and LH-RH nerve terminals might also interact at the level of the lateral part of the external layer of the ME where a small amount of D-β-H activity has been localized (Goldstein *et al.*, 1973).

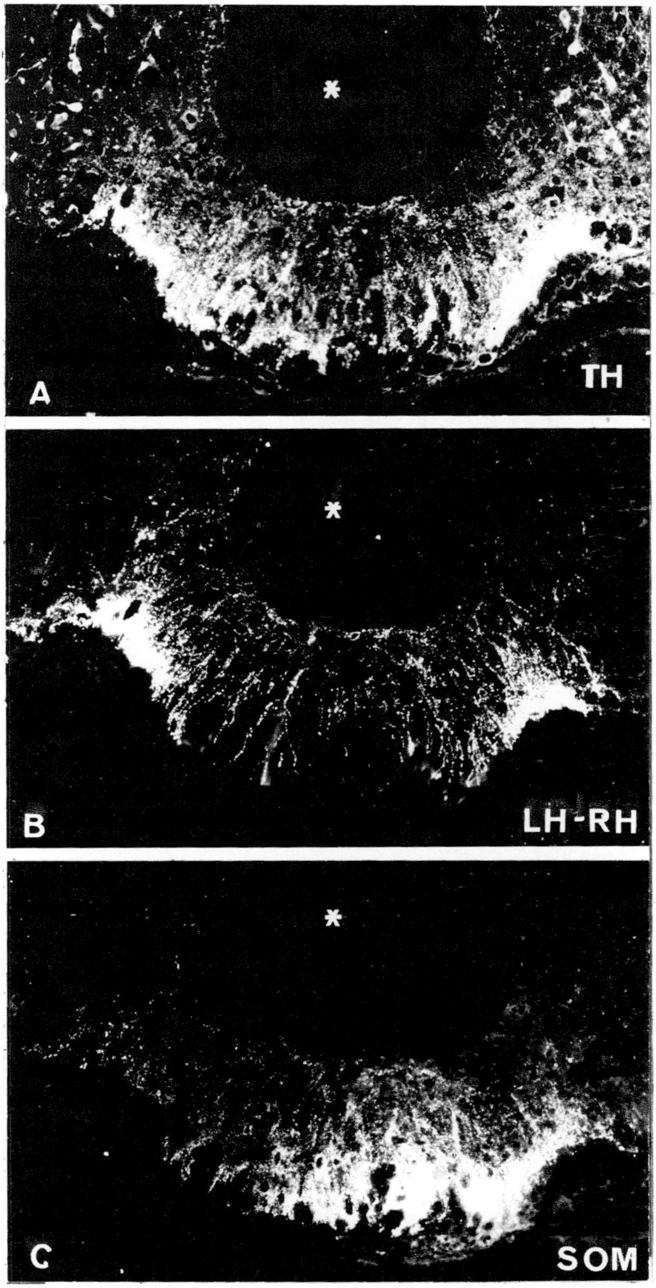

Fig. 54 Immunofluorescence micrographs of the medial part of the median eminence after incubation with antibodies to tyrosine hydroxylase (A), LH-RH (B), and somatostatin (C). TH is present mainly in the lateral part (A), LH-RH almost exclusively in the lateral part (B), and somatostatin mainly in the medial part (C). (×140). Asterisk indicates the third ventricle. The distribution of TH positive nerve terminals corresponds to nerve terminal distribution of the DA A11–A14 cell groups (see Chapter II, Section B,4). (Reproduced, with permission, from Hökfelt *et al.*, 1976c.)

5-HT and melatonin have also been shown capable of inhibiting gonadotropin release following intracerebral injection; their possible sites of action in the CNS have been discussed at great length in Section A,3,b. For the possible site(s) at which sex steroids and monoamine system(s) interact to establish a closed feedback loop see Section B,3.

4. Prolactin

As indicated previously, PRL appears to be predominantly under inhibitory hypothalamic control (see Chapter III, Section A,5,a), and a variety of pharmacological tests have led to the almost inescapable conclusion that it is DA rather than NE which is the inhibitory transmitter. The previous belief that DA acts primarily by releasing (hypothalamic?) PIF which then inhibits PRL release has been questioned on the basis of experiments which showed that the CA may act directly on the gland (see Section A,4,a,i). The presence of receptor sites in the AP which would be physiologically related to the inhibition of PRL secretion has been demonstrated for both DA, apomorphine, and ergocryptine in experiments in which cold DA or the DA agonist drugs blocked the binding of ^{3}H-DA to the receptor.

At present it is still controversial whether DA is the only inhibitor of PRL secretion or that a true PIF exists distinct from DA; the latter would act to release PIF from peptidergic neurons (in the ME?) and might also be secreted into portal blood to inhibit the gland directly. Proof of the physiological significance of the direct action of DA on the AP requires its demonstration in portal blood, a problem which has been faced (Ben-Jonathan *et al.*, 1975), and recently solved (Ben-Jonathan *et al.*, 1977) by Porter and co-workers. Studies aimed at correlating DA titers in the hypophyseal portal blood with spontaneous or experimentally induced changes in endocrine function are awaited with interest.

Evidence is emerging that brain 5-HT, in contrast to DA, plays a stimulatory role on PRL secretion. Although the anatomic site(s) for the interaction of 5-HT and PRL secretion is not established, it may be hypothesized that it occurs mainly in the CNS (see also Section A,4,b). This hypothesis is especially supported by the finding that IVT administration of microgram amounts of 5,7-DHT induced a clear-cut and persistent reduction in plasma PRL levels in male rats (Gil-Ad *et al.*, 1976a).

The use of neurotoxic compounds, such as 5,7-DHT, has recently led to neuroanatomic information in the rat on the medial ascending 5-HT pathway, which primarily innervates the hypothalamus and the POA (Fuxe and Jonsson, 1974; see also Chapter II, Section B,2). It is of note in this context that in the anterior hypothalamus the suprachiasmatic region, a typical area of distribution of 5-HT nerve terminals, is certainly involved in the control of PRL secretion (Pasteels, 1970).

It cannot be excluded *a priori,* however, even though it seems unlikely, that similar to the adrenergic compounds (see above) 5-HT may also act directly on the AP to induce its PRL-releasing effect. Increased serum PRL levels have been reported by Lu and Meites (1973) following systemic administration of 5-HTP to hypophysectomized rats bearing an AP graft and by Lawson and Gala (1975) after 5-HT injection in intact rats. 5-HT also stimulated PRL release from pituitary cell in culture (Fang, 1976). It must be recalled in this context that 5-HT is normally present in the rat AP and the presence of Trp-H suggests that it is formed locally (Saavedra *et al.,* 1975). Possible site(s) of action of monoamines in influencing TSH and MSH secretion have already been discussed in Sections A,5 and A,6 to which the reader is referred.

V

Diagnostic and Therapeutic Implications of the New Concepts on Neurotransmitter–Neurohormone Interactions

A. BRAIN CATECHOLAMINES AND DRUG THERAPY FOR ENDOCRINE DISORDERS

1. L-Dopa, DA Agonist, and Antagonist Drugs as a Test of Anterior Pituitary Function

The wide acceptance of the role played by brain CA's in the control of AP function has led to the utilization of adrenergic drugs as a reliable screening procedure for assessing the functional intactness of the hypothalamopituitary system. Administration of the parent molecules, most probably because of the poor penetrability of the BBB (Axelrod, 1965), gave inconsistent results (see Chapter IV, Section A,1,a), whereas the introduction of permeable precursor molecules such as L-dopa proved to be more rewarding.

In spite of the alleged effectiveness of DA in stimulating or inhibiting release of LH-RH in the experimental animal (see Chapter IV, Section A,3,a), a single dose of L-dopa in normal subjects (Eddy *et al.*, 1971; Hayek and Crawford, 1972), or in patients with amenorrhea and galactorrhea (Zarate *et al.*, 1973a) produced no significant changes in serum FSH and LH levels, although in one study increased serum gonadotropin levels were found in normal female subjects (Stevens and Boutselis, 1972).

Also, in patients with Parkinson's disease, chronic administration of L-dopa did not alter FSH and LH secretion (Lundberg, 1972) (see also Section A,3a). DA was observed to increase serum LH and FSH concentrations in two postmenopausal women (Franchimont, 1971), but the poor penetration of the amine into brain tissue makes the results questionable. In contrast to these findings, a significant fall in LH but not FSH levels, with an immediate rebound, has been noted in normal subjects of both sexes after administration of DA (Leblanc *et al.*, 1976) or L-dopa (Lachelin *et al.*, 1977).

The demonstration by Boyd *et al.* (1970) and Kansal and associates (1972) of L-dopa stimulation of GH secretion has introduced a new method for the investigation of hypothalamic–pituitary function in humans and for the treatment of those conditions associated with hGH dysfunction. The results of experiments in which the effects of L-dopa on GH release were studied have indicated that it is as reliable a stimulus for GH release as are arginine and insulin. Weldon *et al.* (1973), in experiments with many children of short stature who were subjected to insulin, arginine, or L-dopa administration, observed fairly consistent agreement in the results, and similar conclusions were reached by Lin and Tucci (1972) and Porter *et al.* (1972). Peak GH responses to L-dopa appeared at 90 min compared to the peak arginine response at 45 and 60 min. In the opinion of these authors the L-dopa test (Larodopa, 500 mg orally) is advisable for the diagnosis of GH deficiency since it is simple to perform; it does not require an iv infusion, does not involve the risk of hypoglycemia, nor does it require a physician in constant attendance. However, some normal subjects failed to respond to L-dopa and, in others, the increase in GH levels did not occur until 90–120 min after oral administration (Kansal *et al.*, 1972; Sachar *et al.*, 1972; Chakmakjian *et al.*, 1973).

Whether a given oral dose of L-dopa is sufficient to induce a reliable hGH stimulation in a normal individual will depend on many factors, of which absorption and transport across the gastrointestinal wall, rapidity of systemic L-dopa degradation, and degree of penetration through the BBB into brain parenchyma are but a few. Decreased responsiveness to L-dopa was observed after a repeat oral dose of the drug. The fact that addition of a peripheral L-AAAD inhibitor (see below) to the restimulation dose of L-dopa shortened but did not eliminate the refractory interval was taken to indicate that the latter was due to a centrally rather than peripherally located mechanism (Parker and Eddy, 1976).

Animal experiments have indicated that only a small proportion (less than 0.1%) of oral L-dopa is metabolized to DA in the brain (Wurtman *et al.*, 1970). This percentage is significantly increased by the concurrent use of peripheral L-AAAD inhibitors (see Chapter II, Table VII). In this way, peripheral side effects attributed to L-dopa or its metabolites, including tachycardia, postural hypotension, nausea, and vomiting, are reduced because of the lower dose of L-dopa which can be administered.

In the human, the addition of one of these peripheral L-AAAD inhibitors, MK-486 (carbidopa, 50 mg) with an oral dose of 250 mg L-dopa, resulted in a fivefold potentiation of plasma L-dopa and a twofold increase in hGH levels. An analysis of the individual curves suggested that a minimal L-dopa level of 400 ng per millimeter plasma is necessary to evoke an unequivocal hGH response in healthy individuals (Mars and Genuth, 1973). Similar, though lower plasma L-dopa titers (at least 250 ng/ml) have been indicated by Essman *et al.* (1976) for subjects who respond to L-dopa with a rise in GH. Potentiation of the L-dopa effect by MK-486 (50 mg) has also been reported by Frantz *et al.* (1973), who observed GH stimulation of equal degree when 250 mg of L-dopa with MK-486 or 500 mg of L-dopa alone was administered to 6 normal subjects. Administration of carbidopa alone did not alter baseline GH levels (G. M. Brown *et al.,* 1976b). Collectively, it would appear from these results that the use of oral L-dopa as a test for GH secretion is more reliable if supplemented by decarboxylase inhibitors. The L-dopa-related GH peaks are reduced in aged, normal subjects and in unipolar depressed patients (Sachar *et al.,* 1972). In both conditions the deficient hGH response to L-dopa might be due to a central neurochemical disturbance in CA metabolism or/and action (see Chapter II, Section B,13 and Van Praag and Korf, 1975). In contrast, bipolar, manic-depressive patients secrete substantial amounts of hGH in response to L-dopa, emphasizing the potential value of neuroendocrine techniques in studying the nature of hypothalamic dysfunction in the affective disorders (Sachar *et al.,* 1972).

The recognition that apomorphine, which selectively stimulates the DA receptor, is capable of promptly increasing serum hGH levels following systemic administration (Lal *et al.,* 1972, 1973; Brown *et al.,* 1973; Ettigi *et al.,* 1975) led to its use as a stimulatory agent for hGH secretion in normal subjects. Data to date on healthy men, based on blood samples taken 30, 45, and 60 min after apomorphine administration (0.1–1.0 mg), show elevated serum hGH levels (10 ng/ml or more) in the great majority of subjects, while in women some [e.g., 7 out of 24 trials in 13 women (Ettigi *et al.,* 1975)] failed to develop an increase in serum GH of 6 ng/ml or more. This is somewhat surprising in view of the higher GH responsiveness of women than man to other GH-provoking stimuli, such as arginine infusion or ambulation (Merimee *et al.,* 1966; Frantz and Rabkin, 1965). A comparative double-blind crossover study of L-dopa and apomorphine effects in the same subjects led to the conclusion that the latter drug elicited more reliable, prompt, and consistent elevation in GH levels than did L-dopa, with similar frequency and severity of side effects, thus emphasizing the potential usefulness of apomorphine as a stimulatory agent for assessing hGH secretory capacity (Lal *et al.*, 1975b).

A single dose of apomorphine (1.0 or 1.5 mg) to adult men decreased serum PRL but failed to decrease CPZ-induced elevated levels of PRL; similar to that observed for L-dopa, the drug had no effect on serum LH or FSH (Lal *et al.*, 1973). As for L-dopa, the GH response to apomorphine is significantly greater in younger than in older subjects (Maany *et al.*, 1975), which again emphasizes the possibility that "a final common pathway" responsible for GH release, which might involve dopaminergic receptors, may become less sensitive with age.

The availability of CNS-active compounds capable of blocking central dopaminergic receptors, but unlike the classic neuroleptic antipsychotic drugs such as CPZ and haloperidol, free of extrapyramidal, psychotropic, and tranquillizing effects (see Table XI), offers a new tool for the functional exploration of the hypothalamic pituitary system. Among them, metoclopramide (2.5 mg iv) appears to be a safe, reliable, and potent stimulus of PRL secretion (McNeilly *et al.*, 1974; Judd *et al.*, 1976), whose site of action seems to be mainly the CNS. While in normal subjects it induced a prompt and sustained rise in plasma PRL, there was no response to metoclopramide in patients with structural disorders of the hypothalamus (Judd *et al.*, 1976; Yamauchi *et al.*, 1976). Sulpiride, also, induced in normal subjects after single (100 mg po) or chronic (150 mg po daily) administration a quick, marked and sustained increment of serum PRL levels (Mancini *et al.*, 1976) and, according to one report (Mancini *et al.*, 1977), not confirmed by others (Thorner *et al.*, 1974b), also reduced the circulating gonadotropins. Clozapine is another drug in this group whose PRL-releasing effect observed in the rat (Meltzer *et al.*, 1975) should be assessed in man (Sachar *et al.*, 1977).

2. Drug Therapy for GH Hypersecretion: A Neuropharmacological Approach to Treating Acromegaly

For the use of CB-154 or cyproheptadine in the medical treatment of Cushing's disease or Nelson's syndrome see Chapter IV, Sections A,1,a and b.

In the past few years, results of experiments using different stimuli for GH secretion have indicated that in acromegaly GH secretion is often nonautonomous. Patterns of GH responsiveness to glucose suppression and/or provocative stimuli are highly variable in acromegalic subjects as a group, and GH secretion can be altered significantly if several perturbations are attempted (Cryer and Daughaday, 1969; Lawrence *et al.*, 1970).

The surprising variability and the persistence of normal or exaggerated GH responses to provocative stimuli led to the theory that in some patients acromegaly may begin as a primary hypothalamic disturbance with chronically elevated levels of GRF, ultimately resulting in hypertrophy and hyperplasia of pituitary somatotropes and, in some cases, pituitary tumor formation. This view, which challenged the concept that chronic hypersecretion of GH that results in gigantism and acromegaly is independent of a hypothalamic control (Glick *et al.*, 1965; Beck *et al.*, 1966), has been reinforced by the detection of elevated levels of circulating GRF activity in acromegalic plasma (Hagen *et al.*, 1971, 1972b).

If the hypothalamus played a role in the pathogenesis of hypersomatotropism, since the release of GH in the human is either noradrenergically or dopaminergically mediated, it was reasoned that the phenothiazines might act to lower elevated GH titers in active acromegaly. Previous animal studies had shown tht CPZ suppresses insulin-induced release of GH, acting directly on the CNS (Müller *et al.*, 1967a; see also Chapter IV, Section 2,a,i). In fact, Kolodny *et al.* (1971a) initially reported a marked decrease in fasting serum GH levels with improvement in symptoms in one acromegalic patient following a short course of CPZ therapy. In further studies the same authors (Kolodny *et al.*, 1971b) noted decreased serum GH levels and clinical improvement in only 2 out of 4 acromegalics following a longer CPZ treatment (2 weeks–1 month), and, more recently, Dimond *et al.* (1973) reported that out of 8 patients with active acromegaly, mean basal GH levels decreased markedly in only one following a 3- to 6-month period of treatment with high doses of CPZ. Clinical improvement did not occur in any patient, and patterns of GH secretion following administration of glucose, insulin, and arginine varied greatly in individual patients before and throughout treatment. On the basis of their results, they suggested that the magnitude of long-term response to CPZ may be properly evaluated after at least a 1-month trial period. From their studies, CPZ did not appear to be a satisfactory agent for the treatment of acromegaly.

Another antiadrenergic drug which has been tried therapeutically in acromegaly is phentolamine, alone or in association with β-stimulants. In 12 patients, 50 mg of the drug infused over 90 min suppressed basal plasma GH levels by more than 40% in 6 subjects (Nakagawa and Mashimo, 1973); in 10 acromegalics serum GH concentration fell to 67% after 120 min of iv phentolamine administration and to 50% of the mean baseline 1 hr after the addition of isoproterenol, which resulted in β-adrenergic stimulation on α-blockade (Cryer and Daughaday, 1974). The action of phentolamine in the acromegalic patients was attributed to α-adrenergic blockade of excessive α-adrenergic flow into the GRF-secreting hypophysiotropic neurons; the effect of superimposed β-stimulation was attributed to a reestablishment of diminished β-adrenergic activity, which possibly plays a role in excessive GH secretion. Further studies are needed to assess the effectiveness of α-adrenergic blockers as GH suppressors in acromegaly; in any event, the need for constant infusion of the drug makes its therapeutic application impractical.

Clinical observations suggested that carbamazepine, a dibenzazepine derivative, active as an anticonvulsant and antidiuretic agent (Fram *et al.*, 1969), might be of value in the treatment of acromegaly (Braunhofer and Zicha, 1966). However, in 3 acromegalic patients no effect of carbamazepine (3 × 200 mg/day) on arginine-induced GH secretion and on basal GH levels was seen. According to these findings this drug does not appear suitable for the medical management of acromegaly.

In addition to the extreme daily variation in plasma hGH levels (Cryer and Daughaday, 1969), acromegaly is characterized by paradoxical responses to different stimuli. The most usual response to glucose ingestion in acromegaly is a "paradoxical" rise in GH levels (Cryer and Daughaday, 1969). The unexpected variation—physiological or paradoxical—in blood GH concentrations resulting from different stimuli is reflected by the pattern of GH response to dopaminergic drugs. In 8 untreated women with acromegaly, L-dopa (500 mg) administered orally, as a single bolus, produced a marked decrease in plasma hGH levels within 30 min. In 6 patients, GH levels remained persistently low for 120–180 min (Liuzzi *et al.*, 1972). The paradoxical decrease of GH in the plasma of acromegalics following a single dose of L-dopa administration has been subsequently confirmed by many authors, e.g., Sherman *et al.* (1972) in 5 out of 7 patients; Mims *et al.* (1973a) in 5 out of 14 patients; and Guyda *et al.* (1973) in a 5 year-old female child with acromegalic gigantism. Treatment with chronic L-dopa administration (2–3 gm per day) proved to be unsuccessful; the brief inhibitory effect was in fact followed by a rebound increase of plasma GH to basal levels or higher (Chiodini *et al.*, 1974).

In the search for additional dopaminergic compounds capable of lowering hGH levels in acromegaly, it was then shown by the Italian group that the same inhibitory action of L-dopa was shared by apomorphine (Chiodini *et al.*, 1974). The transient inhibitory action on hGH levels (30–45 min) observed in 8 subjects, previously shown to be responsive to L-dopa, correlated well with the short-lived activation of DA receptors by apomorphine (Andén *et al.*, 1967). Further confirmation of the paradoxical effect of DA-like drugs on GH secretion in acromegaly was derived from the use of 2-Br-α-ergocryptine (Sandoz, CB 154), a drug which exerts DA receptor stimulation of long duration (Corrodi *et al.*, 1973) (see Table XI). A single dose of CB-154 (2.5 mg po) decreased plasma GH levels in the 7 patients studied. The fall was initiated within 60 min, and GH levels remained persistently suppressed until the last sampling (270–300 min) (Liuzzi *et al.,* 1974a). Figure 55 compares the GH-lowering effects of L-dopa and CB-154 in a group of "responder" acromegalics. In additional studies, it was shown that chronic CB-154 treatment (2.5 mg four times daily) in 7 subjects, who had exhibited suppressed GH levels to acute CB 154 administration, induced a marked decrease in serial resting GH levels determined at day 30 of the treatment period. Cessation of the treatment induced a prompt resumption of pretreatment hGH values, which were again suppressed following a second course of CB-154 therapy (Fig. 56). In 5 acromegalics who had not reacted to acute CB 154 administration, chronic treatment did not produce appreciable hGH inhibition (Chiodini *et al.,* 1975). Moreover, a single administration of double the dose (5.0 mg) did not produce better results in the unresponsive patients, suggesting different, individual sensitivities to dopaminergic stimulation.

That activation of the dopaminergic mechanism(s) underlies the hGH-lowering effect of CB-154 was suggested by the fact that pimozide (4.0 mg po for 7 days) given to 4 patients before CB-154 administration reduced the hGH-lowering effect (Müller *et al.,* 1976c).

Long-term administration of CB-154, in addition to suppressing circulat-

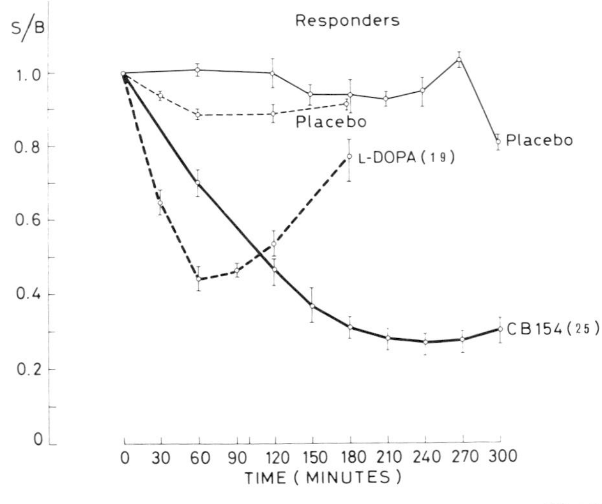

Fig. 55 Plasma hGH after administration of L-dopa (500 mg po) or CB-154 (2.5 mg po) in "responder" acromegalic subjects. In parentheses number of cases. Placebo refers to baseline determinations in L-dopa (----) or CB-154 (——) treated acromegalics. Values are expressed as ratio of suppressed (S) to baseline (B) values ±SE. (Reproduced, with permission, from Liuzzi *et al.,* 1976.)

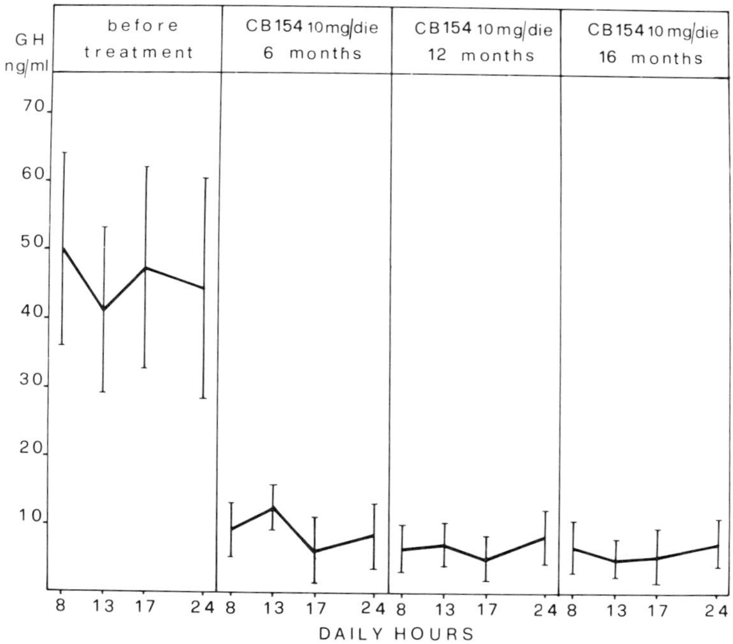

Fig. 56 Plasma hGH levels (mean ± SE) in 7 acromegalic patients, responsive to acute CB-154 (2.5 mg) administration, during chronic treatment with CB-154 (10 mg/day). (Reproduced, with permission from Chiodini *et al.*, 1975.)

ing GH levels, was followed by marked clinical (i.e., reduction in soft tissue thickening and improvement in facial features) and metabolic improvement. In an overt diabetic acromegalic, during the initial 15 days of therapy, chronic CB-154 suppressed hGH levels, reduced blood glucose, and glycosuria disappeared; oral antidiabetic therapy could be stopped (Fig. 57).

The beneficial effects of oral bromocriptine have been widely confirmed (Thorner *et al.*, 1975a; Camanni *et al.*, 1975a; Althoff *et al.*, 1975; Sachdev *et al.*, 1975). In addition, it has been shown that acute oral administration of piribedil, another DA-stimulant drug, lowers GH levels in some acromegalics, while amantadine and D-145 are ineffective (Camanni *et al.*, 1975b). Provided the dose of CB-154 is built up slowly, it is well tolerated; however, any patient given a sufficiently large initial dose may develop nasal stuffiness, nausea, vomiting, and postural hypotension. Micturition syncope has also been described. Normal people are most prone to these side effects; patients with acromegaly and hyperprolac-

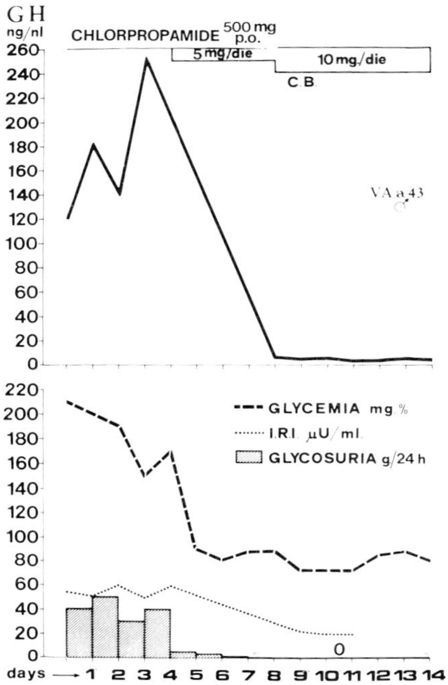

Fig. 57 Plasma hGH and immunoreactive insulin (IRI) levels, blood glucose and glycosuria in an acromegalic subject with overt diabetes before and after institution of CB-154 treatment. (Reproduced, with permission, from Müller *et al.*, 1976c.)

tinemia (see Section A,3) less so; and, finally, the most resistant to the action of CB-154 as well as other DA stimulants are women immediately postpartum, in whom side effects are virtually unknown. Recently, the occurrence of gastrointestinal bleeding due to peptic ulcers of the stomach and duodenum has been reported in 6 out of 96 acromegalic subjects treated chronically with CB-154 (Wass *et al.*, 1976); however, this finding has not been confirmed by other authors using a larger group of patients (del Pozo and Maclay, 1976).

It appears from above data that the selective suppressive effect of bromocriptine on GH (and PRL) levels, the rapidity of its action, and its low toxicity make it particularly suitable for early, mild acromegaly in which it may be important to preserve other pituitary functions, particularly gonadotropin secretion. Bromocriptine may also be used with benefit in the large number of patients who have shown only partial response to other forms of therapy. In view of the present uncertainty about its effect on tumor size, bromocriptine administration appears unwise in patients with a suprasellar extension of the tumor because of the possibility of further expansion and the risk to vision.

The depressing effect of dopaminergic drugs on hGH levels in acromegaly must be considered "paradoxical" on the basis of the knowledge available presently on the physiopathology of GH secretion in this disease (Liuzzi *et al.,* 1976). It is similar to the "paradoxical" effects of suppressant agents on GH secretion during the first two weeks of life (Cornblath *et al.,* 1965), a situation which shares in common with acromegaly high circulating levels of GH.

Although the precise site and mechanism(s) of action of dopaminergic drugs in lowering hGH in acromegaly have not yet been completely clarified some theories on these aspects are presented below. The almost complete dissociation observed in the same patient between responses elicited by dopaminergic drugs and those by the application of stimuli (e.g., insulin hypoglycemia, glucose or arginine infusion) almost with certainty acting on the CNS (see Chapter III, Sections A,3,a, and b) (Liuzzi *et al.,* 1974b) favors an extra-CNS site of action of the dopaminergic stimuli. In line with a pituitary site of action are the reported homogeneity in the hGH response in acromegalics to a combined application of CB-154 testing and TRH (Liuzzi *et al.,* 1974b), a stimulus which acts at the pituitary level (Udeschini *et al.,* 1976), and the observation of Hoyte and Martin (1975), expanded by Faglia *et al.* (1977), that in acromegalics with microadenomata selective adenomectomy suppressed both the paradoxical decrease to the DA agonists and the altered responsiveness to TRH.

Although DA-like compounds are generally believed to act on CNS loci, a direct effect of CA's on the pituitary for PRL secretion has been reported by many groups of workers (see Chapter IV, Section A,4,a). Also, for GH secretion acute suppression of hormone serum levels has been reported in rats bearing transplanted GH-secreting tumors following high doses of L-dopa (Malarkey and Daughaday, 1972). Evidence for a direct pituitary site of action of dopaminergic compounds in acromegaly has been provided by the observation that DA, which does not cross the BBB (see Chapter IV), as does L-dopa, apomorphine, and CB 154, proved to be effective in lowering GH levels when infused into 4 bromocriptine-responsive acromegalics (Massura *et al.,* 1976), an observation confirmed by Verde *et al.* (1976) with a larger groups of patients. DA also inhibited the GH response to TRH (Camanni *et al.,* 1977). Although an action of systemically administered DA at the ME level cannot be excluded, since this area is thought to lie outside the BBB, this is unlikely because of the finding that systemic DA was completely ineffective in elevating plasma GH levels in normal subjects (Verde *et al.,* 1976). Also the ineffectiveness of carbidopa, a peripheral L-AAAD inhibitor (see

Table VII), on GH levels in normal subjects (G. M. Brown *et al.,* 1976b) makes the possibility that DA neurons terminating in the ME may be involved in GH regulation unlikely.

An analysis of the dopaminergic compounds used so far in acromegaly showed that they were not all effective in decreasing plasma hGH levels. Of the five dopaminergic compounds investigated in the study of Camanni *et al.* (1975b), piribedil (see Table XI) was less effective than CB 154 and L-dopa and no activity was shown in "responder" acromegalics with amantadine or D-145, an amantadine derivative. Also amphetamine, which elevates hGH levels in normal subjects, failed to reduce hGH levels in acromegalic patients (Müller *et al.,* 1977a), and equally ineffective was nomifensine, a blocker of DA reuptake (see Table XI) (E. E. Müller, unpublished results). From these findings it would appear that the ability of a given compound to suppress the release of GH from the tumoral or hyperplastic pituitary is related to its ability to stimulate directly the DA receptor site. Apomorphine, CB 154, DA, and lisuride, which share in common the ability of stimulating the DA-sensitive AC of the rat striatum (see Table XI), and L-dopa, which can be converted to DA by the L-AAAD system present in the AP (see Chapter II, Section B,5), were effective, whereas amantadine, D-145, amphetamine, and nomifensine, whose action occurs mainly through presynaptically released DA, were not. Since it is unlikely that the pituitary contains the presynaptic neuronal component through which the activation of the postsynaptic CA receptor occurs physiologically (see Chapter II, Section C,3), only drugs directly affecting the postsynaptic site at the pituitary level would be capable of lowering hGH levels in acromegaly. Figure 58 shows schematically the

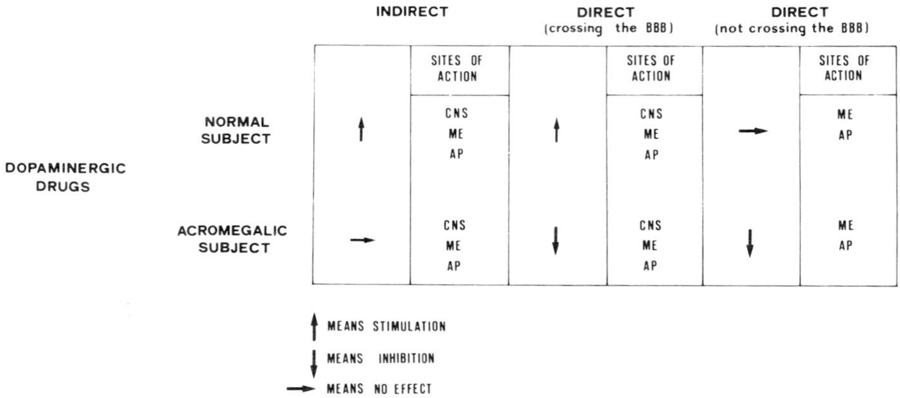

Fig. 58 Effect of indirect or direct dopaminergic drugs on hGH in normal or acromegalic subjects.

effects of direct or indirect DA-agonist drugs on hGH levels in normal or acromegalic "responder" patients.

The view that the release of GH in the human may be serotoninergically mediated (see Chapter IV, Section A,2,b) has led to the use of 5-HT blockers in GH hypersecretion. In one acromegalic woman decreased GH levels were reported on two occasions following administration of methysergide (Bivens *et al.*, 1973b), and during oral glucose-tolerance tests methysergide in one patient and Cy in 4 of 6 patients decreased plasma GH concentration (Feldman *et al.*, 1976). However, in larger groups of patients a single or chronic administration of Cy was found completely ineffective (Winkelmann *et al.*, 1975; Chiodini *et al.*, 1976).

A consistent, but short-lived suppression of GH levels has been observed in 5 out of 12 acromegalics following methergoline administration (Chiodini *et al.*, 1976), a drug reportedly possessing a specific 5-HT receptor-blocking activity (see Table XIV). However, the homogeneity encountered in the same subjects between GH response to methergoline and to CB-154, the ineffectiveness in the same subjects to Cy, and, finally, the suppressive effect of a pimozide pretreatment suggest that the methergoline effect was not due to an anti-5-HT action but probably to a dopaminergic mechanism.

In Table XXXVIII some of the drugs which have been used for suppressing GH hypersecretion in acromegalic subjects are listed.

3. L-Dopa and Related Drugs in States of Hyperprolactinemia

Experimental evidence linking brain CA's to PRL secretion (see Chapter IV, Section A,4,a) and the clinical availability of L-dopa provided the opportunity of studying the effect of this CA precursor on PRL secretion and of the pharmacological approach to hyperprolactinemia states.

Inappropriate secretion of hPRL is recognized as one of the important causes of galactorrhea of diverse origin, e.g., that observed after disruption of the portohypophyseal circulation, after administration of certain drugs, including birth control pills, during puerperium, and in some patients with chromophobe pituitary tumors or with no clinically detectable disease of the pituitary or hypothalamus. Galactorrhea with high prolactin levels is usually associated with hypogonadism, whether the hyperprolactinemia is physiological during the postpartum period or pathological due to pituitary or hypothalamic disease. This complication occurs most often because the high PRL levels interfere either with the hypothalamic LH-RH centers or with the direct effects of the gonadotropins on the gonads, or, in few cases, because of associated destruction of either the pituitary or the hypothalamus (see Besser and Mortimer, 1976).

TABLE XXXVIII

Neuropharmacological Drugs Used in Acromegaly[a]

Class of compound or mechanism of action	Drug	Effect on GH levels[b]	Authors
Phenothiazines	Chlorpromazine	↓	Kolodny *et al.* (1971a)
		→	Dimond *et al.* (1973)
		→	AvRuskin *et al.* (1973)
α-Adrenergic blockers alone or +	Phentolamine alone or +	↓	Nakagawa and Mashimo (1973)
β-stimulants	Isoproterenol	↓	Cryer and Daughaday (1974)
Dibenzazepines	Carbamazepine	→	Braunhofer and Zicha (1966)
DA agonists	DA	↓	Massara *et al.* (1976); Verde *et al.* (1976)
	Apomorphine	↓	Chiodini *et al.* (1974)
	CB 154	↓	Liuzzi *et al.* (1974a); Thorner *et al.* (1975a); Camanni *et al.* (1975a)
	Piribedil	↓	Camanni *et al.* (1975b)
	Lysenil[c]	↓	A. Liuzzi and P. G. Chiodini, unpublished results
	L-Dopa	↓	Liuzzi *et al.* (1972); Sherman *et al.* (1972); Mims *et al.* (1973a); Guyda *et al.* (1973)
	Amantadine	→	Camanni *et al.* (1975b)
	D-145	→	Camanni *et al.* (1975b)
	Amphetamine	→	Müller *et al.* (1977a)
	Nomifensine	→	E. E. Müller, unpublished results
5-HT blockers	Methysergide	↓	Bivens *et al.* (1973b); Feldman *et al.* (1976)
	Cyproheptadine	→	Winkelmann *et al.* (1975); Chiodini *et al.* (1976)
	Methergoline	↓	Chiodini *et al.* (1976)

[a] Key to symbols: →, no effect; ↓, inhibition.
[b] See text for details.
[c] Lisuride hydrogen maleate.

In women with nonpuerperal lactation and elevated serum concentrations of PRL due either to PRL-secreting pituitary tumors (Forbes-Albright syndrome) or functional pituitary disorders, a single dose of L-dopa (500 mg) resulted in a profound fall in serum PRL in 90 min. Sustained suppression of serum PRL levels could not be achieved, however, when the drug was given four times daily, and more prolonged treatment (six weeks) did not change basal serum hPRL or milk production (Malarkey *et al.,* 1971). Turkington (1972b), however, reported favorable results in 11 out of 14 galactorrhea patients treated chronically with L-dopa. Discontinuation of therapy was associated with the return of PRL levels to the initial control values and with resumption of galactorrhea. More recently, Zarate *et al.* (1973b) reported restoration of menses with chronic L-dopa therapy in 9 out of 11 women with inappropriate lactation. Ovulation was confirmed in 4 of them. On the other hand, slight PRL responses to L-dopa have been reported in nonpathological hyperprolactinemia such as in the last trimester of pregnancy (Pujol-Amat *et al.*, 1973) or that resulting from sulpiride administration (L'Hermite *et al.*, 1975).

As already mentioned, L-dopa is converted to DA in the hypothalamus, and its efficacy after administration of an oral dose depends on the concurrence of many factors (see Section A,1). It is therefore not unexpected that the prolactin-inhibiting effect of a dose of L-dopa in man has a short duration of action (Malarkey *et al.*, 1971). Thus, this compound must be administered systemically in order to maintain stable reduction of PRL serum levels. In endocrine patients, its therapeutic application is also limited by the high incidence of side effects and the reported impairment of glucose tolerance (Sirtori *et al.*, 1972). Nevertheless, its use has initiated the development of other drugs capable of similar induction of high dopaminergic activity within the hypothalamus and/or the pituitary.

Similar to L-dopa, apomorphine was found effective in suppressing serum hPRL levels in abnormal hPRL secretory states, whether secondary to assumed hypothalamic disturbance (2 cases) or to hPRL-secreting pituitary adenomas (4 cases) (Martin *et al.*, 1974). Apomorphine, however, is not likely to be a useful therapeutic drug, having a very short duration of action and giving rise to side effects. The same limitations can be applied to DA, which, when infused in 2 patients with amenorrhea–galactorrhea, induced a marked inhibition of the elevated plasma PRL levels (Crosignani *et al.*, 1976).

Ergot alkaloids, known to exhibit antigalactic properties in the experimental animal (see del Pozo and Flückiger, 1973, and Chapter IV, Section A,4,a,i), have provided more enduring benefit in many cases of hyperprolactinemia. Lawrence and Hagen (1972), in a group of patients with nonpuerperal galactorrhea treated with ergonovine maleate, observed cessation of the syndrome within 3 months of therapy initiation. Discontinuation of the drug was not associated with the return of galactorrhea.

The search for a compound with specific antigalactic action and free of the uterotonic and vascular properties of the more traditional ergot alkaloids resulted in the selection of 2-Br-α-ergocryptine (Flückiger and Wagner, 1968; Flückiger, 1972). Lutterbeck *et al.* (1971) reported termination of galactorrhea with CB-154 in 3 patients suffering from a nonpuerperal form of the disease for periods ranging from 2 months to 6 years.

Later, Besser *et al.* (1972) were able to induce with CB-154 suppression of lactation and diminution of the elevated plasma PRL levels in 5 patients with inappropriate lactation and amenorrhea. Similar results were obtained by Copinschi *et al.* (1972) in a male subject with galactorrhea and secondary hypogonadism due to a pituitary tumor; after failure to induce a response with L-dopa, CB-154 markedly depressed serum PRL levels and consistently reduced milk secretion.

Cessation of galactorrhea and suppression of PRL levels are accompanied by regular menstruation in women and return *libido* and potency in men. Cessation of galactorrhea and lowering of serum PRL concentration was also reported by del Pozo *et al.* (1972) in 2 female subjects and more recently (del Pozo *et al.*, 1974) in 9 more cases. In the latter, treatment with CB-154 (3–5 mg/day for 2–8 months) produced a fall in serum hPRL levels accompanied by cessation of milk secretion in 8 cases and restored regular menses and ovulation in 7 amenorrheic women. Satisfactory results with CB-154, as shown by restoration of normal gonadal function, were also reported by Thorner *et al.* (1974a) in a number of hypogonadal men and women. Irrespective of the cause of the elevated PRL levels, the only patients who did not respond appeared to be the minority who were shown to be truly deficient in gonadotropins.

Since bromocriptine is neither teratogenic nor contraceptive, and because in man PRL suppression does not block implantation of the ovum as in the rat, it has been used recently to restore fertility (Thorner *et al.*, 1975b). Although its role still must be clarified, it appears to be effective not only in patients with hyperprolactinemia but also in normoprolactinemic patients with secondary amenorrhea (Seppälä *et al.*, 1976). The results of these and of a large series of analogous studies have been summarized (Editorials, 1975, 1977). In hyperprolactinemia, CB-154 therapy produces a gradual fall in serum PRL. The drug is also effective in suppressing abnormal PRL-dependent galactorrhea, without side effects with the dosages used, and, at the same time, allows the return of normal menstruation or potency when amenorrhea or impotence with low plasma androgens is present. The latter effect would result from the fall of PRL levels which return the cyclic LH-RH hypothalamic center(s) to normal functioning (Boyar *et al.*, 1974; Bohnet *et al.*, 1976) without excluding suppression of the direct PRL effect at the level of the pituitary (Maneckjee *et al.*, 1976) or the ovary (Zarate *et al.*, 1972b).

Recently, a controlled trial of estrogen versus CB-154 administration for suppression of puerperal lactation showed the latter to be more effective as judged by milk flow and the relief of breast pain and congestion. Headache and nausea, sometimes observed in nonpuerperal patients receiving bromocriptine, did not occur. Interruption of treatment after 14 days resulted, however, in a small rebound rise in prolactin levels on day 21, corresponding to the clinical observation of a slight increase in milk leakage, congestion, and discomfort (Walker *et al.*, 1975). Thus, although bromocriptine now appears to be the most effective suppressor of puerperal lactation, long-acting or depot preparations of the drug or ergot derivatives with an even more sustained duration of action would be advantageous. The same conclusion holds true for acromegaly.

In addition to peptide-containing ergot alkaloids, such as CB-154, a number of simple synthetic ergoline derivatives, e.g., lergotrile, lisuride, and methergoline, which lack the peptide side chain of ergot and appear to possess DA-like stimulation properties (Clemens *et al.*, 1975; Horowski and Watchel, 1976; Chiodini *et al.*, 1976), are now undergoing successful clinical trial for the inhibition of puerperal lactation and for the treatment of prolactin-dependent disorders (Lemberger *et al.*, 1974; Crosignani *et al.*, 1977). In addition, other stimulants of DA-receptors, e.g., piribedil (Müller *et al.*, 1977b), or blockers of DA reuptake, such as nomifensine (A. R. Genazzani, unpublished results), appear to be promising in the medical management of excessive PRL secretion. Finally, another approach for the treatment of hyperprolactinemic states, which involves an enhance-

TABLE XXXIX

Drugs Which Inhibit Plasma PRL in States of Hyperprolactinemia

Class of compounds	Drug	Authors
DA AGONISTS	Da	Crosignani *et al.* (1976)
	Apomorphine	Martin *et al.* (1974)
	Piribedil	Müller *et al.* (1977b)
	L-Dopa	Malarkey *et al.* (1971)
	Nomifensine	A. R. Genazzani, unpublished results
Ergot alkaloids		
	Ergonovine	Lawrence and Hagen (1972)
	Methylergonovine	Perez-Lopez *et al.* (1975)
	2-Br-α-ergocryptine	Lutterbeck *et al.* (1971); del Pozo *et al.* (1972, 1974); Besser *et al.* (1972); Copinschi *et al.* 1972)
Ergolines		
	Lergotrile	Lemberger *et al.* (1974)
	Lysenil	R. Horowski, personal communication
	Methergoline[a]	Crosignani *et al.* (1977); Delitala *et al.* (1977).
VITAMINS	Vit. B_6	McIntosh (1976)

[a] Central 5-HT blockage.

ment of dopaminergic neurotransmission, seems to be offered by pyridoxine, which functions in the decarboxylation of L-dopa to DA (see also Chapter II, Section B,6). A chronic pyridoxine treatment (200–600 mg/day) in 3 women with the galactorrhea–amenorrhea syndrome and elevated PRL concentrations induced a return of regular ovulatory menses, cessation of the galactorrhea, and inhibition of PRL levels (McIntosh, 1976).

Table XXXIX lists some of the drugs which have been used in hyperprolactinemic states.

Bibliography

Abe, K., and Hiroshige, T. (1974). *Neuroendocrinology* **14,** 195.

Abrams, R. L., Parker, M. L., Blanco, S., Reichlin, S., and Daughaday, W. H. (1965). *Endocrinology* **78,** 605.

Abrams, R. L., Grumbach, M. M., and Kaplan, S. L. (1971). *J. Clin. Invest.* **50,** 940.

Abramson, F. P., McCaman, M. W., and McCaman, R. E. (1974). *Anal. Biochem.* **57,** 482.

Acheson, G. H., and Moe, G. K. (1946). *J. Pharmacol.* **87,** 220.

Adam, H. M. (1961). *In* "Regional Neurochemistry" (S. S. Kety and J. Elkes, eds.), pp. 293–306. Pergamon, Oxford.

Adam, H. M., and Hey, H. K. A. (1966). *Br. J. Pharmacol. Chemother.* **28,** 137.

Aghajanian, G. K. (1972). *Fed. Proc., Fed. Am. Soc. Exp. Biol.* **31,** 91.

Aghajanian, G. K., and Bloom, F. E. (1967a). *J. Pharmacol. Exp. Ther.* **156,** 23.

Aghajanian, G. K., and Bloom, F. E. (1967b). *J. Pharmacol. Exp. Ther.* **156,** 407.

Aghajanian, G. K., and Bunney, B. S. (1973). *In* "Frontiers in Catecholamine Research" (E. Usdin and S. Snyder, eds.), pp. 643–648. Pergamon, Oxford.

Aghajanian, G. K., and Bunney, B. S. (1974). *In* "Frontiers in Neurology and Neuroscience Research" (P. Seeman and G. M. Brown, eds.), pp. 4–11. University of Toronto, Toronto.

Aghajanian, G. K., and Freedman, D. X. (1968). *In* "Psychopharmacology: A 10-Year Progress Report" (D. H. Efron, ed.), pp. 1185–1193. U.S. Govt. Printing Office, Washington, D.C.

Aghajanian, G. K., and Gallager, D. W. (1975). *Brain Res.* **88,** 221.

Aghajanian, G. K., Bloom, F. E., and Sheard, M. H. (1969). *Brain Res.* **13,** 266.

Agrawal, H. C., Davison, A. N., and Kaczmarek, L. K. (1971). *Biochem. J.* **122,** 759.

Ahlqvist, R. P. (1967). *Ann. N.Y. Acad. Sci.* **139,** 549.

Ahren, K., Fuxe, K., Hamberger, L., and Hökfelt, T. (1971). *Endocrinology* **88,** 1415.

Ajika, K., and Hökfelt, T. (1973). *Brain Res.* **57,** 97.

Ajika, K., and Okinaga, S. (1976). *Abstr. Book Int. Congr. Endocrinol., 5th, 1976* Abstract No. 154.

Ajika, K., Kalra, S. P., Fawcett, C. P., Krulich, L., and McCann, S. M. (1972). *Endocrinology* **90,** 707.

Akmayev, I. G., and Donàth, I. (1966). *Z. Mikrosk.-Anat. Forsch.* **74,** 83.

Akmayev, I. G., Réthelyi, M., and Majorossy, K. (1967). *Acta Biol. Acad. Sci. Hung.* **18,** 187.

Albe-Fessard, D., Raiera, S., and Santiago, N. (1967). *J. Physiol. (Paris)* **59,** 324.

Alberici, M., Rodriguez de Lores Arnaiz, G., and De Robertis, E. (1965). *Life Sci.* **4,** 1951.

Alberti, K. G. M. M., Christensen, S. E., Iversen, J., Sayer-Hansen, K., Christensen, N. J., Prange-Hansen, A., Lundbaek, K., and Ørskov, H. (1973). *Lancet* **2,** 1299.

Albrecht, P., Visscher, M. B., Bittner, J. J., and Halberg, F. (1956). *Proc. Soc. Exp. Biol. Med.* **92,** 703.

Alford, F. P., Bloom, S. R., Nabarro, J. D. N., Hall, R., Besser, G. M., Coy, D. H., Kastin, A. J., and Schally, A. V. (1974). *Lancet* **2,** 974.
Algeri, S., and Cerletii, C. (1972). *Abstr. Book Int. Congr. Pharmacol., 5th, 1972* p. 4.
Alouisi, A., and Weiner, N. (1966). *Proc. Natl. Acad. Sci. U.S.A.* **56,** 1491.
Alpert, L. C., Brawer, J. R., Patel, Y. C., and Reichlin, S. (1976). *Endocrinology* **98,** 255.
Althoff, P.-H., Neubauer, M., Handzel, R., Bechstein, V., and Schöffling, K. (1975). *Ric. Sci. Edu. Perm.* **2,** Suppl. 1, 102.
Altszuler, N., and Hampshire, J. (1976). *58th Annu. Meet. Am. Endocr. Soc., San Francisco* Abstract No. 463.
Amenomori, Y., and Meites, J. (1970). *Proc. Soc. Exp. Biol. Med.* **134,** 492.
Amenomori, Y., Chen, C. L., and Meites, J. (1970). *Endocrinology* **86,** 506.
Amin, A. N., Crawford, T. B. B., and Gaddum, J. H. (1954). *J. Physiol. (London)* **126,** 598.
Amoss, M., Blackwell, R., and Guillemin, R. (1972). *J. Clin. Endocrinol. Metab.* **34,** 434.
Amoss, M. S., Jr., Monahan, M. W., and Verlander, M. S. (1974). *J. Clin. Endocrinol. Metab.* **39,** 203.
Andén, N.-E. (1964). *Acta Pharmacol. Toxicol.* **21,** 260.
Andén, N.-E. (1972). *J. Pharm. Pharmacol.* **24,** 905.
Andén, N.-E. (1974). *Naunyn-Schmiedebergs Arch. Pharmacol.* **283,** 419.
Andén, N.-E., and Bédard, P. (1971). *J. Pharm. Pharmacol.* **23,** 460.
Andén, N.-E., and Magnusson, T. (1964). *Proc. Int. Pharmacol. Meet., 2nd, 1963* Vol. 3, pp. 319–328.
Andén, N.-E., Roos, B. E., and Werdinius, B. (1964a). *Life Sci.* **3,** 149.
Andén, N.-E., Carlsson, A., Dahlström, A., Fuxe, K., Hillarp, N. Å., and Larsson, K. (1964b). *Life Sci.* **3,** 523.
Andén, N.-E., Fuxe, K., and Hökfelt, T. (1966). *J. Pharm. Pharmacol.* **18,** 630.
Andén, N.-E., Rubenson, A., Fuxe, K., and Hökfelt, T. (1967). *J. Pharm. Pharmacol.* **19,** 627.
Andén, N.-E., Corrodi, H., Fuxe, K., and Hökfelt, T. (1968). *Br. J. Pharmacol.* **34,** 1.
Andén, N.-E., Carlsson, A., and Häggendal, J. (1969). *Annu. Rev. Pharmacol.* **9,** 119.
Andén, N.-E., Butcher, S. G., Corrodi, H., Fuxe, K., and Ungerstedt, U. (1970). *Eur. J. Pharmacol.* **11,** 303.
Angeli, A. (1974). *Chronobiologia* **1,** Suppl. 1, 253.
Änggård, E. (1971). *In* "Prostaglandins" (P. Ramwell and J. E. Shaw, eds.), pp. 200–217. N.Y. Acad. Sci., New York.
Annunziato, L., Soscia, A., and Di Renzo, G. F. (1973). *Proc. Congr. Soc. Ital. Pharmacol., 1972* p. 3.
Annunziato, L., Di Renzo, G. F., Lombardi, G., Preziosi, P., and Scapagnini, U. (1974). *Br. J. Pharmacol.* **52,** 442P.
Annunziato, L., Lombardi, G., Di Renzo, G. F., Minozzi, M., Scapagnini, U., and Preziosi, P. (1975a). *In* "Gli Ormoni dell'Ipofisi Anteriore" (E. E. Müller and S. Piazzi, eds.), pp. 315–319. I.S.V.T., Sclavo, Siena.
Annunziato, L., Lombardi, G., Di Renzo, G. F., Scapagnini, U., and Preziosi, P. (1975b). *Abstr. Book Int. Congr. Pharmacol., 6th, 1975* Abstract No. 201.
Ansell, G. B., and Spanner, S. (1970). *In* "Drugs and Cholinergic Mechanisms in the CNS" (E. Heilbronn and A. Winter, eds.), pp. 143–159. Försvarets Forskningsanstalt, Stockholm.
Anton-Tay, F. (1971). *Pineal Gland, Ciba Found. Symp., 1970* pp. 213–220.
Anton-Tay, F., and Wurtman, R. J. (1968). *Science* **159,** 1245.
Anton-Tay, F., and Wurtman, R. J. (1969). *Nature (London)* **221,** 474.
Anton-Tay, F., and Wurtman, R. J. (1970). *Neuroendocrinology* **6,** 265.

Anton-Tay, F., and Wurtman, R. J. (1971). *In* "Frontiers in Neuroendocrinology" (L. Martini and W. F. Ganong, eds.), pp. 45–66. Oxford Univ. Press, London and New York.

Anton-Tay, F., Chou, C., Anton, S., and Wurtman, R. J. (1968). *Science* **162,** 277.

Anton-Tay, F., Diaz, J. L., and Fernandez-Guardiola, A. (1971). *Life Sci.* **10,** 841.

Antony, G. J., Van Vyk, J. J., French, F. S., Weaver, R., Dugger, G. S., Timmons, R. L., and Newsome, J. F. (1969). *J. Clin. Endocrinol. Metab.* **29,** 1238.

Apfelbaum, M. E., and Taleisnik, S. (1976). *J. Endocrinol.* **68,** 127.

Aprison, M. H. (1962). *Recent Adv. Biol. Psychiatry* **4,** 133.

Aprison, M. H., and Werman, R. (1965). *Life Sci.* **4,** 2075.

Aprison, M. H., Shank, R. P., and Davidoff, R. A. (1969). *Comp. Biochem. Physiol.* **28,** 1345.

Arendt, J., Paunier, L., and Sizonenko, P. C. (1975). *J. Clin. Endocrinol. Metab.* **40,** 347.

Ariëns, E. J. (1962). *Proc. Int. Pharmacol. Meet., 1st, 1961* Vol. 7, p. 247.

Arimura, A. (1970). *52nd Annu. Meet. Am. Endocr. Soc., St. Louis* p. 127.

Arimura, A., and Schally, A. V. (1970). *Endocrinology* **87,** 653.

Arimura, A., and Schally, A. V. (1971). *Proc. Soc. Exp. Biol. Med.* **136,** 290.

Arimura, A., and Schally, A. V. (1975). *Hypothal. Horm.* [*Symp.*], *1974* pp. 27–42.

Arimura, A., Saito, M., and Wakabayashi, I. (1969). *51st Annu. Meet. Am. Endocr. Soc., New York* p. 46.

Arimura, A., Matsuo, H., Baba, Y., and Schally, A. V. (1971). *Science* **174,** 511.

Arimura, A., Debeljuk, L., Matsuo, H., and Schally, A. V. (1972a). *Proc. Soc. Exp. Biol. Med.* **139,** 851.

Arimura, A., Dunn, J. D., and Schally, A. V. (1972b). *Endocrinology* **90,** 378.

Arimura, A., Spies, H. G., and Schally, A. V. (1973a). *J. Clin. Endocrinol. Metab.* **36,** 372.

Arimura, A., Sato, H., Kumasaka, T., Worobeck, R. B., Debeljuk, L., Dunn, J., and Schally, A. V. (1973b). *Endocrinology* **93,** 1092.

Arimura, A., Vilchez-Martinez, J. A., Coy, D. H., Coy, E. J., Irotsu, Y., and Schally, A. V. (1974). *Endocrinology* **95,** 1174.

Arimura, A., Sato, H., Dupont, A., Nishi, A., and Schally, A. V. (1975a). *Science* **189,** 1007.

Arimura, A., Sato, H., Coy, D. H., and Schally, A. V. (1975b). *Proc. Soc. Exp. Biol. Med.* **148,** 784.

Arimura, A., Smith, W. D., and Schally, A. V. (1976a). *Endocrinology* **98,** 540.

Arimura, A., Gordin, A., and Schally, A. V. (1976b). *Fed. Proc., Fed. Am. Soc. Exp. Biol.* **35,** Abstr. 3174.

Arnold, R., and Creutzfeldt, W. (1975). *Dtsch. Med. Wochenschr.* **100,** 1014.

Asano, Y. (1971). *Life Sci.* **10,** 883.

Ash, A. S. F., and Tucker, J. F. (1966). *Nature (London)* **209,** 306.

Ashcroft, G. W., and Sharman, D. F. (1960). *Nature (London)* **186,** 1050.

Ashkenazi, R., Holman, R. B., and Vogt, M. (1972). *J. Physiol. (London)* **223,** 255.

Audigier, Y., Virion, A., and Schwartz, J.-C. (1976). *Nature (London)* **262,** 307.

Aùskovà, M., Rezabek, K., Zikan, V., and Semonsky, M. (1974). *Endocrinol. Exp.* **8,** 51.

Avanzino, G. L., Bradley, P. B., and Wolstencroft, J. H. (1966). *Br. J. Pharmacol. Chemother.* **27,** 157.

Averill, R. L. W., and Kennedy, T. H. (1967). *Endocrinology* **81,** 113.

AvRuskin, T. W., Sau, K., Tang, S., and Juan, C. (1973). *J. Clin. Endocrinol. Metab.* **37,** 380.

Awapara, J. (1956). *J. Biol. Chem.* **218,** 571.

Axelrod, J. (1958). *Science* **127,** 754.

Axelrod, J. (1962). *J. Biol. Chem.* **237,** 1657.

Axelrod, J. (1965). *Recent Prog. Horm. Res.* **21,** 597.

Axelrod, J. (1974). *In* "The Neurosciences: Third Study Program" (F. O. Schmitt and F. G. Worden, eds.), pp. 863–876. MIT Press, Cambridge, Massachusetts.

Axelrod, J., and Laroche, M. J. (1959). *Science* **130,** 800.

Axelrod, J., and Tomchick, R. (1958). *J. Biol. Chem.* **233,** 702.

Axelrod, J., and Wurtman, R. J. (1968). *Adv. Pharmacol.* **6a,** 157–166.

Axelrod, J., MacLean, P. D., Albers, R. W., and Weissbach, H. W. (1961). *In* "Regional Neurochemistry" (S. S. Kety and J. Elkes, eds.), pp. 307–311. Pergamon, Oxford.

Ayukawa, S., Takeuchi, T., Sezaki, M., Hara, T., Umezawa, H., and Nagatsu, T. (1968). *J. Antibiot.* **21,** 350.

Azmitia, E. C., Jr., and McEwen, B. S. (1969). *Science* **166,** 1274.

Baile, C. A., and Meinardi, H. (1967). *Br. J. Pharmacol. Chemother.* **30,** 302.

Bain, H., Darte, J. M. M., Keith, W., and Kruyff, E. (1966). *Pediatrics* **38,** 473.

Bak, I. J., Choi, W. B., Hassler, R., Usunoff, K. G., and Wagner, A. (1975). *Adv. Neurol.* **9,** 25–41.

Bakay, L. (1956). "The Blood-Brain Barrier." Thomas, Springfield, Illinois.

Baker, B. L., Dermody, W. C., and Reel, J. R. (1974). *Am. J. Anat.* **139,** 129.

Baker, B. R., and Gibson, R. E. (1971). *J. Med. Chem.* **14,** 315.

Baker, R. D., Downey, B. R., and Brinkley, H. J. (1973). *J. Anim. Sci.* **37,** 1376.

Balagura, S., and Kanner, M. (1971). *Physiol. Behav.* **7,** 251.

Baldessarini, R. J., Walton, K. G., and Borgnan, R. J. (1976). *Neuropharmacology* **15,** 471.

Ball, P., Knuppen, R., Haupt, M., and Breuer, H. (1972). *J. Clin. Endocrinol. Metab.* **34,** 736.

Bapna, J., Neff, N. H., and Costa, E. (1971). *Endocrinology* **89,** 1345.

Barasi, S., and Roberts, M. H. T. (1975). *Br. J. Pharmacol.* **54,** 237P.

Barbeau, A. (1962). *Can. Med. Assoc. J.* **84,** 802.

Barbeau, A. (1972). "Monographs on Human Genetics." Karger, Basel.

Barbeau, A. (1975). *Lancet* **2,** 683.

Barbeau, A., and Donaldson, J. (1974). *Arch. Neurol. (Chicago)* **30,** 52.

Barchas, J. D., and Freedman, D. X. (1963). *Biochem. Pharmacol.* **12,** 1232.

Bargmann, W., Lindner, E., and Andres, K. H. (1967). *Z. Zellforsch. Mikrosk. Anat.* **77,** 282.

Barker, J. L., Crayton, J. C., and Nicoll, R. A. (1971). *Science* **171,** 208.

Barraclough, C. A., and Sawyer, C. H. (1957). *Endocrinology* **61,** 341.

Barraclough, C. A., and Sawyer, C. H. (1959). *Endocrinology* **65,** 563.

Barros D'Sa, A. A. J., Bloom, S. R., and Baron, J. H. (1975). *Lancet* **1,** 886.

Barry, J. (1968). *C.R. Hebd. Seances Acad. Sci.* **266,** 2230.

Barry, J., and Dubois, M.-P. (1973). *Ann. Endocrinol.* **34,** 735.

Barry, J., Dubois, M.-P., Poulain, P., and Leonardelli, J. (1973a). *C.R. Hebd. Seances Acad. Sci., Ser. D* **276,** 3191.

Barry, J., Dubois, M.-P., and Poulain, P. (1973b). *Z. Zellforsch. Mikrosk. Anat.* **146,** 351.

Barry, J., Dubois, M.-P., and Carette, B. (1974). *Endocrinology* **95,** 1416.

Bartholini, G., and Pletscher, A. (1968). *J. Pharmacol. Exp. Ther.* **161,** 14.

Bartholini, G., and Pletscher, A. (1969). *J. Pharm. Pharmacol.* **21,** 323.

Bartholini, G., Birland, W. P., Pletscher, A., and Bates, H. M. (1967). *Nature (London)* **215,** 852.

Bartholini, G., Da Prada, M., and Pletscher, A. (1968). *J. Pharm. Pharmacol.* **20,** 228.

Bartholini, G., Pletscher, A., and Richards, S. G. (1970). *Experientia* **26,** 598.

Bartholini, G., Thoenen, H., and Pletscher, A. (1971). *In* "6-Hydroxydopamine and Catecholamine Neurons" (T. Malmfors and H. Thoenen, eds.), pp. 163–170. North-Holland Publ., Amsterdam.

Bartholini, G., Stadler, H., and Lloyd, K. G. (1973). *Adv. Neurol.* **3,** 233–241.
Barzaghi, F., Baumgartner, H. R., Carruba, M., Mantegazza, P., and Pletscher, A. (1973). *Br. J. Pharmacol.* **48,** 245.
Bassiri, R. M., and Utiger, R. D. (1972). *Endocrinology* **90,** 722.
Bassiri, R. M., and Utiger, R. D. (1974). *Endocrinology* **94,** 188.
Baudry, M., Martres, M. P., and Schwartz, J. C. (1973). *J. Neurochem.* **21,** 1301.
Baudry, M., Martres, M. P., and Schwartz, J. C. (1975). *Nature (London)* **253,** 361.
Bauman, A., Bourgoin, A., Beuda, P., Glowinski, J., and Hamon, M. (1974). *Brain Res.* **66,** 253.
Baumgarten, H. G., and Lachenmayer, L. (1974). *Z. Zellforsch. Mikrosk. Anat.* **147,** 285.
Baumgarten, H. G., and Schlossberger, H. G. (1973). *Serotonin Behav., Proc. Symp., 1972* pp. 209–224.
Baumgarten, H. G., Björklund, A., Lachenmayer, L., Nobin, A., and Stenevi, U. (1971). *Acta Physiol. Scand., Suppl.* **373,** 1.
Baumgarten, H. G., Lachenmayer, L., and Schlossberger, H. G. (1972a). *Z. Zellforsch. Mikrosk. Anat.* **125,** 553.
Baumgarten, H. G., Björklund, A., Holstein, A. F., and Nobin, A. (1972b). *Z. Zellforsch. Mikrosk. Anat.* **129,** 256.
Baxter, B. L. (1969). *Exp. Neurol.* **23,** 220.
Baxter, C. F. (1970). *Handb. Neurochem.* **3,** 289–353.
Baxter, C. F., and Roberts, E. (1961). *J. Biol. Chem.* **236,** 3187.
Baxter, C. F., and Senoner, I. (1964). *Anal. Biochem.* **7,** 55.
Beart, P. M., and Johnston, G. A. R. (1973). *J. Neurochem.* **20,** 319.
Beart, P. M., Uhr, M. L., and Johnston, G. A. R. (1972a). *J. Neurochem.* **19,** 1849.
Beart, P. M., Johnston, G. A. R., and Uhr, M. L. (1972b). *J. Neurochem.* **19,** 1855.
Beattie, C. W., and Soyka, L. F. (1973). *Endocrinology* **93,** 1453.
Beattie, C. W., Rodgers, C. H., and Soyka, L. F. (1973). *Endocrinology* **92,** 276.
Beaulieu, M., Labrie, F., Coy, D. H., and Coy, E. J. (1975). *57th Annu. Meet. Am. Endocr. Soc., New York* p. 102.
Beck, P., Parker, M. L., and Daughaday, W. H. (1966). *J. Clin. Endocrinol. Metab.* **26,** 463.
Behrman, H. R., Orczyk, G. P., and Greep, R. O. (1972). *Prostaglandins* **1,** 245.
Belchetz, P. E., Gredley, G., and Himsworth, R. L. (1976). *Abstr. Book, Int. Congr. Endocrinol., 6th, 1976* Abstract No. 799.
Belin, M. F., and Pujol, J. F. (1972). *C.R. Hebd. Seances Acad. Sci.* **275,** 3017.
Belleau, B., and Burba, J. (1961). *Biochim. Biophys. Acta* **54,** 195.
Ben-Jonathan, N., and Porter, J. C. (1976). *Endocrinology* **98,** 1497.
Ben-Jonathan, N., Mical, R. S., and Porter, J. C. (1975). *Endocrinology* **96,** 375.
Ben-Jonathan, N., Oliver, C., Weiner, H. J., Mical, R. S., and Porter, J. C. (1977). *Endocrinology* **100,** 452.
Benkert, O., Laackmann, G., Souvatzoglou, A., and von Verder, K. (1973). *J. Neural Trans.* **34,** 291.
Bennett, G. W., and Edwardson, J. A. (1975). *J. Endocrinol.* **65,** 33.
Bennett, G. W., Edwardson, J. A., Holland, D., Jeffcoate, S. L., and White, N. (1975). *Nature (London)* **257,** 323.
Benson, B., Sorrentino, S., and Evans, J. S. (1969). *Endocrinology* **84,** 369.
Beretta, C., Ferrini, R., and Glässer, A. H. (1965). *Nature (London)* **207,** 421.
Bergström, S., Carlson, L. A., and Weeks, J. R. (1968). *Pharmacol. Rev.* **20,** 1.
Berl, S., and Waelsch, H. (1958). *J. Neurochem.* **3,** 161.
Bern, H. A. (1966). *Symp. Soc. Exp. Biol.* **20,** 325.
Bernardi, G., Marciani, M. G., Morocutti, C., and Giacomini, P. (1975). *Brain Res.* **92,** 511.

Bernardis, L. L., and Frohman, L. A. (1970). *Neuroendocrinology* **6,** 319.
Berneis, K. H., Pletscher, A., and Da Prada, M. (1970). *Br. J. Pharmacol.* **39,** 382.
Bertler, A., and Rosengren, E. (1959). *Experientia* **15,** 10.
Besser, G. M., and Mortimer, C. H. (1976). *In* "Frontiers in Neuroendocrinology" (L. Martini and W. F. Ganong, eds.), pp. 227–254. Raven Press, New York.
Besser, G. M., Butler, P. W. P., Landon, J., and Rees, L. (1969). *Br. Med. J.* **4,** 528.
Besser, G. M., Culler, D. R., Irvine, W. J., Ratcliffe, J. G., and Landon, J. (1971). *Br. Med. J.* **1,** 374.
Besser, G. M., Parke, L., Edwards, C. R. W., Forsyth, I. A., and McNeilly, A. S. (1972). *Br. Med. J.* **3,** 669.
Besser, G. M., Mortimer, C. H., Carr, D., Schally, A. V., Coy, D. H., Evered, D., Kastin, A. J., Tunbridge, W. M. G., Thorner, M. O., and Hall, R. (1974). *Br. Med. J.* **1,** 352.
Besses, G. S., Burrow, G. N., Spaulding, S. W., and Donabedian, R. K. (1975). *J. Clin. Endocrinol. Metab.* **41,** 985.
Besson, M. J., Cheramy, A., Feltz, P., and Glowinski, J. (1969). *Proc. Natl. Acad. Sci. U.S.A.* **62,** 741.
Besson, M. J., Cheramy, A., Feltz, P., and Glowinski, J. (1971). *Brain Res.* **32,** 407.
Bhagat, B., and Shideman, F. E. (1963). *J. Pharmacol. Exp. Ther.* **140,** 317.
Bhattacharya, A. N., and Marks, B. H. (1969a). *J. Pharmacol. Exp. Ther.* **165,** 108.
Bhattacharya, A. N., and Marks, B. H. (1969b). *Proc. Soc. Exp. Biol. Med.* **130,** 1194.
Bhattacharya, A. N., and Marks, B. H. (1970). *Neuroendocrinology* **6,** 49.
Biggio, G., and Guidotti, A. (1976). *Brain Res.* **107,** 365.
Birge, C. A., Jacobs, L. S., Hammer, C. T., and Daughaday, W. H. (1970). *Endocrinology* **86,** 120.
Birks, R. I., and MacIntosh, F. C. (1961). *Can. J. Biochem. Physiol.* **39,** 787.
Biscoe, T. J., and Straughan, D. W. (1966). *J. Physiol. (London)* **183;** 341.
Bivens, C. H., Lebovitz, H. E., and Feldman, J. M. (1973a). *N. Engl. J. Med.* **289,** 236.
Bivens, C. H., Lebovitz, H. E., and Feldman, J. M. (1973b). *N. Engl. J. Med.* **289,** 870.
Björklund, A., and Falck, B. (1969). *Z. Zellforsch. Mikrosk. Anat.* **93,** 254.
Björklund, A., and Nobin, A. (1973). *Brain Res.* **51,** 193.
Björklund, A., Falck, B., and Rosengren, E. (1967). *Life Sci.* **6,** 2103.
Björklund, A., Enemar, A., and Falck, B. (1968). *Z. Zellforsch. Mikrosk. Anat.* **89,** 590.
Björklund, A., Falck, B., Hromek, F., Owman, C., and West, K. A. (1970). *Brain Res.* **71,** 1.
Björklund, A., Moore, R. Y., Nobin, A., and Stenevi, U. (1973). *Brain Res.* **51,** 171.
Björklund, A., Baumgarten, H. G., and Nobin, A. (1974a). *In* "Serotonin—New Vistas" (E. Costa, G. L. Gessa, and M. Sandler, eds.), Vol. 10, pp. 13–33. Raven Press, New York.
Björklund, A., Falck, B., Nobin, A., and Stenevi, U. (1974b). *Neurosecretion—Final Neuroendocr. Pathway, Int. Symp., 6th, 1973* pp. 209–222.
Björklund, A., Lindvall, O., and Nobin, A. (1975a). *Brain Res.* **89,** 29.
Björklund, A., Baumgarten, H. G., and Reusch, A. (1975b). *J. Neurochem.* **24,** 833.
Black, J. W., and Prichard, B. N. C. (1973). *Br. Med. Bull.* **29,** 163.
Black, J. W., Duncan, W. A. M., and Shanks, R. G. (1965). *Br. J. Pharmacol. Chemother.* **25,** 577.
Black, J. W., Duncan, W. A. M., Durant, C. J., Ganellin, C. R., and Parson, E. M. (1972). *Nature (London)* **236,** 385.
Black, J. W., Duncan, W. A. M., Emmett, J. C., Ganellin, C. R., Hasselbo, T., Parson, E. M., and Wyllie, J. H. (1973). *Agents Actions* **3,** 133.
Blackard, W. G., and Heidingsfelder, S. A. (1968). *J. Clin. Invst.* **47,** 1400.
Blackard, W. G., and Hubbel, G. J. (1970). *Metab., Clin. Exp.* **19,** 545.

Blackard, W. G., and Waddel, C. C. (1969). *Proc. Soc. Exp. Biol. Med.* **131,** 192.
Blackmore, M. A., Dyer, R. G., and Morris, J. F. (1974). *J. Physiol. (London)* **242,** 12P.
Blackwell, R. E., and Guillemin, R. (1973). *Annu. Rev. Physiol.* **35,** 357.
Blackwell, R. E., Amoss, M., Jr., Vale, W., Burgus, R., Rivier, J., Monahan, M., Ling, N., and Guillemin, R. (1973). *Am. J. Physiol.* **224,** 170.
Blake, C. A. (1974). *Proc. Soc. Exp. Biol. Med.* **145,** 716.
Blake, C. A. (1976). *Endocrinology* **98,** 99.
Blake, C. A., Weiner, R. J., and Sawyer, C. H. (1972). *Endocrinology* **90,** 862.
Blanco, S., Schalch, D. S., and Reichlin, S. (1966). *Fed. Proc., Fed. Am. Soc. Exp. Biol.* **25,** 191.
Blaschko, H., Richter, D., and Schlossman, H. (1937). *J. Biochem. (Tokyo)* **31,** 2187.
Blask, D. E., Vaughan, M. K., Reiter, R. J., Johnson, L. Y., and Vaughan, G. M. (1976). *Endocrinology* **99,** 152.
Bliss, E. L., and Zwanziger, J. (1966). *J. Psychiatr. Res.* **4,** 189.
Bliss, E. L., Ailion, J., and Zwanziger, J. (1968). *J. Pharmacol. Exp. Ther.* **164,** 122.
Bloom, F. E. (1968). *In* "Psychopharmacology: A 10-Year Progress Report" (D. H. Efron, ed.), pp. 355–373. US Govt. Printing Office, Washington, D.C.
Bloom, F. E. (1971). *In* "6-Hydroxydopamine and Catecholamine Neurons" (T. Malmfors and H. Thoenen, eds.), pp. 135–150. North-Holland Publ., Amsterdam.
Bloom, F. E. (1972). *Handb. Exp. Pharmakol.* **33,** 46–78.
Bloom, F. E., and Aghajanian, G. K. (1968). *J. Pharmacol.* **159,** 261.
Bloom, F. E., Oliver, A. P., and Salmoiraghi, G. C. (1963). *Int. J. Neuropharmacol.* **2,** 181.
Bloom, F. E., von Baumgarten, R., Oliver, A. P., Costa, E., and Salmoiraghi, G. C. (1964). *Life Sci.* **3,** 131.
Bloom, F. E., Costa, E., and Salmoiraghi, G. C. (1965). *J. Pharmacol. Exp. Ther.* **150,** 244.
Bloom, F. E., Algeri, S., Groppetti, A., Revuelta, A., and Costa, E. (1969). *Science* **166,** 1284.
Bloom, F. E., Hoffer, B. J., Siggins, G. R., Burker, J. L., and Nicoll, R. A. (1972). *Fed. Proc., Fed. Am. Soc. Exp. Biol.* **13,** 101.
Bloom, F. E., Hoffer, B. J., and Siggins, G. R. (1973a). *In* "New Concepts in Neurotransmitter Regulation" (A. J. Mandell, ed.), pp. 223–234. Plenum, New York.
Bloom, F. E., Hoffer, B. J., Nelson, C., Sheu, Y.-s., and Siggins, G. R. (1973b). *Serotonin Behav., Proc. Symp., 1972* pp. 249–261.
Bloom, S. R., Mortimer, C. H., Besser, G. M., Hall, R., Gomez-Pan, A., Roy, V. M., Russel, R. C. G., Coy, D. H., Kastin, A. J., and Schally, A. V. (1974). *Lancet* **2,** 1106.
Boakes, R. J., Bradley, P. B., Briggs, I., and Dray, A. (1969). *Brain Res.* **15,** 529.
Boakes, R. J., Bradley, P. G., Briggs, I., and Dray, A. (1970). *Br. J. Pharmacol.* **40,** 202.
Boakes, R. J., Bradley, P. B., Brooks, N., Candy, J. M., and Wolstencroft, J. H. (1971). *Br. J. Pharmacol.* **41,** 462.
Bogdanove, E. M., and Gay, V. L. (1969). *Endocrinology* **84,** 118.
Bogdanski, D. F., Pletscher, A., Brodie, B. B., and Udenfriend, S. (1956). *J. Pharmacol. Exp. Ther.* **117,** 82.
Bogdanski, D. F., Weissbach, H., and Udenfriend, S. (1957). *J. Neurochem.* **1,** 272.
Bohnet, H. G., Dahlen, H. G., Wuttke, W., and Schneider, H. P. G. (1976). *J. Clin. Endocrinol. Metab.* **42,** 132.
Boissin, J., and Assenmacher, I. (1971). *C.R. Hebd. Seances Acad. Sci.* **273,** 1744.
Bölme, P., Corrodi, H., Fuxe, K., Hökfelt, T., Lidbrink, P., and Goldstein, M. (1974). *Eur. J. Pharmacol.* **28,** 89.
Bolton, N. J., Chadwick, A., and Scanes, C. G. (1974). *J. Physiol. (London)* **238,** 78P.
Bonnet, V., and Bremer, F. (1937). *C.R. Seances Soc. Biol. Ses Fil.* **126,** 1271.

Bonnet, V., and Bremer, F. (1952). *Arch. Int. Physiol.* **60,** 33.

Bonta, I. L., De Vos, C. J., and Hillen, F. C. (1969). *Arch. Int. Pharmacodyn. Ther.* **182,** 391.

Bonta, I. L., De Vos, C. J., Grijsen, H., Hillen, F. C., Noach, E. L., and Sim, A. W. (1971). *Br. J. Pharmacol.* **43,** 514.

Borgeat, P., Chavancy, G., Dupont, A., Labrie, F., Arimura, A., and Schally, A. V. (1972). *Proc. Natl. Acad. Sci. U.S.A.* **69,** 2677.

Borgeat, P., Poirier, G., Chavancy, G., Dupont, A., and Labrie, F. (1973). *In* "Hypothalamic Hypophysiotropic Hormones" (C. Gual and E. Rosemberg, eds.), Int. Congr. Ser. No. 263, pp. 174–181. Excerpta Med. Found., Amsterdam.

Borgeat, P., Labrie, F., Côté, J., Ruch, F., Schally, A. V., Coy, D. H., and Yanahihara, N. (1974). *J. Mol. Cell. Endocrinol.* **1,** 7.

Borison, H. L. (1959). *J. Physiol. (London)* **147,** 172.

Borle, A. B. (1974). *J. Membr. Biol.* **16,** 207.

Boura, A. L. A., and Green, A. F. (1959). *Br. J. Pharmacol. Chemother.* **14,** 536.

Boura, A. L. A., and Green, A. F. (1965). *Annu. Rev. Pharmacol.* **5,** 183.

Bourgoin, S., Morot-Guadry, Y., Glowinski, J., and Hamon, M. (1973). *Eur. J. Pharmacol.* **22,** 209.

Bovet, D., Kohn, R., Marotta, M., and Silvestrini, B. (1958). *Br. J. Pharmacol. Chemother.* **13,** 74.

Bower, A., Hadley, M. E., and Hruby, V. J. (1971). *Biochem. Biophys. Res. Commun.* **45,** 1185.

Bower, A., Hadley, M. E., and Hruby, V. J. (1974). *Science* **184,** 70.

Bowers, C. Y., Schally, A. V., Reynolds, G. A., and Hawley, W. D. (1967). *Endocrinology* **81,** 741.

Bowers, C. Y., Lee, K.-L., and Schally, A. V. (1968). *Endocrinology* **82,** 303.

Bowers, C. Y., Schally, A. V., Enzmann, F., Bøler, J., and Folkers, K. (1970). *Endocrinology* **86,** 1143.

Bowers, C. Y., Friesen, H. G., Hwang, P., Guyda, H. J., and Folkers, K. (1971a). *Biochem. Biophys. Res. Commun.* **45,** 1033.

Bowers, C. Y., Schally, A. V., Kastin, A. J., Arimura, A., Schalch, D. S., Gual, C., Castaneda, A., and Folkers, K. (1971b). *J. Med. Chem.* **14,** 477.

Bowers, C. Y., Schally, A. V., Weil, A., Reynolds, G. A., and Folkers, K. (1971c). *Further Adv. Thyroid Res., Trans. Int. Thyroid Conf., 6th, 1970* **2,** p. 109.

Bowers, C. Y., Friesen, H. G., and Folkers, K. (1973). *Biochem. Biophys. Res. Commun.* **51,** 512.

Bowers, M. B., and Roth, R. H. (1972). *Br. J. Pharmacol.* **44,** 301.

Boyar, R. M., Kapen, S., Finkelstein, J. W., Perlow, M., Sassin, J. F., Fukushima, J. K., Weitzman, E. D., and Hellman, L. (1974). *J. Clin. Invest.* **53,** 1588.

Boyd, A. E., and Jackson, I. M. D. (1975). *Endocrinology* **96,** A-201.

Boyd, A. E., Lebovitz, H. E., and Pfeiffer, J. B. (1970). *N. Engl. J. Med.* **238,** 1425.

Bradbury, M. W. B., Burden, J., Hillhouse, E. W., and Jones, M. T. (1974). *J. Physiol. (London)* **239,** 269.

Bradley, P. B. (1960). *Adrenergic Mech., Ciba Found. Symp., 1960* p. 410–420.

Bradley, P. B. (1968). *Int. Rev. Neurobiol.* **11,** 1–56.

Bradley, P. B., and Elkes, J. (1957). *Brain* **80,** 77.

Bradley, P. B., and Key, B. J. (1958). *Electroencaphalogr. Clin. Neurophysiol.* **10,** 97.

Bradley, P. B., and Marley, E. (1965). *Br. J. Pharmacol. Chemother.* **24,** 659.

Bradley, P. B., and Wolstencroft, J. H. (1962). *Nature (London)* **196,** 840.

Bradley, P. B., and Wolstencroft, J. H. (1965). *Br. Med. Bull.* **21,** 15.

Bradley, P. B., Dhavan, B. N., and Wolstencroft, J. H. (1966). *J. Physiol. (London)* **183,** 658.

Brandon, A. (1955). *Arch. Int. Physiol. Biochim.* **63,** 62.

Braunhofer, J., and Zicha, L. (1966). *Med. Welt* **17,** 1875.

Brazeau, P., Vale, W., Burgus, R., Ling, N., Butcher, M., Rivier, J., and Guillemin, R. (1973). *Science* **179,** 77.

Brazeau, P., Rivier, J., Vale, W., and Guillemin, R. (1974a). *Endocrinology* **94,** 184.

Brazeau, P., Vale, W., Rivier, J., and Guillemin, R. (1974b). *Biochem. Biophys. Res. Commun.* **60,** 1202.

Brazeau, P., Epelbaum, J., Tsang, D., Renaud, L. P., and Martin, J. B. (1976). *58th Annu. Meet. Am. Endocr. Soc., San Francisco* Abstract No. 161.

Breese, G. R., and Traylor, T. D. (1970). *J. Pharmacol. Exp. Ther.* **174,** 413.

Breese, G. R., and Traylor, T. D. (1971). *Br. J. Pharmacol.* **42,** 88.

Breese, G. R., Cooper, B. R., Prange, A. J., Cott, J. M., and Lipton, M. A. (1974). *In* "The Thyroid Axis: Drugs and Behavior" (A. J. Prange, ed.), pp. 115–127. Raven, New York.

Breese, G. R., Cott, J. M., Cooper, B. R., Prange, A. J., Jr., Lipton, M. A., and Plotnikoff, N. P. (1975). *J. Pharmacol. Exp. Ther.* **193,** 11.

Breneman, W. R., and Rathkamp, V. (1973). *Biochem. Biophys. Res. Commun.* **52,** 189.

Brezenoff, H. E., and Jenden, D. J. (1969). *Int. J. Neuropharmacol.* **8,** 593.

Bridgers, W. F. (1969). *Arch. Biochem. Biophys.* **133,** 201.

Briggs, F. N., and Munson, P. L. (1955). *Endocrinology* **57,** 205.

Brimblecombe, R. W., Duncan, W. A. M., Durant, G. J., Granellin, C. R., Parsons, M. E., and Black, J. W. (1975). *Br. J. Pharmacol.* **53,** 435P.

Brittain, R. T., Jack, D., and Ritchie, A. C. (1970). *Adv. Drug Res.* **5,** 197.

Broderick, D. S., Candland, K. L., North, J. A., and Mangum, J. H. (1972). *Arch. Biochem. Biophys.* **148,** 196.

Brodie, B. B., Tomich, E. G., Kuntzman, R., and Shore, P. A. (1957). *J. Pharmacol. Exp. Ther.* **119,** 461.

Brodie, B. B., Maickel, R. P., and Westermann, E. O. (1961). *In* "Regional Neurochemistry" (D. D. Kety and J. Elkes, eds.), pp. 351–361. Pergamon, Oxford.

Brodie, B. B., Kuntzman, R., Hirsch, C. W., and Costa, E. (1962). *Life Sci.* **1,** 81.

Bronzino, J. D., Morgane, P. J., and Stern, W. C. (1972). *Am. J. Physiol.* **223,** 376.

Brown, G. L., and Gillespie, J. S. (1957). *J. Physiol. (London)* **138,** 81.

Brown, G. M. (1973). *In* "Advances in Human Growth Hormone Research" (S. Raiti, ed.), DHEW Publ. No. (NIH) 74-612, p. 218. Center for Disease Control, Atlanta, Georgia.

Brown, G. M., Schalch, D. S., and Reichlin, S. (1971). *Endocrinology* **89,** 694.

Brown, G. M., Krigstein, E., Dankova, J., and Hornykiewicz, O. (1972). *Neuroendocrinology* **10,** 207.

Brown, G. M., Seeman, P., and Lee, T. (1976a). *Endocrinology* **99,** 1407.

Brown, G. M., Garfinkel, P. E., Warsh, J. J., and Stancer, H. C. (1976b). *J. Clin. Endocrinol. Metab.* **43,** 236.

Brown, J. (1962). *Metab., Clin. Exp.* **11,** 1098.

Brown, J. H., and Makman, M. H. (1972). *Proc. Natl. Acad. Sci. U.S.A.* **69,** 539.

Brown, M., and Vale, W. (1975). *Endocrinology* **96,** 1333.

Brown, M., Rivier, J., and Vale, W. (1976). Metabolism **25,** suppl. 1, 1501.

Brown, W. A., Van Voert, M. H., and Ambani, L. M. (1973). *J. Clin. Endocrinol. Metab.* **37,** 463.

Brownstein, M. J. (1975). *Life Sci.* **16,** 1363.

Brownstein, M. J., Holtz, R. W., and Axelrod, J. (1973a). *J. Pharmacol. Exp. Ther.* **186,** 109.

Brownstein, M. J., Saavedra, J. M., and Axelrod, J. (1973b). *J. Mol. Pharmacol.* **9,** 605.

Brownstein, M. J., Saavedra, J. M., Palkovits, M., and Axelrod, J. (1974a). *Brain Res.* **77,** 151.

Brownstein, M. J., Palkovits, M., Saavedra, J. M., Bassiri, R. M., and Utiger, R. D. (1974b). *Science* **185,** 267.

Brownstein, M. J., Kobayashi, R., Palkovits, M., and Saavedra, J. M. (1975a). *J. Neurochem.* **24,** 35.

Brownstein, M. J., Arimura, A., Sata, H., Schally, A. V., and Kizer, J. S. (1975b). *Endocrinology* **96,** 1456.

Brownstein, M. J., Palkovits, M., Tappaz, M. L., Saavedra, J. M., and Kizer, J. S. (1976). *Brain Res.* **117,** 287.

Bryant, G. D., Linzell, J. L., and Greenwood, F. C. (1970). *Hormones* **1,** 26.

Buckler, J. H. M., Bold, A. M., Taberner, M., and London, D. R. (1969). *Br. Med. J.* **3,** 153.

Bulat, M., and Sujek, Z. (1968). *J. Neurochem.* **15,** 383.

Bull, G., and Hemsworth, B. A. (1963). *Nature (London)* **199,** 487.

Bunney, B. S., and Aghajanian, G. K. (1975). *In* "Pre- and Postsynaptic Receptors" (E. Usdin and W. E. Bunney, Jr., eds.), pp. 89–122. Dekker, New York.

Bunney, B. S., Aghajanian, G. K., and Roth, R. H. (1973). *Nature (London), New Biol.* **245,** 123.

Burgus, R., and Guillemin, R. (1970). *Annu. Rev. Biochem.* **39,** 444.

Burgus, R., Dunn, T. F., Desideris, D., Ward, D. N., Vale, W., and Guillemin, R. (1970). *Nature (London)* **226,** 321.

Burgus, R., Butcher, M., Ling, N., Monahan, M., Rivier, J., Fellows, R., Amoss, M., Blackwell, R., Vale, W., and Guillemin, R. (1971). *C.R. Hebd. Seances Acad. Sci., Ser. D* **273,** 1611.

Burgus, R., Butcher, M., Amoss, M., Ling, N., Monahan, M., Rivier, J., Fellows, R., Blackwell, R., Vale, W., and Guillemin, R. (1972). *Proc. Natl. Acad. Sci. U.S.A.* **69,** 278.

Burgus, R., Brazeau, P., and Vale, W. (1973). *In* "Advances in Human Growth Hormone Research" (S. Raiti, ed.), DHEW Publ. No. (NIH) 74-612, pp. 144–158. Center for Disease Control, Atlanta, Georgia.

Burkard, W. F., Gey, K. F., and Pletscher, A. (1964). *Life Sci.* **3,** 27.

Burn, J. H., and Rand, M. J. (1958). *J. Physiol. (London)* **144,** 314.

Burrow, G. N., Spaulding, S. W., Donabedian, R., Van Voert, M., and Ambani, L. (1974). *Adv. Neurol.* **5,** 489–493.

Burrow, G. N., May, P. B., Spaulding, S. W., and Donabedian, R. K. (1976). *Proc. Int. Congr. Endocrinol., 5th, 1976* Abstract No. 802.

Burt, R. D., and Snyder, S. H. (1974). *Proc. Soc. Neurosci., 4th Annu. Meet., St. Louis* Abstract No. 101.

Butcher, R. L., Fugo, N. W., and Collins, W. E. (1972). *Endocrinology* **90,** 1125.

Butcher, S. G., and Butcher, L. L. (1974). *Brain Res.* **71,** 167.

Butler, W. R., Key, L. C., Lu, H. K., Peckman, W. D., and Knobil, E. (1975). *Endocrinology* **96,** 1099.

Buu, N. T., and Van Gelder, N. M. (1974). *Br. J. Pharmacol.* **52,** 401.

Buus Lassen, J., Squires, R. F., Christensen, J. A., and Molander, L. (1975). *Psychopharmacologia* **42,** 21.

Bydgeman, M., and Samuelsson, B. (1964). *Clin. Chim. Acta* **10,** 566.

Bydgeman, M., and Samuelsson, B. (1966). *Clin. Chim. Acta* **13,** 465.

Calas, A. (1977). *In* "Advances in Biochemical Psychopharmacology," (E. Costa and G. L. Gessa, eds.), Vol. 16, pp. 79–88, Raven Press, New York.

Caligaris, L., and Taleisnik, S. (1974). *J. Endocrinol.* **62,** 25.
Caligaris, L., Astrada, J. J., and Taleisnik, S. (1968). *Acta Endocrinol. (Copenhagen)* **59,** 177.
Caligaris, L., Astrada, J. J., and Taleisnik, S. (1974). *J. Endocrinol.* **60,** 205.
Callingham, B. A. (1971). *Pharmacol. J.* **207,** 431.
Camanni, F., and Massara, F. (1974). *Lancet* **1,** 942.
Camanni, F., Massara, F., Belforte, L., and Molinatti, G. M. (1975a). *J. Clin. Endocrinol. Metab.* **40,** 363.
Camanni, F., Massara, F., Fassio, V., Molinatti, G. M., and Müller, E. E. (1975b). *Neuroendocrinology* **19,** 227.
Camanni, F., Massara, F., Belforte, L., Rosatello, A., and Molinatti, G. M. (1977). *J. Clin. Endocrinol. Metab.* **44,** 465.
Caraceni, T. A., Parati, E. A., Cocchi, D., Panerai, A. E., and Müller, E. E. (1977). *J. Clin. Endocrinol. Metab.* **44,** 870.
Cardinali, D. P. (1975). *Neuroendocrinology* **19,** 91.
Cardinali, D. P., and Rosner, J. M. (1971). *Endocrinology* **89,** 301.
Cardinali, D. P., Hyyppä, M. T., and Wurtman, R. J. (1973). *Neuroendocrinology* **12,** 30.
Carlson, H. E., Mariz, I. K., and Daughaday, W. H. (1974). *Endocrinology* **94,** 1709.
Carlsson, A. (1959). *Pharmacol. Rev.* **11,** 490.
Carlsson, A. (1964). *Prog. Brain Res.* **8,** 9.
Carlsson, A. (1977). *In* "Advances in Biochemical Psychopharmacology" (E. Costa and G. L. Gessa, eds.), Vol. 16, pp. 439–441, Raven Press, New York.
Carlsson, A., and Hillarp, N. Å. (1962). *Acta Physiol. Scand.* **55,** 95.
Carlsson, A., and Lindqvist, M. (1962). *Acta Physiol. Scand.* **54,** 87.
Carlsson, A., and Lindqvist, M. (1963). *Acta Pharmacol. Toxicol.* **20,** 140.
Carlsson, A., and Waldeck, B. (1964) *Scand. J. Clin. Lab. Invest.* **16,** 133.
Carlsson, A., and Waldeck, B. (1966). *J. Pharm. Pharmacol.* **18,** 252.
Carlsson, A., Rosengren, E., Bertler, A., and Nilsson, J. (1957). *In* "Psychotropic Drugs" (S. Garattini and V. Ghetti, eds.), pp. 363–372. Elsevier, Amsterdam.
Carlsson, A., Lindqvist, M., and Magnusson, T. (1960). *Adrenergic Mech., Ciba Found. Symp., 1960* pp. 434–439.
Carlsson, A., Corrodi, H., and Waldeck, B. (1963). *Helv. Chim. Acta* **46,** 2271.
Carlsson, A., Fuxe, K., and Hökfelt, T. (1967). *J. Pharm. Pharmacol.* **19,** 481.
Carlsson, A., Kehr, W., Lindqvist, M., Magnusson, T., and Atak, C. V. (1972). *Pharmacol. Rev.* **24,** 371.
Carmel, P. C., Araki, S., and Ferin, M. (1976). *Endocrinology* **99,** 243.
Carr, L. A., and Moore, K. E. (1968). *Neuroendocrinology* **3,** 285.
Carr, L. A., and Moore, K. E. (1970). *Biochem. Pharmacol.* **19,** 2671.
Carrer, H. F., and Taleisnik, S. (1970). *Endocrinology* **86,** 231.
Carroll, B. J. (1971). *Clin. Pharmacol. Ther.* **12,** 743.
Carruba, M. (1974). *In* "Serotonin—New Vistas" (E. Costa, G. L. Gessa, and M. Sandler, eds.), Vol. 10, pp. 103–107. Raven Press, New York.
Carvajal, G., Russek, M., Tapia, R., and Massieu, G. (1964). *Biochem. Pharmacol.* **13,** 1059.
Casentini, A., de Poli, A., Hucovic, S., and Martini, L. (1959). *Endocrinology* **64,** 483.
Cassar, J., Mashiter, K., Joplin, G. F., Rees, I. H., and Gilkes, J. J. H. (1976). *Lancet* **2,** 426.
Cattabeni, F., Eros, T., Galli, C. L., and Tonon, G. C. (1975). *Pharmacol. Res. Commun.* **7,** 421.
Cavagnini, F., and Peracchi, M. (1971). *J. Endocrinol.* **51,** 651.
Cavagnini, F., Pontiroli, A. E., Raggi, U., Peracchi, M., and Malinverni, A. (1973). *Experientia* **29,** 573.

Cavagnini, F., Panerai, A. E., Valenti, F., Bulgheroni, M., Peracchi, M., and Pinto, M. (1975). *J. Clin. Endocrinol. Metab.* **41,** 143.

Cavagnini, F., Raggi, U., Micossi, P., Di Landro, A., and Invitti, C. (1976). *J. Clin. Endocrinol. Metab.* **43,** 306.

Cavagnini, F., Di Landro, A., Invitti, C., Raggi, V., Alessandrini, P., Pinto, M., Girotti, G., and Vigo, P. (1977). *Metabolism* **26,** 193.

Cavanough, M. J., and Zeller, E. A. (1964). *Fed. Proc., Fed. Am. Soc. Exp. Biol.* **26,** 814.

Ceccarelli, B., and Hurbult, W. P. (1975). *In* "Proceedings of the Golgi Centennial Symposium" (M. Santini, ed.), pp. 25–41. Raven, New York.

Cehovic, G. D., Robinson, G. A., and Bass, A. D. (1972). *Abstr. Book, Int. Congr. Endocrinol., 3rd, 1972* p. 83.

Celis, M. E. (1975). *Neuroendocrinology* **18,** 256.

Celis, M. E., and Taleisnik, S. (1971). *Experientia* **27,** 1481.

Celis, M. E., and Taleisnik, S. (1974). *Proc. Soc. Exp. Biol. Med.* **145,** 142.

Celis, M. E., Taleisnik, A., and Walter, R. (1971a). *Proc. Natl. Acad. Sci. U.S.A.* **68,** 1428.

Celis, M. E., Taleisnik, S., and Walter, R. (1971b). *Biochem. Biophys. Res. Commun.* **45,** 564.

Ceresa, F., Angeli, A., Boccuzzi, G., and Perotti, L. (1970). *J. Clin. Endocrinol. Metab.* **31,** 491.

Chakmakjian, Z. H., Marks, J. F., and Fink, C. W. (1973). *Pediatr. Res.* **7,** 21.

Chakraborty, P. K., Reeves, J. J., Arimura, A., and Schally, A. V. (1973). *Endocrinology* **92,** 55.

Chakrin, L. W., and Whittaker, V. P. (1969). *Biochem. J.* **113,** 97.

Chambers, J. W., and Brown, G. M. (1976). *Endocrinology* **98,** 420.

Chan, L. T., Schaal, S. M., and Saffran, M. (1970). *Hypophysiotropic Horm. Hypothal.: Assay Chem., Proc. Workshop Conf., 1969* pp. 253–260.

Chang, C. C., Costa, E., and Brodie, B. B. (1965). *J. Pharmacol. Exp. Ther.* **147,** 303.

Chang, H. T. (1953). *J. Neurophysiol.* **16,** 221.

Chang, M. M., Leeman, S. E., and Niall, H. D. (1971). *Nature (London) New Biol.* **232,** 86.

Chase, J. F. A., and Tubbs, P. K. (1966). *Biochem. J.* **100,** 47.

Chase, L. (1970). *Hypophysiotropic Horm. Hypothal.: Assay Chem., Proc. Workshop Conf., 1969* Discussion of the paper by I. Geschwind, p. 317.

Chase, L., Aurbach, G. D., and Wilber, J. F. (1969). *Proc. 51st Annu. Meet. Am. Endocr. Soc., New York* p. 95.

Chase, T. N. (1974). *In* "Serotonin—New Vistas" (E. Costa, G. L. Gessa, and M. Sandler, eds.), Vol. 11, pp. 377–385. Raven Press, New York.

Chase, T. N., Breese, G. R., and Kopin, I. J. (1967). *Science* **157,** 1461.

Chase, T. N., Woods, A. C., Lipton, M. A., and Morris, C. E. (1974). *Arch. Neurol. (Chicago)* **31,** 55.

Chatterton, R. T., Chien, J., and Ward, D. A. (1974). *Proc. Soc. Exp. Biol. Med.* **145,** 874.

Chen, C. L., Amenomori, Y., Lu, K. H., Voogt, J. L., and Meites, J. (1970). *Neuroendocrinology* **6,** 220.

Chen, C. N., Kalucy, R. S., Hartmann, M. K., Lacey, J. H., Crisp, A. H., Bailey, J. E., Eccleston, E. G., and Coppen, A. (1974). *Br. Med. J.* **4,** 564.

Chen, H. J., and Meites, J. (1975). *Endocrinology* **96,** 10.

Chen, H. J., Mueller, G. P., and Meites, J. (1974). *Endocr. Res. Commun.* **1,** 283.

Cheney, D. L., Lefevre, H. F., and Racagni, G. (1975a). *Neuropharmacology* **14,** 801.

Cheney, D. L., Costa, E., Hanin, I., Trabucchi, M., and Wang, C. T. (1975b). *J. Pharmacol. Exp. Ther.* **192,** 288.

Chen-Shih, J., Eiduson, S., Geler, E., and Costa, E. (1974). *In* "Serotonin—New Vistas" (E. Costa, G. L. Gessa, and M. Sandler, eds.), Vol. 11, pp. 101–104. Raven Press, New York.

Cheung, C. Y., and Weiner, R. I. (1976). *Endocrinology* **99,** 914.

Chihara, K., Kato, Y., Ohgo, S., and Imura, H. (1975). *Neuroendocrinology* **18,** 192.

Chihara, K., Kato, Y., Maeda, K., Matsukura, S., and Imura, H. (1976). *J. Clin. Invest.* **57,** 1393.

Chiocchio, S. R., Cannata, M. A., and Tramezzani, J. H. (1976). *Fed. Proc., Fed. Am. Soc. Exp. Biol.* **110,** 612.

Chiodini, P. G., Liuzzi, A., Botalla, L., Cremascoli, G., and Silvestrini, F. (1974). *J. Clin. Endocrinol. Metab.* **38,** 200.

Chiodini, P. G., Liuzzi, A., Botalla, L., Oppizzi, G., Müller, E. E., and Silvestrini, F. (1975). *J. Clin. Endocrinol. Metab.* **40,** 705.

Chiodini, P. G., Liuzzi, A., Müller, E. E., Botalla, L., Cremascoli, G., Oppizzi, G., Verde, G., and Silvestrini, F. (1976). *J. Clin. Endocrinol. Metab.* **43,** 356.

Chiou, C.-Y., and Sastry, V. B. R. (1968). *Biochem. Pharmacol.* **17,** 805.

Chobsieng, P., Naor, P., Koch, Y., Zor, U., and Lindner, H. R. (1975). *Neuroendocrinology* **17,** 12.

Christ, J. F., and Bak, I. J. (1970). *Z. Mikrosk.-Anat. Forsch.* **81,** 329.

Christensen, S. E., Hansen, Aa. P., Lundback, K., Ørskov, H., and Seyer-Hansen, K. (1975). *Lancet* **1,** 1426.

Christiansen, J., and Squires, R. F. (1974). *J. Pharm. Pharmacol.* **26,** 367.

Chu, N., and Bloom, F. E. (1973). *Science* **179,** 908.

Ciaranello, R. D. (1973). *In* "Frontiers in Catecholamine Research" (E. Usdin and S. Snyder, eds.), pp. 101–105. Pergamon, Oxford.

Ciaranello, R. D., and Barchas, J. D. (1973). *Prog. Brain Res.* **39,** 217.

Clark, R. H., and Baker, B. L. (1964). *Science* **143,** 375.

Clark, W. G., and Cumby, H. E. (1975). *Pharmacologist* **17,** 256.

Clarke, D. D., Wilks, S., Gitlow, S. E., and Franklin, M. J. (1967). *Gas Chromatogr.* **5,** 307.

Clemens, J. A., Sar, M., and Meites, J. (1969). *Endocrinology* **84,** 868.

Clemens, J. A., Shaar, C. J., Smalstig, E. B., Bach, N. J., and Kornfield, E. C. (1974). *Endocrinology* **94,** 1171.

Clemens, J. A., Smalstig, E. B., and Shaar, C. J. (1975). *Acta Endocrinol. Copenhagen* **79,** 230.

Clemens, L. E., Kelch, R. P., and Mardovs, M. E. (1974). *Proc. 56th Annu. Meet. Endocr. Soc., Atlanta* p. 225.

Clementi, F., Ceccarelli, B., Cerati, E., Demonte, M. L., Felici, M., Motta, M., and Pecile, A. (1970). *Endocrinology* **48,** 205.

Coates, P. W., Ashby, E. A., Krulich, L., Dhariwal, A. P. S., and McCann, M. S. (1968). *Proc. 26th Meet. Electron-Microsc. Soc. Am. Arcenaux, Baton Rouge* pp. 232–233.

Cocchi, D., Fraschini, F., Jalanbo, H., and Müller, E. E. (1974). *Endocrinology* **95,** 1649.

Cocchi, D., Di Giulio, A., Groppetti, A., Mantegazza, P., Müller, E. E., and Spano, P. F. (1975a). *Experientia* **31,** 384.

Cocchi, D., Di Giulio, A., Groppetti, A., Mantegazza, P., Müller, E. E., and Spano, P. F. (1975b). *Acta Vitaminolog. Enzymolog.* **29,** 90.

Cocchi, D., Gil-Ad, I., Panerai, A. E., Locatelli, V., and Müller, E. E. (1976). *Life Sci.* **19,** 825.

Cocchi, D., Santagostino, A., Gil-Ad. I., Ferri, S., and Müller, E. E. (1977). *Life Sci.,* **20,** 2041.

Coceani, F., and Wolfe, L. S. (1965). *Can. J. Physiol. Pharmacol.* **43,** 445.

Cocola, F., Udeschini, G., Secchi, C., Panerai, A. E., Neri, P., and Müller, E. E. (1976). *Proc. Soc. Exp. Biol. Med.* **151,** 140.

Cohen, S. R., and Lajtha, A. (1972). *Handb. Neurochem.* **7,** 543–572.

Cohn, M. L., Cohn, M., and Taylor, F. H. (1976). *Brain Res.* **96,** 134.

Colburn, R. W., Goodwin, F. K., Bunney, W. E., and Davis, J. M. (1967). *Nature (London)* **215,** 1395.

Coleman, B., and Glaviano, V. V. (1963). *Science* **139,** 54.

Coleoni, A. H., and Masini, A. M. (1974). *Acta Endocrinol. (Copenhagen)* **75,** 717.

Colin-Jones, D. G., and Himsworth, R. L. (1970). *J. Physiol. (London)* **206,** 397.

Collier, B., and MacIntosh, F. C. (1969). *Can. J. Physiol. Pharmacol.* **47,** 127.

Collins, G. G. S., Pryze-Davies, J., Sandler, M., and Southgate, J. (1970). *Nature (London)* **226,** 642.

Collins, G. G. S. (1965). *J. Pharm. Pharmacol.* **17,** 526.

Collins, G. G. S., and Southgate, J. (1970). *Biochem. J.* **117,** 389.

Collu, R., Fraschini, F., and Martini, L. (1971). *Experientia* **28,** 844.

Collu, R., Fraschini, F., Visconti, P., and Martini, L. (1972). *Endocrinology* **90,** 1231.

Collu, R., Jequier, J.-C., Letarte, J., Leboeuf, G., and Ducharme, J. R. (1973). *Neuroendocrinology* **11,** 183.

Connolly, M. E., ed. (1970). "Catapres in Hypertension." Butterworth, London.

Connor, J. D. (1968). *Science* **160,** 899.

Consolo, S., Garattini, S., Ghielmetti, R., Morselli, P., and Valzelli, L. (1965). *Life Sci.* **4,** 625.

Consolo, S., Ladinsky, H., Peri, G., and Garattini, S. (1972). *Eur. J. Pharmacol.* **18,** 251.

Consolo, S., Ladinsky, H., and Garattini, S. (1974). *J. Pharm. Pharmacol.* **26,** 275.

Constantinidis, J., Geissbuhler, F., Gaillard, J. M., Aubert, C., Hovaguimian, T. H., and Tissot, R. (1975). *Psychopharmacologia* **41,** 201.

Convey, E. M., Tucker, H. A., Smith, V. G., and Zolman, J. (1973). *Endocrinology* **92,** 471.

Cools, A. R. (1974). *Psychopharmacologia (Berl.)* **36,** 17.

Cooper, J. R., Bloom, F. E., and Roth, R. H. (1974). "The Biochemical Book of Neuropharmacology," 2nd ed. Oxford Univ. Press, London and New York.

Cooper, K. E., Cranston, W. I., and Honour, A. J. (1965). *J. Physiol. (London)* **181,** 852.

Copinschi, G., L'Hermite, M., Pasteels, J., and Robyn, C. (1972). *In* "Hormones and Antagonists" (P. O. Hubinot, S. H. Hendels, and P. Preumont, eds.), pp. 128–129. Karger, Basel.

Coppen, A., Shaw, D. M., and Farrell, M. B. (1963). *Lancet* **2,** 527.

Coppen, A., Montgomery S., Peet, M., Bailey, J., Marks, V., and Woods, P. (1974). *Lancet* **2,** 433.

Coppola, J. A. (1971). *In* "Frontiers in Neuroendocrinology" (W. F. Ganong and L. Martini, eds.), pp. 129–143. Oxford Univ. Press, London and New York.

Coppola, J. A., Leonard, R. G., Lippman, W., Perrini, J., and Ringler, I. (1965). *Endocrinology* **77,** 485.

Coppola, J. A., Leonardi, R. G., and Lippman, W. (1966). *Endocrinology* **78,** 225.

Corcoran, A. C., and Page, I. H. (1948). *J. Lab. Clin. Med.* **33,** 1236.

Cordeau, J. P., Moreau, A., Beaulnes, A., and Laurin, C. (1963). *Arch. Ital. Biol.* **101,** 30.

Cornblath, M., Parker, M. L., Reyner, S. H., Forbes, A. E., and Daughaday, W. H. (1965). *J. Clin. Endocrinol. Metab.* **25,** 209.

Corrodi, H., and Hanson, L. C. F. (1966). *Psychopharmacologia* **10,** 116.

Corrodi, H., Fuxe, K., Hammer, W., Sjöqvist, F., and Understedt, U. (1967a). *Life Sci.* **6,** 2557.

Corrodi, H., Fuxe, K., and Hökfelt, T. (1967b). *J. Pharm. Pharmacol.* **19,** 433.

Corrodi, H., Fuxe, K., and Hökfelt, T. (1968). *Life Sci.* **7,** 107.

Corrodi, H., Fuxe, K., and Ungerstedt, U. (1971). *J. Pharm. Pharmacol.* **23,** 989.

Corrodi, H., Fuxe, K., Hökfelt, T., and Lidbrink, B. (1973). *J. Pharm. Pharmacol.* **25,** 409.

Corrodi, H., Farnebo, L.-O., Fuxe, K., and Hamberger, B. (1975). *Eur. J. Pharmacol.* **30,** 172.

Costa, E. (1972). *Pharmacol. Rev.* **24,** 167.

Costa, E. (1973). *Proc. Int. Congr. Pharmacol., 5th, 1972* Vol. 4, pp. 215–226.

Costa, E., and Sandler, M., eds. (1972). "Monoamine-Oxidase—New Vistas," Adv. Biochem. Psychopharmacol, Vol. 5. Raven Press, New York.

Costa, E., Groppetti, A., and Revuelta, A. (1971). *Br. J. Pharmacol.* **41,** 57.

Costa, E., Groppetti, A., and Naimzada, N. K. (1972). *Br. J. Pharmacol.* **44,** 742.

Costa, E., Guidotti, A., Mao, C. C., and Suria, A. (1975). *Life Sci.* **17,** 167.

Costall, B., and Naylor, R. J. (1973). *Psychopharmacologia* **32,** 161.

Couch, E. F., Arimura, A., Schally, A. V., Saito, M., and Sawano, S. (1969). *Endocrinology* **85,** 1084.

Couch, J. R. (1970). *Brain Res.* **19,** 137.

Cox, B., and Potkoniak, C. (1969). *J. Pharmacol.* **35,** 1969.

Cox, B., Kastin, A. J., and Schneider, H. (1976). *Eur. J. Pharmacol.* **36,** 141.

Coy, D. H., Coy, E. J., Arimura, A., and Schally, A. V. (1973a). *Biochem. Biophys. Res. Commun.* **54,** 1267.

Coy, D. H., Vilchez-Martinez, J. A., Coy, E. J., Arimura, A., and Schally, A. V. (1973b). *J. Clin. Endocrinol. Metab.* **37,** 331.

Coy, D. H., Schally, A. V., Vilchez-Martinez, J. A., Coy, E. J., and Arimura, A. (1975). *Hypothal. Horm.* [*Symp.*], *1974* pp. 1–12.

Coyle, J. T., and Henry, D. (1973). *J. Neurochem.* **21,** 61.

Coyle, J. T., and Snyder, S. H. (1969). *Science* **166,** 899.

Cramer, H., Rudolph, J., Consbruch, U., and Kendel, K. (1974). *In* "Serotonin—New Vistas" (E. Costa, G. L. Gessa and M. Sandler, eds.), Vol. 11, pp. 187–191. Raven, New York.

Cramer, O. M., and Barraclough, C. A. (1975). *Endocrinology* **96,** 913.

Cranmer, M. F. (1968). *Life Sci.* **7,** 995.

Crawford, I. L., and Connor, J. D. (1973). *Nature (London)* **244,** 442.

Crawford, J. M., and Curtis, D. R. (1964). *Br. J. Pharmacol. Chemother.* **23,** 313.

Crayton, J. C., and Bloom, F. E. (1969). *Anat. Rec.* **163,** 173.

Creveling, C. R., and Daly, J. W. (1971). *In* "Biogenic Amines and Physiological Membranes in Drug Therapy" (J. A. Biel and L. G. Abood, eds.), Part B, pp. 355–411. Dekker, New York.

Crighton, D. B., Schneider, H. P. G., and McCann, S. M. (1970). *Endocrinology* **87,** 323.

Crosignani, P. G., D'Alberton, A., Peracchi, M., and Reschini, E. (1976). *Lancet* **2,** 975.

Crosignani, P. G., Peracchi, M., D'Alberton, A., Lombroso Finzi, G. C., Trojsi, L., Cammareri, G., Caccamo, A., Attansio, A., and Reschini, E. (1977). *In* "Prolactin and Human Reproduction" (P. G. Crosignani and C. Robyn, eds.) pp. 273–283, Academic Press, New York.

Cross, B. A. (1973). *In* "Frontiers in Neuroendocrinology" (W. F. Ganong and L. Martini, eds.), pp. 137–171. Oxford Univ. Press, London and New York.

Cross, B. A., and Dyer, R. G. (1971). *J. Physiol. (London)* **212,** 467.

Crout, J. R., Muskus, A J., and Trendelenburg, U. (1962). *Br. J. Pharmacol.* **18,** 600.

Crowshaw, K., Jessup, S. J., and Ramwell, P. W. (1967). *Biochem. J.* **103,** 79.

Cryer, P. E., and Daughaday, W. H. (1969). *J. Clin. Endocrinol. Metab.* **29,** 386.

Cryer, P. E., and Daughaday, W. H. (1974). *J. Clin. Endocrinol. Metab.* **39,** 658.
Cuculic, Z., Bost, K., and Himwich, H. E. (1968). *Prog. Brain Res.* **28,** 27.
Cuello, A. C., Weiner, R. J., and Ganong, W. F. (1973). *Brain Res.* **59,** 191.
Cuello, A. C., Scapagnini, U., Licko, V., Preziosi, P., and Ganong, W. F. (1973-1974). *Neuroendocrinology* **13,** 115.
Cuello, A. C., Shoemaker, W. J., and Ganong, W. F. (1974). *Brain Res.* **78,** 57.
Curtis, D. R., and Crawford, J. M. (1969). *Annu. Rev. Pharmacol.* **9,** 209.
Curtis, D. R., and Davis, R. (1962). *Br. J. Pharmacol. Chemother.* **18,** 217.
Curtis, D. R., and De Groot, W. C. (1968). *Brain Res.* **10,** 208.
Curtis, D. R., and Johnston, G. A. R. (1974). *Ergeb. Physiol., Biol. Chem. Exp. Pharmakol.* **69,** 97.
Curtis, D. R., and Koizumi, K. (1961). *J. Neurophysiol.* **24,** 80.
Curtis, D. R., and Watkins, J. C. (1961). *Nature (London)* **191,** 1010.
Curtis, D. R., Phillis, J. W., and Watkins, J. C. (1961). *J. Physiol. (London)* **158,** 296.
Curtis, D. R., Ryall, R. W., and Watkins, J. C. (1966). *Exp. Brain Res.* **2,** 97.
Curtis, D. R., Hösli, L., Johnston, G. A. R., and Johnston, I. H. (1967). *Brain Res.* **5,** 112.
Curtis, D. R., Duggan, A. W., Felix, D., and Johnston, G. A. R. (1970a). *Nature (London)* **226,** 1222.
Curtis, D. R., Felix, D., and McLennan, H. (1970b). *Br. J. Pharmacol.* **40,** 881.
Curtis, D. R., Duggan, A. W., and Johnston, G. A. R. (1970c). *Exp. Brain Res.* **10,** 447.
Curtis, D. R., Duggan, A. W., Felix, D., and Johnston, G. A. R. (1971a). *Brain Res.* **32,** 69.
Curtis, D. R., Duggan, A. W., Felix, D., Johnston, G. A. R., and McLennan, H. (1971b). *Brain Res.* **33,** 57.
Curtis, D. R., Duggan, A. W., and Johnston, G. A. R. (1971c). *Exp. Brain Res.* **12,** 547.
Curtis, D. R., Game, C. J. A., Johnston, G. A. R., McCulloch, R. M., and MacLachlan, R. M. (1972). *Brain Res.* **43,** 242.
Curtis, D. R., Felix, D., Game, C. J. A., and McCulloch, R. M. (1973). *Brain Res.* **51,** 358.
Curtis, D. R., Game, C. J. A., Johnston, G. A. R., and McCulloch, R. M. (1974). *Brain Res.* **70,** 493.
Curzon, G. (1974). *In* "Serotonin—New Vistas" (E. Costa, G. L. Gessa, and M. Sandler), Vol. 10, pp. 263–271. Raven, New York.
Curzon, G., and Green, A. R. (1968). *Life Sci.* **7,** 657.
Curzon, G., and Green, A. R. (1969). *Br. J. Pharmacol.* **37,** 689.
Curzon, G., and Green, A. R. (1970). *Br. J. Pharmacol.* **39,** 653.
Curzon, G., and Green, A. R. (1971). *Br. J. Pharmacol.* **43,** 39.
Dählström, A., and Fuxe, K. (1964a). *Acta Physiol. Scand.* **60,** 293.
Dählström, A., and Fuxe, K. (1964b). *Acta Physiol. Scand.* **62,** Suppl. 232, 1.
Dählström, A., and Fuxe, K. (1965). *Acta Physiol. Scand.* **64,** Suppl. 247, 1.
Dählström, A., and Fuxe, K. (1966). *Acta Endocrinol. Copenhagen* **51,** 301.
Dählström, A., and Häggendal, J. (1966). *Acta Physiol. Scand.* **67,** 278.
Dählström, A., and Häggendal, J. (1970). *In* "Biochemistry of Simple Neuronal Models" (E. Costa and E. Giacobini, eds.), Vol. 2, pp. 65–93. Raven, New York.
Dählström, A., Häggendal, J., and Atak, C. (1973). *Serotonin Behav., Proc. Symp., 1972* pp. 87–96.
Dairman, W., and Udenfriend, S. (1970). *Mol. Pharmacol.* **6,** 350.
Dale, H. H. (1914). *J. Pharmacol.* **6,** 147.
Dallmann, M. F., and Yates, F. E. (1968). *Mem. Soc. Endocrinol.* **17,** 39.
D'Amico, D. J., Patel, B. C., and Klavans, H. L. (1976). *J. Pharm. Pharmacol.* **28,** 454.
Dann, O. T., and Carter, C. E. (1964). *Biochem. Pharmacol.* **13,** 677.
Da Prada, M., and Pletscher, A. (1974). *In* "Serotonin—New Vistas" (E. Costa, G. L. Gessa, and M. Sandler, eds.), Vol. 10, pp. 311–320. Raven, New York.

Daughaday, W. H., and Garland, J. T. (1972). *Growth and Growth Horm., Proc. Int. Symp., 2nd, 1971* Int. Congr. Ser. No. 244, pp. 168–179.

Daughaday, W. H., Heins, J. N., Srivastava, L., and Hammer, C. (1968). *J. Lab. Clin. Med.* **72,** 803.

Daughaday, W. H., Hall, K., Raben, M. S., Salmon, W. D., Van Den Branden, J. L., and Van Vyk, J. J. (1972). *Nature (London)* **235,** 107.

Da Vanzo, J. P., Greig, M. E., and Cronin, M. A. (1961). *Am. J. Physiol.* **201,** 833.

Davies, C., Jacoby, J., Lloyd, H. M., and Meares, J. D. (1974). *J. Endocrinol.* **61,** 411.

Davies, I., Naftolin, F., Ryan, K. J., Fishman, J., and Siu, J. (1975). *Endocrinology* **97,** 554.

Davies, J., and Dray, A. (1976). *Brain Res.* **107,** 623.

Davis, S. L., and Borger, M. L. (1973). *Endocrinology* **92,** 303.

Davis, S. L., Anfinson, M. S., and Klindt, J. (1975). *Proc. 57th Annu. Meet. Am. Endocr. Soc., New York* p. 218.

Davison, A. N., and Kaczamarek, L. K. (1971). *Nature (London)* **234,** 107.

Day, M. D., and Rand, M. J. (1963). *J. Pharm. Pharmacol.* **15,** 221.

Debeljuk, L., Arimura, A., and Schally, A. V. (1972a). *Proc. Soc. Exp. Biol. Med.* **139,** 774.

Debeljuk, L., Arimura, A., and Schally, A. V. (1972b). *Endocrinology* **90,** 1578.

Debeljuk, L., Arimura, A., Shiino, M., Rennels, E. G., and Schally, A. V. (1973). *Endocrinology* **92,** 921.

De Belleroche, J. S., and Bradford, H. F. (1973). *J. Neurochem.* **21,** 441.

Debono, A. G., Donaldson, I., Marsden, C. D., and Parkes, J. D. (1975). *Lancet* **2,** 987.

Debreceni, L., Csonka-Takàs, L., and Csete, B. (1969). *Acta Physiol. Acad. Sci. Hung.* **36,** 425.

De Feudis, F. V. (1975). *Annu. Rev. Pharmacol.* **15,** 105.

De Feudis, F. V., Delgado, J. M. R., and Roth, R. H. (1970). *Brain Res.* **18,** 15.

Deguchi, T. (1973). *Mol. Pharmacol.* **9,** 184.

Deguchi, T., and Axelrod, J. (1972a). *Proc. Natl. Acad. Sci. U.S.A.* **69,** 2208.

Deguchi, T., and Axelrod, J. (1972b). *Proc. Natl. Acad. Sci. U.S.A.* **69,** 2547.

Deguchi, T., and Barchas, J. (1973). *Serotonin Behav., Proc. Symp., 1972* p. 33.

de la Cruz, K. G., and Arimura, A. (1975). *Proc. 57th Annu. Meet. Am. Endocr. Soc., New York* p. 103.

de la Cruz, A., Arimura, A., de la Cruz, K. G., and Schally, A. V. (1976). *Endocrinology* **98,** 490.

Del Castillo, J., and Katz, B. (1954). *J. Physiol. (London)* **124,** 560.

Del Castillo, J., and Katz, B. (1955). *J. Physiol. (London)* **128,** 396.

Del Castillo, J., and Katz, B. (1956). *Progr. Biophys. Chem.* **6,** 121.

Delitala, G., Masala, A., Alagna, S., and Devilla, L. (1976). *J. Clin. Endocrinol. Metab.* **42,** 603.

Delitala, G., Masala, A., Alagna, S., Devilla, L., and Lotti, G. (1977). *Brit. Med. J.* **1,** 744.

Delorme, F., Froment, J. L., and Jouvet, M. (1966). *C.R. Seances Soc. Biol. Ses Fil.* **160,** 2347.

del Pozo, E., and Flückiger, E. (1973). *In* "Human Prolactin" (J. L. Pasteels and C. Robyn, eds.), pp. 291–301. Excerpta Med. Found., Amsterdam.

del Pozo, E., and Maclay, W. P. (1976). *Lancet* **2,** 906.

del Pozo, E., Brun del Re, R., Varga, L., and Friesen, H. G. (1972). *J. Clin. Endocrinol. Metab.* **35,** 768.

del Pozo, E., Varga, L., Wyss, H., Tolis, G., Friesen, H. G., Wenner, R., Vetter, L., and Vettwiller, A. (1974). *J. Clin. Endocrinol. Metab.* **39,** 18.

Dement, W. C., Mitler, M. M., and Henriksen, S. J. (1972). *Rev. Can. Biol.* **31,** 239.

Demos, C. H., Krasner, F., and Groel, J. T. (1964). *Clin. Pharmacol. Ther.* **5,** 721.

Dengler, H. (1962). *Naunyn-Schmiedebergs Arch. Exp. Pathol. Pharmakol.* **243,** 366.

Dent, J. N., and Gupta, B. J. (1967). *Gen. Comp. Endocrinol.* **8,** 273.

De Potter, W. P., De Schaepdryver, A. F., Moerman, E. J., and Smith, A. D. (1969). *J. Physiol. (London)* **204,** 102P.

De Robertis, E. (1967). *Science* **156,** 907.

De Robertis, E., and Bennett, H. S. (1955). *J. Biophys. Biochem. Cytol.* **1,** 47.

De Robertis, E., Rodriguez de Lores Arnaiz, G., and Pellegrino de Iraldi, A. (1962). *Nature (London)* **194,** 794.

De Robertis, E., Rodriguez de Lores Arnaiz, B., Salganicoff, L., Pellegrino de Iraldi, A., and Zieher, L. (1963). *J. Neurochem.* **10,** 225.

De Schaepdryver, A. F., and Preziosi, P. (1959). *Arch. Int. Pharmacodyn. Ther.* **119,** 506.

De Schaepdryver, A. F., Preziosi, P., and Scapagnini, U. (1969a). *Br. J. Pharmacol.* **35,** 460.

De Schaepdryver, A. F., Preziosi, P., and Scapagnini, U. (1969b). *Arch. Int. Pharmacodyn, Ther.* **180,** 11.

Desclin, L. (1950). *Ann. Endocrinol.* **2,** 656.

Deuben, R. R., and Meites, J. (1964). *Endocrinology* **74,** 408.

Dewhurst, W. G., and Marley, E. (1965). *Br. J. Pharmacol. Chemother.* **25,** 682.

De Wied, D. (1967). *Pharmacol. Rev.* **19,** 251.

De Wied, D., and De Jong, W. (1974). *Annu. Rev. Pharmacol.* **14,** 389.

De Wied, D., Smelik, P. G., Moll, J., and Bouman, P. R. (1964). *In* "Major Problems in Neuroendocrinology" (E. Bajusz and G. Jasmin, eds.), pp. 156–176. Williams & Wilkins, Baltimore, Maryland.

De Wied, D., Witter, A., Versteeg, D. H. G., and Mulder, A. H. (1969). *Endocrinology* **85,** 561.

Dhariwal, A. P. S., Antunes-Rodrigues, J., Krulich, L., and McCann, M. S. (1965). *Neuroendocrinology* **1,** 341.

Dhariwal, A. P. S., Antunes-Rodrigues, J., Reeser, F., Chowers, I., and McCann, S. M. (1966). *Proc. Soc. Exp. Biol. Med.* **121,** 8.

Dhariwal, A. P. S., Grosvenor, C. E., Antunes-Rodrigues, J., and McCann, S. M. (1968). *Endocrinology* **82,** 1236.

Dhariwal, A. P. S., Krulich, L., and McCann, S. M. (1969a). *Neuroendocrinology* **4,** 282.

Dhariwal, A. P. S., Russel, S. M., McCann, S. M., and Yates, F. E. (1969b). *Endocrinology* **84,** 544.

Diamond, I., and Kennedy, E. P. (1969). *J. Biol. Chem.* **244,** 3258.

Dibbett, J. A., Boudreau, M. J., and Bruni, J. F. (1974). *Proc. 56th Annu. Meet. Am. Endocr. Soc., Atlanta* p. 186.

Dicker, S. E. (1966). *J. Physiol. (London)* **185,** 429.

Dickerman, S., Klediz, G., Gelato, M., Chen, H. J., and Meites, J. (1974). *Neuroendocrinology* **15,** 10.

Diefenbach, W. M. P., Carmel, P. W., Frantz, A. G., and Ferin, M. (1976). *J. Clin. Endocrinol. Metab.* **43,** 638.

Dierickx, K. (1965). *Naturwissenschaften* **52,** 109.

Dierickx, K. (1967). *Z. Zellforsch. Mikrosk. Anat.* **78,** 114.

Di Giulio, A. M., Arezzini, C., Cocchi, D., Mantegazza, P., Spano, P. F., and Groppetti, A. (1975). *In* "Gli Ormoni dell-Ipofisi Anteriore" (E. E. Müller and S. E. Piazzi, eds.), pp. 387–396. I.S.V.T., Sclavo, Siena.

Dimitrikoudi, M., Hanson-Norty, E., and Jenner, F. A. (1974). *Lancet* **1,** 456.

Dimond, R. C., Brammer, S. R., Atkinson, R. L., Jr., Howard, A. J., and Earl, J. M. (1973). *J. Clin. Endocrinol. Metab.* **36,** 1189.

Dismukes, K., Kuhar, M. J., and Snyder, S. H. (1974). *Brain Res.* **78,** 144.

Dixit, B. N., and Buckley, J. P. (1967). *Life Sci.* **6,** 755.

Dolphin, A., Jenner, P., Marsden, C. D., Pycock, C., and Tarsy, D. (1975). *Psychopharmacologia* **41,** 133.

Domanski, E., Przekop, F., and Skubiszewski, B. (1972). *In* "Symposium: Differentiation and Neuroendocrine Regulation in the Hypothalamo-Hypophysial-Gonadal System," Abstr., p. 48. Berlin.

Domanski, E., Przekop, F., Skubiszewski, B., and Wolinska, E. (1975). *Neuroendocrinology* **17,** 265.

Domer, F. R., and Feldberg, W. (1960). *Br. J. Pharmacol. Chemother.* **15,** 578.

Dominic, J. A., and Moore, K. E. (1969). *Psychopharmacologia* **15,** 96.

Domino, E. F., Yamamoto, K. I., and Dren, A. T. (1968). *Prog. Brain Res.* **28,** 113.

Donoso, A. O. (1972). *Abstr. Book Int. Congr. Endocrinol., 4th, 1972* Int. Congr. Ser. No. 256, p. 220.

Donoso, A. O., and Banzan, A. M. (1976). *J. Neural Trans.* **39,** 95.

Donoso, A. O., Bishop, W., Fawcett, C. P., Krulich, L., and McCann, M. S. (1971). *Endocrinology* **89,** 774.

Donoso, A. O., Brazeau, A. M., and Borzino, M. I. (1976). *J. Endocrinol.* **68,** 171.

Doteuchi, M., Wang, C., and Costa, E. (1974). *Mol. Pharmacol.* **10,** 225.

Douglas, W. W. (1963). *Nature (London)* **197,** 81.

Douglas, W. W. (1968). *Br. J. Pharmacol.* **34,** 451.

Douglas, W. W. (1973). *In* "Drug Effects on Neuroendocrine Regulation" (E. Zimmermann *et al.*, eds.), pp. 21–39. Elsevier, Amsterdam.

Douglas, W. W. (1975). *In* "The Pharmacological Basis of Therapeutics" (L. S. Goodman and A. Gilman, eds.), 5th ed., pp. 590–629. Macmillan, New York.

Douglas, W. W., and Poisner, A M. (1964). *J. Physiol. (London)* **172,** 1.

Douglas, W. W., and Poisner, A. M. (1966). *J. Physiol. (London)* **183,** 236.

Douglas, W. W., Nagasawa, J., and Schultz, R. A. (1971). *Mem. Soc. Endocrinol.* **19,** 353.

Dowd, A. J., Hoffman, D. C., and Speroff, L. (1973). *Proc. 55th Annu. Meet. Am. Endocr. Soc., Chicago* p. 173.

Drayson, A. M. (1974). *Lancet* **1,** 312.

Drucker- Colin, R. R., Spanis, C. W., Hunyadi, J., Sassin, J. F., and McGaugh, J. L. (1975). *Neuroendocrinology* **18,** 1.

Dubois, M.-P. (1975). *Proc. Natl. Acad. Sci. U.S.A.* **72,** 1340.

Dubois, M.-P., Barry, J., and Leonardelli, J. (1974). *C.R. Hebd. Seances Acad. Sci., Ser. D* **279,** 1899.

Duchen, L. W. (1962). *J. Endocrinol.* **25,** 161.

Dudel, J. and Kuffler, S. W. (1961). *J. Physiol. (London)* **155,** 543.

Duffy, M. J., Mulhall, D., and Powell, D. (1975). *J. Neurochem.* **25,** 305.

Dular, R., La Bella, F., Vivian, S., and Eddie, L. (1974). *Endocrinology* **94,** 563.

Dunlop, D., and Shanks, R. G. (1968). *Br. J. Pharmacol. Chemother.* **32,** 201.

Dunn, J. D., Arimura, A., and Scheving, L. E. (1972). *Endocrinology* **90,** 29.

Dunstan, R., and Jackson, D. M. (1976). *Psychopharmacology* **48,** 105.

Dupont, A., and Redding, T. W. (1975). *Proc. 57th Annu. Meet. Am. Endocr. Soc., New York* Abstract No. 85.

Dyball, R. E. J., and Dyer, R. G. (1974). *J. Physiol. (London)* **242,** 14P.

Dyball, R. E. J., Dyer, R. G., and Drewett, R. F. (1974). *Brain Res.* **71,** 140.

Dyer, R. G. (1975). *Hypothal. Horm.* [*Symp.*], *1974* pp. 161–182.

Dyer, R. G., and Dyball, R. E. J. (1974). *Nature (London)* **252,** 486.

Eakins, K. E., and Sanner, J. H. (1972). Prostaglandins: *Prog. Res.* pp. 263–292.

Eccles, J. C. (1946). *J. Neurophysiol.* **9,** 87.

Eccleston, D. (1968). *J. Neurochem.* **15,** 947.

Eddy, R. L., Jones, A. L., Chakmakjian, Z. H., and Silverthorne, M. C. (1971). *J. Clin. Endocrinol. Metab.* **33,** 709.

Editorial. (1975). *Br. Med. J.* **4,** 667.

Editorial. (1977). *Br. Med. J.* **1,** 863.

Edwardson, J. A., and Bennett, G. W. (1974). *Nature (London)* **251,** 426.

Efendić, S., Luft, R., and Sievertsson, H. (1975). *FEBS Lett.* **58,** 302.

Efendić, S., Hökfelt, T., and Luft, R. (1976). *In* "Growth Hormone and Related Peptides" (A. Pecile and E. E. Müller, eds.), pp. 216–221. Excerpta Med. Found. Amsterdam.

Egdahl, R. H. (1960). *Endocrinology* **96,** 200.

Egdahl, R. H., Richards, J. B., and Hume, D. H. (1956). *Science* **123,** 418.

Ehrensing, R. H., Kastin, A. J., Schalch, D. S., Friesen, H. G., Vargas, R., and Schally, A. V. (1974). *Am. J. Psychiatry* **131,** 714.

Ehringer, H., and Hornykiewicz, O. (1960). *Klin. Wochenschr.* **38,** 1236.

Eichorn, J. H., Salzman, E. W., and Silen, W., *Nature (London)* **248,** 238.

Eisenberg, R. M. (1975). *Neuroendocrinology* **17,** 154.

Elde, R., Hökfelt, T., Johansson, P., Efendic, S., and Luft, R. (1976). *Abstr. Book, Int. Congr. Endocrinol., 5th, 1976* Abstract No. 659.

Elefthériou, B. E., and Church, R. L. (1968). *Physiol. Behav.* **3,** 977.

Ellis, S., and Grindeland, R. E. (1973). *In* "Advances in Human Growth Hormone Research" (S. Raiti, ed.), DHEW Publ. No. (NIH) 74-612, pp. 409–433. Center for Disease Control, Atlanta, Georgia.

Ellis, S., Grindeland, R. E., Reilly, T. J., and Yang, S. H. (1976). *In* "Growth Hormone and Related Peptides" (A. Pecile and E. E. Müller, eds.), pp. 75–83. Excerpta Med. Found., Amsterdam.

Eltherington, L. G., and Horita, A. (1960). *J. Pharmacol. Exp. Ther.* **128,** 7.

Emlen, W., Segal, D., and Mandel, A. (1972). *Science* **175,** 79.

Endröczi, E., Schreiber, G., and Lissak, K. (1963). *Acta Physiol. Acad. Sci. Hung.* **24,** 211.

Enemar, A., Falck, B., and Iturriza, F. C. (1967). *Z. Zellforsch. Mikrosk. Anat.* **77,** 325.

Enero, M. A., Langer, S. Z., Rothlin, R. P., and Stefano, F. J. (1972). *Br. J. Pharmacol.* **44,** 672.

Engel, J., Hanson, L. C., Ross, B. E., and Strömbergsson, L. E. (1968). *Psychopharmacologia* **13,** 140.

Engström, G., Strömbom, U., Svensson, T. H., and Waldech, B. (1975). *J. Neural Trans.* **37,** 1.

Enjalbert, A., Carbonell, L., Priam, M., and Kordon, C. (1977). *Europ. J. Pharmacol.* **41,** 243.

Enna, S. J., Kuhar, M. J., and Snyder, S. H. (1975). *Brain Res.* **93,** 168.

Ericson, L. E., Melander, A., Owman, C. H., and Sandler, F. (1974). *Endocrinology* **87,** 915.

Ernst, A. M. (1967). *Psychopharmacologia* **10,** 316.

Erspamer, V. (1953). *Arch. Int. Pharmacodyn. Ther.* **93,** 293.

Erspamer, V. (1966). *Handb. Exp. Pharmakol.* **19,** 132.

Eskay, R. L., Oliver, C., Warberg, J., and Porter, C. J. (1976). *Endocrinology* **98,** 269.

Eskin, J. A., and Shchedrina, R. N. (1971). *Byull. Eksp. Biol. Med.* **72,** 16.

Essman, W. B., Sherman, L., and Kolodny, H. D. (1973). *N. Engl. J. Med.* **289,** 870.

Essman, W. B., Kolodny, H. D., Sherman, L., Rammohan, M., and Deutsch, S. (1976). *58th Annu. Meet. Am. Endocr. Soc., San Francisco* Abstract No. 309.

Etkin, W. (1967). *In* "Neuroendocrinology" (L. Martini and W. F. Ganong, eds.), Vol. 2, pp. 261–289. Academic Press, New York.

Eto, S., McMillin Wood, J., Hutchins, M., and Fleischer, N. (1974). *Am. J. Physiol.* **226,** 1315.

Ettigi, P., Lal, S., Martin, J. B., and Friesen, H. G. (1975). *J. Clin. Endocrinol. Metab.* **40,** 1094.

Everett, G. M. (1974). *In* "Serotonin—New Vistas" (E. Costa, G. L. Gessa, and M. Sandler, eds.), Vol. 10, pp. 261–271. Raven, New York.

Everett, G. M., and Borcherding, J. W. (1970). *Science* **168,** 849.

Everett, J. W. (1954). *Endocrinology* **54,** 685.

Everett, J. W. (1964). *Physiol. Rev.* **44,** 373.

Evetts, K. D., and Iversen, L. L. (1970). *J. Pharm. Pharmacol.* **22,** 540.

Evetts, K. D., Uretsky, N. J., Iversen, L. L., and Iversen, S. D. (1970). *Nature (London)* **225,** 961.

Faglia, G., Beck-Peccoz, P., Travaglini, P., Paracchi, A., Spada, A., and Lewin, A. (1973). *J. Clin. Endocrinol. Metab.* **37,** 338.

Faglia, G., Paracchi, A., Beck-Peccoz, P., Spada, A., and Ferrari, C. (1977). *In* "European Workshop on Treatment of Pituitary Adenomas" (R. Fahlbusch and K. von Werder, eds.) G. Thieme Verlag, Stuttgart (in press).

Fahn, S., and Coté, L. J. (1968). *Brain Res.* **7,** 323.

Falck, B., Hillarp, N.-Å., Thieme, G., and Torp, A. (1962). *J. Histochem. Cytochem.* **10,** 348.

Falck, J. D., Jessup, R., and Ramwell, P. W. (1969). *Science* **163,** 691.

Fang, V. S. (1976). *Fed. Proc., Fed. Am. Soc. Exp. Biol.* **35,** Abstr. 549.

Farnebo, L.-O., and Hamberger, B. (1971). *Br. J. Pharmacol.* **43,** 97.

Farnebo, L.-O., Fuxe, K., Hamberger, B., and Ljungdahl, H. (1970). *J. Pharm. Pharmacol.* **22,** 733.

Fatt, P., and Katz, B. (1953). *J. Physiol. (London)* **131,** 374.

Fehm, H. L., Voigt, K. H., and Pfeiffer, E. F. (1977). *In* "European Workshop on Treatment of Pituitary Adenomas" (R. Fahlbusch and K. von Werder, eds.) G. Thieme Verlag, Stuttgart (in press).

Feierabend, J. F., and Fernström, J. D. (1976). *Fed. Proc., Fed. Am. Soc. Exp. Biol.* **35,** Abstr. 551.

Feigenson, M. E., and Saelens, J. K. (1969). *Biochem. Pharmacol.* **18,** 1479.

Feilden, R., and Green, A. L. (1967). *Br. J. Pharmacol. Chemother.* **30,** 155.

Feinberg, I. (1969). *In* "Sleep—Physiology and Pathology" (A. Kales, ed.), Chapter III. Lippincott, Philadelphia, Pennsylvania.

Fekete, F., Herman, J. P., Kanycska, B., and Palkovits, M. (1976). *Proc. Int. Congr. Endocrinol., 5th, 1976* Abstract No. 153.

Feldberg, W., and Myers, R. D. (1966). *J. Physiol. (London)* **184,** 837.

Feldberg, W., and Saxena, P. N. (1971a). *J. Physiol. (London)* **215,** 23.

Feldberg, W., and Saxena, P. N. (1971b). *J. Physiol. (London)* **219,** 739.

Feldberg, W., and Sherwood, S. L. (1953). *J. Physiol. (London)* **120,** 3P.

Feldberg, W., and Sherwood, S. L. (1954). *J. Physiol. (London)* **123,** 148.

Feldberg, W., and Vogt, M. (1948). *J. Physiol. (London)* **107,** 372.

Feldberg, W., Malcom, J. L., and Darian Smith, I. (1957). *J. Physiol. (London)* **138,** 178.

Feldman, J. M., Plonk, J. W., Bivens, C. H., Lebovitz, H. E., and Handwerger, S. (1975). *Am. J. Med. Sci.* **269,** 333.

Feldman, J. M., Plonk, J. W., and Bivens, C. H. (1976). *Clin. Endocrinol.* **5,** 71.

Fell, L. R., Findlay, J. K., Cumming, I. A., and Goding, J. R. (1973). *Endocrinology* **93,** 487.

Fellman, J. H. (1969). *J. Neurochem.* **16,** 135.

Felten, D. L. (1976). *J. Neural Trans.* **39,** 269.

Fenske, M., and Wuttke, W. (1976). *Brain Res.* **104,** 63.

Ferland, L., Labrie, F., Jobin, M., Arimura, A., and Schally, A. V. (1976a). *Biochem. Biophys. Res. Commun.* **68,** 149.
Ferland, L., Labrie, F., Coy, D. H., Arimura, A., and Schally, A. V. (1976b). *Mol. Cell. Endocrinol.* **4,** 79.
Ferland, L., Labrie, F., Coy, D. H., Coy, E. J., and Schally, A. V. (1976c). *Fertil. Steril.* **26,** 889.
Fernandez, B. E., Dominguez, A. E., and Vidal, N. A. (1973). *Acta Endocrinol. Copenhagen* **73,** 273.
Fernando, J. C. R., Knott, P. J., and Curzon, G. (1976). *J. Neurochem.* **27,** 343.
Fernström, J. D. (1976). *Fed. Proc., Fed. Am. Soc. Exp. Biol.* **35,** 1151.
Fernström, J. D., and Wurtman, R. J. (1971). *Science* **174,** 1023.
Fernström, J. D., and Wurtman, R. J. (1972a). *Metab., Clin. Exp.* **21,** 337.
Fernström, J. D., and Wurtman, R. J. (1972b). *Science* **178,** 414.
Ferrini, R., and Glässer, A. (1965). *Psychopharmacologia* **8,** 271.
Fideleff, H., Debeljuk, L., Laszlo, C. Vargas, C., Mancini, A., and Moguilevsky, J. A. (1974). *Abstr. Book Congr. Panam. Endocrinol. 8th, 1974* p. 103.
Fideleff, H., Aparicio, N. J., Guitelman, A., Debeljuk, L., Mancini, A., and Cramer, C. (1976). *J. Clin. Endocrinol. Metab.* **42,** 1014.
Fillenz, P. (1971). *Trans. R. Soc. London, Ser. B* **261,** 319.
Finch, C. E. (1973). *Brain Res.* **52,** 261.
Finch, C. E., and Hicks, P. E. (1975). *Proc. Br. Pharmacol. Soc.* p. 38.
Finch, C. E., Foster, J. R., and Mirsky, A. E. (1969). *J. Gen. Physiol.* **54,** 690.
Fiorindo, R. P., and Martini, L. (1975). *Neuroendocrinology* **18,** 322.
Fischer, J. E., Horst, W. D., and Kopin, I. J. (1965). *Br. J. Pharmacol. Chemother.* **24,** 477.
Fisher, G. H., Humphries, J., Folkers, K., Pernow, B., and Bowers, C. Y. (1974). *J. Med. Chem.* **17,** 843.
Fishman, J. (1963). *J. Clin. Endocrinol. Metab.* **23,** 207.
Fishman, J., and Norton, B. (1975). *Endocrinology* **96,** 1054.
Fishman, J., Naftolin, F., Davies, I. J., Ryan, K. J., and Petro, Z. (1976). *J. Clin. Endocrinol. Metab.* **42,** 177.
Fishman, M. A., and Peake, G. T. (1970). *Pediatrics* **45,** 973.
Fiszer, S., and De Robertis, E. (1969). *J. Neurochem.* **16,** 1201.
Flament-Durand, J. (1965). *Endocrinology* **77,** 446.
Fleischer, N., and Woods, J. M. (1973). *Endocrinology* **92,** 1555.
Fleischer, N., Lorente, M., Hauger-Klevene, J., and Kalderon, M. (1971). *53rd Annu. Meet. Am. Endocr. Soc., San Francisco* p. 87.
Fleischer, N., Zimmerman, G., Schindler, W., and Hutchins, M. (1972). *Endocrinology* **91,** 1436.
Florey, E. (1961). *J. Physiol. (London)* **156,** 1.
Florey, E., and Elliott, K. A. C. (1961). *Methods Med. Res.* **9,** 196–202.
Floss, H. G., Cassidy, J. M., and Robberts, J. E. (1973). *J. Pharm. Sci.* **62,** 699.
Flower, R. J. (1974). *Pharmacol. Rev.* **26,** 33.
Flückiger, E. (1972). *In* "Prolactin and Carcinogenesis" (A. R. Boyns and K. Griffiths, eds.), pp. 162–171. Alpha Omega Alpha Publ., Cardiff, Wales.
Flückiger, E., and Kovacs, E. (1974). *Experientia* **30,** 1173.
Flückiger, E., and Wagner, E. (1968). *Experientia* **24,** 1130.
Folkers, K., Enzmann, F., Böler, J., Bowers, C. Y., and Schally, A. V. (1969). *Biochem. Biophys. Res. Commun.* **37,** 1236.
Fonnum, F., Grotova, I., Rinvik, E., Storm-Mathisen, J., and Walberg, F. (1974). *Brain Res.* **71,** 77.

Foster, J. P., Holland, D. T., Jeffcoate, S. L., and Crighton, D. B. (1974). *J. Endocr.* **61,** lxiii.

Fowler, L. J., and John, R. A. (1972). *Biochem. J.* **130,** 569.

Fram, H., Smejkal, E., and Kratzenstein, R. (1969). *Verh. Dtsch. Ges. Inn. Med.* **75,** 777.

Franchimont, P. (1971). *In* "Frontiers in Neuroendocrinology" (L. Martini and W. F. Ganong, eds.), pp. 331–358. Oxford Univ. Press, London and New York.

Franchimont, P., and Legros, J. J. (1975). *Hypothal. Horm. [Symp.], 1974* pp. 311–324.

Frankel, R. J., and Jenkins, J. S. (1975). *Acta Endocrinol. Copenhagen* **78,** 209.

Frantz, A. G. (1973). *In* "Frontiers in Neuroendocrinology" (W. F. Ganong and L. Martini, eds.), pp. 337–374. Academic Press, New York.

Frantz, A. G., and Rabkin, M. T. (1965). *J. Clin. Endocrinol. Metab.* **25,** 1470.

Frantz, A. G., Kleinberg, D. L., and Noel, G. L. (1972). *Recent Prog. Horm. Res.* **28,** 527–573.

Frantz, A. G., Suh, H. K., and Noel, G. L. (1973). *In* "Frontiers in Catecholamine Research" (E. Usdin and S. Snyder, eds.), pp. 843–847. Pergamon, Oxford.

Fraschini, F. (1969). *Prog. Endocrinol., Proc. Int. Congr. Endocrinol., 3rd, 1968* Int. Congr. Ser. No. 184, pp. 637–644.

Fraschini, F., and Martini, L. (1970). *In* "The Hypothalamus" (L. Martini, M. Motta, and F. Fraschini, eds.), pp. 529–549. Academic Press, New York.

Fraschini, F., Mess, B., Piva, F., and Martini, L. (1968). *Science* **159,** 1104.

Fraschini, F., Collu, R., and Martini, L. (1971). *Pineal Gland, Ciba Found. Symp., 1970* pp. 259–273.

Freeman, M. E., and Neill, J. D. (1972). *Endocrinology* **90,** 1292.

Freinkel, N., Metzger, B. E., Harris, E., Robinson, S., and Mager, M. (1972). *N. Engl. J. Med.* **287,** 841.

French, E. D., Garcia, J. F., and George, R. (1974). *Proc. West. Pharmacol. Soc.* **17,** 159.

Friedman, A. H., and Walker, C. A. (1968). *J. Physiol. (London)* **197,** 77.

Friedman, A. H., and Walker, C. A. (1969). *J. Physiol. (London)* **202,** 133.

Friedman, E., Friedman, J., and Gershon, S. (1973). *Science* **182,** 831.

Friesen, A. J. D., Kemp, J. W., and Woodbury, D. M. (1965). *J. Pharmacol. Exp. Ther.* **148,** 312.

Friesen, H. G., Guyda, H., Hwang, P., Tyson, J. E., and Barbeau, A. (1972). *J. Clin. Invest.* **51,** 706.

Frohman, L. A., and Bernardis, L. L. (1968). *Endocrinology* **82,** 1125.

Frohman, L. A., Bernardis, L. L., and Kant, K. J. (1968). *Science* **162,** 580.

Frohman, L. A., Maran, J. W., and Dhariwal, A. P. S. (1971). *Endocrinology* **88,** 1483.

Frohman, L. A., Müller, E. E., and Cocchi, D. (1973). *Horm. Metab. Res.* **5,** 21.

Fujimori, M., and Himwich, H. E. (1971). *Fed. Proc., Fed. Am. Soc. Exp. Biol.* **30,** Abstr. 541.

Fujino, M., Kobayashi, S., Obayashi, M., Shinagawa, S., Fukuda, T., Kitada, C., Nakayama, R., Yamazaki, I., White, W. F., and Rippel, R. H. (1972). *Biochem. Biophys. Res. Commun.* **49,** 863.

Fuller, R. W., and Perry, K. W. (1974). *Fed. Proc., Fed. Am. Soc. Exp. Biol.* **33,** 255.

Fuller, R. W., Snoddy, H. D., and Molloy, B. B. (1976). *Life Sci.* **19,** 337.

Furchgott, R. F. (1967). *Ann. N.Y. Acad. Sci.* **139,** 553.

Furchgott, R. F. (1972). *In* "Catecholamines" (F. Blaschko and E. Muscholl, eds.), pp. 283–335. Springer-Verlag, Berlin and New York.

Furchgott, R. F., and Sanchez-Garcia, P. (1968). *J. Pharmacol. Exp. Ther.* **163,** 98.

Furr, B. J. A., Onoura, G. I., Bonney, R. C., and Cunningham, F. J. (1973). *J. Endocrinol.* **59,** 495.

Furukawa, T. (1955). *Osaka Shiritsu Daigaka Igaku Zasshi* **4,** 293.
Fuxe, K. (1963). *Acta Physiol. Scand.* **58,** 383.
Fuxe, K. (1964). *Z. Zellforsch. Mikrosk. Anat.* **61,** 710.
Fuxe, K. (1965a). *Acta Physiol. Scand.* **247,** Suppl. 64, 38.
Fuxe, K. (1965b). *Z. Zellforsch. Mikrosk. Anat.* **65,** 573.
Fuxe, K., and Hillarp, N.-Å. (1964). *Life Sci.* **3,** 1403.
Fuxe, K., and Hökfelt, T. (1966). *Acta Physiol. Scand.* **66,** 245.
Fuxe, K., and Hökfelt, T. (1967). *Neurosecretion, Int. Symp., 4th, 1966* pp. 165–177.
Fuxe, K., and Hökfelt, T. (1969). *In* "Frontiers in Neuroendocrinology" (W. F. Ganong and L. Martini, eds.), pp. 47–96. Oxford Univ. Press, London and New York.
Fuxe, K., and Hökfelt, T. (1970). *In* "The Hypothalamus" (L. Martini, M. Motta, and F. Fraschini, eds.), pp. 123–138. Academic Press, New York.
Fuxe, K. and Jonsson, G. (1974). *In* "Serotonin—New Vistas" (E. Costa, G. L. Gessa, and M. Sandler, eds.), Vol. 11, pp. 1–12. Raven, New York.
Fuxe, K., Hökfelt, T., and Nilsson, O. (1969). *Neuroendocrinology* **5,** 107.
Fuxe, K., Goldstein, M., Hökfelt, T., and Joh, T. H. (1971). *Prog. Brain Res.* **34,** 127.
Fuxe, K., Hökfelt, T., Jonsson, G., Levine, S., Lidbrink, P., and Löfstrom, A. (1973a). *In* "Brain-Pituitary-Adrenal Interrelationships" (A. Brodish and E. S. Redgate, eds.), pp. 239–269. Karger, Basel.
Fuxe, K., Hökfelt, T., Jonsson, G., and Löfström, A. (1973b). *In* "Frontiers in Catecholamine Research" (E. Usdin and S. Snyder, eds.), pp. 787–794. Pergamon, Oxford.
Fuxe, K., Hökfelt, T., Jonsson, G., and Löfström, A. (1974). *Neurosecretion—Final Neuroendocr. Pathway, Int. Symp. 6th, 1973* pp. 269–275.
Fuxe, K., Hökfelt, T., Bolme, P., Goldstein, M., Johansson, O., Jonsson, G., Lidbrink, P., Ljungdahl, A., and Sachs, C. (1975a). *In* "Central Action of Drugs in Blood Pressure Regulation" (D. S. Davies and J. L. Reid, eds.), pp. 8–22. Pitman, London.
Fuxe, K., Agnati, L. F., Corrodi, H., Everitt, B. J., Hökfelt, T., Löfström, A., and Ungerstedt, U. (1975b). *Adv. Neurol.* **9,** 223–242.
Fuxe, K., Hökfelt, T., Agnati, L., Goldstein, M., and Johanson, O. (1976). *1st Meet. Eur. Soc. Neurochem., 1976* Abstract No. 57.
Gaddum, J. H. (1959). *Pharmacol. Rev.* **11,** 241.
Gaddum, J. H., and Schild, H. (1934). *J. Physiol. (London)* **83,** 1.
Gaddum, J. H., and Vogt, M. (1956). *Br. J. Pharmacol. Chemother.* **11,** 175.
Gaitonde, M. K. (1970). *Handb. Neurochem.* **3,** 225–287.
Gàl, E. M. (1974). *In* "Serotonin—New Vistas" (E. Costa, G. L. Gessa, and M. Sandler, eds.), Vol. 11, pp. 1–11. Raven, New York.
Gàl, E. M., and Drewes, P. A. (1962). *Proc. Soc. Exp. Biol. Med.* **110,** 368.
Gala, R. R., Janson, P. A., and Kuo, E. Y. (1972). *Proc. Soc. Exp. Biol. Med.* **140,** 569.
Galindo, A. (1969). *Brain Res.* **14,** 763.
Galindo, A., Krnjevic, K., and Schwartz, S. (1967). *J. Physiol. (London)* **192,** 359.
Gallagher, D. W., and Aghajanian, G. K. (1976). *Neuropharmacology* **15,** 149.
Gallagher, D. W., Sanders-Bush, E., Aghajanian, G. K., and Sulser, F. (1975). *Brain Res.* **93,** 111.
Gallagher, T. F., Yoshida, K., Roffwarg, H. D., Fukushima, D. K., Weitzman, E. D., and Hellman, L. (1973). *J. Clin. Endocrinol. Metab.* **36,** 1058.
Gallo, R. V., Rabii, J., and Möberg, G. P. (1975). *Endocrinology* **97,** 1096.
Ganong, W. F. (1963). *Adv. Neuroendocrinol., Proc. Symp., 1961* pp. 92–149.
Ganong, W. F. (1974). *In* "The Neurosciences: Third Study Program" (F. O. Schmitt and F. G. Worden, eds.), pp. 549–563. MIT Press, Cambridge, Massachusetts.

Ganong, W. F. (1975). *Hypothal. Horm. [Symp.], 1974* pp. 237–248.
Ganong, W. F., and Lorenzen, L. (1967). *In* "Neuroendocrinology" (L. Martini and W. F. Ganong, eds.), Vol. 2, pp. 583–640. Academic Press, New York.
Ganong, W. F., Wise, B. L., Shackelford, R., Boryczka, A. T., and Zipf, B. (1965). *Endocrinology* **76,** 526.
Ganong, W. F., Scapagnini, U., Cuello, A. C., and Shoemaker, W. J. (1973). *55th Annu. Meet. Am. Endocr. Soc., Chicago* p. 81.
Ganong, W. F., Reid, I. A., Lovinger, R. D., and Kaplan, S. L. (1975). Abstr. Book *Int. Congr. Pharmacol., 6th, 1975* p. 93.
Ganong, W. F., Kramer, N., Salmon, J., Reid, I. A., Lovinger, R., Scapagnini, U., Boryczka, A. T., and Schackelford, R. (1976). *Neuroscience* **1,** 167.
Garattini, S., and Valzelli, L. (1965). "Serotonin." Elsevier, Amsterdam.
Garattini, S., Lamesta, L., Mortari, A., Palma, V., and Valzelli, L. (1961). *J. Pharm. Pharmacol.* **13,** 385.
Gargiulo, G., and Nisticò, G. (1975). *Acta Med. Vet.* **21,** 137.
Garguilo, G., and Nisticò, G. (1977). *J. Anat.* (in press).
Geffen, L. B., Livett, B. G., and Rush, R. A. (1969). *J. Physiol. (London)* **204,** 593.
Geiger, R., Loning, W., Wissmann, H., and Enzmann, F. (1971). *Biochem. Biophys. Res. Commun.* **45,** 767.
George, R., Haslett, W. L., and Jenden, D. J. (1964). *Int. J. Neuropharmacol.* **3,** 541.
George, W. J., Polson, J. B., O'Toole, A. G., and Goldberg, N. D. (1970). *Proc Natl. Acad. Sci.* **66,** 398.
Gerald, H. C., and Maickel, R. P. (1972). *Br. J. Pharmacol.* **44,** 462.
Gerich, J. E., Lorenzi, M., Schneider, V., and Forsham, P. H. (1974a). *J. Clin. Endocrinol. Metab.* **39,** 1057.
Gerich, J. E., Lorenzi, M., Schneider, V., Karam, J. H., Rivier, J., Guillemin, R., and Forsham, P. H. (1974b). *N. Engl. J. Med.* **291,** 544.
Gerich, J. E., Lovinger, R., and Grodsky, G. (1975). *Endocrinology* **96,** 746.
Gershberg, H., Fry, E. G., Brobeck, J. R., and Long, C. N. H. (1950). *Yale J. Biol. Med.* **23,** 32.
Gershenfeld, H. M. (1971). *Science* **171,** 1252.
Gessa, G. L., and Tagliamonte, A. (1974). *In* "Serotonin—New Vistas" (E. Costa, G. L. Gessa, and M. Sandler, eds.), Vol. 11, pp. 119–131. Raven, New York.
Ghinghelli, F. (1961). *Arch. Sci. Biol. (Bologna)* **45,** 127.
Giacalone, E., and Valzelli, L. (1969). *Pharmacology* **2,** 171.
Gianutsos, G., Drawbaugh, R. B., Hynes, M. D., and Lal, H. (1974). *Life Sci.* **14,** 887.
Gibson, S., McGeer, E. G., and McGeer, P. L. (1969). *J. Neurochem.* **16,** 1491.
Gil-Ad, I., Zambotti, F., Carruba, M. O., Vicentini, L., and Müller, E. E. (1976a). *Proc. Soc. Exp. Biol. Med.* **151,** 512.
Gil-Ad, I., Cocchi, D., Panerai, A. E., Locatelli, V., Mantegazza, P., and Müller, E. E. (1976b). *Neuroendocrinology* **21,** 368.
Gilfoil, T. M., Hart, E. R., and Marrazzi, A. S. (1960). *Fed. Proc., Fed. Am. Soc. Exp. Biol.* **19,** 262.
Gillessen, D., Felix, A. M., Lergier, W., and Stüder, R. O. (1970). *Helv. Chim. Acta* **53,** 63.
Ginther, O. J., and Wentworth, B. C. (1974). *Am. J. Vet. Res.* **35,** 79.
Glick, S. M., Roth, J., Walow, R. S., and Berson, S. A. (1965). *Recent Prog. Horm. Res.* **21,** 241.
Glover, V. A. S., and Potter, L. T. (1971). *J. Neurochem.* **18,** 571.
Glowinski, J. (1972). *In* "Perspectives in Neuropharmacology" (S. H. Snyder, ed.), pp. 349–404. Oxford Univ. Press, London and New York.
Glowinski, J., and Axelrod, J. (1965). *J. Pharmacol. Exp. Ther.* **149,** 43.

Glowinski, J., and Axelrod, J. (1966). *Pharmacol. Rev.* **18,** 775.
Glowinski, J., and Baldessarini, R. J. (1966). *Pharmacol. Rev.* **18,** 1201.
Glowinski, J., and Iversen, L. L. (1966). *J. Neurochem.* **13,** 655.
Glowinski, J., Iversen, L. L., and Axelrod, J. (1966a). *J. Pharmacol. Exp. Ther.* **151,** 385.
Glowinski, J., Axelrod, J., and Iversen, L. L. (1966b). *J. Pharmacol. Exp. Ther.* **153,** 30.
Gnodde, H. P., and Schuiling, G. A. (1976). *Neuroendocrinology* **20,** 212.
Godowski, Z. Z. (1948). *Br. Med. J.* **1,** 46.
Goebelsmann, U., Midgley, A. R., Jr., and Jaffe, R. B. (1969). *J. Clin. Endocrinol. Metab.* **29,** 1222.
Goldfien, A., and Ganong, W. F. (1961). *Am. J. Physiol.* **202,** 205.
Goldstein, M. (1964). *Int. J. Neuropharmacol.* **3,** 37.
Goldstein, M., and Weiss, Z. (1965). *Life Sci.* **4,** 261.
Goldstein, M., Lauber, E., and McKereghan, M. R. (1964). *Biochem. Pharmacol.* **3,** 1103.
Goldstein, M., Gang, H., and Anagnoste, B. (1967). *Biochem. Pharmacol.* **13,** 1103.
Goldstein, M., Fuxe, K., and Hökfelt, T. (1972). *Pharmacol. Rev.* **24,** 293.
Goldstein, M., Anagnoste, B., Friedman, L. S., Roffman, M., Ebstein, R. P., Park, D. H., Fuxe, K., and Hökfelt, T. (1973). *In* "Frontiers in Catecholamine Research" (E. Usdin and S. Snyder, eds.), pp. 69–78. Pergamon, Oxford.
Golter, T. D., Reeves, J. J., O'Mary, C. C., Arimura, A., and Schally, A. V. (1972). *J. Anim. Sci.* **34,** 914.
Gomez-Pan, A., Albinus, M., Reed, J. D., Shaw, B., Hall, R., Besser, G. M., Coy, D. H., Kastin, A. J., and Schally, A. V. (1975). *Lancet* **1,** 888.
Gonzales-Barcena, D., Kastin, A. J., Schalch, D. S., Torres-Zamora, M., Perez-Pasten, E., Kato, A., and Schally, A. V. (1973). *J. Clin. Endocrinol. Metab.* **36,** 117.
Gonzales-Barcena, D., Kastin, A. J., Clinton Miller, M., III, Schalch, D. S., Coy, D. H., Schally, A. V., and Escalante-Herrera, A. (1975). *Lancet* **2,** 1126.
Goodale, D. B., and Moore, K. E. (1975). *Neuropharmacology* **14,** 585.
Goodall, M. C., and Alton, H. (1972). *Biochem. Pharmacol.* **21,** 2401.
Goodman, L. S., and Gilman, A., eds. (1975). "The Pharmacological Basis of Therapeutics," 5th ed. Macmillan, New York.
Goodrick, C. L. (1971). *J. Gerontol.* **26,** 58.
Goodwin, J. S., Katz, R. I., and Kopin, I. J. (1969). *Nature (London)* **221,** 556.
Gordon, A. E., and Meldrum, B. S. (1970). *Biochem. Pharmacol.* **19,** 3042.
Gordon, R., Spector, S., Sjoerdsma, A., and Udenfriend, S. (1966). *J. Pharmacol. Exp. Ther.* **153,** 440.
Goridis, C., and Neff, N. H. (1971). *Neuropharmacology* **10,** 557.
Goth, A. (1973). *In* "Histamine and Antihistamines" (M. Schacter, ed.), Int. Encycl. Pharmacol. Ther., Sect. 74, Vol. 1, pp. 25–43. Pergamon, Oxford.
Goujet, M. A., Simon, P., Chermat, R., and Boissier, J. R. (1975). *Psychopharmacologia* **45,** 87.
Grabner, K., and Lembeck, K. (1960). *Arch. Exp. Pathol. Pharmakol.* **240,** 157.
Grabowska, M., Antikiewicz, L., Maj, J., and Michaluk, J. (1973). *Pol. J. Pharmacol. Pharm.* **25,** 29.
Grabowska, M., Antikiewicz, L., and Michaluck, J., (1974). *J. Pharm. Pharmacol.* **26,** 74.
Gräf, K.-J., Neümann, F., and Horowski, R. (1976). *Endocrinology* **98,** 598.
Gräf, K.-J., Horowski, R., and El Etreby, M. F. (1977). *Acta Endocrinol. (Copenhagen)* **85,** 267.
Graham, L. T., and Aprison, M. H. (1966). *Anal. Biochem.* **15,** 487.
Graham, L. T., Shank, R. P., Werman, R., and Aprison, M. H. (1967). *J. Neurochem.* **14,** 465.

Grahame-Smith, D. G. (1971). *J. Neurochem.* **18,** 1053.
Grahame-Smith, D. G. (1973). *Serotonin Behav., Proc. Symp., 1972* pp. 5–7.
Grahame-Smith, D. G. (1974). *In* "Serotonin—New Vistas" (E. Costa, G. L. Gessa and M. Sandler, eds.), Vol. 10, pp. 83–91. Raven, New York.
Grana, E., and Lilla, L. (1959). *Br. J. Pharmacol. Chemother.* **14,** 501.
Grandison, L., and Meites, J. (1975). *Fed. Proc., Fed. Am. Soc. Exp. Biol.* **34,** Abstr. 195.
Grandison, L., and Meites, J. (1976). *Fed. Proc., Fed. Am. Soc. Exp. Biol.* **35,** Abstr. 356.
Grandison, L., Gelato, M., and Meites, J. (1974). *Proc. Soc. Exp. Biol. Med.* **145,** 1236.
Grant, E. C. G., and Pryse-Davies, J. (1968). *Br. Med. J.* **3,** 770.
Grant, G., Vale, W., and Guillemin, R. (1972). *Biochem. Biophys. Res. Commun.* **46,** 28.
Grant, S. D., Forsham, P. H., and Di Raimondo, V. C. (1965). *N. Engl. J. Med.* **273,** 1115.
Green, A. R., and Grahame-Smith, D. G. (1974). *Neuropharmacology* **13,** 949.
Green, A. R., and Grahame-Smith, D. G. (1976). *Nature (London)* **260,** 487.
Green, J. D., and Harris, G. W. (1949). *J. Physiol. (London)* **108,** 359.
Green, J. P. (1970). *Handb. Neurochem.* **4,** 221–250.
Green, M. D., Simon, M. L., and Lomax, P. (1975). *Life Sci.* **16,** 1293.
Green, M. D., Cox, B., and Lomax, P. (1976). *Neuropharmacology* **15,** 321.
Greer, M. A. (1951). *Proc. Soc. Exp. Biol. Med.* **77,** 603.
Greig, M. E., Walk, R. A., and Gibbons, A. J. (1959). *J. Pharmacol. Exp. Ther.* **127,** 110.
Griffiths, E. C. (1976). *Hormone Res.* **7,** 179.
Grillo, M. A., and Palay, S. L. (1962). *Proc. Int. Congr. Electron Microsc., 5th, 1962* Vol. 2, Artic. U-1.
Grimm, Y., and Reichlin, S. (1972). *Abstr. Book Int. Congr. Endocrinol., 4th, 1972* Int. Congr. Ser. No. 184, p. 85.
Grimm, Y., and Reichlin, S. (1973). *Endocrinology* **93,** 626.
Grimm-Jorgensen, Y., McKelvy, J. F., and Jackson, I. M. D. (1975). *Nature (London)* **254,** 620.
Groppetti, A., Algeri, S., Cattabeni, F., Di Guilio, A. M., Galli, C. M., Ponzio, F., and Spano, P. F. (1977). *J. Neurochem.* **28,** 193.
Grosvenor, C. E., McCann, S. M., and Nallar, R. (1965). *Endocrinology* **76,** 883.
Gruneberg, H. (1952). *In* "The Genetics of the Mouse" (H. Gruneberg, ed.), pp. 122–129. Nijhoff, The Hague.
Gual, C. (1973). *In* "Frontiers in Neuroendocrinology" (W. F. Ganong and L. Martini, eds.), pp. 89–131. Academic Press, New York.
Gudelsky, G. A., and Moore, K. E. (1976). *J. Neural Trans.* **38,** 95.
Gudelsky, G. A., Thornburg, J. E., and Moore, K. E. (1975). *Life Sci.* **16,** 1331.
Guidotti, A., Badiani, G., and Pepeu, G. (1972). *J. Neurochem.* **19,** 431.
Guillemin, R. (1955). *Endocrinology* **56,** 248.
Guillemin, R. (1964). *Recent Prog. Horm. Res.* **20,** 89–130.
Guillemin, R. (1971a). *Adv. Metab. Disord.* **5,** 1.
Guillemin, R. (1971b). *1st Annu. Meet. Soc. Neurosci., Washington* p. 70.
Guillemin, R. (1971c). *Vitam. Horm. (N.Y.)* **29,** 1.
Guillemin, R., and Rosenberg, B. (1955). *Endocrinology 57,* 559.
Guillemin, R., and Schally, A. V. (1963). *Adv. Neuroendocrinol., Proc. Symp., 1961* pp. 314–328.
Guillemin, R., Sakiz, E., and Ward, D. N. (1965). *Proc. Soc. Exp. Biol. Med.* **118,** 1132.
Guldberg, G. C., Aschroft, G. W., and Crawford, T. B. B. (1966). *Life Sci.* **5,** 1571.
Gumulka, W., Samanin, R., Valzelli, L., and Consolo, S. (1971). *J. Neurochem.* **18,** 533.
Gunaga, K. P., and Menon, K. M. J. (1973). *Biochem. Biophys. Res. Commun.* **54,** 440.
Gunaga, K. P., Kawano, A., and Menon, K. M. J. (1974). *Neuroendocrinology* **16,** 273.

Guyda, H., Robert, F., Colle, E., and Hardy, J. (1973). *J. Clin. Endocrinol. Metab.* **36,** 531.

Haas, H. L. (1974). *Brain Res.* **76,** 363.

Haas, H. L., and Hösli, L. (1973). *Brain Res.* **52,** 399.

Haas, H. L., Anderson, E. G., and Hösli, L. (1973). *Brain Res.* **51,** 269.

Haefely, W., Thoenen, H., and Hurlimann, A. (1962). *Med. Pharmacol. Exp.* **7,** 245.

Hagen, T. C., Lawrence, A. M., and Kirsteins, L. (1971). *J. Clin. Endocrinol. Metab.* **33,** 448.

Hagen, T. C., Lawrence, A. M., and Kirsteins, L. (1972a). *Metab., Clin. Exp.* **21,** 603.

Hagen, T. C., Lawrence, A. M., Kacerian, R. E., and Kirsteins, L. (1972b). *Proc. Soc. Exp. Biol. Med.* **141,** 536.

Hagen, T. C., Guansing, A. R., and Sill, A. J., (1976). *Neuroendocrinology* **21,** 255.

Häggendal, J. (1970). *New Aspects Storage Release Mech. Catecholamines, Bayer-Symp., 2nd, 1969* pp. 100–109.

Häggendal, J., and Lindqvist, M. (1964). *Acta Physiol. Scand.* **60,** 351.

Häggendal, J., and Malmfors, T. (1965). *Acta Physiol. Scand.* **64,** 58.

Haigler, E. D., Jr., Pittman, J. A., Jr., Hersham, J. M., and Baugh, C. M. (1971). *J. Clin. Endocrinol. Metab.* **33,** 573.

Haïgler, H. J., and Aghajanian, G. K. (1974). *J. Neural Trans.* **35,** 257.

Håkanson, R., and Sundler, F. (1971a). *Biochem. Pharmacol.* **20,** 3223.

Håkanson, R., and Sundler, F. (1971b). *J. Histochem. Cytochem.* **19,** 693.

Halàsz, B. (1969). *In* "Frontiers in Neuroendocrinology" (W. F. Ganong and L. Martini, eds.), pp. 307–342. Oxford Univ. Press, London and New York.

Halàsz, B., and Pupp, L. (1965). *Endocrinology* **77,** 553.

Halàsz, B., Pupp, L., and Uhlarik, S. (1968). *J. Endocrinol.* **25,** 147.

Haley, T. J. (1957). *In* "Psychotropic Drugs" (S. Garattini and V. Ghetti, eds.), p. 313. Elsevier, Amsterdam.

Hall, M. M. (1970). Ph.D. Thesis, University of Ohio, Athens.

Hall, R., and Besser, G. M. (1975). *Hypothal. Horm.* [*Symp.*], *1974* pp. 299–309.

Hall, R., Amos, J., Garry, R., and Buxton, R. L. (1970). *Br. Med. J.* **2,** 274.

Hall, R., Besser, G. M., Schally, A. V., Coy, D. H., Evered, D., Goldie, D. J., Kastin, A. J., McNeilly, A. S., Mortimer, C. H., Turnbridge, W. M. G., Phenekos, C., and Weightman, D. (1973). *Lancet* **2,** 581.

Hall, T. R., Chadwick, A., Bolton, N. J., and Scanes, C. G. (1975). *Gen. Comp. Endocrinol.* **25,** 298.

Hamon, M., and Glowinski, J. (1974). *Life Sci.* **15,** 1533.

Hamon, M., Bourgoin, S., Morot-Gaudry, Y., Héry, F., and Glowinski, J. (1974). *In* "Serotonin—New Vistas" (E. Costa, G. L. Gessa, and M. Sandler, eds.), Vol. 11, pp. 153–162. Raven, New York.

Handwerger, S., Plonk, J. W., Lebovitz, H. E., Bivens, C. H., and Feldman, J. M. (1975). *Horm. Metab. Res.* **7,** 214.

Hanin, I. (1974). "Choline and Acetylcholine: Handbook of Chemical Assay Methods." Raven, New York.

Hanin, I., and Jenden, D. J. (1969). *Biochem. Pharmacol.* **18,** 837.

Hanin, I., and Schubert, J. (1974). *J. Neurochem.* **23,** 819.

Hanin, I., Jenden, D. J., and Cho, A. K. (1966). *Mol. Pharmacol.* **2,** 352.

Hanin, I., Cheney, D. L., Trabucchi, M., Massarelli, R., Wang, C. T., and Costa, E. (1973). *J. Pharmacol. Exp. Ther.* **187,** 68.

Hanson, L. C. F. (1965). *Psychopharmacologia* **6,** 100.

Hardman, J. G. (1974). *In* "Textbook of Endocrinology" (R. H. Williams, ed.), 5th ed., pp. 869–879. Saunders, Philadelphia, Pennsylvania.

Hardy, R. W. (1970). *Epilepsia* **11,** 179.
Harms, P. G., Ojeda, S. R., and McCann, S. M. (1973). *Science* **181,** 760.
Harms, P. G., Ojeda, S. R., and McCann, S. M. (1974). *Endocrinology* **94,** 1459.
Harms, P. G., Ojeda, S. R., and McCann, S. M. (1976). *Endocrinology* **98,** 318.
Harrington, R. J., Scapagnini, U., and Moberg, G. P. (1973). *J. Anim. Sci.* **37,** 313.
Harris, G. W. (1937). *Proc. R. Soc. London, Ser. B* **122,** 374.
Harris, G. W., and Jacobsohn, D. (1952). *Proc. R. Soc. London, Ser. B* **139,** 263.
Harris, G. W., Jacobsohn, D., and Kahlson, G. (1952). *Ciba Found. Colloq. Endocrinol.* [*Proc.*] **4,** 186–193.
Harris, M., Hopkin, J. M., and Neal, M. J. (1973). *Br. J. Pharmacol.* **47,** 229.
Hartmann, E. (1967). "The Biology of Dreaming." Thomas, Springfield, Illinois.
Hartmann, E., Chung, R., Draskoczy, P. R., and Schildkraut, J. J. (1971a). *Nature (London)* **233,** 425.
Hartmann, E., Chung, R., and Chien, C.-P. (1971b). *Psychopharmacologia* **19,** 114.
Hartwig, W., Kasperlik-Zaluska, A., Wilczyńska, J., and Migdalska, B. (1976). *N. Engl. J. Med.* **295,** 394.
Harwood, C. T., and Bigelow, W. M. (1960). *Pharmacologist* **2,** 93.
Harwood, C. T., and Mason, J. W. (1957). *Endocrinology* **60,** 239.
Havlicek, V., Rezek, M., and Friesen, H. (1975). *Abstr. Book, Int. Congr. Pharmacol., 6th 1975* Abstract No. 1260.
Havlicek, V., Rezek, M., Friesen, H., Martin, J. B., and Brazeau, P. (1976). *Abstr. Book, Int. Congr. Endocrinol., 5th, 1976* Abstract No. 860.
Hawkins, J. E., and Saret, L. H. (1957). *Clin. Chim. Acta* **2,** 481.
Hayashi, T. (1959). *J. Physiol. (London)* **145,** 570.
Hayek, A., and Crawford, J. D. (1972). *J. Clin. Endocrinol. Metab.* **34,** 764.
Hayes, J. R., Johnson, D. G., Koerker, D., and Williams, R. H. (1975). *Endocrinology* **96,** 1374.
Hebb, C. (1970). *Annu. Rev. Physiol.* **32,** 165.
Hebb, C. (1972a). *In* "Cholinesterases and Anticholinesterases Agents" (G. B. Koelle, ed.), pp. 55–88. Springer-Verlag, Berlin and New York.
Hebb, C. (1972b). *Physiol. Rev.* **52,** 918.
Hebb, C., Krnjevic, K., and Silver, A. (1964). *J. Physiol. (London)* **171,** 504.
Hechter, O. (1949). *Fed. Proc., Fed. Am. Soc. Exp. Biol.* **8,** 70.
Hedge, G. A. (1972). *Endocrinology* **91,** 925.
Hedge, G. A., and De Wied, D. (1971). *Endocrinology* **88,** 1257.
Hedge, G. A., and Smelik, P. G. (1968). *Science* **159,** 891.
Hedge, G. A., Yates, M. B., Marcus, R., and Yates, F. E. (1966). *Endocrinology* **79,** 328.
Hedqvist, P. (1970). *Acta Physiol. Scand.* **79,** Suppl. 345.
Hedqvist, P. (1973a). *In* "Prostaglandins" (P. Ramwell and J. E. Shaw, eds.), Vol. 1, pp. 101–131. Plenum, New York.
Hedqvist, P. (1973b). *Adv. Biosci.* **9,** 461.
Hefco, E., Krulich, L., and Aschenbrenner, J. E. (1975). *Endocrinology* **97,** 1234.
Heidingsfelder, S. A., and Blackard, W. G. (1968). *Metab., Clin. Exp.* **17,** 1019.
Hendley, E., and Snyder, S. (1968). *Nature (London)* **220,** 1330.
Henning, N., and Rubenson, A. (1971). *Acta Pharmacol. Toxicol.* **29,** 145.
Henriksen, S., Jacobs, B., Dement, W., and Barchas, J. (1973). *In* "Frontiers in Catecholamine Research" (E. Usdin and S. Snyder, eds.), pp. 759–762. Pergamon, Oxford.
Henriksen, S., Dement, W., and Barchas, J. (1974). *In* "Serotonin—New Vistas" (E. Costa, G. L. Gessa, and M. Sandler, eds.), Vol. 11, pp. 169–179. Raven, New York.

Henry, J. L., Krnjevic, K., and Morris, M. E. (1974). *Fed. Proc., Fed. Am. Soc. Exp. Biol.* **33,** 548.
Hernández-Peón, R. (1962). *Electroencephalogr. Clin. Neurophysiol.* **14,** 423.
Hernández-Peón, R., O'Flaherty, J. J., and O'Flaherty-Mazzuchelli, A. L. (1967). *Brain Res.* **4,** 243.
Hersham, J. M., and Pittman, J. A., Jr. (1971). *Ann. Intern. Med.* **74,** 481.
Hertelendy, F., Machlin, L., and Kipnis, D. M. (1969). *Endocrinology* **84,** 192.
Hertelendy, F., Peake, G. T., and Todd, H. (1971). *Biochem. Biophys. Res. Commun.* **44,** 253.
Hertting, G., Axelrod, J., and Whitby, L. G. (1961). *J. Pharmacol. Exp. Ther.* **134,** 146.
Héry, F., Rouer, E., and Glowinski, J. (1972). *Brain Res.* **43,** 445.
Hess, S. M., Connamacher, R. H., Ozaki, M., and Udenfriend, S. (1961). *J. Pharmacol. Exp. Ther.* **134,** 129.
Hidaka, H. (1973). *In* "Frontiers in Catecholamine Research" (E. Usdin and S. Snyder, eds.), pp. 87–90. Pergamon, Oxford.
Hidaka, H., Nagasawa, A., and Takeda, A. (1973). *J. Clin. Endocrinol. Metab.* **37,** 145.
Hill, R. G., and Simmonds, M. A. (1973). *Br. J. Pharmacol.* **48,** 1.
Hill, R. G., Simmonds, M. A., and Straughan, D. W. (1972). *Nature (London), New Biol.* **240,** 51.
Hill, R. G., Simmonds, M. A., and Straughan, D. W. (1973). *Br. J. Pharmacol.* **49,** 37.
Hillarp, N.-Å., Fuxe, K., and Dählström, A. (1966). *Pharmacol. Rev.* **18,** 727.
Hillhouse, E. W., Burden, J., and Jones, M. T. (1975). *Neuroendocrinology* **17,** 1.
Hilliard, J., Schally, A. V., and Sawyer, C. H. (1971). *Endocrinology* **88,** 730.
Hills, A. G., Forsham, P. H., and Finch, C. A. (1948). *Blood* **3,** 755.
Hill-Samli, M., and MacLeod, R. M. (1974). *Endocrinology* **95,** 1189.
Hill-Samli, M., and MacLeod, R. M. (1975). *Proc. Soc. Exp. Biol. Med.* **149,** 511.
Himsworth, R. L. (1970). *J. Physiol. (London)* **206,** 411.
Himsworth, R. L., Carmel, P. W., and Frantz, A. G. (1972). *Endocrinology* **91,** 217.
Himwich, H. E., and Rinaldi, F. (1957). *In* "Brain Mechanisms and Drugs Action" (W. S. Fields, ed.), pp. 15–44. Thomas, Springfield, Illinois.
Hinkle, P. M., and Tashjian, A. H., Jr. (1975). *Endocrinology* **97,** 324.
Hinsey, J. C. (1937). *Cold Spring Harbor Symp. Quant. Biol.* **5,** 269.
Hiroshige, T., and Abe, K. (1973). *In* "Neuroendocrine Control" (M. Yagi and T. Yoshida, eds.), pp. 205–288. Univ. of Tokyo Press, Tokyo.
Hiroshige, T., and Sakakura, M. (1971). *Neuroendocrinology* **7,** 25.
Hiroshige, T., Kunita, H., Ogura, C., and Itoh, S. (1968). *Jpn. J. Physiol.* **18,** 606.
Hiroshige, T., Abe, K., Wada, S., and Kaneko, M. (1973). *Neuroendocrinology* **11,** 306.
Hirsch, G. H., and Moore, K. E. (1968). *Neuroendocrinology* **3,** 398.
Hirvonen, E., Ranta, T., and Seppälä, M. (1976). *J. Clin. Endocrinol. Metab.* **42,** 1024.
Hishikawa, Y., Cramer, H., and Kuhlo, W. (1969). *Exp. Brain Res.* **7,** 84.
Hodges, J. R., and Vellucci, S. V. (1975). *Br. J. Pharmacol.* **53,** 555.
Hodgkin, A. L., and Martin, K. (1965). *J. Physiol. (London)* **179,** 26P.
Hoffer, B. J., Siggins, G. R., and Bloom, F. E. (1969). *Science* **166,** 1418.
Hoffer, B. J., Siggins, G. R., Oliver, A. P., and Bloom, F. E. (1971). *Ann. N.Y. Acad. Sci.* **185,** 531.
Hoffman, G., Felten, D. L., and Sladek, J. R., Jr. (1976). *Am. J. Anat.* **147,** 501.
Hökfelt, T. (1967a). *Z. Zellforsch. Mikrosk. Anat.* **79,** 110.
Hökfelt, T. (1967b). *Brain Res.* **5,** 121.
Hökfelt, T. (1968). *Z. Zellforsch. Mikrosk. Anat.* **91,** 1.
Hökfelt, T. (1970). *Brain Res.* **22,** 147.

Hökfelt, T. (1974). *Fed. Proc., Fed. Am. Soc. Exp. Biol.* **33,** 2177.
Hökfelt, T., and Fuxe, K. (1972). *Neuroendocrinology* **9,** 100.
Hökfelt, T., and Ungerstedt, U. (1969). *Acta Physiol. Scand.* **76,** 415.
Hökfelt, T., Fuxe, K., Goldstein, M., and Johansson, O. (1974a). *Brain Res.* **66,** 235.
Hökfelt, T., Efendić, S., Johansson, O., Luft, R., and Arimura, A. (1974b). *Brain Res.* **80,** 165.
Hökfelt, T., Johansson, O., Efendić, S., Luft, R., and Arimura, A. (1975a). *Experientia* **31,** 852.
Hökfelt, T., Fuxe, K., Johansson, O., Jeffcoate, S., and White, N. (1975b). *Eur. J. Pharmacol.* **34,** 389.
Hökfelt, T., Fuxe, K., Johansson, O., Jeffcoate, S., and White, N. (1975c). *Neurosci. Lett.* **1,** 133.
Hökfelt, T., Efendic, S., Hellerström, C., Johansson, O., Luft, R., and Arimura, A. (1975d). *Acta Endocrinol. (Copenhagen)* **80,** Suppl. 200, 5.
Hökfelt, T., Kellerth, J. O., Nilsson, G., and Pernow, B. (1975e). *Science* **190,** 889.
Hökfelt, T., Kellerth, J. O., Nilsson, G., and Pernow, B. (1975f). *Brain Res.* **100,** 235.
Hökfelt, T., Elde, R., Johansson, O., Luft, R., and Arimura, A. (1975g). *Neurosci. Lett.* **1,** 231.
Hökfelt, T., Elde, R., Johansson, O., Luft, R., Nilsson, G., and Arimura, A. (1976a). *Neuroscience* **1,** 131.
Hökfelt, T., Meyerson, B., Nilsson, G., Pernow, B., and Sachs, C. (1976b). *Brain Res.* **104,** 181.
Hökfelt, T., Johansson, O., Fuxe, K., Löfström, A., Goldstein, M., Park, D., Ebstein, R. E., Fraser, H., Jeffcoate, S., Efendić, S., Luft, R., and Arimura, A. (1976c). *Proc. Int. Congr. Pharmacol., 6th, 1975* Vol. 3, pp. 93–104.
Hollander, C. S., Mitsuma, T., Shenkman, L., Woolf, P., and Gershengorn, M. C. (1972). *Science* **175,** 209.
Holman, R. B., Shillito, E. E., and Vogt, M. (1971). *Br. J. Pharmacol.* **43,** 685.
Holman, R. B., Elliott, G. R., and Barchas, J. D. (1975). *Annu. Rev. Med.* **26,** 499–519.
Holmes, S. W. (1970). *Br. J. Pharmacol.* **38,** 653.
Holmes, S. W., and Horton, E. W. (1967). *J. Physiol. (London)* **191,** 134.
Holmstead, B., and Tham, R. (1952). *Acta Physiol. Scand.* **45,** 152.
Holton, P. (1959). *J. Physiol. (London)* **149,** 35P.
Holton, P. (1960). *In* "Polypeptides Which Affect Smooth Muscles and Blood Vessels (M. Shachter, ed.) pp. 192–194. Pergamon, Oxford.
Holtz, P. (1939). *Naturwissenschaften* **27,** 724.
Holtz, P., and Palm, D. (1964). *Pharmacol. Rev.* **16,** 113.
Holzbauer, M., and Sharman, D. F. (1972). *Handb. Exp. Pharmakol.* **33,** 110–185.
Holzbauer, M., and Vogt, M. (1956). *J. Neurochem.* **1,** 8.
Holzbauer, M., and Youdim, M. B. H. (1972). *Br. J. Pharmacol.* **44,** 355P.
Hong, E., Sancilio, L. F., Vargas, R., and Pardo, E. (1969). *Eur. J. Pharmacol.* **6,** 274.
Horn, A. S., and Snyder, S. (1971). *Proc. Natl. Acad. Sci. U.S.A.* **68,** 2325.
Hornykiewicz, O. (1966). *Pharmacol. Rev.* **18,** 925.
Hornykiewicz, O. (1973). *Br. Med. Bull.* **29,** 172.
Horowski, R., and Wachtel, H. (1976). *Eur. J. Pharmacol.* **36,** 373.
Horowski, R., Neumann, F., and Gräf, K.-J. (1975). *J. Pharm. Pharmacol.* **27,** 532.
Horton, E. W. (1964). *Br. J. Pharmacol. Chemother.* **22,** 189.
Horton, E. W. (1969). *Physiol. Rev.* **49,** 122.
Horton, E. W. (1972). "Prostaglandins." Springer-Verlag, Berlin and New York.
Horton, E. W., and Main, I. H. M. (1965a). *Int. J. Neuropharmacol.* **4,** 65.

Horton, E. W., and Main, I. H. M. (1965b). *J. Physiol. (London)* **179,** 18P.
Horton, E. W., and Main, I. H. M. (1967). *Br. J. Pharmacol. Chemother.* **30,** 582.
Horton, E. W., and Meldrum, B. S. (1973). *Br. J. Pharmacol.* **49,** 52.
Hösli, L., and Tebécis, A. K. (1970). *Exp. Brain Res.* **11,** 111.
Houssay, B.-A., Biasotti, A., and Sammartino, R. (1935). *C.R. Seances Soc. Biol. Ses Fil.* **120,** 725.
Howe, A. (1973). *J. Endocrinol.* **59,** 385.
Howe, A., and Thody, A. J. (1969). *J. Physiol. (London)* **203,** 159.
Howe, A., and Thody, A. J. (1970). *J. Endocrinol.* **46,** 201.
Howell, S. L., and Montague, W. (1971). *FEBS Lett.* **18,** 293.
Hoyte, K. M., and Martin, J. B. (1975). *J. Clin. Endocrinol. Metab.* **41,** 656.
Hsu, T-H., Hsu, C-K., and Gann, D. S. (1976). *Endocrinology* **99,** 115.
Hubbard, J. I. (1970). *Prog. Biophys. Mol. Biol.* **21,** 33.
Hubbard, J. I., and Kwanbunbumpen, S. (1968). *J. Physiol. (London)* **194,** 407.
Hughes, J., Smith, T., Morgan, B., and Forthergill, L. (1975). *Life Sci.* **16,** 1753.
Huibodro-Toro, J. P., Scotti de Carolis, A., and Longo, V. G. (1975). *Pharmacol. Biochem. Behav.* **3,** 235.
Hukovic, S., and Muscholl, E. (1962). *Naunyn Schmiedebergs Arch. Exp. Pathol. Pharmakol.* **244,** 81.
Hunt, P., Kannengiesser, M.-K., and Raynaud, J.-P. (1974). *J. Pharm. Pharmacol.* **26,** 371.
Hyyppä, M., and Wurtman, J. (1973). *Prog. Brain Res.* **39,** 210.
Ichiyama, A., Nakamura, S., Nishizula, Y., and Hayaishi, O. (1970). *J. Biol. Chem.* **245,** 1699.
Ikeda, M., Fahien, L. A., and Udenfriend, S. (1966). *J. Biol. Chem.* **241,** 4452.
Illner, P., Steiner, R. A., and Gale, C. C. (1976). *Fed. Proc., Fed. Am. Soc. Exp. Biol.* **35,** Abstr. 3171.
Imura, H., Kato, Y., Ikeda, M., Morimoto, M., and Yawata, M. (1971). *J. Clin. Invest.* **50,** 1069.
Imura, H., Nakai, Y., and Yoshimi, T. (1973). *J. Clin. Endocrinol. Metab.* **36,** 204.
Invitti, C., Cavagnini, F., Di Landro, A., and Pinto, M. (1976). *Abstr. Book, Int. Congr. Endocrinol., 5th, 1976* Abstract No. 661.
Irie, M., and Tsushima, T. J. (1972). *J. Clin. Endocrinol. Metab.* **35,** 97.
Israël, M., and Gautron, J. (1969). *Symp. Int. Soc. Cell Biol.* **8,** 137.
Israël, M., Gautron, J., and Lesbats, B. (1969). *C.R. Hebd. Seances Acad. Sci.* **266,** 278.
Ito, T. (1974). *Neuroendocrinology* **16,** 312.
Ito, M., Highstein, S. M., and Tsuchiya, T. (1968). *Brain Res.* **17,** 520.
Ito, H., Momose, G., Katayama, T., Takagishi, H., Ito, L., Nakajima, H., and Takei, Y. (1971). *J. Clin. Endocrinol. Metab.* **32,** 857.
Iturriza, F. C. (1964). *Gen. Comp. Endocrinol.* **4,** 492.
Iturriza, F. C. (1966). *Gen. Comp. Endocrinol.* **6,** 19.
Iturriza, F. C. (1969). *Gen. Comp. Endocrinol.* **12,** 417.
Iversen, D. S., and Iversen, L. L. (1975). *In* "Handbook of Psychobiology 1975," pp. 153–200. Academic Press, New York.
Iversen, L. L. (1965). *Adv. Drug Res.* **2,** 5.
Iversen, L. L. (1967). "The Uptake and Storage of Noradrenaline in Sympathetic Nerves." Cambridge Univ. Press, London and New York.
Iversen, L. L. (1971a). *Br. J. Pharmacol.* **41,** 571.
Iversen, L. L. (1971b). *In* "Biogenic Amines and Physiological Membranes in Drug Therapy" (J. A. Biel and L. G. Abood, eds.), Part B, pp. 259–327. Dekker, New York.

Iversen, L. L. (1972). *In* "Perspectives in Neuropharmacology" (S. Snyder, ed.), pp. 75–111. Oxford Univ. Press, London and New York.
Iversen, L. L. (1973). *Br. Med. Bull.* **29,** 130.
Iversen, L. L. (1974). *Biochem. Pharmacol.* **23,** 1927.
Iversen, L. L. (1975). *Science* **188,** 1084.
Iversen, L. L., and Glowinski, J. (1966). *J. Neurochem.* **13,** 671.
Iversen, L. L., and Neal, M. J. (1968). *J. Neurochem.* **15,** 1141.
Iversen, L. L., and Salt, P. J. (1970). *Br. J. Pharmacol.* **40,** 828.
Iversen, L. L., and Salt, P. J. (1972). *Nature (London), New Biol.* **238,** 91.
Iversen, L. L., and Schon, F. E. (1973). *In* "New Concepts in Neurotransmitter Regulation" (A. J. Mandell, ed.), pp. 153–193. Plenum, New York.
Iversen, L. L., and Uretsky, N. J. (1970). *Brain Res.* **24,** 364.
Iversen, L. L., and Uretsky, N. J. (1971). *In* "6-Hydroxydopamine and Catecholamine Neurons" (T. Malmfors and H. Thoenen, eds.), pp. 171–186. North-Holland Publ., Amsterdam.
Iversen, L. L., Mitchell, J. F., and Srinivasan, V. (1971). *J. Physiol. (London)* **212,** 519.
Iversen, L. L., Horn, A. S., and Miller, R. J. (1975). *In* "Pre- and Post-Synaptic Receptors" (E. Usdin and W. E. Bunney, Jr., eds.), pp. 207–243. Dekker, New York.
Iversen, L. L., Jessell, T., and Kanazawa, I. (1976). *Nature (London)* **264,** 81.
Iwata, T., and Ishii, S. (1969). *Neuroendocrinology* **5,** 140.
Jackson, I. M. D., and Reichlin, S. (1974). *Endocrinology* **95,** 854.
Jackson, I. M. D., Papapetrou, P., and Reichlin, S. (1974). *Proc. Am. Thyroid Assoc., 50th,* T-1 (Abstr.).
Jacobowitz, D., and Kostrzewa, R. (1971). *Life Sci.* **10,** 1329.
Jacobowitz, D. M. (1973). *Prog. Brain Res.* **39,** 199.
Jacobs, B. L. (1974). *Psychopharmacologia* **39,** 81.
Jacobs, L. S., and Daughaday, W. H. (1973). *In* "Human Prolactin" (J. L. Pasteels and C. Robyn, eds.), pp. 189–205. Excerpta Med. Found., Amsterdam.
Jacobs, L. S., Snyder, P. J., Wilber, J. F., Utiger, R. D., and Daughaday, W. H. (1971). *J. Clin. Endocrinol. Metab.* **33,** 996.
Jacobsen, E. (1968). *Psychopharmacol. Dimensions Perspect.* pp. 175–213.
Jacoby, J. H., Greenstein, M., Sassin, J. F., and Weitzman, E. D. (1974). *Neuroendocrinology* **14,** 95.
Jaffe, I. A. (1972). *Adv. Biochem. Psychopharmacol.* **4,** 217–226.
Jalfre, M., Ruch-Monachon, M.-A., and Haefely, W. (1974). *Adv. Biochem. Psychopharmacol.* **10,** 121–134.
Janowsky, D. S., and Davis, J. M. (1970). *Life Sci.* **9,** 525.
Janssen, P. A. J. (1965). *Int. Rev. Neurobiol.* **8,** 221.
Janssen, P. A. J., Niemegeers, C. J. E., Schellekens, K. H. L., Lenaerts, F. M., and Wauquier, A. (1975). *Arzneim.-Forsch.* **25,** 1287.
Jasper, H. H., and Koyama, I. (1969). *Can. J. Physiol. Pharmacol.* **47,** 889.
Jasper, H. H., Khan, R. T., and Elliott, K. A. C. (1965). *Science* **147,** 1448.
Javoy, F., and Glowinski, J. (1971a). *Int. Soc. Neurochem. Meet., 3rd, 1971* p. 169.
Javoy, F., and Glowinski, J. (1971b). *J. Neurochem.* **18,** 1305.
Javoy, F., and Glowinski, J. (1972). *Br. J. Pharmacol.* **44,** 742.
Javoy, F., Glowinski, J., and Kordon, C. (1968). *Eur. J. Pharmacol.* **4,** 103.
Javoy, F., Agid, Y., and Glowinski, J. (1973). *In* "Frontiers in Catecholamine Research" (E. Usdin and S. Snyder, eds.), pp. 747–749. Pergamon, Oxford.
Jeffcoate, S. L., and Holland, D. T. (1974a). *J. Endocrinol.* **61,** lxii.
Jeffcoate, S. L., and Holland, D. T. (1974b). *J. Endocrinol.* **62,** 333.

Jeffcoate, S. L., Holland, D. T., Fraser, H. M., and Gunn, A. (1973). *Nature (London)* **244,** 161.

Jeffcoate, S. L., Sharpe, P. J., Fraser, H. M., Holland, D. T., and Gunn, A. (1974a). *J. Endocrinol.* **62,** 85.

Jeffcoate, S. L., Fraser, H. M., Holland, D. T., and Gunn, A. (1974b). *Acta Endocrinol. (Copenhagen)* **75,** 625.

Jenden, D. J. (1973). *In* "Gas-Chromatography-Mass Spectrometry in Neurobiology" (E. Costa and B. Holmstead, eds.), pp. 69–81. Raven, New York.

Jenkinson, D. H. (1973). *Br. Med. Bull.* **29,** 142.

Jequier, E., Robinson, D. S., Lovenberg, W., and Sjoerdsma, A. (1969). *Biochem. Pharmacol.* **18,** 1071.

Johansen, K. (1973). *Metab. Clin. Exp.* **22,** 773.

Johke, T. (1970). *Endocrinol. Jpn.* **17,** 393.

Johnson, A. M., Loew, D. M., and Vigouret, J. M. (1976). *Br. J. Pharmacol.* **76,** 59.

Johnson, E. S., Roberts, M. H. T., Sobieszek, A., and Straughan, D. W. (1969a). *Neuropharmacology* **8,** 549.

Johnson, E. S., Roberts, M. H. T., and Straughan, D. W. (1969b). *J. Physiol. (London)* **203,** 261.

Johnson, G. A., Kim, E. G., Platz, P. A., and Mickelson, M. M. (1968). *Biochem. Pharmacol.* **17,** 403.

Johnson, G. A., Boukma, S. J., and Kim, E. G. (1970). *J. Pharmacol. Exp. Ther.* **171,** 80.

Johnson, G. A., Kim, E. G., and Boukma, S. J. (1972). *J. Pharmacol. Exp. Ther.* **180,** 539.

Johnson, J. L., and Aprison, M. H. (1971). *Brain Res.* **26,** 141.

Johnston, G. A. R. (1975). *In* "Handbook of Psychopharmacology" (L. L. Iversen, D. S. Iversen, and S. H. Snyder, eds.), Vol. 4, pp. 59–81. Plenum, New York.

Johnston, G. A. R., and Mitchell, J. F. (1971). *J. Neurochem.* **18,** 2441.

Johnston, G. A. R., Triffett, A. C. K., and Wunderlick, J. A. (1968a). *Anal. Chem.* **40,** 1837.

Johnston, G. A. R., Curtis, D. R., De Groat, W. D., and Duggan, A. W. (1968b). *Biochem. Pharmacol.* **17,** 2488.

Johnston, G. A. R., Krogsgaard-Larsen, P., and Stephenson, A. (1975). *Nature (London)* **258,** 627.

Jonas, H. A., Burger, H. G., Cumming, I. A., Findlay, J. K., and de Kretser, D. M. (1975). *Endocrinology* **96,** 384.

Jones, M. T., Gillham, B., and Hillhouse, E. W. (1976a). *Abstr. Book, Int. Congr. Endocrinol., 5th, 1976* Abstract No. 302.

Jones, M. T., Hillhouse, E., and Burden, J. (1976b). *In* "Frontiers in Neuroendocrinology" (L. Martini and W. F. Ganong, eds.), Vol. 4, pp. 195–226. Raven, New York.

Jones, M. T., Hillhouse, E. W., and Burden, J. (1976c). *J. Endocrinol.* **69,** 1.

Jones, S. F., and Kwanbunbumpen, S. (1970a). *J. Physiol. (London)* **207,** 31.

Jones, S. F., and Kwanbunbumpen, S. (1970b). *J. Physiol. (London)* **207,** 51.

Jonsson, G., Fuxe, K., and Hökfelt, T. (1972). *Brain Res.* **40,** 271.

Jordan, D., Ponsin, G., and Mornex, R. (1976). *Abstr. Book, Int. Congr. Endocrinol., 5th, 1976* Abstract No. 151.

Jørgensen, C. B., and Larsen, L. O. (1967). *In* "Neuroendocrinology" (L. Martini and W. F. Ganong, eds.), Vol. 2, pp. 485–528. Academic Press, New York.

Joseph, S. A., Scott, D. E., Vaala, S. S., Knigge, K. M., and Krobisch-Dudley, G. (1973). *Acta Endocrinol. (Copenhagen)* **74,** 215.

Jouvenaz, G. H., Nugteren, D. G., Beerthuis, R. K., and Van Dorp, D. A. (1970). *Biochim. Biophys. Acta* **202,** 231.

Jouvet, M. (1962). *Arch. Ital. Biol.* **100,** 125.

Jouvet, M. (1967). *Physiol. Rev.* **17,** 117.
Jouvet, M. (1969). *Science* **163,** 32.
Jouvet, M. (1972). *Ergeb. Physiol., Biol. Chem. Exp. Pharmakol.* **64,** 166.
Jouvet, M. (1973). *In* "Basic Sleep Mechanisms" (O. J. Petre-Quadens and J. D. Schlag, eds.), pp. 207–232. Academic Press, New York.
Jouvet, M., and Delorme, F. (1965). *C.R. Seances Soc. Biol. Ses Fil.* **159,** 895.
Jouvet, M., and Pujol, J. F. (1974). *Adv. Biochem. Psychopharmacol.* **11,** 199–209.
Judd, H. L., Rebar, R., Vandenberg, G., and Yen, S. S. C. (1974). *J. Clin. Endocrinol. Metab.* **38,** 8.
Judd, S. J., Lazarus, L., and Smythe, G. (1976). *J. Clin. Endocrinol. Metab.* **43,** 313.
Juorio, A. V., and Vogt, M. (1967). *J. Physiol. (London)* **189,** 489.
Justine-Besançon, L., Thominet, M., Laville, C., and Margarit, J. (1967). *C.R. Hebd. Seances Acad. Sci.* **265,** 1253.
Justo, G., Motta, M., and Martini, L. (1975). *Experientia* **31,** 598.
Jutisz, M., Kerdelhué, B., Berauld, A., and Paloma de la Llosa, M. (1972). *Gonadotropins, Int. Symp., 1971* pp. 64–71.
Kadota, K., and Kanaseki, T. (1969). *J. Cell Biol.* **42,** 202.
Kadzielawa, K., and Widy-Tyszkiewicz, R. (1970). *Electroencephalogr. Clin. Neurophysiol.* **28,** 259.
Kahlson, G. E., Rosengren, E., and Thunberg, O. (1963). *J. Physiol. (London)* **169,** 467.
Kakimoto, Y., and Armstrong, M. D. (1962). *J. Biol. Chem.* **237,** 422.
Kalra, P. S., Kalra, S. P., Krulich, L., Fawcett, C. P., and McCann, S. M. (1972). *Endocrinology* **90,** 1168.
Kalra, S. P. (1975). *Neuroendocrinology* **18,** 333.
Kalra, S. P., and McCann, S. M. (1973). *Endocrinology* **93,** 356.
Kalra, S. P., and McCann, S. M. (1974). *Neuroendocrinology* **15,** 79.
Kamberi, I. A., and Kobayashi, Y. (1970). *J. Neurochem.* **17,** 261.
Kamberi, I. A., Mical, R. S., and Porter, J. C. (1969). *Science* **166,** 388.
Kamberi, I. A., Mical, R. S., and Porter, J. C. (1970a). *Experientia* **26,** 1150.
Kamberi, I. A., Mical, R. S., and Porter, J. C. (1970b). *Endocrinology* **87,** 1.
Kamberi, I. A., Mical, R. S., and Porter, J. C. (1971a). *Endocrinology* **88,** 1012.
Kamberi, I. A., Mical, R. S., and Porter, J. C. (1971b). *Endocrinology* **88,** 1288.
Kamberi, I. A., Mical, R. S., and Porter, J. C. (1971c). *Endocrinology* **88,** 1294.
Kanai, T., and Szerb, J. C. (1965). *Nature (London)* **205,** 80.
Kaneko, T., Saito, S., Oka, H., Oda, T., and Yanaihara, N. (1973). *Metab., Clin. Exp.* **22,** 77.
Kansal, P. C., Buse, J., Talbert, O. R., and Buse, M. G. (1972). *J. Clin. Endocrinol. Metab.* **34,** 99.
Kaplanski, J., and Smelik, P. G. (1973a). *Res. Commun. Chem. Pathol. Pharmacol.* **5,** 263.
Kaplanski, J., and Smelik, P. G. (1973b). *Acta Endocrinol. (Copenhagen)* **73,** 651.
Kaplanski, J., Dorst, W., and Smelik, P. G. (1972). *Eur. J. Pharmacol.* **20,** 238.
Kaplanski, J., Nyakas, C., van Delft, A. M. L., and Smelik, P. G. (1973–1974). *Neuroendocrinology* **13,** 123.
Kara, J. N., Karki, T., and Tala, E. (1961). *Acta Pharmacol. Toxicol.* **18,** 255.
Karczmar, A. G., Longo, V. G., and Scotti de Carolis, A. (1970). *Physiol. Behav.* **5,** 175.
Karim, S. M. M., and Hillier, K. (1972). *Prostaglandins: Prog. Res.* pp. 1–46.
Karobath, M. E. (1974). *Eur. J. Pharmacol.* **28,** 376.
Karobath, M. E., Diaz, J.-L., and Huttunen, M. O. (1971). *Eur. J. Pharmacol.* **14,** 393.
Karobath, M. E., Diaz, J.-L., and Huttunen, M. (1972). *Biochem. Pharmacol.* **21,** 1245.
Kastin, A. J., and Barbeau, A. (1972). *Can. Med. Assoc. J.* **107,** 1079.

Kastin, A. J., and Ross, G. T. (1964). *Endocrinology* **95,** 187.

Kastin, A. J., and Schally, A. V. (1971). *Recent Adv. Endocrinol., Proc. Pan-Am. Congr. Endocrinol., 7th, 1970* Excerpta Med. Found. Int. Congr. Ser. No. 238, p. 311.

Kastin, A. J., Schally, A. V., Gual, C., Midgley, A. R., Jr., Bowers, C. Y., and Diaz-Infante, A. (1969). *J. Clin. Endocrinol. Metab.* **29,** 1046.

Kastin, A. J., Schally, A. V., and Viosca, S. (1971). *Proc. Soc. Exp. Biol. Med.* **137,** 1437.

Kastin, A. J., Schally, A. V., Gual, C., and Arimura, A. (1972a). *J. Clin. Endocrinol. Metab.* **34,** 753.

Kastin, A. J., Schally, A. V., Gual, C., Glick, S., and Arimura, A. (1972b). *J. Clin. Endocrinol. Metab.* **35,** 326.

Kastin, A. J., Gual, C., and Arimura, A. (1972c). *Recent Prog. Horm. Res.* **28,** 201.

Kastin, A. J., Viosca, S., Nair, R. M. G., and Miller, M. C. (1972d). *Endocrinology* **91,** 1323.

Kastin, A. J., Ehrensing, R. H., Schalch, D. S., and Anderson, M. S. (1972e). *Lancet* **2,** 740.

Kastin, A. J., Barbeau, A., Ehrensing, R. H., Plotnikoff, N. P., and Schally, A. V. (1973a). *Adv. Neurol.* **5,** 225–229.

Kastin, A. J., Miller, M. C., Ferrell, L., and Schally, A. V. (1973b). *Physiol. Behav.* **10,** 399.

Kastin, A. J., Nissel, C., Redding, T. W., Nair, R. M., and Schally, A. V. (1974). *Neuroendocrinology* **16,** 36.

Kastin, A. J., Plotnikoff, N. P., Hall, R., and Schally, A. V. (1975). *Hypothal. Horm. [Symp.], 1974,* pp. 261–268.

Kataoka, K. (1962). *Jpn. J. Physiol.* **12,** 81.

Kataoka, K., Ramwell, P. W., and Jessup, S. J. (1967). *Science* **157,** 1187.

Kataoka, K., Bak, I. J., Hassler, R., Kim, J. S., and Wagner, A. (1974). *Exp. Brain Res.* **19,** 217.

Kato, Y., Dupré, J., and Beck, J. C. (1973). *Endocrinology* **93,** 135.

Kato, Y., Chihara, K., Ohgo, S., and Imura, H. (1974a). *Endocrinology* **95,** 1608.

Kato, Y., Nakai, Y., Imura, H., Chihara, K., and Ohgo, S. (1974b). *J. Clin. Endocrinol. Metab.* **38,** 695.

Kato, Y., Chihara, K., Maeda, K., Ohgo, S., Okanishi, Y., and Imura, H. (1975). *Endocrinology* **96,** 1114.

Kato, Y., Chihara, K., Ohgo, S., Iwasaki, Y., Abe, H., and Imura, K. (1976a) *Life Sci.* **19,** 441.

Kato, Y., Iwasaki, Y., Abe, H., Imura, F., Hirata, F., and Hayaishi, O. (1976b). *Abstr. Book Int. Congr. Endocrinol., 5th, 1976* Abstract No. 28.

Katz, B. (1966). "Nerve, Muscle and Synapse." McGraw-Hill, New York.

Katz, B. (1969). "The Sherrington Lectures X: The Release of Neural Transmitter Substances." Liverpool Univ. Press, Liverpool.

Katz, B., Chase, T. N., and Kopin, I. J. (1969). *Science* **162,** 466.

Kawakami, M., and Sakuma, Y. (1974). *Neuroendocrinology* **15,** 290.

Kawakami, M., Negoro, H., Kimura, F., Higuchi, T., and Asai, T. (1975). *Neuroendocrinology* **19,** 137.

Kebabian, J. W., Petzold, G., and Greengard, P. (1972). *Proc. Natl. Acad. Sci. U.S.A.* **69,** 2145.

Keberle, H., and Faigle, J. W. (1972). *In* "Spasticity—A Topical Survey" (W. Birkmayer, ed.), pp. 90–93. Huber, Bern.

Kehr, W. (1977). *Eur. J. Pharmacol.* **41,** 261.

Keller, H. H., Bartholini, G., and Pletscher, A. (1973). *Eur. J. Pharmacol.* **23,** 183.

Keller, H. H., Bartholini, G., and Pletscher, A. (1974). *Nature (London)* **248,** 528.

Kelly, J. S., and Beart, P. M. (1975). *In* "Handbook of Psychopharmacology" (L. L. Iversen, S. D. Iversen, and S. H. Snyder, eds.), Vol. 4, pp. 29–209. Plenum, New York.

Kendall, J. W., Rees, L. H., and Kramer, R. (1971). *Endocrinology* **88,** 1503.

Kerdelhué, B., Jutisz, M. Gillessen, D., and Studer, R. O. (1973). *Biochim. Biophys. Acta* **297,** 540.

Key, B. J. (1975). *Neuropharmacology* **14,** 41.

Key, B. J., and Marley, E. (1961). *J. Physiol. (London)* **155,** 39P.

Khaunina, R. A. (1968). *Farmakol. Toksikol. (Moscow)* **31,** 202.

Khazan, H., Sulman, F. G., and Winnik, H. Z. (1961). *Proc. Soc. Exp. Biol. Med.* **106,** 579.

Killam, K. F., and Bain, J. A. (1957). *J. Pharmacol. Exp. Ther.* **199,** 255.

Kim, J. S., Bak, I. J., Hassler, R., and Okada, Y. (1971a). *Brain Res.* **14,** 95.

Kim, J. S., Sherman, L., Kolodny, H. D., Benjamin, F., and Singh, A. (1971b). *Clin. Res.* **15,** 718.

Kimura, H., and Kuriyama, K. (1975). *J. Neurochem.* **24,** 903.

Kimura, H., and MacLeod, R. M. (1975a). *57th Annu. Meet. Endocr. Soc., New York* p. 94.

Kimura, H., and MacLeod, R. M. (1975b). *Abstr. Book Int. Congr. Pharmacol., 6th, 1975* p. 88.

King, A. B., and Thomas, J. A. (1968). *J. Pharmacol. Exp. Ther.* **159,** 18.

King, D., and Jewett, R. E. (1971). *J. Pharmacol. Exp. Ther.* **177,** 188.

Kirpekar, S. M., and Puig, M. (1971). *Br. J. Pharmacol.* **43,** 359.

Kirshner, N. (1957). *J. Biol. Chem.* **226,** 821.

Kitay, J. I., and Altschule, M. D. (1954). "The Pineal Gland." Harvard Univ. Press, Cambridge, Massachusetts.

Kitchen, J. H. (1974). *Neuroendocrinology* **15,** 240.

Kitchen, J. H., Ruf, K. B., and Young Lai, F. H. (1974). *J. Reprod. Fertil.* **40,** 249.

Kizer, J. S., Palkovits, M., Zivin, J. A., Brownstein, M. J., Saavedra, J. M., and Kopin, I. J. (1974). *Endocrinology* **95,** 799.

Kizer, J. S., Arimura, A., Schally, A. V., and Brownstein, M. J. (1975a). *Endocrinology* **96,** 523.

Kizer, J. S., Kopin, I. J., and Zivin, J. A. (1975b). *Br. J. Pharmacol.* **54,** 243P.

Kizer, J. S., Zivin, J. A., Saavedra, J. M., and Brownstein, M. J. (1975c). *J. Neurochem.* **24,** 779.

Kizer, J. S., Zivin, J. A., Jacobowitz, D. M., and Kopin, I. J. (1975d). *Endocrinology* **96,** 1230.

Kizer, J. S., Palkovits, M., and Brownstein, M. J. (1976a). *Endocrinology* **98,** 311.

Kizer, J. S., Muth, E., and Jacobowitz, D. M. (1976b). *Endocrinology* **98,** 886.

Kizer, J. S., Palkovits, M., Kopin, I. J., Saavedra, J. M., and Brownstein, M. J. (1976c). *Endocrinology* **98,** 743.

Kizer, J. S., Palkovits, M., and Brownstein, M. (1976d). *Brain Res.* **108,** 363.

Klavans, H. L., and Rubovits, R. (1972). *Neurology* **22,** 107.

Klein, D. C. (1973). *Serotonin Behav., Proc. Symp., 1972* pp. 109–119.

Klein, D. C., and Weller, J. L. (1970). *Science* **169,** 1093.

Klein, D. C., Berg, G. R., and Weller, J. L. (1970). *Science* **168,** 979.

Knapp, S., and Mandell, A. J. (1972a). *Life Sci.* **16,** 761.

Knapp, S., and Mandell, A. J. (1972b). *Science* **177,** 1209.

Knigge, K. M., and Joseph, S. A. (1974). *Acta Endocrinol. (Copenhagen)* **76,** 209.

Knigge, K. M., and Schock, D. (1975). *Neuroendocrinology* **19,** 277.

Knigge, K. M., Joseph, S. A., Scott, D. E., and Jacobs, J. J. (1971). *In* "The Neuroendocrinology of Human Reproduction" (H. C. Mack and A. I. Sherman, eds.), pp. 6–22. Thomas, Springfield, Illinois.

Knigge, K. M., Joseph, S. A., Silverman, A. J., and Vaala, S. (1973). *In* "Drug Effects on Neuroendocrine Regulation" (E. Zimmermann *et al.*, eds.), pp. 7–20. Elsevier, Amsterdam.

Knigge, K. M., Hoffman, G., Scott, D. E., and Sladek, J. R. (1977). *Brain Res.* **120,** 393.

Knobil, E. (1974). *Recent Prog. Horm. Res.* **30,** 1.

Knobil, E., and Hotchkiss, J. (1964). *Annu. Rev. Physiol.* **26,** 47.

Knoche, H., Alpes, H., and Moellman, H. (1969). *Experientia* **25,** 515.

Knott, P. J., and Curzon, G. (1972). *Nature (London)* **239,** 452.

Knowles, F. (1965). *Arch. Anat. Microsc. Morphol. Exp.* **54,** 343.

Knowles, F., and Kumar, T. C. A. (1969). *Philos. Trans. R. Soc. London, Ser. B* **256,** 257.

Kobayashi, H., Matsui, T., and Ishii, S. (1970). *Int. Rev. Cytol.* **29,** 281.

Kobayashi, R. M., Palkovits, M., Kopin, I. J., and Jacobowitz, D. M. (1974). *Brain Res.* **77,** 269.

Koch, Y., Lu, K. H., and Meites, J. (1970). *Endocrinology* **87,** 673.

Koch, Y., Goldhaber, G., Fireman, T., Tor, V., Shani, J., and Tal, E. (1977). *Endocrinology* **100,** 1476.

Koe, B. K., and Weissman, A. (1966). *J. Pharmacol. Exp. Ther.* **154,** 499.

Koella, W. P., and Czicman, J. S. (1963). *Am. J. Physiol.* **204,** 873.

Koella, W. P., and Czicman, J. S. (1966). *Am. J. Physiol.* **211,** 926.

Koella, W. P., Feldstein, A., and Czicman, J. S. (1968). *Electroencephalogr. Clin. Neurophysiol.* **25,** 481.

Koelle, G. B. (1961). *Nature (London)* **190,** 208.

Koelle, G. B. (1970). *In* "The Pharmacological Basis of Therapeutics" (L. S. Goodman and A. Gilman, eds.), 4th ed., pp. 402–422. Macmillan, New York.

Koelle, G. B., and Friedenwald, J. S. (1949). *Proc. Soc. Exp. Biol. Med.* **70,** 617.

Koerker, D. J., Ruch, W., Chideckel, E., Palmer, J., Goodner, C. J., Ensinck, J., and Gale, C. C. (1974). *Science* **184,** 482.

Kokka, N., Garcia, J. F., and Elliott, H. W. (1973). *Prog. Brain Res.* **39,** 347.

Kolodny, H. D., Sherman, L., Singh, S., Kim, S., and Benjamin, F. (1971a). *N. Engl. J. Med.* **284,** 819.

Kolodny, H. D., Sherman, L., Singh, A., Benjamin, F., and Kim, S. (1971b). *53rd Annu. Meet. Endocr. Soc., San Francisco* p. 213.

Konishi, S., and Otsuka, M. (1974). *Brain Res.* **65,** 397.

Konstantinova, M. (1967). *Z. Zellforsch. Mikrosk. Anat.* **83,** 549.

Kopin, I. J. (1968). *Annu. Rev. Pharmacol.* **8,** 377.

Kopin, I. J. (1971). *Med. Res.* **5,** 329–353.

Kopin, I. J. (1972). *In* "Perspectives in Neuropharmacology" (S. H. Snyder, ed.), pp. 340–348. Oxford Univ. Press, London and New York.

Kopin, I. J. (1973). *In* "New Concepts in Neurotransmitter Regulation" (A. J. Mandell, ed.), pp. 21–31. Plenum, New York.

Kopin, I. J., and Gordon, R. (1962). *J. Pharmacol. Exp. Ther.* **138,** 351.

Kopin, I. J., and Gordon, E. K. (1963). *J. Pharmacol. Exp. Ther.* **140,** 207.

Kopin, I. J., Pare, C. M. B., Axelrod, J., and Weissbach, H. (1961). *J. Biol. Chem.* **236,** 3072.

Kopin, I. J., Fischer, J. E., Musacchio, J. M., Horst, W. D., and Weise, V. K. (1965). *J. Pharmacol. Exp. Ther.* **147,** 186.

Kopin, I. J., Breese, G. R., Krauss, K. R., and Weise, V. K. (1968). *J. Pharmacol. Exp. Ther.* **161,** 271.

Kordon, C. (1971). *Neuroendocrinology* **7,** 202.

Kordon, C., and Glowinski, J. (1969). *Endocrinology* **85,** 924.

Kordon, C., Javoy, F., Vassent, G., and Glowinski, J. (1968). *Eur. J. Pharmacol.* **4,** 169.

Kordon, C., Blake, C. A., Terkel, J., and Sawyer, C. H. (1973–1974). *Neuroendocrinology* **13,** 213.

Koslow, S. H. (1974). *Adv. Biochem. Psychopharmacol.* **11,** 95–100.

Koslow, S. H., Racagni, G., and Costa, E. (1974). *Neuropharmacology* **13,** 1123.

Kostowski, W., Giacalone, E., Garattini, S., and Valzelli, L. (1969). *Eur. J. Pharmacol.* **7,** 170.

Kostrzewa, R. M. (1975). *Res. Commun. Chem. Pathol. Pharmacol.* **11,** 567.
Kostrzewa, R. M., and Jacobowitz, D. M. (1974). *Pharmacol. Rev.* **26,** 199.
Kostrzewa, R. M., Kastin, A. J., and Spirtes, M. A. (1975). *Pharmacol. Biochem. Behav.* **3,** 1017.
Kotani, M., Onaya, T., and Yamada, T. (1973). *Endocrinology* **92,** 288.
Kovacs, E., and Flückiger, E. (1974). *Experientia* **30,** 1174.
Koyama, I. (1972). *Can. J. Physiol. Pharmacol.* **50,** 740.
Kremzner, L. T. (1966). *Anal. Biochem.* **15,** 270.
Krey, L. C., Butler, W. R., and Knobil, E. (1975). *Hypothal. Horm.* [*Symp.*], *1974* pp. 255–260.
Krieger, D. T. (1973). *J. Clin. Endocrinol. Metab.* **36,** 277.
Krieger, D. T. (1975). *Neuroendocrinology* **17,** 62.
Krieger, D. T. (1976). *N. Engl. J. Med.* **295,** 394.
Krieger, D. T., and Gewirtz, G. P. (1974). *J. Clin. Endocrinol. Metab.* **39,** 46.
Krieger, D. T., and Krieger, H. P. (1970). *Endocrinology* **87,** 179.
Krieger, D. T., and Rizzo, F. (1969). *Am. J. Physiol.* **217,** 1703.
Krieger, D. T., Silverberg, A. I., Rizzo, F., and Krieger, H. P. (1968). *Am. J. Physiol.* **215,** 959.
Krieger, D. T., Amorosa, L., and Linick, F. (1975). *N. Engl. J. Med.* **293,** 893.
Krieger, D. T., Howanitz, P. J., and Frantz, A. G. (1976). *J. Clin. Endocrinol. Metab.* **42,** 260.
Krieger, H. P., and Krieger, D. T. (1970). *Am. J. Physiol.* **218,** 1632.
Krnjević, K. (1964). *Int. Rev. Neurobiol.* **7,** 41.
Krnjević, K. (1965). *Br. Med. Bull.* **21,** 10.
Krnjevic, K. (1970). *Nature (London)* **228,** 119.
Krnjević, K., and Morris, M. (1973). *Soc. Neurosci.* p. 349.
Krnjević, K., and Morris, M. E. (1974). *Can. J. Physiol. Pharmacol.* **52,** 736.
Krnjević, K., and Phillis, J. W. (1963a). *Br. J. Pharmacol. Chemother.* **20,** 471.
Krnjević, K., and Phillis, J. W. (1963b). *J. Physiol. (London)* **165,** 274.
Krnjević, K., and Phillis, J. W. (1963c). *J. Physiol. (London)* **166,** 296.
Krnjević, K., and Phillis, J. W. (1963d). *J. Physiol. (London)* **166,** 328.
Krnjević, K., and Schwartz, S. (1966). *Nature (London)* **211,** 1372.
Krnjević, K., and Schwartz, S. (1967). *Exp. Brain Res.* **3,** 320.
Krnjević, K., and Silver, A. (1965). *J. Anat.* **99,** 711.
Krnjević, K., and Silver, A. (1966). *J. Anat.* **100,** 63.
Krogsgaard-Larsen, P., and Johnston, G. A. R. (1975). *J. Neurochem.* **25,** 797.
Kronheim, S., Berelowitz, M., and Pimstone, B. L. (1976). *Clin. Endocrinol.* **5,** 619.
Krulich, L. (1975). *Life Sci.* **17,** 1141.
Krulich, L., and Marchlewska-Koj, A. (1976). *Fed. Proc., Fed. Am. Soc. Exp. Biol.* **35,** Abstr. 555.
Krulich, L., Dhariwal, A. P. S., and McCann, S. M. (1967). *Abstr., Int. Symp. Growth Horm., 1st, 1967* Excerpta Med. Found. Int. Congr. Ser. No. 142, p. 32.
Krulich, L., Dhariwal, A. P. S., and McCann, S. M. (1968). *Endocrinology* **83,** 783.
Krulich, L., Quijada, M., Illner, P., and McCann, S. M. (1971) *Abstr. Book Int. Congr. Physiol. Sci. 25th, 1971* Vol. 9, p. 326.
Krulich, L., Illner, P., Fawcett, C. P., Quijada, M., and McCann, S. M. (1972). *Growth Growth Horm., Proc. Int. Symp., 2nd, 1971* Excerpta Med. Found. Int. Congr. Ser. No. 244, pp. 306–316.
Krulich, L., Quijada, L., Hefco, E., and Sundberg, D. K. (1974). *Endocrinology* **95,** 9.
Krulich, L., Hefco, E., and Aschenbrenner, J. E. (1975). *Endocrinology* **96,** 107.
Kuhar, M. J., Shaskan, E. G., and Snyder, S. (1971). *J. Neurochem.* **18,** 33.

Kuhar, M. J., Aghajanian, G. K., and Roth, R. H. (1974). *Adv. Biochem. Psychopharmacol.* **10,** 287–295.

Kulig, B. M. (1975). *Neuropharmacology* **14,** 489.

Kumeda, H., Huchimura, H., Kawabata, T., Maeda, Y., Okamoto, O., Kawa, A., and Kanehisa, T. (1974). *J. Endocrinol.* **62,** 161.

Kuntzman, R., Shore, P. A., Bogdanski, D., and Brodie, B. B. (1961). *J. Neurochem.* **6,** 226.

Kuppen, R., Lubrich, W., Haupt, M., Amerlahn, U., and Breuer, H. (1969). *Hoppe-Seyler's Z. Physiol. Chem.* **359,** 1067.

Kuriyama, K., Haber, B., Sisken, B., and Roberts, E. (1966). *Proc. Natl. Acad. Sci. U.S.A.* **55,** 846.

Kuriyama, K., Sisken, B., Haber, B., and Roberts, E. (1968). *Brain Res.* **9,** 165.

Kuroshima, A., Arimura, A., Bowers, C. Y., and Schally, A. V. (1966). *Endocrinology* **78,** 216.

Kwa, H. G., and Verhofstad, F. (1967). *J. Endocrinol.* **39,** 455.

Kwaitkowski, H. (1943). *J. Physiol. (London)* **102,** 32.

Labhsetwar, A. P. (1971). *Nature (London)* **229,** 203.

Labhsetwar, A. P., and Zolovick, A. (1973). *Nature (London), New Biol.* **246,** 55.

Laborit, H., Jouany, J. M., and Niaussat, P. (1959). *C.R. Seances Soc. Biol. Ses Fil.* **153,** 549.

Labrie, F., Beraud, G., Gauthier, M., and Lemay, A. (1971). *J. Biol. Chem.* **246,** 1902.

Labrie, F., Barden, N., Porrier, G., and De Lean, A. (1972). *Proc. Natl. Acad. Sci. U.S.A.* **69,** 283.

Labrie, F., Gauthier, M., Pelletier, G., Borgeat, P., Lemay, A., and Gouge, J.-J. (1973). *Endocrinology* **93,** 903.

Labrie, F., Borgeat, P., Ferland, L., Lemay, A., Dupont, A., Lemaire, S., Pelletier, G., Barden, N., Drouin, J., De Lean, A., and Bélanger, A. (1975). *Hypothal. Horm.* [*Symp.*], *1974* pp. 109–123.

Labrie, F., Savary, M., Coy, D. H., Coy, E. J., and Schally, A. V. (1976a). *Endocrinology* **98,** 289.

Labrie, F., Borgeat, P., Barden, N., Beaulieu, M., Ferland, L., Drouin, J., De Lean, A., and Morin, O. (1976b). *In* "Subcellular Mechanisms in Reproductive Neuroendocrinology" (F. Naftolin, K. S. Ryan, and I. Davies, eds.). Elsevier, Amsterdam (in press).

Lachelin, G. C. L., Leblanc, H., and Yen, S. S. C. (1977). *J. Clin. Endocrinol. Metab.* **44,** 728.

Ladinsky, H., Consolo, S., and Garattini, S. (1974). *Life Sci.* **14,** 1351.

Laguzzi, R., Petitjean, F., Pujol, J. F., and Jouvet, M. (1971). *C.R. Seances Soc. Biol. Ses Fil.* **165,** 1649.

Laguzzi, R., Petitjean, F., Pujol, J. F., and Jouvet, M. (1972). *Brain Res.* **48,** 295.

Lajtha, A. (1957). *J. Neurochem.* **1,** 216.

Lal, S., de la Vega, C. E., Sourkes, T. L., and Friesen, H. G. (1972). *Lancet* **1,** 661.

Lal, S., de la Vega, C. E., Sourkes, T. L., and Friesen, H. G. (1973). *J. Clin. Endocrinol. Metab.* **37,** 719.

Lal, S., Tolis, G., Martin, J. B., Brown, G. M., and Guyda, H. (1975a). *J. Clin. Endocrinol. Metab.* **41,** 827.

Lal. S., Martin, J. B., de la Vega, C. E., and Friesen, H. G. (1975b). *Clin. Endocrinol.* **4,** 277.

Lamar, C. (1970). *J. Neurochem.* **17,** 165.

Lamberts, S. W. J., and Birkenhägen, J. C. (1976a). *J. Endocrinol.* **70,** 315.

Lamberts, S. W. J., and Birkenhägen, J. C. (1976b). *Lancet* **2,** 811.

Lamberts, S. W. J., and Birkenhägen, J. C. (1976c). *Abstr. Book Int. Congr. Endocrinol., 5th, 1976* Abstract No. 492.

Lands, A. M., Arnold, A., McAuliff, J. P., Luduena, F. P., and Brown, T. G., Jr. (1967). *Nature (London)* **214,** 597.
Langer, S. Z. (1970). *J. Physiol. (London)* **208,** 515.
Langer, S. Z. (1973). *In* "Frontiers in Catecholamine Research" (E. Usdin and S. Snyder, eds.), pp. 543–549. Pergamon, Oxford.
Langer, S. Z., Adler, E., Enero, M. A., and Stefano, F. J. E. (1971). *Abstr. Book Congr. Physiol. Sci. 25th, 1971* p. 335.
Laron, Z., Pertzelan, A., and Mannheimer, S. (1966). *Isr. J. Med. Sci.* **2,** 152.
Laverty, R. (1973). *Br. Med. Bull.* **29,** 152.
Laverty, R., and Sharman, D. F. (1965). *Br. J. Pharmacol. Chemother.* **24,** 759.
Laverty, R., and Taylor, K. M. (1968). *Anal. Biochem.* **22,** 269.
Lawrence, A. M., and Hagen, T. C. (1972). *N. Engl. J. Med.* **287,** 150.
Lawrence, A. M., Goldfine, I. D., and Kirsteins, L. (1970). *J. Clin. Endocrinol. Metab.* **31,** 239.
Lawson, D. M., and Gala, R. R. (1975). *Endocrinology* **96,** 313.
Lawton, E. E., and Sawyer, C. H. (1968). *Endocrinology* **82,** 831.
Leblanc, H., Lachelin, G. C. L., Abu-Fadil, S., and Yen, S. S. C. (1976). *J. Clin. Endocrinol. Metab.* **43,** 668.
Lebovitz, H. E., Skyler, J. S., and Boyd, A. E. (1974). *Adv. Neurol.* **5,** 461–469.
Lederis, K. (1967). *Neurosecretion, Int. Symp., 4th, 1966* pp. 155–164.
Lederis, K., and Livingston, A. (1968). *J. Physiol. (London)* **196,** 34P.
Lederis, K., and Livingston, A. (1970). *J. Physiol. (London)* **210,** 187.
Lee, T. P., Kuo, J. F., and Greenzard, P., (1972). *Proc. Natl. Acad. Sci.* **69,** 3287.
Legait, H., and Legait, E. (1962). *Mem. Soc. Endocrinol.* **12,** 165–170.
Legros, J. J., and Breulet, M. (1975). *Lancet* **2,** 663.
Lehmann, A. (1964). *Aggressologie* **5,** 311.
Leibowitz, S. F. (1973). *Brain Res.* **63,** 440.
Leinweber, F. J. (1968). *Mol. Pharmacol.* **4,** 337.
Lemarchand-Béraud, T., Rappaport, G., Raymond, M., and Berthier, C. (1974). *Abstr. Book Hypothal. Horm. [Symp.] 1974.*
Lemberger, L., Crabtree, R., Lemens, J., Dyke, R., and Woodburn, R. (1974). *J. Clin. Endocrinol. Metab.* **39,** 579.
Lengvàri, I., and Hàlasz, B. (1973). *Neuroendocrinology* **11,** 191.
Leonardelli, J., Barry, J., and Dubois, M.-P. (1973). *C.R. Hebd. Seances Acad. Sci., Ser. D* **276,** 2043.
Leonardelli, J., Dubois, M. P., and Poulain, P. (1974). *Neuroendocrinology* **15,** 69.
Leppäluoto, J., Männisto, P., Ranta, T., and Lonnoila, M. (1976). *Acta Endocrinol. (Copenhagen)* **81,** 455.
Leslie, G. B. (1965). *Nature (London)* **208,** 1291.
Levitt, M., Spector, S., Sjoerdsma, A., and Udenfriend, S. (1965). *J. Pharmacol. Exp. Ther.* **148,** 1.
Levitt, M., Gibb, J. W., Daly, J. W., Lipton, M., and Udenfriend, S. (1967). *Biochem. Pharmacol.* **16,** 1313.
L'Hermite, P., Denayer, P., Degueldre, M., Goldstein, J., Vauhaelst, L., and Robyn, C. (1975). *Ric. Sci. Educ. Perm.* Suppl. 1, Abstr. 149.
Libertun, C., and McCann, S. M. (1973). *Endocrinology* **92,** 1714.
Libertun, C., and McCann, S. M. (1974). *Proc. Soc. Exp. Biol. Med.* **147,** 498.
Libertun, C., and McCann, S. M. (1976). *Neuroendocrinology* **20,** 110.
Lichtensteiger, W. (1969). *J. Physiol. (London)* **203,** 675.
Lichtensteiger, W., Mutzner, V., and Langeman, N. (1967). *J. Neurochem.* **14,** 489.
Lidbrink, P., Jonsson, G., and Fuxe, K. (1974). *Brain Res.* **67,** 439.
Liew, M.-C., and Laverty, R. (1975). *Eur. J. Pharmacol.* **33,** 165.

Lightman, S., and Iversen, L. L. (1969). *Br. J. Pharmacol.* **37,** 638.
Lin, R. C., Ngai, S. H., and Costa, E. (1969). *Science* **166,** 237.
Lin, T., and Tucci, J. R. (1972). *Clin. Res.* **20,** 431.
Lipinski, J. F., Schaumburg, H. H., and Baldessarini, R. J. (1973). *Brain Res.* **52,** 403.
Lippa, A. S., Antelman, S. M., Fahringer, E. E., and Redgate, E. S. (1973). *Nature (London), New Biol.* **241,** 24.
Lippman, W., Leonardi, R., Ball, J., and Coppola, A. (1967). *J. Pharmacol. Exp. Ther.* **156,** 258.
Lister, R. C., Underwood, L. E., Marshall, R. N., Friesen, H. G., and Van Vyk, J. J. (1974). *J. Clin. Endocrinol. Metab.* **39,** 1148.
Liu, T.-S., Jackson, G. L., and Gorski, J. (1976). *Endocrinology* **98,** 151.
Liuzzi, A., Chiodini, P. G., Botalla, L., Cremascoli, G., and Silvestrini, F. (1972). *J. Clin. Endocrinol. Metab.* **35,** 941.
Liuzzi, A., Chiodini, P. G., Botalla, L., Cremascoli, G., Müller, E. E., and Silvestrini, F. (1974a). *J. Clin. Endocrinol. Metab.* **38,** 910.
Liuzzi, A., Chiodini, P. G., Botalla, L., Silvestrini, F., and Müller, E. E. (1974b). *J. Clin. Endocrinol. Metab.* **39,** 871.
Liuzzi, A., Panerai, A. E., Chiodini, P. G., Secchi, C., Cocchi, D., Botalla, L., Silvestrini, F., and Müller, E. E. (1976). *In* "Growth Hormone and Related Peptides" (A. Pecile and E. E. Müller, eds.), pp. 236–251. Excerpta Med. Found., Amsterdam.
Livett, B. G. (1973). *Br. Med. Bull.* **29,** 93.
Lloyd, K. G., and Bartholini, G. (1974). *Adv. Biochem. Psychopharmacol.* **10,** 305–310.
Lloyd, K. G., and Hornykiewicz, O. (1974). *In* "Frontiers in Neurology and Neuroscience Research" (P. Seeman and G. M. Brown, eds.), pp. 26–35. Univ. of Toronto Press, Toronto.
Locatelli, V., Panerai, A. E., Cocchi, D., Gil-Ad, I., Mantegazza, P., and Müller, E. E. (1977). *Neuroendocrinology* in press.
Lockhart-Ewart, R. B., and Taylor, K. W. (1971). *Biochem. J.* **124,** 815.
Lofgren, F. (1960). *Acta Morphol. Neerl. Scand.* **3,** 55.
Logothetis, J., Harner, R., Morrell, F., and Torres, F. (1959). *Neurology* **9,** 352.
Loh, H. H., Hitzemann, R. J., and Way, E. L. (1973). *Life Sci.* **12,** 33.
Long, C. N. H. (1952). *Recent Prog. Horm. Res.* **7,** 75.
Long, C. N. H., and Fry, E. G. (1945). *Proc. Soc. Exp. Biol. Med.* **59,** 67.
Longo, V. G. (1966). *Pharmacol. Rev.* **18,** 965.
Longo, V. G., Scotti de Carolis, A., Liuzzi, A., and Massotti, M. (1974). *Adv. Biochem. Psychopharmacol.* **10,** 109–114.
Lorenzen, L. C., and Ganong, W. F. (1967). *Endocrinology* **80,** 889.
Lovenberg, W. (1972). *Abstr. Book Int. Congr. Pharmacol., 5th, 1972* p. 250.
Lovenberg, W., Jequier, E., and Sjoerdsma, A. (1968). *Adv. Pharmacol.* **6,** Part A, 21–36.
Lovinger, R. D., Boryczka, A. T., Shackelford, R., Kaplan, S. L., Ganong, W. F., and Grumbach, M. M. (1974a). *Endocrinology* **95,** 943.
Lovinger, R. D., Connors, M. H., Kaplan, S. L., Ganong, W. F., and Grumbach, M. M. (1974b). *Endocrinology* **95,** 1317.
Lovinger, R. D., Rose, J., Boryczka, A. T., Shackelford, R., Kaplan, S. L., Ganong, W. F., and Grumbach, M. M. (1975). *Ric. Sci. Educ. Perm.,* Suppl. 1, 59.
Lu, H. K., and Meites, J. (1971). *Proc. Soc. Exp. Biol. Med.* **137,** 480.
Lu, H. K., and Meites, J. (1972). *Endocrinology* **91,** 868.
Lu, H. K., and Meites, J. (1973). *Endocrinology* **93,** 152.
Lucke, C., and Glick, S. (1971). *J. Clin. Endocrinol. Metab.* **32,** 729.
Lucke, C., Höffken, B., and von zur Mühlen, A. (1975). *J. Clin. Endocrinol. Metab.* **41,** 1082.

Luedecke, G., Mueller, D., and Patino, C. (1976). *Abstr. Book Int. Congr. Endocrinol., 5th, 1976* p. 272.

Lufkin, E. G., Greene, H. L., and Meck, J. R. (1971). *J. Lab. Clin. Med.* **78,** 820.

Luft, R., Efendic, S., Hökfelt, T., Johansson, O., and Arimura, A. (1974). *Med. Biol.* **52,** 428.

Lundberg, P. O. (1972). *Acta Neurol. Scand.* **48,** 427.

Lutterbeck, P. M., Pryor, J. S., Varga, L., and Wenner, R. (1971). *Br. Med. J.* **3,** 228.

Luttge, W. G., Chronister, R. B., and Hall, N. R. (1973). *Life Sci.* **12,** 419.

Lymangrover, J. R., and Brodish, A. (1973). *Neuroendocrinology* **12,** 225.

Lymangrover, J. R., and Brodish, A. (1973–1974). *Neuroendocrinology* **13,** 234.

Lyons, W. R., and Page, E. (1935). *Proc. Soc. Exp. Biol. Med.* **32,** 1049.

Maany, I., Frazer, A., and Mendels, J. (1975). *J. Clin. Endocrinol. Metab.* **40,** 164.

McCann, S. M., and Brobeck, J. R. (1954). *Proc. Soc. Exp. Biol. Med.* **87,** 318.

McCann, S. M., and Fruit, A. (1957). *Proc. Soc. Exp. Biol. Med.* **96,** 566.

McCann, S. M., and Porter, J. C. (1969). *Physiol. Rev.* **49,** 240.

McCann, S. M., Dhariwal, A. P. S., and Porter, J. C. (1968). *Annu. Rev. Physiol.* **30,** 589.

McCann, S. M., Fawcett, C. P., and Krulich, L. (1974). *In* "Endocrine Physiology" (S. M. McCann, ed.), pp. 31–65. Butterworth, London.

McCaul, J. A., Cassel, K. J., and Stern, G. M. (1974). *Lancet* **1,** 735.

McDermott, W. V., Fry, E. C., Brobeck, J. R., and Long, C. N. H. (1950). *Yale J. Biol. Med.* **23,** 52.

McEwen, B. S., Magnus, C., and Wallach, G. (1972a). *Endocrinology* **90,** 217.

McEwen, B. S., Zigmond, R. L., and Gerlach, J. L. (1972b). *In* "The Structure and Function of Nervous Tissue" (G. H. Bourne, ed.), Vol. 5, pp. 250–292. Academic Press, New York.

McFarland, D., and Weiner, A. (1965). *Life Sci.* **4,** 1587.

McGeer, E. G. (1973). *Serotonin Behav., Proc. Symp., 1972* pp. 55–59.

McGeer, E. G., and McGeer, P. L. (1962). *J. Biochem. Physiol.* **40,** 114.

McGeer, E. G., McGeer, P. L., and Peters, D. A. (1967). *Life Sci.* **6,** 2221.

McGeer, P. L., McGeer, E. G., Fibiger, H. G., and Wickson, W. (1971). *Brain Res.* **35,** 308.

Machlin, L. J., Takahashi, Y., Horino, M., Hertelendy, F., Gordon, R. S., and Kipnis, D. M. (1968). *Proc. Int. Symp. Growth Horm. 1st, 1967* Excerpta Med. Found. Int. Congr. Ser. No. 158, pp. 292–305.

Machlin, L. J., Jacobs, L. S., Cirulis, N., Kimes, R., and Miller, R. (1974). *Endocrinology* **95,** 1350.

McIndoe, J. H., and Turkington, R. V. (1973). *J. Clin. Invest.* **52,** 1972.

McIntosh, E. N. (1976). *J. Clin. Endocrinol. Metab.* **42,** 1193.

McIntyre, H. B., and Odell, W. D. (1974). *Neuroendocrinology* **16,** 8.

McIsaac, W. M., and Estevez, V. (1966). *Biochem. Pharmacol.* **15,** 1625.

McKelvy, J. F., and Grimm-Jørgensen, Y. (1975). *Hypothal. Horm.* [*Symp.*], *1974* pp. 13–25.

McKinney, W. T., Jr. Prange, A. J., Jr., Majchowicz, E., and Schlesinger, K. (1971). *Dis. Nerv. Syst.* **32,** 308.

McLean, B. K., and Nikitovitch-Winer, M. B. (1975). *Endocrinology* **97,** 763.

McLean, P. O. (1957). *Arch. Neurol. Psychiatry* **78,** 113.

McLennan, H. (1975). *In* "Handbook of Psychopharmacology" (L. L. Iversen, S. D. Iversen, and S. H. Snyder, eds.), Vol. 4, pp. 211–228. Plenum, New York.

MacLeod, R. M., and Krieger, D. T. (1976). *58th Annu. Meet. Endocr. Soc., San Francisco* Abstract No. 317.

MacLeod, R. M., and Lehemeyer, J. E. (1970). *Proc. Natl. Acad. Sci. U.S.A.* **67,** 1172.

Macleod, R. M., and Lehemeyer, J. E. (1972). *Lactogenic Horm., Ciba Found. Symp., 1971* pp. 53–76.

MacLeod, R. M., and Lehemeyer, J. E. (1974). *Endocrinology* **94,** 1077.

MacLeod, R. M., Kimura, H., and Login, I. (1976). *In* "Growth Hormone and Related Peptides" (A. Pecile and E. E. Müller, eds.), pp. 443–453. Excerpta Med. Found., Amsterdam.

McMenamy, R. H., and Oncley, J. L. (1958). *J. Biol. Chem.* **233,** 1436.

McNeilly, A. S., Thorner, M. O., Volans, G., and Besser, G. M. (1974). *Br. Med. J.* **2,** 729.

Macon, J., Sokoloff, L., and Glowinski, J. (1971). *J. Neurochem.* **18,** 323.

MacPhail, E. M., and Miller, N. E. (1968). *J. Comp. Physiol. Psychol.* **65,** 499.

Madras, B. K., Cohen, E. L., Fernstrom, J. D., Larin, F., Munro, H. N., and Wurtman, R. J. (1973). *Nature (London)* **244,** 34.

Madras, B. K., Cohen, E. L., Munro, H. N., and Wurtman, R. J. (1974). *Adv. Biochem. Psychopharmacol.* **11,** 143–151.

Maeda, K., Kato, Y., Ohgo, S., Chihara, K., Yoshimoto, Y., Yamaguchi, N., Kuromaru, S., and Imura, H. (1975). *J. Clin. Endocrinol. Metab.* **40,** 501.

Maeda, K., Kato, Y., Yamaguchi, N., Chihara, K., Ohgo, S., Iwasaki, Y., Yoshimoto, Y., Moridera, K., Kuromaru, S., and Imura, H. (1976). *Acta Endocrinol. (Copenhagen)* **81,** 1.

Magnusson, T., Carlsson, A., Fisher, G. H., Chang, D., and Folkers, K. (1976). *J. Neural Trans.* **38,** 89.

Mahler, H. R., and Cordes, E. H. (1966). *In* "Biological Chemistry," pp. 653–713. Harper, New York.

Maickel, R. P., and Miller, F. P. (1968). *Adv. Pharmacol.* **6a,** 71–77.

Maickel, R. P., Westermann, E. D., and Brodie, B. B. (1961). *J. Pharmacol. Exp. Ther.* **134,** 167.

Maitre, L., Waldmeier, P. C., Baumann, P. A., and Staehelin, M. (1974). *Adv. Biochem. Psychopharmacol.* **10,** 297–304.

Makara, G. B., and Stork, E. (1974). *Neuroendocrinology* **16,** 178.

Makara, G. B., and Stork, E. (1975). *Neuroendocrinology* **18,** 213.

Makino, T. (1973). *Am. J. Obstet. Gynecol.* **115,** 606.

Malacara, J. M., and Reichlin, S. (1972). *Growth Growth Horm., Proc. Int. Symp., 2nd, 1971* Excerpta Med. Found. Int. Congr. Ser. No. 244, pp. 299–305.

Malacara, J. M., Seyler, L. E., Jr., and Reichlin, S. (1972). *J. Clin. Endocrinol. Metab.* **34,** 271.

Malarkey, W. B., and Daughaday, W. H. (1972). *Endocrinology* **91,** 1314.

Malarkey, W. B., Jacobs, L. S., and Daughaday, W. H. (1971). *N. Engl. J. Med.* **285,** 1160.

Malarkey, W. B., Cyrus, J., and Paulson, G. W. (1974). *J. Clin. Endocrinol. Metab.* **39,** 229.

Malven, P. V. (1975a). *Endocrinology* **97,** 808.

Malven, P. V. (1975b). *Neuroendocrinology* **19,** 81.

Mancini, A. M., Guitelman, A., Vargas, C. A., Debeljuk, L., and Aparicio, N. J. (1976). *J. Clin. Endocrinol. Metab.* **42,** 181.

Mancini, A. M., Guitelman, A., Debeljuk, L., and Vargas, C. A. (1977). *J. Endocrinol.* (in press).

Mandell, A. J., and Morgan, M. (1970). *Nature (London)* **227,** 75.

Mandell, M. P., Mandell, A. J., and Jacobson, A. (1965). *Recent Adv. Biol. Psychiatry* **7,** 115.

Maneckjee, R., Srinath, B. R., and Moudgal, N. R. (1976). *Nature (London)* **262,** 507.

Mangan, J. L., and Whittaker, V. P. (1966). *Biochem. J.* **98,** 128.

Mannarino, E., Kirshner, N., and Nashold, B. S. (1963). *J. Neurochem.* **10,** 373.

Mannisto, M., and Tuomisto, J. (1976). *Abstr. Book Int. Congr. Endocrinol., 5th, 1976* Abstract No. 912.

Mao, C. C., Guidotti, A., and Costa, E. (1974a). *Brain Res.* **79,** 510.
Mao, C. C., Guidotti, A., and Costa, E. (1974b). *Mol. Pharmacol.* **10,** 736.
Marantz, R., and Rechtschaffen, A. (1967). *Percept. Mot. Skills* **25,** 805.
Marchbanks, R. M. (1968). *Biochem. J.* **110,** 533.
Marchbanks, R. M. (1969). *Biochem. Pharmacol.* **18,** 1763.
Marchbanks, R. M. (1972). *In* "The Structure and Function of Nervous Tissue" (G. H. Bourne, ed.), Vol. 4, pp. 20–33. Academic Press, New York.
Marchbanks, R. M., and Israël, M. (1973). *Biochem. J.*
Marchlewska-Koj, A., and Krulich, L. (1975). *Fed. Proc., Fed. Am. Soc. Exp. Biol.* **34,** 252.
Marcus, E. M., Watson, C. W., and Goldman, P. L. (1966). *Arch. Neurol. (Chicago)* **15,** 521.
Marczynski, T. J., Yamaguchi, N., Ling, G. M., and Grodzinska, L. (1964). *Experientia* **20,** 435.
Marks, B. H., Hall, M. M., and Bhattacharya, A. N. (1970). *Prog. Brain Res.* **32,** 57.
Marley, E., and Blackwell, B. (1970). *Adv. Pharmacol. Chemother.* **8,** 185–239.
Marley, E., and Nisticò, G. (1972). *Br. J. Pharmacol.* **46,** 619.
Marley, E., and Nisticò, G. (1975a). *Br. J. Pharmacol.* **53,** 193.
Marley, E., and Nisticò, G. (1975b). *Br. J. Pharmacol.* **55,** 459.
Marley, E., and Seller, T. J. (1972). *Br. J. Pharmacol.* **44,** 413.
Marley, E., and Seller, T. J. (1974). *Br. J. Pharmacol.* **51,** 347.
Marley, E., and Stephenson, J. D. (1972). *Handb. Exp. Pharmakol.* [N. S.] **33,** 463–537.
Marley, E., and Whelan, J. E. (1976). *Br. J. Pharmacol.* **56,** 133.
Mars, H., and Genuth, S. M. (1973). *Clin. Pharmacol. Ther.* **14,** 390.
Martin, J. B. (1972). *Endocrinology* **91,** 107.
Martin, J. B. (1973a). *N. Engl. J. Med.* **288,** 1384.
Martin, J. B. (1973b). *In* "Advances in Human Growth Hormone Research" (S. Raiti, ed.), DHEW Publ. No. (NIH) 74-612, pp. 223–249. Center for Disease Control, Atlanta, Georgia.
Martin, J. B. (1974). *56th Annu. Meet. Endocr. Soc., Atlanta* p. 248.
Martin, J. B. (1976). *In* "Frontiers in Neuroendocrinology" (L. Martin and W. F. Ganong, eds.), Vol. 4, pp. 129–168. Raven Press, New York.
Martin, J. B., Lal, S., Tolis, G., and Friesen, H. G. (1974). *J. Clin. Endocrinol. Metab.* **39,** 180.
Martin, J. B., Tannenbaum, G., Willoughby, J. O., Renaud, L. P., and Brazeau, P. (1975a). *Hypothal. Horm.* [*Symp.*], *1974* pp. 217–235.
Martin, J. B., Audet, J., and Saunders, A. (1975b). *Endocrinology* **96,** 839.
Martin, M. M., and Hellman, D. E. (1964). *J. Clin. Endocrinol. Metab.* **24,** 253.
Martin, S., York, D. H., and Kraicer, J. (1973). *Endocrinology* **92,** 1084.
Martin, W. R., Sloan, J. W., and Christian, S. T. (1971). *Histochemie* **25,** 32.
Martini, L., and Pecile, A. (1962). *J. Endocrinol.* **24,** 11.
Martini, L., Pecile, A., Saito, S., and Tani, F. (1960). *Endocrinology* **66,** 501.
Martres, M. P., Baudry, M., and Schwartz, J. C. (1975). *Brain Res.* **83,** 261.
Mason, J. (1959). *Recent Prog. Horm. Res.* **15,** 345.
Massara, F., and Camanni, F. (1972). *J. Endocrinol.* **54,** 195.
Massara, F., and Strumia, E. (1970). *J. Endocrinol.* **47,** 95.
Massara, F., Camanni, F., and Molinatti, G. M. (1973a). *Folia Endocrinol.* **26,** 476.
Massara, F., Camanni, F., and Molinatti, G. M. (1973b). *Horm. Metab. Res.* **5,** 454.
Massara, F., Camanni, F., Belforte, L., and Molinatti, G. M. (1976). *Lancet* **1,** 485.
Matsukara, S., West, C. D., Ichikawa, Y., Jubiz, W., Harada, G., and Tyler, F. H. (1971). *J. Lab. Clin. Med.* **77,** 490.
Matsuo, H., Fujimoto, Y., and Tasuno, T. (1966). *Biochem. Biophys. Res. Commun.* **22,** 69.

Matsuo, H., Baba, Y., Nair, R. M. G., Arimura, A., and Schally, A. V. (1971a). *Biochem. Biophys. Res. Commun.* **43,** 1334.

Matsuo, H., Arimura, A., Nair, R. M. G., and Schally, A. V. (1971b). *Biochem. Biophys. Res. Commun.* **45,** 822.

Matsuzaki, M. (1969). *Brain Res.* **13,** 247.

Matussek, N., and Patschke, U. (1963). *Med. Exp.* **11,** 81.

Matz, R., Rick, W., Oh, D., Thompson, H., and Gershon, S. (1974). *Curr. Ther. Res.* **16,** 687.

Maynert, E. W., and Levi, R. (1964). *J. Pharmacol. Exp. Ther.* **143,** 90.

Medina, M., and Shore, P. A. (1966). *Biochem. Pharmacol.* **15,** 1627.

Meites, J. (1966). *In* "Neuroendocrinology" (L. Martini and W. F. Ganong, eds.), Vol. 1, pp. 669–707. Academic Press, New York.

Meites, J. (1970). *Hypophysiotropic Horm. Hypothala.: Assay Chem., Proc. Workshop Conf., 1969* Comments, pp. 144–146.

Meites, J. (1973). *In* "Human Prolactin" (J. L. Pasteels and C. Robyn, eds.), pp. 105–118. Excerpta Med. Found., Amsterdam.

Meites, J., and Clemens, J. A. (1972). *Vitam. Horm. (N.Y.)* **30,** 165.

Meites, J., and Nicoll, C. S. (1966). *Annu. Rev. Physiol.* **28,** 57.

Meites, J., Talwalker, P. K., and Nicoll, C. S. (1960). *Proc. Soc. Exp. Biol. Med.* **103,** 298.

Meites, J., Nicoll, C. S., and Talwalker, P. K. (1963). *Adv. Neuroendocrinol., Proc. Symp., 1961* pp. 238–277.

Meites, J., Lu, K. H., Wuttke, W., Welsch, G. W., Nagasawa, H., and Quadri, S. K. (1972). *Recent Prog. Horm. Res.* **28,** 471.

Melander, A. (1969). *Acta Endocrinol. (Copenhagen)* **62,** 565.

Melander, A. (1970). *Acta Endocrinol. (Copenhagen)* **65,** 371.

Melander, A., and Sandler, F. (1972). *Endocrinology* **90,** 188.

Meldrum, B. S. (1975). *Int. Rev. Neurobiol.* **17,** 1.

Meltzer, H. Y., Daniels, S., and Fang, V. S. (1975). *Life Sci.* **17,** 339.

Meltzer, H. Y., Fang, V. S., and Daniels, S. (1976). *Fed. Proc., Fed. Am. Soc. Exp. Biol.* **35,** Abstr. 554.

Mendelson, W. B., Jacobs, L. S., Reichman, J. D., Othmer, E., Cryer, P. E., Trivedi, B., and Daughaday, W. H. (1975). *J. Clin. Invest.* **56,** 690.

Mendoza, D., Arimura, A., and Schally, A. V. (1973). *Endocrinology* **92,** 1153.

Merimee, T. J., Burgess, J. A., and Rabinowitz, D. (1966). *J. Clin. Endocrinol. Metab.* **26,** 791.

Merritt, J. H., Schultz, E. J., and Wykes, A. A. (1965). *Biochem. Pharmacol.* **13,** 1364.

Mess, B., Heizer, A., Tóth, A., and Tima, L. (1971). *Pineal Gland, Ciba Found. Symp., 1970* pp. 229–240.

Metcalf, G. (1974). *Nature (London)* **252,** 310.

Meyer, V., and Knobil, K. (1967). *Endocrinology* **80,** 163.

Meyers, F. H., Jawetz, E., and Goldfien, A. (1974). "Review of Medical Pharmacology," 4th ed. Lange Med. Publ., Los Altos, California.

Meyerson, B. J., and Sawyer, C. H. (1968). *Endocrinology* **83,** 170.

Michaelson, I. A., and Whittaker, V. P. (1963). *Biochem. Pharmacol.* **12,** 203.

Middler, S. A. (1976). *N. Engl. J. Med.* **295,** 394.

Millard, S. A., Costa, E., and Gàl, E. M. (1972). *Brain Res.* **40,** 545.

Miller, F. P., Cox, R. H., Jr., Snodgrass, W. R., and Maickel, R. P. (1970). *Biochem. Pharmacol.* **19,** 435.

Miller, R. J., and Iversen, L. L. (1974). *Naunyn-Schmiedebergs Arch. Exp. Pathol. Pharmakol.* **282,** 213.

Miller, R. J., and Kelly, P. H. (1975). *Nature (London)* **255,** 163.
Miller, R. J., Horn, A. S., and Iversen, L. L. (1974). *Mol. Pharmacol.* **10,** 759.
Milligan, J. V., and Kraicer, J. (1971). *Endocrinology* **89,** 766.
Milligan, J. V., Kraicer, J., Illner, P., and Fawcett, P. (1972). *Can. J. Physiol. Pharmacol.* **50,** 613.
Mills, J., Kattan, R., Slater, I. H., and Fuller, R. W. (1968). *J. Med. Chem.* **11,** 95.
Milton, A. S., and Wendlandt, S. (1970). *J. Physiol. (London)* **207,** 76P.
Milton, A. S., and Wendlandt, S. (1971a). *J. Physiol. (London)* **217,** 33P.
Milton, A. S., and Wendlandt, S. (1971b). *J. Physiol. (London)* **218,** 325.
Mims, R. B., Stein, R. B., and Bethune, J. E. (1973a). *J. Clin. Endocrinol. Metab.* **37,** 34.
Mims, R. B., Scott, C. L., Modebe, M. O., and Bethune, J. E. (1973b). *J. Clin. Endocrinol. Metab.* **37,** 660.
Mims, R. B., Scott, C. L., Modebe, O., and Bethune, J. E. (1975). *J. Clin. Endocrinol. Metab.* **40,** 256.
Minozzi, M., Faggiano, M., Lombardi, G., Carella, G., Criscuolo, T., and Scapagnini, U. (1975). *Neuroendocrinology* **17,** 147.
Mioduszewski, R., Grandison, L., and Meites, J. (1976). *Proc. Soc. Exp. Biol. Med.* **151,** 44.
Mira, E. (1960). *Proc. Int. Congr. Histochem. Cytochem., 1st, 1960* p. 147.
Mishkinsky, J., Dikstein, S., and Sulman, F. G. (1968). *Endocrinology* **82,** 611.
Mitchell, A. J., Smyrl, R., Hutchins, M., Schindler, W. J., and Critchlow, V. (1972). *Neuroendocrinology* **10,** 31.
Mitnick, M. A., and Reichlin, S. (1971). *Science* **172,** 1241.
Mittler, J. C., Arimura, A., and Schally, A. V. (1970). *Proc. Soc. Exp. Biol. Med.* **133,** 1321.
Miyai, K., Onishi, T., Hosokawa, M., Ishabashi, K., and Kumahara, Y. (1974). *J. Clin. Endocrinol. Metab.* **39,** 391.
Möberg, G. P., Scapagnini, U., De Groot, J., and Ganong, W. F. (1971). *Neuroendocrinology* **7,** 11.
Moir, A. T. B., and Eccleston, D. (1968). *J. Neurochem.* **15,** 1093.
Molrar, J., Arimura, A., and Kastin, A. J. (1976). *Fed. Proc., Fed. Am. Soc. Exp. Biol.* **35,** 782.
Monachon, M. A., Burkard, W. P., Jalfre, M., and Haefely, W. (1972). *Naunyn-Schmiedebergs Arch. Pharmacol.* **274,** 192.
Monahan, M., Rivier, J., Burgus, R., Amoss, M., Blackwell, R., Vale, W., and Guillemin, R. (1971). *C.R. Hebd. Seances Acad. Sci., Ser. D* **273,** 508.
Montagu, K. A. (1957). *Nature (London)* **180,** 244.
Montanari, R., and Stockham, M. A. (1962). *Br. J. Pharmacol. Chemother.* **18,** 337.
Monti, J. M., Sala, M. A., Otegui, J. T., Benedetti, W. L., and Grino, E. (1970). *Experientia* **26,** 1263.
Montoya, L., Seibel, M. J., and Wilber, J. F. (1975). *Endocrinology* **96,** 1413.
Moore, K. E., and Larivière, E. W. (1964). *Biochem. Pharmacol.* **13,** 1098.
Moore, K. E., and Phillipson, D. T. (1975a). *Br. J. Pharmacol.* **53,** 453P.
Moore, K. E., and Phillipson, D. T. (1975b). *J. Neurochem.* **25,** 289.
Moore, K. E., and Thornburg, J. E. (1975). *Adv. Neurol.* **9,** 93–104.
Mora, S., Loizzo, A., and Longo, V. G. (1976). *Pharmacol. Biochem. Behav.* **4,** 279.
Moran, J. F., and Sourkes, T. L. (1963). *J. Biol. Chem.* **238,** 3006.
Morgan, C. M., and Hadley, M. E. (1976). *Neuroendocrinology* **21,** 10.
Morris, D., and Grewaal, D. S. (1969). *Life Sci.* **8,** 511.
Mortimer, C. H., Carr, D., Lind, T., Bloom, S. R., Mallison, C. N., Schally, A. V., Tunbridge, W. M. G., Yeomans, L., Coy, D. H., Kastin, A. J., Besser, G. M., and Hall, R. (1974). *Lancet* **1,** 697.

Mortimer, C. H., Besser, G. M., and McNeilly, A. S. (1975). *Hypothal. Horm.* [*Symp.*], *1974* pp. 325–335.

Moruzzi, G., and Magoun, H. W. (1949). *Electroencephalogr. Clin. Neurophysiol.* **1,** 455.

Moss, R. L., and Foreman, M. M. (1976). *Neuroendocrinology* **20,** 176.

Moss, R. L., and McCann, S. M. (1973). *Science* **181,** 177.

Moss, R. L., Kelly, M., and Riskind, P. (1975a). *Brain Res.* **89,** 265.

Moss, R. L., Dudley, C., Foreman, M., and McCann, S. M. (1975b). *Hypothal. Horm.* [*Symp.*], *1974* pp. 269–278.

Motovilov, P. E., and Kozhevnikov, S. P. (1968). *Pharmakol. Toksikol. (Moscow)* **31,** 205.

Mountjoy, C. Q., Weller, M., Hall, R., Price, J. S., Hunter, P., and Dewar, J. H. (1974). *Lancet* **1,** 958.

Mowles, T. F., Ashkanazy, B., Mix, E., Jr., and Sheppard, H. (1971). *Endocrinology* **89,** 484.

Moyersoons, F., Evraerd, A., Dauby, J., and Giurgea, C. (1969). *Arch. Int. Pharmacodyn. Ther.* **179,** 388.

Muller, G. P., Chen, H. J., and Meites, J. (1973). *Proc. Soc. Exp. Biol. Med.* **144,** 613.

Mueller, G. P., Simpkins, J., Meites, J., and Moore, K. E. (1976a). *Neuroendocrinology* **20,** 121.

Mueller, G. P., Twohy, C. P., Chen, H. T., Advis, J. P., and Meties, J. (1976b). *Life Sci.* **18,** 715.

Mueller, R. A., Thoenen, H., and Axelrod, J. (1969a). *Science* **163,** 468.

Mueller, R. A., Thoenen, H., and Axelrod, J. (1969b). *J. Pharmacol. Exp. Ther.* **169,** 74.

Muenter, M. D., and Tyce, G. F. (1971). *Mayo Clin. Proc.* **46,** 231.

Mulder, A. H., and Smelik, P. G. (1974). *In* "Hormones, Homeostasis and the Brain," p. 15. Elsevier, Amsterdam.

Müller, E. E. (1973). *Neuroendocrinology* **11,** 338.

Müller, E. E. (1976). *Proc. Int. Congr. Pharmacol., 6th, 1975* Vol. 3, pp. 131–145.

Müller, E. E., and Pecile, A. (1965). *Proc. Soc. Exp. Biol. Med.* **119,** 1191.

Müller, E. E., Saito, T., Arimura, A., and Schally, A. V. (1967a). *Endocrinology* **80,** 109.

Müller, E. E., Sawano, S., Arimura, A., and Schally, A. V. (1967b). *Endocrinology* **80,** 471.

Müller, E. E., Dal Prà, P., and Pecile, A. (1968). *Endocrinology* **83,** 893.

Müller, E. E., Pecile, A., Felici, M., and Cocchi, D. (1970). *Endocrinology* **86,** 1376.

Müller, E. E., Schally, A. V., and Cocchi, D. (1971a). *Proc. Soc. Exp. Biol. Med.* **137,** 489.

Müller, E. E., Cocchi, D., and Forni, A. (1971b). *Life Sci.* **10,** 1057.

Müller, E. E., Miedico, D., Giustina, G., and Cocchi, D. (1971c). *Endocrinology* **88,** 345.

Müller, E. E., Giustina, G., Miedico, D., and Cocchi, D. (1972a). *Growth Growth Horm., Proc. Int. Symp., 2nd, 1971* Excerpta Med. Found. Int. Congr. Ser. No. 244, pp. 283–299.

Müller, E. E., Cocchi, D., and Mantegazza, P. (1972b). *Am. J. Physiol.* **223,** 945.

Müller, E. E., Panerai, A., Cocchi, D., Frohman, L. A., and Mantegazza, P. (1973a). *Experientia* **29,** 874.

Müller, E. E., Frohman, L. A., and Cocchi, D. (1973b). *Am. J. Physiol.* **224,** 1210.

Müller, E. E., Cocchi, D., Jalanbo, H., and Udeschini, G. (1973c). *55th Annu. Meet. Endocr. Soc., Chicago* p. 248.

Müller, E. E., Pecile, A., Cocchi, D., and Olgiati, V. R. (1974a). *Am. J. Physiol.* **226,** 1100.

Müller, E. E., Brambilla, F., Cavagnini, F., Peracchi, M., and Panerai, A. (1974b). *J. Clin. Endocrinol. Metab.* **39,** 1.

Müller, E. E., Panerai, A. E., Cocchi, D., Gil-Ad, I., Rossi, G. L., and Olgiati, V. R. (1976a). *58th Annu. Meet. Endocr. Soc., San Francisco,* p. 120.

Müller, E. E., Udeschini, G., Secchi, C., Zambotti, F., Vicentini, L., Panerai, A. E., Cocola, F., and Mantegazza, P. (1976b). *Acta Endocrinol. (Copenhagen)* **82,** 71.

Müller, E. E., Gil-Ad, I., Panerai, A. E., Secchi, C., Liuzzi, A., Chiodini, P. G., and Silvestrini, F. (1976c). *In* "The Basic and Clinical Aspects of Hypothalamic Hormones" (A. L. Charro Salgado, R. Fernandez Durango, and J. G. Lopéz del Campo, eds.), pp. 211–231. Excerpta Med. Found., Amsterdam.

Müller, E. E., Liuzzi, A., Cocchi, D., Panerai, A. E., Oppizzi, G., Locatelli, V., Silvestrini, F. Mantegazza, P., and Chiodini, P. G. (1977a). *In* "Advances in Biochemical Psychopharmacology" (E. Costa and G. L. Gessa, eds.), Vol. 16, pp. 127–138. Raven Press, New York.

Müller, E. E., Genazzani, A. R., Murru, S., and Fioretti, P. (1977b). *Acta Endocrinol. (Copenhagen)* (in press).

Mulloy, A. L., and Möberg, G. P. (1975). *Fed. Proc., Fed. Am. Soc. Exp. Biol.* **34,** 251.

Murphy, J. T., Dreifuss, J. J., and Gloor, P. (1968). *Am. J. Physiol.* **214,** 1443.

Musacchio, J. M., Kopin, I. J., and Snyder, S. H. (1964). *Life Sci.* **3,** 769.

Musacchio, J. M., Bhagat, B., Jackson, C. J., and Kopin, I. J. (1966). *J. Pharmacol. Exp. Ther.* **152,** 293.

Muscholl, E. (1972). *In* "Catecholamines" (H. Blaschko and E. Muscholl, eds.), pp. 618–660. Springer-Verlag, Berlin and New York.

Myers, R. D. (1964a). *In* "Thirst: First International Symposium on Thirst and Regulation of Body Water" (M. J. Wayner, ed.), pp. 533–549 Pergamon, New York.

Myers, R. D. (1964b). *Can. J. Physiol.* **18,** 6.

Naftolin, F., Morishita, H., Davies, I. J., Todd, R., Ryan, K. J., and Fishman, J. (1975). *Biochem. Biophys. Res. Commun.* **65,** 905.

Nagatsu, T., Levitt, M., and Udenfriend, S. (1964). *Biochem. Biophys. Res. Commun.* **14,** 543.

Nair, R. M. G., Barret, J. F., Bowers, C. Y., and Schally, A. V. (1970). *Biochemistry* **9,** 1103.

Nair, R. M. G., Redding, T. W., and Schally, A. V. (1971a). *Biochemistry* **10,** 3621.

Nair, R. M. G., Kastin, A. J., and Schally, A. V. (1971b). *Biochem. Biophys. Res. Commun.* **43,** 1376.

Nair, R. M. G., Kastin, A. J., and Schally, A. V. (1972). *Biochem. Biophys. Res. Commun.* **47,** 1420.

Nair, R. M. G., de Villier, C., Barnes, M., and Wilbur, D. L. (1976). *58th Annu. Meet. Endocr. Soc., San Francisco* Abstract No. 125.

Nakagawa, K., and Mashimo, K. (1973). *J. Clin. Endocrinol. Metab.* **37,** 238.

Nakahara, T., Uchimura, H., Hirano, M., Saito, T., and Ito, M. (1976). *Brain Res.* **117,** 351.

Nakai, Y., Imura, H., Yoshimi, T., and Matsuokara, S. (1973). *Acta Endocrinol. (Copenhagen)* **74,** 263.

Nakai, Y., Imura, H., Sakura, H., Kurahachi, H. and Yoshimi, T. (1974). *J. Clin. Endocrinol. Metab.* **38,** 446.

Naumenko, E. V. (1965). *Izv. Sib. Otd. Akad. Nauk. SSSR* **3,** 143.

Naumenko, E. V. (1967). *Endocrinology* **80,** 69.

Naumenko, E. V. (1968). *Brain Res.* **11,** 1.

Naumenko, E. V. (1969). *Neuroendocrinology* **5,** 81.

Naumenko, E. V., Maslova, L. N., and Popova, N. K. (1973). *Byull. Eksp. Biol. Med.* **76,** 512.

Neal, M. J., and Iversen, L. L. (1969). *J. Neurochem.* **16,** 1245.

Neff, N. H., and Costa, E. (1966). *Life Sci.* **5,** 951.

Neff, N. H., and Tozer, T. N. (1968). *Adv. Pharmacol.* **6a,** 97–109.

Neff, N. H., Tozer, T. N., and Brodie, B. B. (1964). *Pharmacologist* **6,** 162.
Neff, N. H., Tozer, T. N., and Brodie, B. B. (1967). *J. Pharmacol. Exp. Ther.* **158,** 214.
Neff, N. H., Ngai, S. H., Wang, C. T., and Costa, E. (1969). *Mol. Pharmacol.* **5,** 90.
Neff, N. H., Spano, P. F., Groppetti, A., Wang, C. T., and Costa, E. (1971). *J. Pharmacol. Exp. Ther.* **176,** 701.
Neill, D. B., and Linn, C. L. (1975). *Physiol. Behav.* **14,** 617.
Neill, J. D. (1970). *Endocrinology* **87,** 1192.
Neill, J. D., and Reichert, L. E., Jr. (1971). *Endocrinology* **88,** 548.
Nett, T. M., Akbar, A. M., Niswender, G. D., Hedlund, M. T., and White, W. F. (1973). *J. Clin. Endocrinol. Metab.* **36,** 880.
Ng, L. K. Y., Chase, T. N., Colburn, R. W., and Kopin, I. J. (1972). *Neurology* **22,** 688.
Nichols, C. W., Jacobowitz, D., and Hottenstein, M. (1967). *Invest. Ophthalmol.* **6,** 642.
Nickel, E., and Potter, L. T. (1970). *Brain Res.* **23,** 95.
Nickers, L., and Sze, P. Y. (1975). *Brain Res.* **93,** 123.
Nicoll, C. S. (1971). *In* "Frontiers in Neuroendocrinology" (L. Martini and W. F. Ganong, eds.), pp. 291–330. Oxford Univ. Press, London and New York.
Nicoll, C. S. (1972). *Gen. Comp. Endocrinol.* **3,** Suppl. 3.
Nicoll, C. S., Fiorindo, R. P., McKennee, C. T., and Parsons, J. A. (1970). *Hypophysiotropic Horm. Hypothal.: Assay Chem. Proc. Workshop Conf., 1969* pp. 115–150.
Nicoll, R. A. (1975). *In* "Handbook of Psychopharmacology" (L. L. Iversen, S. D. Iversen, and S. H. Snyder, eds.), Vol. 4, pp. 229–263. Plenum, New York.
Nikodijevic, B., Daly, J. W., and Creveling, C. R. (1969). *Biochem. Pharmacol.* **18,** 1577.
Nilson, J. A., and Pycock, C. J. (1976). *Br. J. Pharmacol.* **56,** 77.
Nilsson, G., Hökfelt, T., and Pernow, B. (1974). *Med. Biol.* **52,** 424.
Nilsson, K. O., Thorrel, J. I., and Hökfelt, B. (1974). *Acta Endocrinol. (Copenhagen)* **76,** 24.
Nilsson, K. O., Thorrel, J. I., and Hökfelt, B. (1975). *Acta Endocrinol. (Copenhagen)* **80,** 229.
Nisticò, G., and Marley, E. (1973). *Neuropharmacology* **12,** 1009.
Nisticò, G., and Marley, E. (1976). *Neuropharmacology* **12,** 737.
Nisticò, G., and Preziosi, P. (1969). *Pharmacol. Res. Commun.* **1,** 363.
Nisticò, G., and Preziosi, P. (1970). *Lancet* **2,** 213.
Nisticò, G., Gargiulo, G., De Girolamo, G.,and Rotiroti, D. (1975). *Boll. Soc. Ital. Biol. Sper.* **51,** 1086.
Nisticò, G., Stephenson, J. D., and Preziosi, P. (1976). *Eur. J. Pharmacol.* **36,** 459.
Niswender, G. D., Midgley, A. R., Jr., Monroe, S. E., and Reichert, L. E. (1968). *Proc. Soc. Exp. Biol. Med.* **128,** 807.
Niswender, G. D., Chen, C. L., Midgley, A. R., Jr., Meites, J., and Ellis, S. (1969). *Proc. Soc. Exp. Biol. Med.* **130,** 793.
Nixon, R. L., Janowsky, D. S., and Davis, J. M. (1974). *Res. Commun. Chem. Pathol. Pharmacol.* **7,** 233.
Noel, G. L., Such, H. K., and Frantz, A. G. (1973). *J. Clin. Endocrinol. Metab.* **36,** 1255.
Noel, G. L., Dimond, R. C., Wartofski, L., Earll, J. M., and Frantz, A. G. (1974). *J. Clin. Endocrinol. Metab.* **39,** 6.
Nokin, J., Vekemans, M., L'Hermite, M., and Robyn, C. (1972). *Br. Med. J.* **3,** 561.
Norman, R. L., Resko, J. A., and Spies, H. G. (1976). *Endocrinology* **99,** 59.
North, R. B., Harik, S. I., and Snyder, S. H. (1973). *Brain Res.* **63,** 435.
Nukada, T., Sakurai, T., and Imaizumi, R. (1963). *Jpn. J. Pharmacol.* **13,** 124.
Nyback, H. (1971). *Acta Pharmacol. Toxicol.* **30,** 372.
Nyback, H., and Sedvall, G. (1969). *Eur. J. Pharmacol.* **5,** 245.
Obata, K., and Highstein, S. M. (1970). *Brain Res.* **18,** 538.

Obata, K., and Takeda, K. (1969). *J. Neurochem.* **16,** 1043.
Obata, K., Ito, M., Ochi, R., and Sato, N. (1967). *Exp. Brain Res.* **4,** 43.
Odell, W. D., and Daughaday, W. H., eds. (1971). "Principles of Competitive Protein Binding Assays." Lippincott, Philadelphia, Pennsylvania.
Ogawa, N., Friesen, H. G., Martin, J. B., and Brazeau, P. (1976). *58th Annu. Meet. Endocr. Soc., San Francisco* Abstract No. 196.
Ojeda, S. R., Harms, P. G., and McCann, S. M. (1974a). *Endocrinology* **94,** 1650.
Ojeda, S. R., Harms, P. G., and McCann, S. M. (1974b). *Endocrinology* **95,** 1694.
Ojeda, S. R., Wheaton, J. E., and McCann, S. M. (1975). *Neuroendocrinology* **17,** 283.
Okada, F. (1971). *Life Sci.* **10,** 77.
Oliver, C., Eskay, R. L., Ben-Jonathan, N., and Porter, J. C. (1974a). *Endocrinology* **95,** 540.
Oliver, C., Charvet, J. P., Codaccioni, J. L., and Vague, J. (1974b). *J. Clin. Endocrinol. Metab.* **39,** 406.
Oliver, C., Ben-Jonathan, N., Mical, R. S., and Porter, J. C. (1975). *Endocrinology* **97,** 1138.
Olson, L., and Fuxe, K. (1972). *Brain Res.* **43,** 289.
Onaya, T., and Hashizume, K. (1976). *Neuroendocrinology* **20,** 47.
Onaya, T., Hashizume, K., Sato, A., and Kojima, A. (1975). *57th Annu. Met. Endocr. Soc., New York* p. 301.
Ondo, J. G. (1974). *Science* **186,** 738.
Ondo, J. G., and Kitay, I. (1972). *Neuroendocrinology* **9,** 72.
Ondo, J. G., and Pass, K. A. (1976). *Endocrinology* **98,** 1248.
Ondo, J. G., Pass, K. A., and Baldwin, R. (1976). *Neuroendocrinology* **21,** 79.
Ong, H. H., Creveling, C. R., and Daly, J. W. (1969). *J. Med. Chem.* **12,** 458.
Orci, L., and Unger, R. H. (1975). *Lancet* **2,** 1243.
Ordy, J. M., Samorajski, T., and Schroeder, D. (1966). *J. Pharmacol. Exp. Ther.* **152,** 445.
Ormston, B. J., Kilborn, J. R., Garry, R., Amos, J., and Hall, R. (1971). *Br. Med. J.* **2,** 199.
Orr, E., and Quay, W. B. (1975). *Endocrinology* **96,** 941.
O'Steen, K. W. (1964). *Endocrinology* **74,** 885.
Otsuka, M. (1976). *Proc. Int. Congr. Pharmacol., 6th, 1975* Vol. 2, pp. 39–47.
Otsuka, M., and Konishi, S. (1976). *Nature (London)* **264,** 83.
Otsuka, M., Konishi, S., and Takahashi, T. (1972). *Proc. Jpn. Acad.* **48,** 342.
Otsuka, M., Konishi, S., and Takahashi, T. (1975). *Fed. Proc., Fed. Am. Soc. Exp. Biol.* **34,** 1922.
Owman, C., Hakanson, R., and Sundler, F. (1973). *Fed. Proc., Fed. Am. Soc. Exp. Biol.* **32,** 1785.
Palay, S. L. (1956). *J. Biophys. Biochem. Cytol.* **2,** Suppl., 193.
Palay, S. L. (1957). *In* "Ultrastructure and Cellular Chemistry of Neural Tissue" (H. Waelsch, ed.), pp. 31–49. Harper (Hoeber), New York.
Palka, A., Coyer, D., and Critchlow, V. (1969). *Neuroendocrinology* **5,** 33.
Palkovits, M. (1973). *Brain Res.* **59,** 449.
Palkovits, M., and Jacobowitz, D. M. (1974). *J. Comp. Neurol.* **157,** 29.
Palkovits, M., Brownstein, M., Saavedra, J. M., and Axelrod, J. (1974a). *Brain Res.* **77,** 137.
Palkovits, M., Arimura, A., Brownstein, M., Schally, A. V., and Saavedra, J. M. (1974b). *Endocrinology* **95,** 554.
Palkovits, M., Kobayashi, R. M., Kizer, J. S., Jacobovitz, D. M., and Kopin, I. J. (1975). *Neuroendocrinology* **18,** 144.
Panerai, A. E., Cocchi, D., Gil-Ad, I., Locatelli, V., Rossi, G. L., and Müller, E. E. (1977a). *J. Endocrinol.* **72,** 301.

Panerai, A. E., Salerno, F., Manneschi, M. Cocchi, D., and Müller, E. E. (1977b). *J. Clin. Endocrinol. Metab.* **45,** 134.
Panksepp, J., Jalowiec, J. E., Morgane, P. J., Zolovich, A. J., and Stern, W. C. (1973). *Exp. Neurol.* **41,** 233.
Papeschi, F. (1974). *Neuropharmacology* **13,** 131.
Pare, C. M. B. (1963). *Lancet* **2,** 527.
Parker, C. R., and Barnea, A. (1976). *58th Annu. Meet. Endocr. Soc., San Francisco* Abstract No. 123.
Parker, D. C., and Rossman, L. G. (1973). *In* "Advances in Human Growth Hormone Research" (S. Raiti, ed.), DHEW Publication No. (NIH) 74–612, pp. 294–312. Center for Disease Control, Atlanta, Georgia.
Parker, D. C., Rossman, L. G., and Vanderlaan, E. F. (1973). *J. Clin. Endocrinol. Metab.* **36,** 1119.
Parker, K. M., and Eddy, R. L. (1976). *J. Clin. Endocrinol. Metab.* **42,** 1188.
Parker, L. N. (1972). *J. Neurochem.* **19,** 1611.
Parkes, J. D., Debono, A. G., and Mardsen, C. D. (1976). *Lancet* **1,** 484.
Parkhie, M. R., and Johnson, H. D. (1971). *Neuroendocrinology* **8,** 1.
Pashkis, K. E., Cantarow, A., and Boyle, D. (1949). *J. Clin. Endocrinol. Metab.* **9,** 658.
Pasteels, J. L. (1961). *C.R. Hebd. Seances Acad. Sci., Ser. D* **253,** 3074.
Pasteels, J. L. (1970). *Hypothala. Proc. Workshop Conf., 1969* pp. 385–399.
Paton, W. D. M. (1953). *Lect. Sci. Basis Med.* **3,** 105–126.
Paton, W. D. M., and Zaimis, E. J. (1943). *Nature (London)* **161,** 178.
Paton, W. D. M., and Zaimis, E. J. (1952). *Pharmacol. Rev.* **4,** 219.
Patry, G., and Naquet, R. (1971). *Can. J. Physiol. Pharmacol.* **49,** 568.
Paulsen, E. C., and Hess, S. M. (1963). *J. Neurochem.* **10,** 453.
Payton, B. W., and Shand, D. G. (1966). *Br. J. Pharmacol. Chemother.* **28,** 23.
Peake, G. T. (1973). *In* "Frontiers in Neuroendocrinology" (W. F. Ganong and L. Martini, eds.), pp. 173–208. Academic Press, New York.
Peake, G. T., Steiner, A. L., and Daughaday, W. H. (1972). *Endocrinology* **90,** 212.
Pearse, A. G. E. (1968). *Proc. R. Soc. London, Ser. B* **170,** 71.
Pearse, A. G. E. (1969). *J. Histochem. Cytochem.* **17,** 303.
Pelletier, G., Labrie, F., Puviani, R., Arimura, A., and Schally, A. V. (1974a). *Endocrinology* **93,** 314.
Pelletier, G., Labrie, F., Arimura, A., and Schally, A. V. (1974b). *Am. J. Anat.* **140,** 445.
Pelletier, G., Leclerc, R., Dube, D., Labrie, F., Puviani, R., Arimura, A., and Schally, A. V. (1975). *Am. J. Anat.* **142,** 397.
Pelletier, G., Labrie, F., Kastin, A. J., Coy, D. H., and Schally, A. V. (1977). In preparation.
Peng, T.-C., Six, K. M., and Munson, P. L. (1970). *Endocrinology* **86,** 202.
Pepeu, G., and Bartholini, A. (1968). *Eur. J. Pharmacol.* **4,** 254.
Pepeu, G., Garau, L., and Mulas, M. L. (1974). *Adv. Biochem. Psychopharmacol.* **10,** 247–252.
Perez-Lopez, F. R., Delvoye, P., Roncero, M. C., and Robyn, C. (1975). *Acta Endocrinol. (Copenhagen)* **79,** 644.
Peringer, E., Jenner, P., Donaldson, I. M., Marsden, C. D., and Miller, R. (1976). *Neuropharmacology* **15,** 463.
Pearlmutter, A. F., Rapino, E., and Saffran, M. (1974). *Neuroendocrinology* **15,** 106.
Pearlmutter, A. F., Rapino, E., and Saffran, M. (1975). *Endocrinology* **97,** 1336.
Perry, T. L. (1971). *J. Neurochem.* **18,** 521.
Perry, T. L., Urquhart, N., Hansen, S., and Kennedy, J. (1974). *J. Neurochem.* **23,** 443.
Peskar, B., and Spector, S. (1973). *Science* **179,** 1340.
Peters, D. A. V., McGeer, P. L., and McGeer, E. G. (1968). *J. Neurochem.* **14,** 1431.

Petitjean, F., and Jouvet, M. (1970). *C.R. Seances Soc. Biol.* **164,** 2288.
Pfaff, D. W. (1973). *Science* **182,** 1148.
Pfeifer, A. K., and Galambos, E. (1967). *J. Pharm. Pharmacol.* **19,** 400.
Phillis, J. W. (1965). *Br. Med. Bull.* **21,** 26.
Phillis, J. W. (1976). *Experientia* **32,** 593.
Phillis, J. W., and Limacher, J. J. (1974). *Brain Res.* **69,** 158.
Phillis, J. W., and Tebecis, A. K. (1967). *Comp. Biochem. Physiol.* **23,** 541.
Phillis, J. W., Tebecis, A. K., and York, D. H. (1968). *Br. J. Pharmacol. Chemother.* **33,** 426.
Pieri, L., Pieri, M., and Haefely, W. (1974). *Nature (London)* **252,** 586.
Pirch, J. H., and Norton, S. (1965). *J. Pharmacol. Exp. Ther.* **155,** 506.
Piva, F., and Borrell, J. (1976). *Abstr. Book Int. Congr. Endocrinol., 5th, 1976* Abstract 156.
Pixner, D. B. (1967). *J. Physiol. (London)* **189,** 15P.
Pletscher, A. (1973). *In* "Frontiers in Catecholamine Research" (E. Usdin and S. Snyder, eds.), pp. 27–37. Pergamon, Oxford.
Pletscher, A., Brossi, A., and Gey, K. F. (1962). *Int. Rev. Neurobiol.* **4,** 275.
Pletscher, A., Burckard, W. P., and Gey, K. F. (1964). *Biochem. Pharmacol.* **13,** 385.
Pletscher, A., Da Prada, M., Bartholini, G., Burckard, W. P., and Bruderer, H. (1965). *Life Sci.* **4,** 2301.
Pletscher, A., Gey, K. F., and Burckard, W. P. (1966). *Handb. Exp. Pharmakol.* [N.S.] **19,** 593–735.
Pletscher, A., Da Prada, M., Burckard, W. P., and Tranzer, J. P. (1968). *Adv. Pharmacol.* **6b,** 55–69.
Plonk, J. W., and Feldman, J. M. (1975). *Metab., Clin. Exp.* **24,** 1035.
Plonk, J. W., and Feldman, J. M. (1976). *J. Clin. Endocrinol. Metab.* **42,** 291.
Plonk, J. W., Bivens, C. H., and Feldman, J. M.(1974). *J. Clin. Endocrinol. Metab.* **38,** 836.
Plotnikoff, N. P., Kastin, A. J., Anderson, M. S., and Schally, A. V. (1971). *Life Sci.* **10,** 1279.
Plotnikoff, N. P., Kastin, A. J., Anderson, M. S., and Schally, A. V. (1972a). *Proc. Soc. Exp. Biol. Med.* **140,** 811.
Plotnikoff, N. P., Prange, A. J., Breese, G. R., Anderson, M. S., and Wilson, I. C. (1972b). *Science* **178,** 417.
Plotnikoff, N. P., Kastin, A. J., Anderson, M. S., and Schally, A. V. (1973). *Neuroendocrinology* **11,** 67.
Plotnikoff, N. P., Kastin, A. J., and Minard, F. (1974). *Neuroendocrinology* **14,** 271.
Plummer, N. A., Thody, A. J., Burton, J. L., Goolamali, S. K., Shuster, S., Cole, E. N., and Boyns, A. R. (1975). *J. Clin. Endocrinol. Metab.* **41,** 380.
Poch, G. R., and Kopin, I. J. (1966). *Biochem. Pharmacol.* **15,** 210.
Podolsky, S., and Leopold, N. A. (1973). *In* "Drug Effects on Neuroendocrine Regulation" (E. Zimmermann *et al.*, eds.). pp. 225–235. Elsevier, Amsterdam.
Podolsky, S., and Leopold, N. A. (1976). *In* "Growth Hormone and Related Peptides" (A. Pecile and E. E. Müller, eds.), pp. 252–260. Excerpta Med. Found., Amsterdam.
Poignant, J. C., Gressier, H., Petitjean, M., Regnier, G., and Canevari, R. (1975). *Experientia* **31,** 1204.
Poirier, L. J., and Sourkes, T. L. (1965). *Brain* **88,** 181.
Polak, J. M., Grimelius, L., Pearse, A. G. E., Bloom, S. R., and Arimura, A. (1975). *Lancet* **1,** 1220.
Pollard, H., Bishoff, S., and Schwartz, J. C. (1974). *J. Pharmacol. Exp. Ther.* **190,** 88.
Pomerantz, G., and Sorrentino, S., Jr. (1973). *Neuroendocrinology* **12,** 354.
Pontiroli, A. E., Cavagnini, F., Viberti, G. C., Peracchi, M., Tognetti, A., and Pozza, G. (1975a). *In* "Gli Ormoni dell'Ipofisi Anteriore" (E. E. Müller and S. E. Piazzi, eds.). pp. 343–349. I.S.V.T. Sclavo, Siena.

Pontiroli, A. E., Viberti, G. C., Tognetti, A., and Pozza, G. (1975b). *Acta Endocrinol. (Copenhagen)* **70,** Suppl. 29, 289.
Popa, G. T., and Fielding, U. (1930). *J. Anat.* **65,** 88.
Popa, G. T., and Fielding, U. (1933). *J. Anat.* **67,** 227.
Popova, N. K., Maslova, L. N., and Naumenko, E. V. (1972). *Brain Res.* **47,** 61.
Porter, B. A., Rosenfield, R. L., and Lawrence, A. M. (1972). *Pediatr. Res.* **6,** 350.
Porter, C. C. (1971). *Fed. Proc., Fed. Am. Soc. Exp. Biol.* **30,** 871.
Porter, C. C., Watson, L. S., Titus, D. C., Totaro, J. A., and Byer, S. S. (1962). *Biochem. Pharmacol.* **11,** 1067.
Porter, J. C., Mical, R. S., Kamberi, I. A., and Grazia, Y. R. (1970a). *Endocrinology* **87,** 197.
Porter, J. C., Goldman, B. D., and Wilber, J. F. (1970b). *Hypophysiotropic Horm. Hypothalamus: Assay Chem., Proc. Workshop Conf., 1969* pp. 282–297.
Porter, J. C., Vale, W., Burgus, R., Mical, R. S., and Guillemin, R. (1971). *Endocrinology* **89,** 1054.
Porter, J. C., Mical, R. S., and Cramer, O. M. (1971–1972). *Gynecol. Invest.* **2,** 13.
Portig, P. J., and Vogt, M. (1969). *J. Physiol. (London)* **204,** 687.
Potter, L. T. (1968). *In* "The Interaction of Drugs and Subcellular Components on Animal Cells" (P. N. Campbell, ed.), pp. 293–304. Churchill, London.
Potter, L. T. (1970). *J. Physiol. (London)* **206,** 145.
Potter, L. T. (1971). Quoted in *Neurosci. Res. Program, Bull.* **8,** 382.
Potter, L. T. (1972). *In* "The Structure and Function of Nervous Tissue" (G. H. Bourne, ed.), Vol. 4, pp. 105–128. Academic Press, New York.
Potter, L. T., and Axelrod, J. (1963). *J. Pharmacol. Exp. Ther.* **142,** 291.
Potter, L. T., Glover, V. A. S., and Saelens, J. K. (1968). *J. Biol. Chem.* **243,** 3864.
Powell, D., Leeman, S., Tregear, G. W., Niall, H. D., and Potts, J. T. (1973). *Nature (London), New Biol.* **241,** 252.
Powell, E. D. U., Frantz, A. G., Rabkin, M. T., and Field, R. A. (1966). *New Engl. J. Med.* **275,** 922.
Powers, B., and Valenstein, E. S. (1972). *Science* **175,** 1003.
Pradhan, S. N., and Dutta, S. N. (1971). *Int. Rev. Neurobiol.* **14,** 173.
Prange, A. J., Wilson, I. C., Lara, P. P., Alltop, L. B., and Breese, G. R. (1972). *Lancet* **2,** 999.
Prange-Hansen, Aa, Ørskov, H., Seyer-Hansen, K., and Lundbaek, K. (1973). *Br. Med. J.* **3,** 523.
Precht, W., and Yoshida, M. (1971). *Brain Res.* **32,** 229.
Preziosi, P., Scapagnini, U., and Nisticò, G. (1968). *Biochem. Pharmacol.* **17,** 1309.
Puig, M., Bartholini, G., and Pletscher, A. (1974). *Naunyn Schmiedberg's Arch. Pharmacol.* **281,** 443.
Pujol, J. F., Mouret, J., Jouvet, M., and Glowinski, J. (1968). *Science* **159,** 112.
Pujol-Amat, P., Gamissans, O., Calaf, J., Benito, E., Perez-Lopez, F. R., L'Hermite, P., and Robyn, C. (1973). *Excerpta Med. Found. Int. Congr. Ser. D* **308,** 316.
Putkonen, P. T. S. (1974). *Med. Biol.* **52,** 193.
Quadri, S. K., and Meites, J. (1973). *Proc. Soc. Exp. Biol. Med.* **142,** 837.
Quadri, S. K., Kledzek, G. A., and Meites, J. (1973). *Neuroendocrinology* **11,** 248.
Quadri, S. K., Pierson, C., and Spies, H. G. (1976). *Fed. Proc., Fed. Am. Soc. Exp. Biol.* **35,** Abstr. 552.
Quastel, J. H. (1962). *In* "Neurochemistry" (K. A. C. Elliott, I. H. Page, and J. H. Quastel, eds.), pp. 431–451. Thomas, Springfield, Illinois.
Quay, W. B. (1965). *Life Sci.* **4,** 379.
Quay, W. B. (1968). *Am. J. Physiol.* **215,** 1448.

Quay, W. B. (1969). *J. Neuro-Visc. Relat.* **9,** 212.

Quijada, M., Krulich, L., Fawcett, C. P., Sundberg, D. K., and McCann, S. M. (1971). *Fed. Proc., Fed. Am. Soc. Exp. Biol.* **30,** 197.

Quijada, M., Illner, P., Krulich, L., and McCann, S. M. (1973–1974). *Neuroendocrinology* **13,** 151.

Rabinowitz, D., Merimee, T. J., Nelson, J. K., Schultz, R. B., and Burgess, J. A. (1968). *Proc. Int. Symp. Growth Horm., 1st, 1967* Excerpta Med. Found. Int. Congr. Ser. No. 158, pp. 105–115.

Ramirez, V. D., and Kordon, C. (1975). *Hypothal. Horm.* [*Symp.*], *1974* pp. 57–74.

Ramwell, P. W., and Shaw, J. E. (1963a). *J. Physiol. (London)* **169,** 51P.

Ramwell, P. W., and Shaw, J. E. (1963b). *Life Sci.* **6,** 419.

Ramwell, P. W., and Shaw, J. E. (1966). *Am. J. Physiol.* **211,** 125.

Ramwell, P. W., Shaw, J. E., and Jessup, R. (1966). *Am. J. Physiol.* **211,** 998.

Ramwell, P. W., Shaw, J. E., Clarke, G. B., Grostic, M. F., Kaiser, D. G., and Pike, J. E. (1968). *Prog. Chem. Fats Other Lipids* **9,** 231–273.

Randic, M., Siminoff, R., and Straughan, D. W. (1964). *Exp. Neurol.* **9,** 236.

Rapaport, A. (1969). *Gerontologia* **13,** 288.

Rapaport, B., Refetoff, S., Fang, V. S., and Friesen, H. G. (1973). *J. Clin. Endocrinol. Metab.* **36,** 256.

Rasmussen, H. (1970). *Science* **170,** 404.

Rasmussen, H. (1974). *In* "Textbook of Endocrinology" (R. H. Williams, ed.), 5th ed., pp. 1–30. Saunders, Philadelphia, Pennsylvania.

Rastogi, R. B., and Singhal, R. L. (1974a). *Brain Res.* **81,** 253.

Rastogi, R. B., and Singhal, R. L. (1974b). *J. Pharmacol. Exp. Ther.* **191,** 72.

Raud, H. R., and Odell, W. D. (1971). *Endocrinology* **88,** 991.

Raziano, J., Cowchock, S., Ferin, M., and Vande Wiele, R. L. (1971). *Endocrinology* **88,** 1516.

Recant, L., Forsham, P. H., and Thorn, G. W. (1948). *Fed. Proc., Fed. Am. Soc. Exp. Biol.* **7,** 99.

Rechtschaffen, A., Lovell, R. A., Freedman, D. X., Whitehead, W. E., and Aldrich, M. (1973). *Serotonin Behav., Proc. Symp., 1972* pp. 401–418.

Redding, T. W., and Coy, E. J. (1974). *56th Annu. Meet. Endocr. Soc., Atlanta* p. 154.

Redding, T. W., and Schally, A. V. (1969a). *Proc. Soc. Exp. Biol. Med.* **131,** 415.

Redding, T. W., and Schally, A. V. (1969b). *Proc. Soc. Exp. Biol. Med.* **131,** 420.

Redding, T. W., and Schally, A. V. (1971). *Endocrinology* **89,** 1075.

Redding, T. W., and Schally, A. V. (1972). *Neuroendocrinology* **9,** 250.

Redding, T. W., Bowers, C. Y., and Schally, A. V. (1966a). *Acta Endocrinol. (Copenhagen)* **51,** 391.

Redding, T. W., Bowers, C. Y., and Schally, A. V. (1966b). *Endocrinology* **79,** 229.

Redding, T. W., Schally, A. V., Arimura, A., and Matsuo, H. (1972). *Endocrinology* **90,** 764.

Redding, T. W., Kastin, A. J., Gonzales-Barcena, D., Coy, D. H., Schalch, D. S., and Schally, A. V. (1973a). *J. Clin. Endocrinol. Metab.* **37,** 626.

Redding, T. W., Kastin, A. J., Nair, R. M., and Schally, A. V. (1973b). *Neuroendocrinology* **11,** 92.

Redding, T. W., Kastin, A. J., Gonzales-Barcena, D., Coy, D. H., Hirotsu, Y., Ruelas, J., and Schally, A. V. (1974). *Neuroendocrinology* **16,** 119.

Rees, L., Butler, P. W. P., Gosling, C., and Besser, G. M. (1970). *Nature (London)* **228,** 565.

Reeves, C. (1966). *Psychol. Bull.* **65,** 321.

Reeves, J. J., Arimura, A., and Schally, A. V. (1970). *J. Anim. Sci.* **31,** 933.

Reeves, J. J., Arimura, A., and Schally, A. V. (1971). *Biol. Reprod.* **4,** 88.

Reeves, J. J., Arimura, A., Schally, A. V., Krogt, C., Beck, T. W., and Casey, J. M. (1972). *J. Anim. Sci.* **35,** 84.

Refetoff, S., Fang, V. S., Rapaport, B., and Friesen, H. G. (1974). *J. Clin. Endocrinol. Metab.* **38,** 450.

Regnier, G., Canevari, R., and Laubie, M. (1968). *J. Med. Chem.* **11,** 1151.

Reichlin, S. (1966). *In* "Neuroendocrinology" (L. Martini and W. F. Ganong, eds.), Vol. 1, pp. 445–536. Academic Press, New York.

Reichlin, S. (1974a). *In* "Textbook of Endocrinology" (R. H. Williams, ed.), 5th ed., pp. 774–831. Saunders, Philadelphia, Pennsylvania.

Reichlin, S. (1974b). *Handb. Physiol., Sect. 7: Endocrinol.* **4,** 405–447.

Reichlin, S., and Mitnick, M. (1973a). *In* "Frontiers in Neuroendocrinology" (W. F. Ganong and L. Martini, eds.), pp. 61–88. Academic Press, New York.

Reichlin, S., and Mitnick, M. (1973b). *In* "Hypothalamic Hypophysiotropic Hormones" (C. Gual and E. Rosemberg, eds.), p. 124–135. Excerpta Med. Found., Amsterdam.

Reichlin, S., Martin, J. B., Mitnick, A., Boshans, R. L., Grimm, Y., Bollinger, J., Gordon, J., and Malacara, J. (1972). *Recent Prog. Horm. Res.* **28,** 229.

Reichlin, S., Saperstein, R., Jackson, I. M. D., Boyd, A. E., III, and Patel, Y. (1976). *Annu. Rev. Physiol.* **38,** 389.

Reid, J. D., and Shepherd, D. M. (1963). *Life Sci.* **2,** 5.

Reid, W. D. (1970). *Br. J. Pharmacol.* **40,** 483.

Reigle, T. G., Avni, J., Platz, P. A., Schildkraut, J. J., and Plotnikoff, N. P. (1974). *Psychopharmacologia* **37,** 1.

Reis, D. J., Weinbren, M., and Corvelli, A. (1968). *J. Pharmacol. Exp. Ther.* **164,** 135.

Reis., D. J., Cowelli, A., and Conners, J. (1969). *J. Pharmacol. Exp. Ther.* **167,** 328.

Reis, D. J., Moorhead, D. T., and Meslino, N. (1970). *Arch. Neurol. (Chicago)* **22,** 31.

Reiter, R. J. (1974). *In* "Endocrine Physiology" (S. M. McCann, ed.), pp. 277–308. Butterworth, London.

Reiter, R. J., and Fraschini, F. (1969). *Neuroendocrinology* **5,** 219.

Reiter, R. J., and Sorrentino, S. (1971). *Contraception* **4,** 385.

Reivich, M., and Glowinski, J. (1967). *Brain* **90,** 633.

Renaud, L. P. (1975). *Br. J. Pharmacol.* **55,** 277P.

Renaud, L. P., Martin, J. B., and Brazeau, P. (1975). *Nature (London)* **255,** 233.

Réthelyi, M., and Halàsz, B. (1970). *Exp. Brain Res.* **11,** 145.

Rice, R. W., and Critchlow, V. (1976). *Endocrinology,* **99,** 970.

Richards, J. G., and Tranzer, J. P. (1974). *Neurosecretion—Final Neuroendocr. Pathway, Int. Symp., 6th, 1973* pp. 246–259.

Riddel, D., and Szerb, J. C. (1971). *J. Neurochem.* **18,** 989.

Riddle, O., Bates, R. W., and Dykshorn, S. W. (1933). *Am. J. Physiol.* **105,** 191.

Rieke, W. O. (1958). *Anat. Rec.* **130,** 363.

Rigler, G. L., Peake, G. T., Wise, P. M., and Ratner, A. (1976). *58th Annu. Meet. Endocr. Soc., San Francisco* Abstract No. 309.

Rinaldi, F., and Himwich, H. C. (1955). *Dis. Nerv. Syst.* **16,** 133.

Rinne, U. K., and Sonninen, V. (1968). *Experientia* **24,** 177.

Rippel, R. H., Johnson, E. S., White, W. F., Fujino, M., Yamazaki, I., and Nakayama, R. (1973). *Endocrinology* **93,** 1449.

Rivier, C., and Vale, W. (1974). *Endocrinology* **95,** 978.

Rivier, C., Brown, M., and Vale, W. (1977). *Endocrinology* **100,** 751.

Rivier, J. (1974). *J. Am. Chem. Soc.* **96,** 2986.

Robberecht, P., Deschodt-Lanckman, M., De Neff, P., and Christophe, J. (1975). *Biochem. Biophys. Res. Commun.* **67,** 315.

Roberts, E., Wein, J., and Simonsen, D. G. (1964). *Vitam. Horm. (N.Y.)* **22,** 503.

Roberts, M. H. T., and Straughan, D. W. (1967). *J. Physiol. (London)* **193,** 269.
Roberts, P. J. (1973). *Brain Res.* **49,** 451.
Robinson, O. P. W. (1973). *Adv. Neurol.* **3,** 97–104.
Robyn, C., Delvoye, P., Nokin, J., Vekemans, M., Badawi, M., Perez-Lopez, F. R., and L'Hermite, M. (1973). *In* "Human Prolactin" (J. L. Pasteels and C. Robyn, eds.), pp. 167–188. Excerpta Med. Found., Amsterdam.
Rocha-Silva, M. (1959). *Br. J. Pharmacol. Chemother.* **14,** 243.
Rodger, N. W., Beck, J. C., Burgus, R., and Guillemin, R. (1969). *Endocrinology* **84,** 1373.
Rodriguez, E. M. (1972). *Brain-Endocr. Interact., Median Eminence: Struct. Funct., Int. Symp., 1971* pp. 319–334.
Rodriguez de Lores Arnaiz, G., and De Robertis, E. (1962). *J. Neurochem.* **9,** 503.
Ronzoni, E., and Reichlin, S. (1950). *Am. J. Physiol.* **160,** 490.
Roos, B. E. (1962). *Life Sci.* **1,** 25.
Roos, B. E. (1965). *J. Pharm. Pharmacol.* **17,** 820.
Roos, B. E., and Werdinius, B. (1963). *Life Sci.* **2,** 92.
Root, A. W., Snyder, P. J., Revani, I., Di George, A. M., and Utiger, R. D. (1973). *J. Clin. Endocrinol. Metab.* **36,** 103.
Rosencrans, J. A. (1968). *Fed. Proc., Fed. Am. Soc. Exp. Biol.* **27,** 540.
Rosencrans, J. A. (1970). *Eur. J. Pharmacol.* **9,** 170.
Rosenthal, J., Escobar-Jimenez, F., Raptis, S., and Pfeiffer, E. F. (1976). *Lancet* **1,** 772.
Rosenthal, N. R., and Mason, J. W. (1961). *Psychopharmacologia* **2,** 55.
Ross, S. B., and Haljasmaa, O. (1964). *Acta Endocrinol. (Copenhagen)* **21,** 215.
Roth, G. I., Walton, P. L., and Yamamoto, W. S. (1970). *Brain Res.* **23,** 223.
Roth, J., Glick, M., Yalow, R. S., and Berson, S. A. (1963). *Science* **140,** 987.
Roth, J., Glick, S. M., Yalow, R. S., and Berson, S. A. (1964). *Diabetes* **13,** 355.
Roux, M. (1962). *C.R. Seances Soc. Biol. Ses Fil.* **156,** 1664.
Royce, P. C., and Sayers, G. (1960). *Proc. Soc. Exp. Biol. Med.* **103,** 447.
Rubin, A. L., Levin, S. R., Bernstein, R. I., Tyrrel, J. B., Noacco, C., and Forsham, P. H. (1973). *J. Clin. Endocrinol. Metab.* **37,** 160.
Rubin, R. P. (1970). *Pharmacol. Rev.* **22,** 389.
Rubin, R. T., Kales, A., Adler, R., Fagan, T., and Odell, W. (1971). *Science* **175,** 196.
Rubinstein, L., and Sawyer, C. H. (1970). *Endocrinology* **86,** 988.
Ruch, W., Koerker, D. J., Carino, M., Johnsen, S. D., Webster, B. R., Ensink, J. W., Goodner, C. J., and Gale, C. C. (1973). *In* "Advances in Human Growth Hormone Research" (S. Raiti, ed.), DHEW Publication No. (NIH) 74-612, pp. 271–289.
Rufener, C., and Dreifuss, J. J. (1970). *Brain Res.* **22,** 402.
Rumsfeld, H. W., Jr., and Porter, J. C. (1959). *Arch. Biochem. Biophys.* **82,** 473.
Russell, S. M., Dhariwal, A. P. S., McCann, S. M., and Yates, F. E. (1969). *Endocrinology* **85,** 512.
Ryall, R. W. (1964). *J. Neurochem.* **11,** 131.
Ryall, R. W. (1975). *In* "Handbook of Psychopharmacology" (L. L. Iversen, S. D. Iversen, and S. H. Snyder, eds.), Vol. 4, pp. 83–128. Plenum, New York.
Saavedra, J. M., Brownstein, M., and Axelrod, J. (1973). *J. Pharmacol. Exp. Ther.* **186,** 508.
Saavedra, J. M., Palkovits, M., Brownstein, M. J., and Axelrod, J. (1974a). *Nature (London)* **248,** 692.
Saavedra, J. M., Palkovits, M., Brownstein, M. J., and Axelrod, J. (1974b). *Brain Res.* **77,** 157.
Saavedra, J. M., Palkovits, M., Kizer, J. S., Brownstein, M. J., and Zivin, J. A. (1975). *J. Neurochem.* **25,** 257.
Sachar, E. J., Mushrush, G., Perlow, M., Weitzman, E. D., and Sassin, J. (1972). *Science* **178,** 1304.

Sachar, E. J., Gruen, P. H., Altman, N. S., Halpern, F. S., and Frantz, A. G. (1976). *In* "Hormones, Behavior and Psychopathology" (E. J. Sachar, ed.), pp. 161–176. Raven, New York.

Sachar, E. J., Gruen, P. H., Altman, N., Langer, G., Halpern, F. S., and Liefer, M. (1977). *In* "Neuroregulators and Psychiatric Disorders" (E. Usdin, D. A. Hamburg and J. D. Barchas, eds.), pp. 242–249, Oxford University Press, Oxford.

Sachdev, Y., Turnbridge, W. M. G., Weightman, D. R., Gomez-Pan, A., Duns, A., and Hall, R. (1975). *Lancet* **2,** 1164.

Sachs, C., and Jonsson, G. (1972). *J. Neurochem.* **19,** 1561.

Saelens, J. K., Schoen, M. S., and Kovacsics, G. B. (1967). *Biochem. Pharmacol.* **16,** 1043.

Saffran, M., and Rowell, P. (1969). *Endocrinology* **85,** 652.

Saffran, M., and Schally, A. V. (1955). *Can. J. Biochem. Physiol.* **33,** 408.

Saffran, M., Schally, A. V., and Benfey, B. G. (1955). *Endocrinology* **57,** 439.

Saito, K., Konishi, S., and Otsuka, M. (1975). *Brain Res.* **97,** 177.

Sakuma, M., and Knobil, E. (1970). *Endocrinology* **86,** 890.

Sala, M. A., Otegui, J. T., Benedetti, W. L., Monti, J. M., and Grino, E. (1971). *J. Neurovisc. Relat.* **32,** 241.

Salmoiraghi, G. C., and Stefanis, C. N. (1967). *Int. Rev. Neurobiol.* **10,** 1.

Salmoiraghi, G. C., and Steiner, F. A. (1963). *J. Neurophysiol.* **26,** 581.

Salmoiraghi, G. C., Costa, E., and Bloom, F. E. (1965). *Annu. Rev. Pharmacol.* **5,** 213.

Samli, M. H., and Geschwind, I. I. (1968). *Endocrinology* **82,** 225.

Samuelsson, B. (1972). *In* "Prostaglandins and Fertility Control" (S. Bergstrom, K. Green, and B. Samuelsson, eds.), Vol. 2, pp. 1–17. Karolinska Institutet, Stockholm.

Samuelsson, B., Hamberg, B., and Sweely, C. C. (1970). *Anal. Biochem.* **38,** 301.

Sanders-Bush, E., and Sulser, F. (1970). *J. Pharmacol. Exp. Ther.* **175,** 419.

Sanders-Bush, E., Gallager, D. A., and Sulser, F. (1974). *Adv. Biochem. Psychopharmacol.* **10,** 185–194.

Sandler, M., Goodwin, B. L., Leask, B. G. S., and Ruthven, C. R. J. (1973). *Lancet* **1,** 612.

Sandman, R. P. (1962). *Anal. Biochem.* **3,** 158.

Sandow, J., Arimura, A., and Schally, A. V. (1972). *Endocrinology* **90,** 1315.

Sano, K., Mayanagi, Y., Sekino, H., Ogashuwa, M., and Ishiÿma, B. (1970). *J. Neurosurg.* **33,** 689.

Sassin, J. F., Frantz, A. G., Kapen, S., and Weitzman, E. D. (1973). *J. Clin. Endocrinol. Metab.* **37,** 436.

Sato, T., Hirono, M., Jyujo, T., Iesaka, T., Taya, K., and Igarashi, M. (1975a). *Endocrinology* **96,** 45.

Sato, T., Jyujo, T., Kawarai, Y., and Asai, T. (1975b). *Am. J. Obstet. Gynecol.* **122,** 637.

Sawano, S., Arimura, A., Bowers, C. Y., and Schally, A. V. (1967). *Endocrinology* **81,** 1410.

Sawyer, C. H. (1952). *Anat. Rec.* **112,** 385.

Sawyer, C. H., Markee, J. E., and Hollinshead, W. H. (1947). *Endocrinology* **41,** 395.

Sawyer, C. H., Hilliard, J., Kanematsu, S., Scaramuzzi, R., and Blake, C. A. (1974). *Neuroendocrinology* **15,** 328.

Scales, B., and Cosgrove, M. B. (1970). *J. Pharmacol. Exp. Ther.* **175,** 338.

Scanes, C. G. (1974). *Neuroendocrinology* **15,** 1.

Scapagnini, U. (1974). *In* "The Neurosciences: Third Study Program" (F. O. Schmitt and F. G. Worden, eds.), p. 565. MIT Press, Cambridge, Massachusetts.

Scapagnini, U., and Preziosi, P. (1972). *Arch. Int. Pharmacodyn. Ther.* **196,** Suppl., 205.

Scapagnini, U., and Preziosi, P. (1973a). *Prog. Brain Res.* **39,** 171.

Scapagnini, U., and Preziosi, P. (1973b). *Neuropharmacology* **12,** 57.

Scapagnini, U., Preziosi, P., and De Schapdryver, A. (1969). *Pharmacol. Res. Commun.* **1,** 63.

Scapagnini, U., Van Loon, G. R., Möberg, G. P., and Ganong, W. F. (1970). *Eur. J. Pharmacol.* **11,** 266.
Scapagnini, U., Möberg, G. P., Van Loon, G. R., De Groot, J., and Ganong, W. F. (1971). *Neuroendocrinology* **7,** 90.
Scapagnini, U., Van Loon, G. R., Möberg, G. P., Preziosi, P., and Ganong, W. F. (1972). *Neuroendocrinology* **10,** 155.
Scapagnini, U., Annunziato, L., and Preziosi, P. (1974a). *J. Pharmacol. (Paris)* **5,** 118.
Scapagnini, U., Annunziato, L., and Di Renzo, G. F. (1974b). *Proc. Acad. Sci. Med. Chir.* **128,** 89.
Scapagnini, U., Annunziato, L., Lombardi, G., Oliver, C., and Preziosi, P. (1975). *Neuroendocrinology* **18,** 272.
Scapagnini, U., Annunziato, L., Di Renzo, G. F., Lombardi, G., and Preziosi, P. (1976a). *Neuroendocrinology* **20,** 243.
Scapagnini, U., Raxas, M., D'Agata, R., Annunziato, L., and Preziosi, P. (1976b). *Abstr. Book Int. Congr. Endocrinol., 5th, 1976* Abstract No. 800.
Scatton, B., Cheramy, A., Besson, M. J., and Glowinski, J. (1970). *Eur. J. Pharmacol.* **13,** 131.
Schacter, M., ed. (1973). "International Encyclopedia of Pharmacology and Therapeutics," Vol. I. Pergamon, Oxford.
Schain, R. J. (1962). *World Neurol.* **3,** 706.
Schalch, D. S. (1967). *J. Lab. Clin. Med.* **69,** 256.
Schalch, D. S., and Reichlin, S. (1966). *Endocrinology* **79,** 275.
Schalch, D. S., Gonzales-Barcena, D., Kastin, A. J., Schally, A. V., and Lee, L. A. (1972). *J. Clin. Endocrinol. Metab.* **35,** 609.
Schally, A. V., and Bowers, C. Y. (1964). *Metab., Clin. Exp.* **13,** 1190.
Schally, A. V., and Redding, T. W. (1967). *Proc. Soc. Exp. Biol. Med.* **126,** 320.
Schally, A. V., Saffran, M., and Zimmerman, B. (1958). *Biochem. J.* **70,** 97.
Schally, A. V., Lipscomb, H. S., and Guillemin, R. (1962). *Endocrinology* **71,** 164.
Schally, A. V., Kuroshima, A., Ishida, Y., Redding, T. W., and Bowers, C. Y. (1965). *Proc. Soc. Exp. Biol. Med.* **118,** 350.
Schally, A. V., Arimura, A., Bowers, C. Y., Kastin, A. J., Sawano, S., and Redding, T. W. (1968a). *Recent Prog. Horm. Res.* **24,** 497.
Schally, A. V., Sawano, S., Müller, E. E., Arimura, A., Bowers, C. Y., Redding, T. W., and Steelman, S. L. (1968b). *Proc. Int. Symp. Growth Horm., 1st, 1967* Excerpta Med. Found. Int. Congr. Ser. No. 158, pp. 185–203.
Schally, A. V., Sawano, S., Arimura, A., Barrett, J. F., Wakabayashi, I., and Bowers, C. Y. (1969a). *Endocrinology* **84,** 1493.
Schally, A. V., Itoh, Z., Carter, W. H., Saito, M., Sawano, S., Arimura, A., and Bowers, C. Y. (1969b). *Gen. Comp. Endocrinol.* **12,** 176.
Schally, A. V., Bowers, C. Y., Redding, T. W., and Barrett, J. F. (1969c). *J. Biol. Chem.* **244,** 4077.
Schally, A. V., Baba, Y., Nair, R. M. G., and Bennet, C. D. (1971a). *J. Biol. Chem.* **246,** 6647.
Schally, A. V., Nair, R. M. G., Redding, T. W., and Arimura, A. (1971b). *J. Biol. Chem.* **246,** 7230.
Schally, A. V., Arimura, A., Kastin, A. J., Matsuo, H., Baba, Y., Redding, T. W., Nair, R. M. G., Debeljuk, L., and White, W. F. (1971c). *Science* **173,** 1036.
Schally, A. V., Baba, Y., Arimura, A., and Redding, T. W. (1971d) *Biochem. Biophys. Res. Commun.* **42,** 50.
Schally, A. V., Arimura, A., Baba, Y., Nair. R. M. G., Matsuo, H., Redding, T. W., Debeljuk, L., and White, W. F. (1971e). *Biochem. Biophys. Res. Commun.* **43,** 393.

Schally, A. V., Baba, Y., Redding, T. W., Matsuo, H., and Arimura, A. (1971f). *Neuroendocrinology* **8,** 347.

Schally, A. V., Arimura, A., Wakabayashi, I., Redding, T. W., Dickerman, E., and Meites, J. (1972a). *Experientia* **28,** 205.

Schally, A. V., Kastin, A. J., and Arimura, A. (1972b). *Am. J. Obstet. Gynecol.* **114,** 423.

Schally, A. V., Arimura, A., and Kastin, A. J. (1973a). *Science* **179,** 341.

Schally, A. V., Kastin, A. J., Arimura, A., Coy, D. H., Coy, E. J., Debeljuk, L., and Redding, T. W. (1973b). *J. Reprod. Fertil., Suppl.* **20,** 119.

Schally, A. V., Arimura, A., Debeljuk, L., Redding, T. W., and Reeves, J. J. (1973c). *In* "Hypothalamic Hypophysiotropic Hormones" (C. Gual and E. Rosemberg, eds.), pp. 3–11. Excerpta Med. Found., Amsterdam.

Schally, A. V., Dupont, A., Arimura, A., Redding, T. W., and Linthicum, G. L. (1975). *Fed. Proc., Fed. Am. Soc. Exp. Biol.* **34,** 584.

Schally, A. V., Dupont, A., Arimura, A., Takahara, S., Redding, T. W., Clemens, A., and Shaar, C. (1976). *Acta Endocrinol. (Copenhagen)* **82,** 1.

Schally, A. V., Redding, T. W., Arimura, A., Dupont, A., and Linthicum, G. L. (1977). *Endocrinology* **100,** 681.

Scharrer, B. (1964a). *Z. Zellforsch. Mikrosk. Anat.* **62,** 125.

Scharrer, B. (1964b). *Z. Zellforsch. Mikrosk. Anat.* **64,** 301.

Scharrer, B. (1967). *Am. Zool.* **7,** 161.

Scharrer, B. (1969). *Med. Coll. Va. Q.* **5,** 27.

Scharrer, B. (1972). *Prog. Brain Res.* pp. 7–18.

Scharrer, B., and Weitzman, M. (1970). *Aspects Neuroendocrinol., Int. Symp. Neurosecretion, 5th, 1969* pp. 1–23.

Scharrer, E. (1966). *Res. Publ., Assoc. Res. Nerv. Ment. Dis.* **43,** 1–35.

Scheving, L. E., Harrison, W. H., Gordon, P., and Pauly, J. E. (1968). *Am. J. Physiol.* **214,** 166.

Schildkraut, J., Schanberg, S. M., and Kopin, I. J. (1966). *Life Sci.* **5,** 1479.

Schildkraut, J., Draskoczy, P. R., and Lo, P. S. (1971). *Science* **172,** 587.

Schneider, H. P. G., and McCann, S. M. (1969). *Endocrinology* **85,** 121.

Schneider, H. P. G., and McCann, S. M. (1970). *Endocrinology* **87,** 330.

Schneider, H. P. G., Crighton, D. B., and McCann, S. M. (1969). *Neuroendocrinology* **5,** 271.

Schoenfeld, R., and Uretsky, N. (1972). *Eur. J. Pharmacol.* **19,** 115.

Schofield, J. G. (1970). *Nature (London)* **215,** 1382.

Schorderet, M. (1975). *Experientia* **31,** 1325.

Schou, M. (1968). *In* "Psychopharmacology: A Review of Progress, 1957–1967" (D. H. Efrom *et al.*, eds.), PHSD No. 1036, pp. 701–718. U.S. Govt. Printing Office, Washington, D.C.

Schreiber, V., Eckertova, A., Franc, Z., Koci, J., Rybak, M., and Kmentova, V. (1961). *Experientia* **17,** 264.

Schreiber, V., Rybak, M., Eckertova, A., Jirgl, V., Koci, J., Franc, Z., and Kmentova, V. (1962). *Experientia* **18,** 338.

Schubert, J., Sparf, B., and Sundwall, A. (1969). *J. Neurochem.* **16,** 695.

Schubert, J., Sparf, B., and Sundwall, A. (1970). *J. Neurochem.* **17,** 461.

Schwartz, J. C. (1975). *Life Sci.* **17,** 502.

Scott, A. P., and Lowry, P. J. (1974). *Biochem. J.* **139,** 593.

Scott, D. E., and Knigge, K. M. (1970). *Z. Zellforsch. Mikrosk. Anat.* **105,** 1.

Scott, R., Espiner, E. A., and Donald, R. A. (1977). *N. Engl. J. Med.* **296,** 57.

Sedvall, G. C. (1964). *Acta Physiol. Scand.* **60,** 39.

Sedvall, G. C., Weise, V. K., and Kopin, I. J. (1968). *J. Pharmacol. Exp. Ther.* **159,** 274.
Seeman, P., and Lee, T. (1975). *Science* **188,** 1217.
Segal, D. S., and Mandell, A. J. (1974). *In* "The Thyroid Axis: Drugs and Behavior" (A. J. Prange, ed.), pp. 129–133. Raven, New York.
Segal, D. S., Sullivan, J. L., Kuczenski, R. T., and Mandell, A. J. (1971). *Science* **173,** 847.
Segales, T., Erill, S., and Domino, E. F. (1969). *Clin. Pharmacol. Ther.* **10,** 522.
Seiden, G., and Brodish, A. (1971). *Neuroendocrinology* **8,** 145.
Seppälä, M., Ranta, T., and Hirvonen, E. (1976). *Lancet* **1,** 1154.
Sétàlo, G., Vigh, S., Schally, A. V., Arimura, A., and Flerkò, B. (1975). *Endocrinology* **96,** 135.
Shaar, C. J., and Clemens, J. A. (1972). *Endocrinology* **90,** 285.
Shaar, C. J., and Clemens, A. J. (1976). *Fed. Proc., Fed. Am. Soc. Exp. Biol.* **35,** Abstr. 546.
Shalyapina, V. G. (1967). *Probl. Endokrinol.* **13,** 102.
Shamanesh, M., and Jeffcoate, S. L. (1976). *J. Endocrinol.* **68,** 89.
Shank, R. P., and Aprison, M. H. (1970). *J. Neurochem.* **17,** 1461.
Shank, R. P., Aprison, M. H., and Baxter, C. F. (1973). *Brain Res.* **52,** 301.
Sharman, D. F. (1966). *Br. J. Pharmacol. Chemother,* **28,** 153.
Sharman, D. F., and Smith, S. E. (1962). *J. Neurochem.* **9,** 403.
Shaw, J. E., and Ramwell, P. W. (1968). *Am. J. Physiol.* **215,** 262.
Shaw, J. E., and Ramwell, P. W. (1969). *Methods Biochem. Anal.* **17,** 325.
Shaw, K. M., Lees, A. J., Hayes, S., Ross, E. J., Stern, G. M., and Thompson, B. D. (1976). *Lancet* **1,** 194.
Sheard, M. H., and Aghajanian, G. K. (1968). *J. Pharmacol. Exp. Ther.* **163,** 425.
Shenkman, L., Sanghvi, I., Schopsin, B., Kataoka, K., Imai, Y., Gershon, S., and Hollander, C. S. (1973). *55th Annu. Meet. Endocr. Soc., Chicago* p. 138.
Sherman, G. P., Kim, S., Benjamin, F., and Kolodny, K. D. (1971). *N. Engl. J. Med.* **284,** 72.
Sherman, L., Kolodny, H. D., Singh, A., Deutsch, S., and Benjamin, F. (1972). *Clin. Res.* **20,** 867.
Sheu, Y.-s., Chu, N. S., Nelson, J. P., and Bloom, F. E. (1971). *Proc. West. Pharmacol. Soc.* **14,** 133.
Shields, P. J., and Eccleston, D. (1973). *J. Neurochem.* **20,** 881.
Shiino, M., Arimura, A., Schally, A. V., and Rennels, E. G. (1972). *Histochimie* **128,** 152.
Shirsishi, T., and Mager, M. (1974). *Fed. Proc., Fed. Am. Soc. Exp. Biol.* **33,** 408.
Shoemaker, W. J. (1973). *In* "Frontiers in Catecholamine Research" (E. Usdin and S. Snyder, eds.), pp. 815–817. Pergamon, Oxford.
Shore, P. A. (1962). *Pharmacol. Rev.* **14,** 531.
Shore, P. A., Burkhalter, A., and Cohn, V. P. (1959). *J. Pharmacol. Exp. Ther.* **127,** 182.
Shore, P. A., Busfield, D., and Alpers, H. S. (1964). *J. Pharmacol. Exp. Ther.* **146,** 194.
Shute, C. C. D., and Lewis, P. R. (1966). *Br. Med. Bull.* **22,** 221.
Sievertsson, H., Chang, J. K., Bogentoft, B. L., Currie, B. L., Folkers, K., and Bowers, C. Y. (1971). *Biochem. Biophys. Res. Commun.* **44,** 1566.
Siggins, G. R., Hoffer, B. J., and Bloom, F. E. (1969). *Science* **165,** 1018.
Siggins, G. R., Hoffer, B. J., and Bloom, F. E. (1971). *Ann. N.Y. Acad. Sci.* **180,** 302.
Sih, C. J., and Takeguchi, C. A. (1973). *In* "The Prostaglandins" (P. W. Ramwell, ed.), Vol. 1, pp. 83–100. Plenum, New York.
Siler, T. M., VandeBergh, G., Yen, S. S. C., Brazeau, P., Vale, W., and Guillemin, R. (1973). *J. Clin. Endocrinol. Metab.* **37,** 632.
Silver, A. (1967). *Int. Rev. Neurobiol.* **10,** 57.
Silverman, A. J. (1976). *Endocrinology* **99,** 30.

Silverman, A. J., and Knigge, K. M. (1972). *Neuroendocrinology* **10,** 71.
Silverman, A. J., Knigge, K. M., and Peck, W. A. (1972). *Neuroendocrinology* **9,** 123.
Silvestrini, F., Liuzzi, A., Chiodini, P. G., Cremascoli, G., Botalla, L., and Müller, E. E. (1975). *In* "Gli Ormoni dell'Ipofisi Anteriore" (E. E. Müller and S. E. Piazzi, eds.), pp. 229–241. I.S.V.T., Sclavo, Siena.
Simmonds, M. A., and Uretsky, N. J. (1970). *Br. J. Pharmacol.* **40,** 630.
Simon, M. L., and George, L. (1975). *Neuroendocrinology* **17,** 125.
Simonovic, I., Motta, M., and Martini, L. (1974). *Endocrinology* **95,** 1373.
Sinha, Y. N., Gupta, M. L., and Bhargava, K. P. (1969). *Eur. J. Pharmacol.* **5,** 235.
Sinha, Y. N., Selby, F. W., Lewis, U. J., and Vanderlaan, W. P. (1972). *Endocrinology* **91,** 1045.
Sirett, N. E., and Purves, H. D. (1974). *Neuroendocrinology* **14,** 257.
Sirtori, C., Bohne, P., and Azarnoff, D. L. (1972). *N. Engl. J. Med.* **287,** 729.
Sitaram, N., Wyatt, R. J., Dawson, S., and Gillin, C. (1976). *Science* **191,** 1281.
Smalstig, E. B., Sawyer, B. D., and Clemens, J. A. (1974). *Endocrinology* **95,** 123.
Smelik, P. G. (1967). *Neuroendocrinology* **2,** 247.
Smith, A. D. (1973). *Br. Med. Bull.* **29,** 123.
Smith, A. D., and Winkler, H. (1972). *In* "Catecholamines" (H. Blaschko and E. Muscholl, eds.), pp. 538–617. Springer-Verlag, Berlin and New York.
Smith, A. F. (1975). *Neuroendocrinology* **19,** 363.
Smith, C. B. (1966). *J. Pharmacol. Exp. Ther.* **151,** 207.
Smith, D. S., Jälfors, U., and Berànek, R. (1970). *J. Cell Biol.* **46,** 199.
Smith, G. P., and Root, A. W. (1971). *Neuroendocrinology* **8,** 235.
Smith, I. (1960). "Chromatographic and Electrophoretic Techniques," Vols. 1 and 2. Wiley (Interscience), New York.
Smith, J. C., Cavallito, C. J., and Foldes, F. F. (1967). *Biochem. Pharmacol.* **16,** 2438.
Smith, R. L., Maikel, R. P., and Brodie, B. B. (1973). *J. Pharmacol. Exp. Ther.* **139,** 185.
Smythe, G. A., and Lazarus, L. (1973a). *Nature (London)* **244,** 1973.
Smythe, G. A., and Lazarus, L. (1973b). *Endocrinology* **93,** 147.
Smythe, G. A., and Lazarus, L. (1974). *J. Clin. Invest.* **54,** 116.
Smythe, G. A., Brandstater, J. F., and Lazarus, L. (1975a). *Neuroendocrinology* **17,** 245.
Smythe, G. A., Compton, P. J., and Lazarus, L. (1975b). *J. Clin. Endocrinol. Metab.* **40,** 714.
Smythe, G. A., Compton, P. J., and Lazarus, L. (1976). *In* "Growth Hormone and Related Peptides" (A. Pecile and E. E. Müller, eds.), pp. 222–235. Excerpta Med. Found., Amsterdam.
Snyder, P. J., and Utiger, R. D. (1972). *J. Clin. Endocrinol. Metab.* **34,** 380.
Snyder, S. H., and Bennett, J. P. Jr (1976). *Ann. Rev. Physiol.* **38,** 153.
Snyder, S. H., and Coyle, J. T. (1968). *J. Pharmacol. Exp. Ther.* **165,** 78.
Snyder, S. H., and Taylor, K. M. (1972). *In* "Perspectives in Neuropharmacology" (S. H. Snyder, ed.), pp. 43–73. Oxford Univ. Press, London and New York.
Snyder, S. H., Axelrod, J., and Zweig, M. (1965). *Biochem. Pharmacol.* **14,** 831.
Snyder, S. H., Baldessarini, R. J., and Axelrod, J. (1966). *J. Pharmacol. Exp. Ther.* **153,** 544.
Snyder, S. H., Kuhar, M. J., Green, A. I., Coyle, J. T., and Shaskan, E. G. (1970). *Int. Rev. Neurobiol.* **13,** 127.
Sørensen, R., Svendsen, K., and Schou, M. (1974). *Lancet* **1,** 865.
Sorrentino, S. (1968). *Anat. Rec.* **160,** 432.
Sorrentino, S., Jr., Schalch, D. S., and Reiter, R. J. (1972). *Growth and Growth Horm., Proc. Int. Symp., 2nd, 1971* Excerpta Med. Found. Int. Congr. Ser. No. 244, pp. 330–348.

Sourkes, T. G. (1954). *Arch. Biochem. Biophys.* **51,** 444.
Spector, S., Shore, P. A., and Brodie, B. B. (1960). *J. Pharmacol. Exp. Ther.* **128,** 15.
Spector, S., Sjoerdsma, A., Zaltman-Nirenberg, P., Levitt, M., and Udenfriend, S. (1963). *Science* **139,** 1299.
Spector, S., Sjoerdsma, A., and Udenfriend, S. (1965). *J. Pharmacol. Exp. Ther.* **147,** 86.
Spies, H. G., and Norman, R. L. (1973). *Prostaglandins* **4,** 131.
Spirtes, M. A., Plotnikoff, N. P., and Kastin, A. J. (1976). *Proc. Int. Congr. Pharmacol., 6th, 1975* pp. 121–129.
Spooner, C. E., and Winters, W. D. (1965). *Experientia* **21,** 256.
Squires, R. (1974). *Adv. Biochem. Psychopharmacol.* **10,** 207–211.
Stachura, M. E., Dhariwal, A. P. S., and Frohman, L. A. (1972a). *Endocrinology* **91,** 1071.
Stachura, M. E., Dhariwal, A. P. S., and Frohman, L. A. (1972b). *Abstr. Book Int. Congr. Endocrinol., 4th, 1972* p. 83.
Stadler, H., Lloyd, K. G., Gadea-Cira, M., and Bartholini, G. (1973). *Brain Res.* **55,** 476.
Stanbury, J. B., Kassenaar, A. A. H., Meyer, J. W. A., and Terpstra, J. (1956). *J. Clin. Endocrinol. Metab.* **16,** 735.
Stärke, K. (1971). *Naturwissenschaften* **58,** 420.
Stärke, K. (1972). *Naunyn-Schmiedberg's Arch. Pharmacol.* **274,** 18.
Stavinoha, W. B., and Ryan, L. C. (1965). *J. Pharmacol. Exp. Ther.* **150,** 231.
Stein, D., Jouvet, M., and Pujol, J. F. (1974). *Brain Res.* **72,** 360.
Steiner, A. L., Peake, G. T., Utiger, R., and Kipnis, D. M. (1969). *J. Clin. Invest.* **48,** 80a.
Steiner, A. L., Peake, G. T., Utiger, R. D., Karl, I. E., and Kipnis, D. M. (1970). *Endocrinology* **86,** 1354.
Steiner, F. A., and Pieri, L. (1969). *Prog. Brain Res.* **31,** 191.
Steiner, F. A., Rup, K., and Akerk, K. (1969). *Brain Res.* **12,** 74.
Steiner, J., Cassar, J., Mashiter, K., Daves, I., Russel Fraser, T., and Breckenridge, A. (1976). *Br. Med. J.* **1,** 1186.
Sterman, M. B., and Clemente, C. D. (1962). *Exp. Neurol.* **6,** 103.
Stern, P., and Hadzovic, M. (1973). *Arch. Int. Pharmacodyn. Ther.* **202,** 259.
Stern, W. C., and Morgane, P. J. (1973). *Biol. Psychiatry* **6,** 301.
Stevens, V. C. (1969). *J. Clin. Endocrinol. Metab.* **29,** 904.
Stevens, V. C., and Boutselis, J. (1972). *Obstet. Gynecol.* **39,** 633.
Stewart, J. M., Getto, C. J., Neldner, K., Reeve, E. B., Krivoy, W. A., and Zimmermann, E. (1976). *Nature (London)* **262,** 784.
Stille, G., and Hippius, H. (1971). *Pharmakopsychiatr. Neuro-Psychopharmakol.* **4,** 182.
Stoeckhart, R., Jansen, H. G., and Kreike, A. J. (1972). *Z. Zellforsch. Mikrosk. Anat.* **131,** 99.
Stolk, J. M., and Rech, R. H. (1968). *J. Pharmacol. Exp. Ther.* **163,** 75.
Stolk, J. M., Barchas, J., Dement, W., and Schanberg, S. (1969). *Pharmacologist* **11,** 258.
Stone, C. L., Wenger, H. C., Ludden, C. T., Stravorski, J. M., and Ross, C. A. (1961). *J. Pharmacol. Exp. Ther.* **131,** 73.
Storm-Mathisen, J. (1975). *Brain Res.* **84,** 409.
Strada, S. J., Sander-Bush, E., and Sulser, F. (1970). *Biochem. Pharmacol.* **19,** 2621.
Strada, S. J., Klein, D. C., Weller, J., and Weiss, B. (1972). *Endocrinology* **90,** 1470.
Straughan, D. W., Neal, M. J., Simmonds, M. A., Collins, G. G. S., and Hill, R. G. (1971). *Nature (London)* **233,** 352.
Streefkerk, J. G. (1967). *In* Ph.D. Thesis, University of Amsterdam.
Stumpf, W. E., and Sar, M. (1973). *Prog. Brain Res.* **39,** 53.
Subramanian, M. G., and Gala, R. R. (1976). *Endocrinology* **98,** 842.
Sulaiman, W. R., and Johnson, R. H. (1973). *Br. Med. J.* **2,** 329.

Sulser, F., and Bass, D. (1968). *Public Health Serv., Publ.* **1836.**

Sulser, F., and Dingell, J. V. (1968). *Biochem. Pharmacol.* **17,** 634.

Sundberg, D. K., Krulich, L., Fawcett, C. P., Illner, P., and McCann, S. M. (1973). *Proc. Soc. Exp. Biol. Med.* **142,** 1097.

Suzuki, T., Ikeda, H., Narita, S., Shibata, O., Waki, S., and Egashiza, K. (1973). *Anat. J. Exp. Physiol.* **58,** 139.

Suzuki, T., Abe, K., and Hiroshige, T. (1975). *Neurochemistry* **17,** 75.

Svensson, T. H. (1973a). *Eur. J. Pharmacol.* **23,** 232.

Svensson, T. H. (1973b). *J. Pharm. Pharmacol.* **25,** 75.

Svensson, T. H., Bunney, B. S., and Aghajanian, G. K. (1975). *Brain Res.* **92,** 291.

Sweeley, C. C., Elliott, W. H., Fries, I., and Ryhage, R. (1966). *Anal. Chem.* **38,** 1549.

Szabo, M., and Frohman, L. A. (1975). *Endocrinology* **96,** 955.

Szabo, M., and Frohman, L. A. (1976). *Endocrinology* **98,** 1451.

Szara, S. (1956). *Experientia* **12,** 441.

Sze, P. Y., Nickers, L., and Towle, A. C. (1976). *J. Neurochem.* **26,** 169.

Tagliamonte, A., Tagliamonte, P., Corsini, G. U., Mereu, G. P., and Gessa, G. L. (1973). *J. Pharm. Pharmacol.* **25,** 101.

Tait, S. A. S., Tait, J. F., Okamoto, M., and Flood, C. (1967). *Endocrinology* **81,** 1213.

Takahara, J., Arimura, A., and Schally, A. V. (1974a). *Endocrinology* **95,** 462.

Takahara, J., Arimura, A., and Schally, A. V. (1974b). *Endocrinology* **95,** 1490.

Takahara, J., Arimura, A., and Schally, A. V. (1974c). *Proc. Soc. Exp. Biol. Med.* **146,** 831.

Takahara, J., Arimura, A., and Schally, A. V. (1975). *Acta Endocrinol. (Copenhagen)* **78,** 428.

Taleisnik, S., and Celis, M. E. (1974). *Proc. Int. Congr. Endocrinol., 4th, 1972* Excerpta Med. Found. Int. Congr. Ser. No. 273, p. 100.

Taleisnik, S., and Orias, R. (1965). *Am. J. Physiol.* **208,** 293.

Taleisnik, S., and Tomatis, M. E. (1967a). *Endocrinology* **81,** 819.

Taleisnik, S., and Tomatis, M. E. (1967b). *Am. J. Physiol.* **212,** 157.

Taleisnik, S., and Tomatis, M. E. (1968). *Neuroendocrinology* **3,** 303.

Taleisnik, S., Orias, R., and De Olmos, J. (1966). *Proc. Soc. Exp. Biol. Med.* **122,** 325.

Taleisnik, S., De Olmos, J., Orias, R., and Tomatis, M. E. (1967). *J. Endocrinol.* **39,** 485.

Taleisnik, S., Tomatis, M. E., and Celis, M. E. (1972). *Neuroendocrinology* **10,** 235.

Taleisnik, S., Celis, M. E., and Tomatis, M. E. (1973–1974). *Neuroendocrinology* **13,** 327.

Tang, L. C., and Cotzias, G. C. (1976). *Arch. Neurol. (Chicago)* **33,** 131.

Tarsy, D., and Baldessarini, R. J. (1974). *Neuropharmacology* **13,** 927.

Tarsy, D., Pycock, C., Meldrum, B., and Marsden, C. D. (1975). *Brain Res.* **89,** 160.

Tarver, J., Bautz, G., and Horst, W. D. (1975). *Fed. Proc., Fed. Am. Soc. Exp. Biol.* **34,** 283.

Tashjian, A. H., Jr., Barowsky, N. J., and Jensen, D. K. (1971). *Biochem. Biophys. Res. Commun.* **43,** 516.

Tate, S. S., Ross, L. L., and Meister, A. (1973). *Proc. Natl. Acad. Sci. U.S.A.* **70,** 1447.

Taylor, K. M., and Snyder, S. H. (1972). *J. Neurochem.* **19,** 1343.

Taylor, K. M., Gfeller, E., and Snyder, S. H. (1972). *Brain Res.,* **41,** 171.

Taylor, R. J., Jr., Stubbs, C. S., Jr., and Ellenbogen, L. (1968). *Biochem. Pharmacol.* **17,** 1779.

Teitelbaum, P., and Epstein, A. N. (1962). *Psychol. Rev.* **69,** 74.

Telegdy, G., and Vermes, I. (1975). *Neuroendocrinology* **18,** 16.

Teresawa, E., Bridson, W. E., Davenport, J. W., and Goy, R. W. (1975). *Neuroendocrinology* **18,** 345.

Terry, L. C., Willoughby, J. O., Brazeau, P., and Martin, J. B. (1976). *Science* **192,** 565.

Thierry, A. M., Fekete, M., and Glowinski, J. (1968). *Eur. J. Pharmacol.* **4,** 384.

Thoa, N. B., Wooten, G. F., Axelrod, J., and Kopin, I. J. (1972). *Proc. Natl. Acad. Sci. U.S.A.* **69,** 520.

Thody, A. J. (1974). *Neuroendocrinology* **16,** 323.

Thody, A. J., Shuster, S., Plummer, N. A., Bogie, W., Leigh, R. J., Goolamali, S. K., and Smith, A. G. (1974). *J. Clin. Endocrinol. Metab.* **38,** 491.

Thoenen, H. (1972). *In* "Perspectives in Neuropharmacology" (S. H. Snyder, ed.), pp. 301–338. Oxford Univ. Press, London and New York.

Thoenen, H., and Tranzer, J. P. (1968). *Naunyn-Schmiedeberg's Arch. Pharmacol. Exp. Pathol.* **281,** 271.

Thoenen, H., Haefely, W., Gey, K. F., and Huerlimann, A. (1965). *Life Sci.* **4,** 2033.

Thompson, C. J., Los, M., and Horton, E. W. (1970). *Life Sci.* **9,** 983.

Thorner, M. O., McNeilly, A. S., Hagan, C., and Besser, G. M. (1974a). *Br. Med. J.* **2,** 419.

Thorner, M. O., Besser, G. M., Hagan, C., and McNeilly, A. S. (1974b). *J. Endocrinol.* **61,** lxxxii.

Thorner, M. O., Chait, A., Aitken, M., Benker, G., Bloom, S. M., Mortimer, C. H., Sanders, P., Stuart-Mason, A., and Besser, G. M. (1975a). *Br. Med. J.* **1,** 299.

Thorner, M. O., Besser, G. M., Jones, A., Dacie, J., and Jones, A. E. (1975b). *Clin. Endocrinol.* **4,** 694.

Thorner, M. O., Wass, J. A. H., Jones, A., Bloom, S. R., and MacLeod, R. M. (1976). *Abstr. Book Int. Congr. Endocrinol., 5th, 1976* Abstract No. 860.

Thornton, V. F., and Geschwind, I. I. (1974). *Gen. Comp. Endocrinol.* **23,** 294.

Thum, C. H., Meissner, C., Beischer, W., Schroder, K. E., and Pfeiffer, E. F. (1975). *Acta Endocrinol. (Copenhagen)* **78,** 69.

Tima, L., and Flerkò, B. (1974). *Neuroendocrinology* **15,** 346.

Tima, L., Trentini, G. P., and Mess, B. (1973). *Neuroendocrinology* **12,** 149.

Tindal, J. S., and Knaggs, G. S. (1972). *J. Endocrinol.* **52,** 253.

Tissot, R., Bartholini, G., and Pletscher, A. (1969). *Arch. Neurol. (Chicago)* **20,** 187.

Tiwary, C. M., Frias, J. L., and Rosenbloom, A. L. (1972). *Lancet* **2,** 1086.

Toivola, P. T. K., and Gale, C. C. (1970). *Neuroendocrinology* **6,** 210.

Toivola, P. T. K., and Gale, C. C. (1972). *Endocrinology* **90,** 895.

Toivola, P. T. K., and Gale, C. C. (1973). *Fed. Proc., Fed. Am. Soc. Exp. Biol.* **32,** 265.

Tiovola, P. T. K., Gale, C. C., Goodner, C. J., and Werrbach, J. H. (1972). *Hormones* **3,** 193.

Torda, C. (1968). *Brain Res.* **10,** 200.

Torii, S., Mitsumori, K., Inubushi, S., and Yanagisawa, I. (1973). *Psychopharmacologia* **29,** 65.

Torök, B. (1954). *Acta Morphol. Acad. Sci. Hung.* **4,** 83.

Trabucchi, M., Cheney, D. L., Racagni, G., and Costa, E. (1974). *Nature (London)* **249,** 664.

Trabucchi, M., Cheney, D. L., Racagni, G., and Costa, E. (1975). *Brain Res.* **85,** 130.

Trabucchi, M., Spano, P. F., Tonon, G. C., and Frattola, L. (1976a). *Life Sci.* **19,** 225.

Trabucchi, M., Longoni, R., Fresia, P., and Spano, P. F. (1976b). *Life Sci.* **17,** 1551.

Tranzer, J. P., and Thoenen, H. (1972). *Experientia* **24,** 484.

Tranzer, J. P., Da Prada, M., and Pletscher, A. (1966). *Nature (London)* **212,** 1574.

Tregear, G. W., Niall, H. D., Potts, J. T., Leeman, S. E., and Chang, M. M. (1971). *Nature (London), New Biol.* **232,** 87.

Trendelenburg, U. (1957). *Circ. Res.* **5,** 105.

Trendelenburg, U. (1972). *In* "Catecholamines" (H. Blaschko and E. Muscholl, eds.), pp. 337–362. Springer-Verlag, Berlin and New York.

Trentini, G. P., Tima, L., and Mess, B. (1973). *Horm. Res.* **4,** 349.

Trentini, G. P., Tima, L., De Gaetani, C. F., and Mess, B. (1974). *Steroids Lipids Res.* **5,** 262.

Tsafriri, A., Koch, Y., and Lindner, H. R. (1973). *Prostaglandins* **3,** 461.
Tsai, C. C., and Yen, S. S. C. (1971). *J. Clin. Endocrinol. Metab.* **33,** 917.
Tsukui, T., Kanno, Y., Koizumi, K., Hashizume, K., Onaya, T., and Yamada, T. (1974). *Proc. Soc. Exp. Biol. Med.* **145,** 346.
Tullner, W. W., and Hertz, R. (1964). *Proc. Soc. Exp. Biol. Med.* **16,** 837.
Tuomisto, J., and Männisto, P. (1973). *Lancet* **1,** 836.
Tuomisto, J., Rante, T., Saarinen, A., Männisto, P., and Leppäluoto, J. (1973). *Lancet* **2,** 510.
Tuomisto, J., Ranta, T., Männisto, P., Saarinen, A., and Leppäluoto, J. (1975). *Eur. J. Pharmacol.* **30,** 221.
Turkington, R. W. (1972a). *Lactogenic Horm., Ciba Found. Symp., 1971* pp. 169–183.
Turkington, R. W. (1972b). *J. Clin. Endocrinol. Metab.* **34,** 306.
Twarog, B. P., and Page, I. H. (1953). *Am. J. Physiol.* **175,** 157.
Tyrrel, J. B., Lorenzi, M., Gerich, J. E., and Forsham, P. H. (1975). *J. Clin. Endocrinol. Metab.* **40,** 1125.
Uchimura, H., Kumeda, H., Kawabata, Y., Maeda, K., Okamoto, O., Ogawa, T., and Inamori, Y. (1972). *Annu. Meet. Jpn. Endocrinol. Soc.* p. 12.
Udenfriend, S., Weissbach, H., and Clark, C. T. (1955). *J. Biol. Chem.* **215,** 337.
Udenfriend, S., Zaltman-Nirenberg, P., and Nagatsu, T. (1965). *Biochem. Pharmacol.* **14,** 837.
Udenfriend, S., Creveling, C. R., Gordon, R., and Spector, S. (1966). *Mol. Pharmacol.* **2,** 95.
Udeschini, G., Cocchi, D., Panerai, A. E., Gil-Ad, I., Rossi, G. L., Chiodini, P. G., Liuzzi, A., and Müller, E. E. (1976). *Endocrinology* **98,** 807.
Uemura, H., and Kobayashi, H. (1971). *Endocrinol. Jpn.* **18,** 91.
Uhr, M. L., and Sneddon, M. K. (1972). *J. Neurochem.* **19,** 1495.
Ulrich, R. S., and Yuwiler, A. (1973). *Endocrinology* **92,** 611.
Ulrich, R. S., Yuwiler, A., and Geller, E. (1975). *Neuroendocrinology* **19,** 259.
Unger, R. H., and Orci, L. (1975). *Lancet* **1,** 14.
Ungerstedt, U. (1968). *Eur. J. Pharmacol.* **5,** 107.
Ungerstedt, U. (1971a). *Acta Phsyiol. Scand., Suppl.* **367,** 1.
Ungerstedt, U. (1971b). *In* "6-Hydroxydopamine and Catecholamine Neurons" (T. Malmfors and H. Thoenen, eds.), pp. 101–127. North-Holland Publ., Amsterdam.
Uretsky, A. M., Gaisinskaya, M. Y., Raisin, M. S., and Brande, I. Y. (1971). *Byull. Eksp. Biol. Med.* **70,** 1388.
Uretsky, N. J., and Iversen, L. L. (1969). *Brain Res.* **24,** 485.
Uretsky, N. J., and Iversen, L. L. (1970). *J. Neurochem.* **17,** 269.
Uretsky, N. J., Simmonds, M. A., and Iversen, L. L. (1971). *J. Pharmacol. Exp. Ther.* **176,** 489.
Vaala, S., and Knigge, K. M. (1974). *Neuroendocrinology* **15,** 147.
Vale, W., Burgus, R., Dunn, T. F., and Guillemin, R. (1967). *Proc. Soc. Exp. Biol. Med.* **125,** 210.
Vale, W., Burgus, R., Dunn, T. F., and Guillemin, R. (1971). *Hormones* **2,** 193.
Vale, W., Burgus, R., Dunn, T. F., and Guillemin, R. (1972a). *Endocrinology* **91,** 562.
Vale, W., Brazeau, P., Grant, G., Nussey, A., Burgus, R., Rivier, J., Ling, N., and Guillemin, R. (1972b). *C.R. Hebd. Seances Acad. Sci.* **275,** 2913.
Vale, W., Brazeau, P., Rivier, J., and Guillemin, R. (1973a). *In* "Advances in Human Growth Hormone Research" (S. Raiti, ed.), DHEW Publication No. (NIH) 74-612, pp. 159–182. Center for Disease Control, Atlanta, Georgia.
Vale, W., Blackwell, R., Grant, G., and Guillemin, R. (1973b). *Endocrinology* **93,** 26.

Vale, W., Grant, G., and Guillemin, R. (1973c). *In* "Frontiers in Neuroendocrinology" (W. F. Ganong and L. Martini, eds.), pp. 375–413. Academic Press, New York.
Valverde, R. C., Chieffo, C. V., and Reichlin, S. (1972). *Endocrinology* **91,** 982.
Valverde, R. C., Chieffo, C. V., and Reichlin, S. (1973). *Life Sci.* **12,** 327.
Van Delft, A. M. L., Kaplanski, J., and Smelik, P. G. (1973). *J. Endocrinol.* **59,** 465.
Vande Wiele, R. L., and Dyrenfurth, I. (1973). *Pharmacol. Rev.* **25,** 189.
Vane, J. R. (1960). *Adrenergic Mech., Ciba Found. Symp., 1960* pp. 356–372.
Vane, J. R. (1966). *Pharmacol. Rev.* **18,** 317.
Vane, J. R. (1971). *Nature (London), New Biol.* **231,** 232.
Van Gelder, N. M. (1968). *J. Neurochem.* **15,** 747.
Van Gelder, N. M. (1969). *J. Neurochem.* **16,** 1355.
Van Gelder, N. M., Sherwin, A. L., and Rasmussen, T. (1972). *Brain Res.* **40,** 385.
Van Loon, G. R. (1970). Ph.D. Thesis, University of California, San Francisco.
Van Loon, G. R. (1973). *In* "Frontiers in Neuroendocrinology" (W. F. Ganong and L. Martini, eds.), pp. 209–247. Oxford Univ. Press, London and New York.
Van Loon, G. R., Nicholson, W. E., and Liddle, G. W. (1971a). *Fed. Proc., Fed. Am. Soc. Exp. Biol.* **30,** 533.
Van Loon, G. R., Hilger, L., King, A. B., Boryczka, A. T., and Ganong, W. F. (1971b). *Endocrinology* **88,** 1404.
Van Loon, G. R., Scapagnini, U., Möberg, G. P., and Ganong, W. F. (1971c). *Endocrinology* **89,** 1464.
Van Loon, G. R., Scapagnini, U., Cohen, R., and Ganong, W. F. (1971d). *Neuroendocrinology* **8,** 257.
Van Loon, G. R., Kamble, A., Sole, M. J., and Green, S. (1976). *Abstr. Book. Int. Congr. Endocrinol., 5th, 1976* Abstract No. 664.
Van Maanen, T. H., and Smelik, P. G. (1968). *Neuroendocrinology* **3,** 177.
Van Praag, H. M., and Korf, J. (1973). *Serotonin Behav., Proc. Symp., 1972* pp. 457–468.
Van Praag, H. M., and Korf, J. (1975). *Pharmakopschia* **8,** 322.
Van Thienhoven, A., and Schally, A. V. (1972). *Gen. Comp. Endocrinol.* **19,** 594.
Vaughan, G. M., Pelham, R. W., Pang, S. F., Loughlin, L. L., Wilson, K. M., Sandock, K. L., Vaughan, M. K., Koslow, S. H., and Reiter, R. J. (1976). *J. Clin. Endocrinol. Metab.* **42,** 752.
Vaughan, M. K., Benson, B., and Norris, J. T. (1970). *J. Endocrinol.* **47,** 397.
Vaughan, M. K., Reiter, R. J., and Vaughan, G. M. (1971). *J. Endocrinol.* **51,** 787.
Vaughan, T., Wyatt, R. J., and Green, R. (1972). *Psychophysiology* **9,** 96.
Veber, D. F., Bennet, C. D., Milkowski, J. D., Gal, G., Denkewalter, R. G., and Hirschmann, R. (1971). *Biochem. Biophys. Res. Commun.* **45,** 235.
Vekemans, M., Delvoye, P., L'Hermite, M., and Robyn, C. (1972). *C.R. Hebd. Seances Acad. Sci.* **275,** 2247.
Verde, G., Oppizzi, G., Colussi, G., Cremascoli, G., Botalla, L., Müller, E. E., Silvestrini, F., Chiodini, P. G., and Liuzzi, A. (1976). *Clin. Endocrinol.* **5,** 419.
Vermes, I., and Telegdy, G. (1972). *Acta Physiol. Acad. Sci. Hung.* **42,** 49.
Vermes, I., and Telegdy, G. (1973). *Acta Physiol. Acad. Sci. Hung.* **43,** 105.
Vermes, I., Telegdy, G., and Lissàk, K. (1972). *Acta Physiol. Acad. Sci. Hung.* **41,** 95.
Vermes, I., Molnàr, D., and Telegdy, G. (1973a). *Acta Physiol. Acad. Sci. Hung.* **43,** 27.
Vermes, I., Telegdy, G., and Lissàk, K. (1973b). *Acta Physiol. Acad. Sci. Hung.* **43,** 33.
Vermes, I., Molnàr, D., and Telegdy, G. (1973c). *Acta Physiol. Acad. Sci. Hung.* **43,** 239.
Vernikos-Danellis, J. (1968). *In* "Pharmacology of Hormonal Polypeptides and Proteins" (N. Back, L. Martini, and R. Paoletti, eds.), p. 175. Plenum, New York.
Vernikos-Danellis, J., and Marks, B. H. (1962). *Endocrinology* **70,** 525.

Vernikos-Danellis, J., Berger, P., and Barchas, J. D. (1973). *Prog. Brain Res.* **39,** 301.
Vijayan, E., and McCann, S. M. (1976). *Abstr. Book. Int. Congr. Endocrinol., 5th, 1976* Abstract No. 29.
Vilchez-Martinez, J. A., Arimura, A., and Schally, A. V. (1975). *Proc. Soc. Exp. Biol. Med.* **148,** 913.
Vinokur, A., and Utiger, R. D. (1974). *Science* **185,** 265.
Vodian, M. A., and Nicoll, C. S. (1975). *Ric. Sci. Educ. Perm.* 56.
Vogt, M. (1944). *J. Physiol. (London)* **103,** 317.
Vogt, M. (1951). *J. Physiol. (London)* **113,** 129.
Vogt, M. (1954). *J. Physiol. (London)* **123,** 451.
Vogt, M. (1959). *Pharmacol. Rev.* **11,** Part 2, 483.
Volkman, P. H., and Heller, A. (1971). *Science* **173,** 839.
von Euler, C., and Holmgren, B. (1956). *J. Physiol. (London)* **131,** 125.
von Euler, U. S. (1959). *Pharmacol. Rev.* **11,** 262.
von Euler, U. S. (1972). *Handb. Exp. Pharmakol.* [N. S.] **33,** 186–230.
von Euler, U. S., and Gaddum, J. H. (1931). *J. Physiol. (London)* **72,** 74.
von Euler, U. S., and Hillarp, N.-Å. (1956). *Nature (London)* **177,** 44.
von Hungen, K., Roberts, S., and Hill, D. F. (1974). *Nature (London)* **252,** 588.
von Voigtandler, P. F., and Moore, K. E. (1973). *Neuropharmacology* **12,** 451.
von Werder, K., Van Loon, G. R., Yatsu, F., and Forsham, P. H. (1970). *Klin. Wochenschr.* **48,** 1454.
von zur Mühlen, A. D., Hesch, D., Embrich, D., and Creutzfeldt, W. (1970). *Dtsch. Med. Wochenschr.* **95,** 2623.
Voogt, J. L., and Carr, L. A. (1975). *Endocrinology* **97,** 891.
Voogt, J. L., and Meites, J. (1971). *Endocrinology* **88,** 286.
Votava, Z. (1967). *Annu. Rev. Pharmacol.* **7,** 223.
Waelsch, H. (1955). *In* "Biochemistry of the Developing Nervous System" (H. Waelsch, ed.), pp. 187–207. Academic Press, New York.
Wagner, K. (1971). *In* "6-Hydroxydopamine and Catecholamine Neurons" (T. Malmfors and H. Thoenen, eds.), pp. 277–278. North-Holland Publ., Amsterdam.
Wakabayashi, K., Kamberi, I. A., and McCann, S. M. (1969). *Endocrinology* **85,** 1046.
Wakabayashi, K., Antunes-Rodriguez, J., Tamaoki, B.-I., and McCann, S. M. (1972). *Endocrinology* **90,** 690.
Walker, S., Hibbard, B. M., Groom, G., Griffiths, K., and Davies, R. H. (1975). *Lancet* **2,** 842.
Wallach, D. P. (1961). *Biochem. Pharmacol.* **5,** 232.
Walsh, D. A., and Sallach, H. J. (1966). *J. Biol. Chem.* **241,** 4068.
Warberg, J., Eskay, R. L., and Porter, J. C. (1976). *Endocrinology* **98,** 1135.
Wass, J. A. H., Thorner, M. O., Besser, G. M., Morris, D., Stuart-Mason, A., Liuzzi, A., and Chiodini, P. G. (1976). *Lancet* **2,** 851.
Watkins, B. E., McKey, D. W., Meites, J., and Riegle, G. D. (1975). *Neuroendocrinology* **19,** 331.
Waton, N. G. (1956). *Br. J. Pharmacol. Chemother.* **11,** 119.
Weeke, J., Prange-Hansen, A., and Lundaek, K. (1975). *J. Clin. Endocrinol. Metab.* **41,** 168.
Weichert, R. F. (1970). *Am. J. Med.* **49,** 232.
Weight, F. F., and Salmoiraghi, G. C. (1966). *J. Pharmacol. Exp. Ther.* **153,** 420.
Weight, F. F., and Salmoiraghi, G. C. (1968). *Adv. Pharmacol.* **6a,** 395–413.
Weil-Malherbe, H., and Bone, A. D. (1957). *Nature (London)* **180,** 1050.
Weil-Malherbe, H., Whitby, L. G., and Axelrod, J. (1961a). *In* "Regional Neurochemistry" (S. S. Kety and J. Elkes, eds.), pp. 284–292. Pergamon, Oxford.

Weil-Malherbe, H., Posner, H. S., and Bowles, G. R. (1961b). *J. Pharmacol. Exp. Ther.* **132,** 278.

Weiner, R. J. (1973). *Prog. Brain Res.* **39,** 165.

Weiner, R. J. (1975). *Hypothal. Horm.* [*Symp.*], *1974* pp. 249–253.

Weiner, R. J., Shryne, J. E., Gorsky, R. A., and Sawyer, C. H. (1972). *Endocrinology* **90,** 867.

Weinshilboum, R. M., Thoa, N. B., Johnson, D. G., Kopin, I. J., and Axelrod, J. (1971). *Science* **174,** 1349.

Weiss, P. (1970). *In* "The Neurosciences: Second Study Program" (F. O. Schmitt, ed.), p. 840. Rockefeller Univ. Press, New York.

Weissman, A. (1973). *Serotonin Behav., Proc. Symp., 1972* pp. 235–248.

Weissman, B. A., Daly, J. W., and Skolnick, P. (1975). *Endocrinology* **97,** 1559.

Weitzman, E. D., Rapport, M. M., McGregor, P., and Jacoby, J. (1968). *Science* **160,** 1361.

Weitzman, E. D., McGregor, P., Moore, C., and Jacoby, J. (1969). *Life Sci.* **8,** 751.

Weitzman, E. D., Fukushima, D., Nogeire, C., Roffwarg, H., Gallagher, T. F., and Hellman, L. (1971). *J. Clin. Endocrinol. Metab.* **33,** 14.

Welch, A. S., and Welch, B. L. (1968). *Proc. Natl. Acad. Sci. U.S.A.* **60,** 478.

Weldon, V. V., Gupta, S. K., Haymond, M. W., Pagliara, A. S., Jacobs, L. S., and Daughaday, W. H. (1973). *J. Clin. Endocrinol. Metab.* **36,** 42.

Weller, C. P. (1963). *Bull. Res. Counc. Isr., Sect. E* **10,** 3.

Welsch, C. W., Squires, M. D., Cassel, E., Chen, C. L., and Meites, J. (1971). *Am. J. Physiol.* **221,** 1714.

Welsh, J. H. (1968). *Adv. Pharmacol.* **6a,** 171–188.

Werdinius, B. (1966). *J. Pharmacol.* **18,** 546.

Werman, R., and Aprison, M. H. (1968). *In* "Structure and Function of Inhibitory Neuronal Mechanisms" (C. von Euler, S. Sköglund, and U. Sodenberg, eds.), pp. 473–486. Pergamon, Oxford.

Werrbach, J. H., Gale, C. C., Goodner, C. J., and Conway, M. J. (1970). *Endocrinology* **86,** 77.

Westermann, E. D., Maickel, R. P., and Brodie, B. B. (1962). *J. Pharmacol. Exp. Ther.* **138,** 208.

Wetterberg, L., Arendt, J., Paunier, L., Sizonenko, P. C., van Donselaar, W., and Heyden, T. (1976). *J. Clin. Endocrinol. Metab.* **42,** 185.

Wheaton, J. E., Krulich, L., and McCann, S. M. (1975). *Endocrinology* **97,** 30.

White, T. (1959). *J. Physiol. (London)* **149,** 34.

White, T. (1960). *J. Physiol. (London)* **152,** 299.

White, T. (1965). *Experientia* **21,** 132.

White, T. (1966). *Handb. Exp. Pharmakol.* **18,** Part 1, 789–796.

White, W. F. (1970a). *Hypophysiotropic Horm. Hypothal.: Assay Chem. Proc. Workshop Conf., 1969* p. 248.

White, W. F. (1970b). *In* "Mammalian Reproduction" (G. Gibian and E. J. Platz, eds.), p. 84. Springer-Verlag, Berlin and New York.

White, W. F., Hedlund, M. T., Weber, G. F., Rippel, R. H., Johnson, E. S., and Wilber, J. F. (1974). *Endocrinology* **94,** 1422.

Whittaker, V. P. (1963). *Handb. Exp. Pharmakol.* [N.S.] **15,** 1–39.

Whittaker, V. P. (1965). *Prog. Biophys.* **15,** 39.

Whittaker, V. P. (1969a). *Handb. Neurochem.* **2,** 327–364.

Whittaker, V. P. (1969b). *Prog. Brain Res.* **31,** 211.

Whittaker, V. P. (1969c). *In* "The Structure and Function of Nervous Tissue (G. H. Bourne, ed.), Vol. 3, p. 1. Academic Press, New York.

Whittaker, V. P., Michaelson, I. A., and Kirkland, J. A. (1964). *Biochem. J.* **90,** 293.

Wilber, G. F., Peake, G. T., and Utiger, R. D. (1969). *Endocrinology* **84,** 758.

Wilber, J. F., Nagel, T., and White, W. F. (1971). *Endocrinology* **89,** 1419.

Wilbur, D. L., and Nair, R. M. G. (1976). *58th Annu. Meet. Endocr. Soc., San Francisco* Abstract No. 160.

Wilbur, D. L., Worthington, W. C., Jr., and Markwald, R. R. (1974). *Am. J. Anat.* **141,** 139.

Wilbur, D. L., Worthington, W. C., Jr., and Markwald, R. R. (1975). *Neuroendocrinology* **19,** 12.

Wilson, D. E., Phillips, C., and Levine, R. A. (1971). *Gastroenterology* **61,** 201.

Wilson, I. C., Lara, P. P., and Prange, A. J. (1973a). *Lancet* **2,** 43.

Wilson, I. C., Prange, A. J., Lara, P. P., Alltop, L. B., Stikeleather, R. A., and Lipton, M. A. (1973b). *Arch. Gen. Psychiatry* **29,** 15.

Wilson, M. C., Steiner, A. L., Dhariwal, A. P., and Peake, G. T. (1974). *Neuroendocrinology* **15,** 313.

Winkelmann, W., Schorn, H., Hadam, W. R., Heesen, D., and Mies, R. (1975). *Ric. Sci. Educ. Perm.* Suppl. 1, 99.

Wislocki, G. B. (1937). *Anat. Rec.* **69,** 361.

Wislocki, G. B., and King, L. S. (1936). *Am. J. Anat.* **58,** 421.

Witorsch, R. J., and Brodish, A. (1972a). *Endocrinology* **90,** 552.

Witorsch, R. J., and Brodish, A. (1972b). *Endocrinology* **90,** 1160.

Wolfe, L. S., Coceani, F., and Pace-Asciak, C. (1967). *Prostaglandins, Proc. Nobel Symp., 2nd, 1966* pp. 265–276.

Wong, D. T., Bymaster, F. P., Horn, J. S., and Molloy, B. B. (1975). *J. Pharmacol. Exp. Ther.* **193,** 804.

Wonnacott, S., and Marchbanks, R. M. (1975). *Biochem. Soc. Trans., 53rd Meet.* pp. 102–106.

Woodruff, G. N. (1971). *Comp. Gen. Pharmacol.* **2,** 439.

Woods, A. C., and Chase, T. N. (1973). *Lancet* **2,** 513.

Woolf, P. D., Jacobs, L. S., Donofrio, R., Burday, S. Z., and Schalch, D. S. (1974). *J. Clin. Endocrinol. Metab.* **38,** 71.

Worthington, W. C., Jr. (1955). *Bull. Johns Hopkins Hosp.* **97,** 343.

Wurtman, R. J. (1966). "Catecholamines." Little, Brown, Boston, Massachusetts.

Wurtman, R. J. (1970). *In* "The Neurosciences: Second Study Program" (F. O. Schmitt, ed.), pp. 530–538. Rockefeller Univ. Press, New York.

Wurtman, R. J. (1971). *Neurosci. Res. Program, Bull.* **9,** 214.

Wurtman, R. J. (1973). *Fed. Proc., Fed. Am. Soc. Exp. Biol.* **32,** 1769.

Wurtman, R. J., and Fernström, J. D. (1972). *In* "Perspectives in Neuropharmacology" (S. H. Snyder, ed.), pp. 143–193. Oxford Univ. Press, London and New York.

Wurtman, R. J., Axelrod, J., and Potter, L. T. (1964). *J. Pharmacol. Exp. Ther.* **143,** 314.

Wurtman, R. J., Axelrod, J., Vesell, E. S., and Ross, G. T. (1968a). *Endocrinology* **82,** 584.

Wurtman, R. J., Axelrod, J., and Kelly, D. E. (1968b). "The Pineal." Academic Press, New York.

Wurtman, R. J., Shein, H. M., Axelrod, J., and Larin, F. (1969). *Proc. Natl. Acad. Sci. U.S.A.* **62,** 749.

Wurtman, R. J., Chou, C., and Rose, C. (1970). *J. Pharmacol. Exp. Ther.* **174,** 351.

Wuttke, W., and Meites, J. (1970). *Proc. Soc. Exp. Biol. Med.* **135,** 648.

Wuttke, W., and Meites, J. (1972). *Pfluegers Arch.* **337,** 71.

Wuttke, W., Cassel, E., and Meites, J. (1971). *Endocrinology* **88,** 737.

Wuttke, W., Gelato, M., and Meites, J. (1972). *Brain-Endocr. Interact. Median Eminence: Struct. Func., Int. Symp., 1971* pp. 267–279.

Wyatt, R. J., Kupfer, D. J., Scott, J., Robinson, D. S., and Snyder, F. (1969). *Psychopharmacologia* **15,** 236.

Wyatt, R. J., Fram, D. H., Kupfer, D. J., and Snyder, F. (1971a). *Arch. Gen. Psychiatry* **24,** 145.

Wyatt, R. J., Zarcone, V., Engelman, K., Dement, W. C., Snyder, F., and Sjoerdsma, A. (1971b). *Electroencephalogr. Clin. Neurophysiol.* **30,** 505.

Wykes, A. A., and Landez, J. H. (1967). *Fed. Proc., Fed. Am. Soc. Exp. Biol.* **26,** No. 2.

Yamada, Y. (1975). *Neuroendocrinology* **18,** 263.

Yamaguchi, N., Marczinski, T. J., and Lingi, M. (1963). *Electroencephalogr. Clin. Neurophysiol.* **15,** 145.

Yamamura, H. I., Kuhar, M. J., Greenberg, D., and Snyder, S. H. (1974). *Brain Res.* **66,** 541.

Yamauchi, J., Takahara, J., and Pfuji, T. (1976). *Abstr. Book. Int. Congr. Endocrinol., 5th, 1976* Abstract No. 30.

Yarbrough, G. G. (1976). *Nature (London)* **263,** 523.

Yarbrough, G. G., Buxbaum, D. M., and Sanders-Bush, E. (1972). *Biochem. Pharmacol.* **21,** 1667.

Yasuda, N., and Greer, M. A. (1976). *Neuroendocrinology* **22,** 48.

Yawata, M., and Fukase, M. (1968). *J. Clin. Endocrinol. Metab.* **28,** 1079.

Yen, S. S. C., Tsai, C. Naftolin, F., Vandenberg, G., and Ajabor, L. (1972). *J. Clin. Endocrinol. Metab.* **34,** 671.

Yen, S. S. C., Siler, T. M., and De Vane, A. W. (1974a). *N. Engl. J. Med.* **290,** 935.

Yen, S. S. C., Vandenberg, G., and Siler, T. M. (1974b). *J. Clin. Endocrinol. Metab.* **39,** 170.

York, D. H. (1970). *Brain Res.* **20,** 233.

Yoshimura, M., Ochi, Y., Miyazachi, T., Shiomi, K., and Hachiya, T. (1973). *Endocrinol. Jpn.* **20,** 135.

Young, A. B., Oster-Granite, M. L., Herndon, R. M., and Snyder, S. H. (1974). *Brain Res.* **73,** 1.

Zaimis, E. (1970). *In* "Catapres in Hypertension" (M. E. Connolly, ed.), pp. 9–22. Butterworth, London.

Zarate, A., Canales, E., Schally, A. V., Ayala-Valdes, L., and Kastin, A. J. (1972a). *Fertil. Steril.* **23,** 672.

Zarate, A., Canales, E. S., Soria, J., Ruiz, F., and McGregor, C. (1972b). *Am. J. Obstet. Gynecol.* **112,** 1130.

Zarate, A., Canales, E., Soria, J., Maneiro, P. J., and MacGregor, C. (1973a). *Neuroendocrinology* **12,** 362.

Zarate, A., Canales, E., Jacobs, L. S., Maneiro, P., Soria, J., and Daughaday, W. H. (1973b). *Fertil. Steril.* **24,** 340.

Zarate, A., Valdes-Vallina, F., Gonzales, A., Perez-Ubierma, C., Canales, C., and Schally, A. V. (1973c). *Fertil. Steril.* **24,** 485.

Zelenà, J. (1969). *Symp. Int. Soc. Cell Biol.* **8,** 73.

Zeller, E. A., Barsky, J., Fouts, J. R., Kirchheimer, W. F., and Van Orden, L. S. (1952). *Experientia* **8,** 349.

Zetler, G. (1956). *Naunyn-Schmiedebergs Arch. Exp. Pathol. Pharmakol.* **228,** 513.

Zetler, G. (1970). *Aspects Neuroendocrinol., Int. Symp. Neurosecretion, 5th, 1969* pp. 287–295.

Zimmerman, E. A., Hsu, K. C., Ferin, M., and Kozlowski, G. P. (1974). *Endocrinology* **95,** 1.

Zivkovic, B., Guidotti, A., and Costa, E. (1973). *Brain Res.* **57,** 522.

Zolovick, A., and Labhsetwar, A. P. (1973). *Nature (London)* **245,** 158.

Zor, U., Kaneko, T., Schneider, H. P. G., McCann, S. M., and Field, J. B. (1969a). *J. Clin. Invest.* **48,** 93a.

Zor, U., Kaneko, T., Schneider, H. P. G., McCann, S. M., Lowe, I. P., Bloom, G., Borland, B., and Field, J. B. (1969b). *Proc. Natl. Acad. Sci. U.S.A.* **63,** 918.

Zor, U., Kaneko, T., Schneider, H. P. G., McCann, S. M., and Field, J. B. (1970). *J. Biol. Chem.* **245,** 1883.

Zschaeck, L. L., and Wurtman, R. J. (1973). *Neuroendocrinology* **11,** 144.

Subject Index

(For abbreviated chemical compound names refer to Abbreviation List on p. v)

B

D

H

N

O

P

Q

R

T